AF595363

Advances in Molecular Simulation

Advances in Molecular Simulation

Editor

Małgorzata Borówko

MDPI • Basel • Beijing • Wuhan • Barcelona • Belgrade • Manchester • Tokyo • Cluj • Tianjin

Editor
Małgorzata Borówko
Department of Theoretical
Chemistry, Institute of Chemical
Sciences, Faculty of Chemistry,
Maria Curie-Skłodowska
University in Lublin
Poland

Editorial Office
MDPI
St. Alban-Anlage 66
4052 Basel, Switzerland

This is a reprint of articles from the Special Issue published online in the open access journal *International Journal of Molecular Sciences* (ISSN 1422-0067) (available at: https://www.mdpi.com/journal/ijms/special_issues/m_simulation).

For citation purposes, cite each article independently as indicated on the article page online and as indicated below:

LastName, A.A.; LastName, B.B.; LastName, C.C. Article Title. *Journal Name* **Year**, *Volume Number*, Page Range.

ISBN 978-3-0365-2710-9 (Hbk)
ISBN 978-3-0365-2711-6 (PDF)

Contents

About the Editor

Małgorzata Borówko is a professor at Maria-Skłodowska University in Lublin (Poland), an institution with which she has been affiliated since 1976. For almost twenty-five years she was the head of the Department of Modeling Physico-Chemical Processes here. She is currently employed at the Department of Theoretical Chemistry.

She received M. Sc. (1976) and PhD. (1980) degrees in chemistry from Maria Curie-Skłodowska University. In 1989 she completed her residency, and in 2001 was awarded with the highest academic degree in Poland—the title of national professor of chemistry—by the President of the Republic of Poland.

Her areas of research include statistical thermodynamics, computer modeling, interface and colloid science, adsorption and chromatography. She has carried out numerous research projects together with colleagues working at UMCS and scientists from research centers in various countries (Czech Republic, Mexico, Spain, Ukraine, USA).

Małgorzata Borówko is the author or coauthor over 120 original scientific articles and several chapters in books and reviews. She was Editor of the monograph "Computational methods in Surface and Colloid Science", Surfactant Science Series, vol. 89 (M. Dekker, New York, 2000). She belonged to the Advisory Board of the Journal of Interface and Colloid Science.

Her scientific achievements have been honored with awards of the Ministry of Education and Science from the Polish Government. In 2001, she received the Langmuir Lecture Award from the Division of Colloid and Surface Chemistry of the American Chemical Society. Her Langmuir Lecture, entitled "Monte Carlo simulation of adsorption", was presented at the 222nd National Meeting of ACS in Chicago.

Her teaching activity concerned such issues as statistical thermodynamics, quantum chemistry, surface science and applications of computer science in natural sciences.

Preface to "Advances in Molecular Simulation"

Rapid advances are taking place in the application of molecular simulations to the study of complex systems. Molecular simulations are widely used in physics, chemistry, biology, material science, engineering, and even medicine. Each study conducted in a purely theoretical manner can be treated as a simulation. The behavior of a considered system can be modeled using different computational approaches. First, the "usual" theories can be utilized for this purpose. Among them are simple phenomenological theories, advanced quantum methods, and sophisticated statistical–thermodynamic calculations. Second, we can apply various special procedures that allow us to mimic real systems and their evolution over time, namely molecular dynamics, the Monte Carlo method, cellular automata, or genetic algorithms. The most popular simulation techniques of this type are molecular dynamics (MD) and the Monte Carlo method (MC). Impressive progress has been made in the development of different versions of these methods.

A detailed description of molecular interactions is always crucial in simulations. Various force fields are developed and permanently improved to estimate the interaction potentials between atoms or coarse-grained particles. The traditional force field provides a specific function for potential energy and a set of its best-fitted parameters. The potential is fit to the results of quantum mechanical calculations and, usually, to certain experimental measurements. The accuracy of molecular interaction description can be improved by the use of machine learning potentials.

During a typical molecular simulation, the trajectory of the system in the configuration space can be generated stochastically, based on the laws of mechanics, or both. The rules for accepting these states take into account changes in the energy of the system. Simulations can be carried out in different statistical ensembles, including the extended ensembles used in multicanonical simulations or the parallel tempering method. A great variety of enhanced sampling techniques have been successfully implemented in molecular modeling. Numerous methods for the analysis of simulation data have been also proposed; for example, the thermodynamic integration, the weighted histogram method, finite-size scaling analysis, and many others.

The exceptional popularity of molecular dynamics results, among other things, from the availability of many ready-made software packages that provide all major MD codes with a set of accompanying programs performing most steps of the preparation for a simulation. The user-friendly environment allows writing of the simulation scripts without the need for a deep knowledge of the underlying procedures. Nevertheless, extracting reliable and useful information from the simulation always requires a lot of care. The results should be interpreted in light of all available theoretical and experimental data for the system under study.

The ever-increasing power of computers and the development of modern programming methods make molecular simulations a very effective tool for studying more and more complex systems. The main purpose of this Special Issue is to present a wide range of molecular simulation methods and their applications in various fields.

The first section consists of two review articles concerning the methodology of molecular simulations. The survey written by Roy and Kovalenko covers the applications of the statistical mechanics-based three-dimensional reference interaction site model with the Kovalenko–Hirata closure molecular solvation theory in different simulations. They show that this approach can be an essential part of a multiscale modeling framework for the simulation of biochemical systems involving liquid media of complex composition.

Next, Cao and Tian present an excellent overview of the application of "dividing and conquering" and "caching" principles in the development of molecular modeling algorithms. They describe algorithms for the coarse-graining, the enhanced sampling, and the local free energy landscape approach. Differences, connections, and potential interactions among these three algorithmic lines are discussed.

The second part is a collection of original scientific articles devoted to the modeling of different molecular systems. These works can be divided into three groups.

The first series of articles deals with the methodology of theoretical calculations. Mun, Sakai, and Kim propose the quantum mechanical approach for strongly ionizing media. They calculate time evolution operators, employing a time-dependent unitary transformation method within the Kramers–Henneberger framework. The method can be applied to atoms or molecules exposed to strong laser fields. In turn, Shtyrov et al. developed a simplified methodology to study spectral properties of rhodopsins. This approach allows estimation of the direct effect of a charged or polar residue substitution on the first absorption band maxima without extensive quantum-mechanical calculations using only a rhodopsin three-dimensional structure and the data provided in the article. The article written by Lai, Jia, and Chi shows new developments in the open-source Monte Carlo simulation tool gMicroMC (Med. Phys. 2020, 47, 1958). They use this method for the transport simulation of protons and heavy ions and the concurrent transport of radicals in the presence of DNA. Herranz et al. present the simulator-descriptor suite (Simu-D) and use it to model numerous polymer-based systems under extreme conditions of concentration, confinement, and nanofiller content. The simulator involves different MC algorithms, including localized, chain-connectivity-altering, identity-exchange, and cluster moves in various statistical ensembles. The simulator includes various MC algorithms, namely localized changing links of the chain, identity exchange, and cluster movements in statistical teams, while the descriptor identifies similarity of computer-generated configurations with respect to reference crystals in two or three dimensions. It has been shown that the Simu-D can be a useful tool in the studies of entropy- and energy-driven phase transition, adsorption, and self-organization of polymer systems.

The second group of articles covers a broad spectrum of applications of molecular simulations in biomedical research. The article by Ahmad et al. is devoted to the modeling of antibody–antigen interactions. They investigate TLR4 antibody-binding epitope using the computational-driven approach. This computational technique involves the construction of interaction networks, epitope prediction, molecular docking, MD simulations, and binding-free-energy calculations. The method can be a promising tool for the development of new monoclonal antibodies. Choi et al. demonstrate the binding mechanism of selected ligands to the coronavirus 3-chymotrypsin-like-protease through docking and MD simulations. The interactions with SARS-CoV-2 proteases and binding energy were calculated. The study can be helpful for the development of novel and effective, antiviral compounds. The article by Rzecki et al. is on the new computational approach for the comparison of antimicrobial effectiveness among different types of Gemini surfactants. In this study, MD simulations are used to investigate the incorporation of these agents into the model bacteria membrane. The development of new antimicrobials is one of the most serious problems facing medicine today.

The last group of works concerns simulations of various physical phenomena in complex systems under different thermodynamic conditions. Eun presents the MD simulation study of the osmosis-driven water transport through nanochannels. Kobayashi et al. propose a possible mechanism of turbulence suppression in surfactant aqueous solution based on the performed large-scale dissipative particle dynamics simulation. Adsorption on ligand-tethered particles is also investigated using coarse-grained MD simulations (Borówko and Staszewski). Patrykiejew presents the results of MC simulation of the phase behavior of two-dimensional systems of Janus-like particles on a triangular lattice. Moreover, Szeleszczuk et al. investigate the isosymmetric structural phase transition of chlorothiazide using periodic Density Functional Theory calculations and ab initio MD simulations. They prove that DFT calculations enabled the prediction of the transitions of this type. Finally, the structural and dynamical properties of colloids with competing interactions confined in slit pores were investigated using MC and MD simulations (Serna, Góźdź, Noya).

We hope that this Special Issue reflects the power of molecular simulation as an effective research tool in many fields and can provide an impetus for further fruitful studies.

Finally, I wish to express my sincere thanks to each of the authors for their contributions.

Małgorzata Borówko

Editor

International Journal of
Molecular Sciences

MDPI

Review

Biomolecular Simulations with the Three-Dimensional Reference Interaction Site Model with the Kovalenko-Hirata Closure Molecular Solvation Theory

Dipankar Roy [1,*] and Andriy Kovalenko [1,2,3,*]

1 10-203 Donadeo Innovation Centre for Engineering, Department of Mechanical Engineering, University of Alberta, Edmonton, AB T6G 1H9, Canada
2 Department of Biological Sciences, University of Alberta, Edmonton, AB T6G 2E9, Canada
3 Nanotechnology Research Centre, National Research Council of Canada, Edmonton, AB T6G 2M9, Canada
* Correspondence: droy1@ualberta.ca (D.R.); andriy.kovalenko@ualberta.ca (A.K.)

Abstract: The statistical mechanics-based 3-dimensional reference interaction site model with the Kovalenko-Hirata closure (3D-RISM-KH) molecular solvation theory has proven to be an essential part of a multiscale modeling framework, covering a vast region of molecular simulation techniques. The successful application ranges from the small molecule solvation energy to the bulk phase behavior of polymers, macromolecules, etc. The 3D-RISM-KH successfully predicts and explains the molecular mechanisms of self-assembly and aggregation of proteins and peptides related to neurodegeneration, protein-ligand binding, and structure-function related solvation properties. Upon coupling the 3D-RISM-KH theory with a novel multiple time-step molecular dynamic (MD) of the solute biomolecule stabilized by the optimized isokinetic Nosé–Hoover chain thermostat driven by effective solvation forces obtained from 3D-RISM-KH and extrapolated forward by generalized solvation force extrapolation (GSFE), gigantic outer time-steps up to picoseconds to accurately calculate equilibrium properties were obtained in this new quasidynamics protocol. The multiscale OIN/GSFE/3D-RISM-KH algorithm was implemented in the Amber package and well documented for fully flexible model of alanine dipeptide, miniprotein 1L2Y, and protein G in aqueous solution, with a solvent sampling rate ~150 times faster than a standard MD simulation in explicit water. Further acceleration in computation can be achieved by modifying the extent of solvation layers considered in the calculation, as well as by modifying existing closure relations. This enhanced simulation technique has proven applications in protein-ligand binding energy calculations, ligand/solvent binding site prediction, molecular solvation energy calculations, etc. Applications of the RISM-KH theory in molecular simulation are discussed in this work.

Keywords: molecular solvation theory; three-dimensional reference interaction site model; Kovalenko-Hirata closure; biomolecular simulation; multiple time step MD; protein-ligand binding; biomolecular solvation

Citation: Roy, D.; Kovalenko, A. Biomolecular Simulations with the Three-Dimensional Reference Interaction Site Model with the Kovalenko-Hirata Closure Molecular Solvation Theory. *Int. J. Mol. Sci.* **2021**, *22*, 5061. https://doi.org/10.3390/ijms22105061

Academic Editor: Małgorzata Borówko

Received: 9 April 2021
Accepted: 10 May 2021
Published: 11 May 2021

1. Introduction

The developments of molecular simulations started first with statistical methods like Monte-Carlo simulations (MC) to address the time-progression of multi-particle systems. The use of macroscopic spheres to simulate atomic motions dates backs to early 1940. The work on elastic collision in phase transition by Alder and Wainwright is attributed as the first realistic simulation [1]. The progress in molecular dynamics (MD) simulations from that point is phenomenal thanks to ever evolving computer architectures and development of efficient algorithms. While the initial applications of the MD simulations were aimed at material science applications, they covered biophysics and biomolecules very fast. The applications in biomolecular systems are numerous: X-ray structure processing, protein folding, and receptor-ligand interactions, to name a few [2–4]. Performance and accuracy of MD simulations were verified by comparison with experimental data obtained from diffraction and NMR experiments. An essential component of MD simulation, the force field is developed by fitting against high-level quantum chemical calculations [5–7]. The other variant of the force field, the Kirkwood-Buff (KB) force fields designed through application of the KB theory in calculating densities and other physical properties of multicomponent solutions, has shown immense potential for use in molecular simulations, both with all atom and united atom settings [8–10]. The deviations of MD simulations from experimental results are attributed to the shortcomings of the force field(s) used, as well as inadequate simulation time frame. It is imperative to point to the local minima problem faced by MC and MD methods for potential energy surfaces (PESs) with multiple minima separated by large energy barriers [11–13]. A plethora of research has been devoted to overcome such issues [14,15]. The explicit solvent simulations using MD techniques are the most adequate ones for modeling biologically important molecules which often requires specific environments. Explicit solvation simulations are quintessential for solvation free energy as well as receptor-ligand binding energy calculations. All these theoretical and computational techniques essentially deal with multiple interactions present in liquid environment (e.g., solution). These interactions involve solvent-solvent and solute-solvent interactions. The solute-solvent interaction further breaks down into the electrostatic and non-polar components. To complete the interaction terms in order to calculate solvation energy, solute polarization and deformation energies are important factors. The last term becomes more significant for binding studies in biomolecular simulations. Incorporating all these intra- and intermolecular terms in solvation process modeling is a daunting task, and justifies development of several theoretical methods to address molecular solvation.

The differences in molecular properties between isolated systems and continuum calculations are often the results of different scalabilities of the systems under consideration. The "gold-standard" quantum mechanical (QM) methods can achieve accuracy up to one-tenth of a kcal/mol, but limited to systems with small sizes [16–19]. Different continuum solvation models (e.g., PCM and its variants, SMD, COSMO) are calibrated against experimental solvation energy databases of small molecules, and are problematic for absolute solvation energy calculations of systems beyond the chemical classes covered in calibration databases (viz. transition metal containing systems) [20–24]. The difficulty in achieving accurate solvation free energy prediction can be attributed to the absence of specific solute-solvent interactions, limited (or most of the time, absent) sampling of the solute conformal steps, etc. Application of quantum chemical calculations of biomolecules (protein, DNA/RNA) are restricted due to system size resulting in a large number of basis functions required to describe such systems. The ONIOM methodology as well as QM/MM and QM/MM/MD techniques provide respite to this handicap by offering a computationally more amenable scenario where site(s) of importance are treated with high level QM calculations while the rest of the system is treated with molecular mechanics potentials for specialized applications [25–28].

The applicability of different theoretical methods to solvation dynamics of systems of different sizes and dimensions is quite compartmentalized. Thus, a theoretical model that spans over a large scale of computational requirements with reasonable accuracy and speed is desirable, and much research activity is devoted toward this goal. The reference interaction site model (RISM) is based on first principle statistical mechanics, with proven applications in the field of van der Waals fluid, biomolecules, material science, and drug development [29–31]. The theoretical framework of RISM is suitable to couple with MD-engines and QM self-consistent-field (SCF) iterations [32–35]. This makes the RISM formalism an excellent candidate from the perspective of building materials of desired properties, as the theory provides understanding of all the underlying interactions between different constituent fragments. The RISM theory with the integral equation formalism was developed and used for solvation structure and energetics calculations, although the potential of this theory goes beyond regular solvation energy calculations and expands to molecular partitioning, physical-chemical property calculation, and molecular simulations [36–48]. The key feature of the RISM theory is that it can provide reasonably accurate result rapidly, a feature that made this the theory an essential part of the multiscale modeling framework (Figure 1).

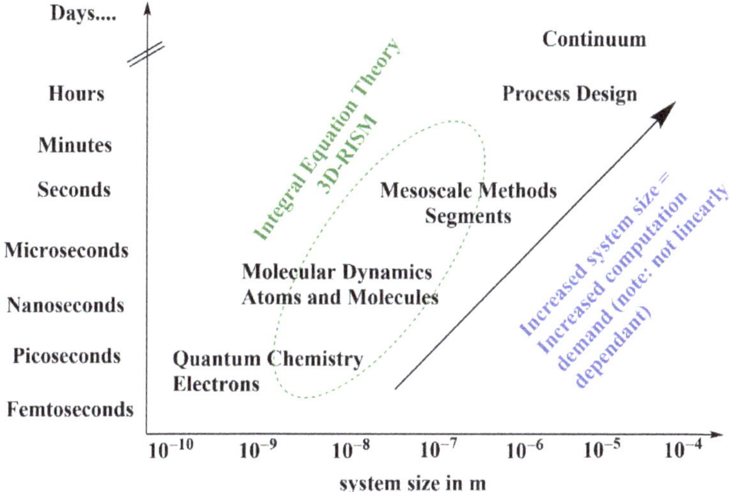

Figure 1. Computational simulation scale and versality of the 3D-RISM theory.

2. Theoretical Background

The foundation of the RISM theory is credited to the seminal works of Chandler and coworkers [49–56]. This theory grew enormously over past forty years. The key theoretical aspects are outlined in this section. Further theoretical backgrounds are provided for individual applications in the respective sections. For a solute of arbitrary shape, the 3-dimensional (3D-) version of the RISM theory provides a probability distribution of all possible interaction sites (γ) of solvent molecules around the solute at position **r** which is a product of the average number density (ρ_γ) in the bulk solution and the normalized density distribution, $g_\gamma(r)$ (Figure 2). The density enhancement and/or depletion ($g_\gamma(r) > 1$ and/or $g_\gamma(r) < 1$) relative to the average density at a point in solution bulk where $g_\gamma(r) \rightarrow 1$ is provided by the average number density. The total correlation function of solvent sites in 3D is related to the 3D direct correlation function $c_\gamma(r)$ and site-site bulk susceptibility function for α-solvent sites around a solute by Equation (1).

$$h_\gamma(r) = \sum_\alpha \int dr' c_\alpha(r - r')\chi_{\alpha\gamma}(r') \quad (1)$$

Additionally, $g_\gamma(r) = h_\gamma(r) + 1$ and $c_\gamma(r) \sim -u_\gamma(r)/(k_B T)$, where T is temperature and k_B is the Boltzmann constant. The bulk susceptibility function χ is an essential input to the 3D-RISM integral equation, and is constructed from the intramolecular correlation function $\omega_{\alpha\gamma}$ from the dielectrically consistent RISM (DRISM) [57]:

$$\chi_{\alpha\gamma}(r) = \omega_{\alpha\gamma}(r) + \omega_{\alpha\gamma}(r)\,\rho_\gamma\,h_{\alpha\gamma}(r) \quad (2)$$

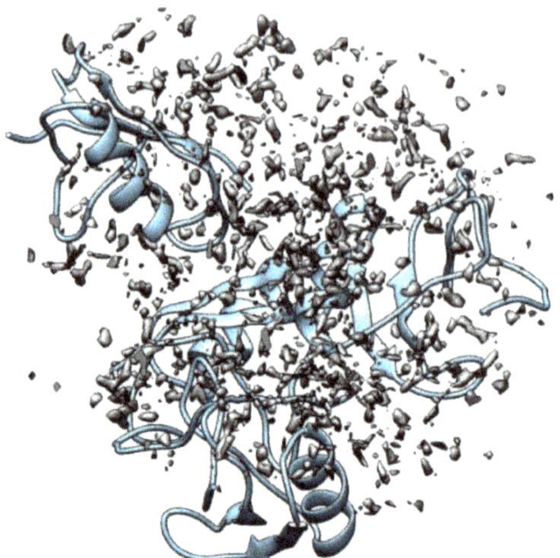

Figure 2. The normalized distribution of the water oxygen sites around the scorpion toxin protein (PDB ID: 1AHO) computed using the 3D-RISM-KH theory and the modified TIP3P water model. The protein backbone is colored in cyan.

The intramolecular correlation function can be expressed in reciprocal k-space via terms of a zeroth-order Bessel function:

$$\omega_{\alpha\gamma}(r) = j_0(kl_{\alpha\gamma}) \tag{3}$$

The intra- and inter-species correlation functions are renormalized through an analytical dielectric bridge function for solvents with high dielectric constant value, thus ensuring that all inter- and intra-species interactions are considered for a few solvent layers around a solute (or cosolvent, etc.). The renormalized form of the dielectric correction is written in terms of zeroth- and first-order Bessel functions over the position of each atom $r_\alpha = (x_\alpha, y_\alpha, z_\alpha)$ with partial charge q_α of site α on species with respect to its molecular origin:

$$\chi_{\alpha\gamma}(k) = j_0(kx_\alpha)j_0(ky_\alpha)j_1(kz_\alpha)h_c(k)j_0(kx_\gamma)j_0(ky_\gamma)j_1(kz_\gamma) \tag{4}$$

The envelope function $h_c(k)$ determines the dielectric constant of the solution using a non-oscillatory form with amplitude A falling off rapidly at wavevectors k larger than the characteristic size l of the liquid. The characteristic length is important for DRISM calculations, to avoid spurious non-physical distribution functions.

$$h_c(k) = A\exp\left(-l^2k^2/4\right) \tag{5}$$

$$A = \frac{1}{\rho_{\text{polar}}}\left(\frac{\varepsilon}{y} - 3\right) \tag{6}$$

For a mixed solvent scenario, the total number density of polar species and solution dielectric susceptibly y can be used in combination with Equations (4)–(6) to apply 3D-RISM formalism as:

$$\rho_{\text{polar}} = \sum_{s\in\text{polar}} \rho_s \tag{7}$$

$$y = \frac{4\pi}{9k_BT}\sum_{s\in\text{polar}} \rho_s(d_s)^2 \tag{8}$$

A closure function is required to integrate an infinite chain of correlation diagrams generated from the direct and total correlation function. The functional form of such a closure function in unknown, and several approximated forms were reported over time for simplified computations. Closure functions differ from each other in the mathematical form of the bridging function used in the construct. The Kovalenko-Hirata (KH) closure is among the best closure relations till date in terms of both numerical stability and reasonable accuracy [58,59]. The mathematical form of the KH closure is given as:

$$g_\gamma(r) = \begin{cases} \exp(-u_\gamma(r)/(k_BT) + h_\gamma(r) - c_\gamma(r)) & \text{for} \quad g_\gamma(r) \leq 1 \\ 1 - u_\gamma(r)/(k_BT) + h_\gamma(r) - c_\gamma(r) & \text{for} \quad g_\gamma(r) > 1 \end{cases} \tag{9}$$

The overall form of the KH closure can be explained as a coupling of the mean spherical approximation for the regions of density enrichment ($g_\gamma(r) > 1$) with the hypernetted chain approximation for the region of density depletion ($g_\gamma(r) < 1$). The excess chemical potential and the solvation free energy is obtained from the analytical form of the KH closure as:

$$\begin{gathered} \mu_{\text{solv}} = \sum_\gamma \int_V dr\, \Phi_\gamma(r) \\ \Phi_\gamma(r) = \rho_\gamma k_B T \left[\tfrac{1}{2} h_\gamma^2(r) \Theta(-h_\gamma(r)) - c_\gamma(r) - \tfrac{1}{2} h_\gamma(r) c_\gamma(r) \right] \end{gathered} \tag{10}$$

The $\Phi_\gamma(r)$ is the Heaviside step function. Important thermodynamic parameters are derived from the excess chemical potential for solute sites (u) and solvent sites (v) as:

$$\Delta\mu = \Delta\varepsilon^{uv} + \Delta\varepsilon^{vv} - T\Delta s_V \tag{11}$$

The entropy (ΔS_V) and partial molar volume (PMV, $\tilde{V}$) are calculated as:

$$\Delta S_V = -\frac{1}{T}\left(\frac{\partial \Delta\mu}{\partial T}\right)_V \tag{12}$$

$$\tilde{V} = k_B T \chi_T \left(1 - \sum_\gamma \rho_\gamma \int dr\, c_\gamma(r)\right) \tag{13}$$

The errors in 3D-RISM calculations have several origins. Firstly, the internal pressure calculated in the 3D-RISM molecular solvation theory is wrong. A few correction schemes were developed to counter this error [60,61]. Another source of errors arises from the choice of the Lennard-Jones potential used for calculating interaction potentials. A careful calibration is warranted while selecting a force field for a specific application. For example, the computational framework of 3D-RISM failed to converge for polar protic hydrogen atom (e.g., water) with conventional force fields, as the hydrogen atoms has no van der Walls parameters assigned. This problem is circumvented by using a non-zero van der Wall's terms for hydrogens [62,63]. The KH closure is known to shift the strongly associated peaks while broadening them simultaneously; interestingly, this provides an adequately correct solvation structure.

3. Biomolecular Simulations with the 3D-RISM-KH Molecular Solvation Theory

The center of simulations for biophysics related problems are structure-function features of protein and nucleic acids. The structural landscape of biomolecular folding is a high demand research field. Recent achievements in achieving millisecond time scale simulation of protein structure opened further developments in order to explore the entire folding landscape of proteins of reasonable sizes [64–66]. The molecular dynamics simulation with the 3D-RISM-KH theory was first incorporated in the AMBER MD simulation suite, using the Sander engine as well as standalone unit for single point solvation free energy calculations [32]. The standard Sander implementation was modified to support long time scale simulations using damped Langevin dynamics for a canonical ensemble to address the instability of the multiple time step MD (MTS-MD). This is achieved by combining two simulation cycles for two different parts of the system (MTS-MD). The outer time steps are obtained from 3D-RISM-KH calculation. For each inner step, the effective solvation force is used to extrapolate solvation force coordinates, based on the outer time steps. The force matrix $\{F\}^{(k)}$ working on each solute atom is approximated as a linear combination of forces at N previous steps, at any given time step t_k:

$$\{F\}^{(k)} = \sum_{i=1}^{N} a_{ki}\{F\}^{(i)}, \; i \in 3D-RISM \text{ steps} \tag{14}$$

The weighted coefficients a_{ki} for a given time step are obtained from the best projection of N previous steps. These non-conservative potentials provided a smooth transition between steps. However, strong coupling through the Nosé-Hoover chain of thermostats impeded structural transitions. The next generation of development provided the advanced solvation force extrapolation scheme (ASFE). This development used the optimized isokinetic Nosé-Hoover thermostat (OIN) for each atom by imposing kinetic energy constraints. The fast-dynamics (solute-solute) and slow dynamics (solute-solvent via 3D-RISM) are separated in the ASFE implementation. The accuracy of extrapolation was estimated by relative mean square deviation of the extrapolated effective solvation forces from their original values calculated from converged 3D-RISM-KH for the outer time step. The applicability of the novel formalism containing two separate time cycles for MD simulation of solute-solvent systems were validated against conformational space of alanine dipeptide in water. The subsequent developments, generalized solvation force extrapolation (GSFE), used rotational transformation of the relative coordinates for each atom in order to smoothen the force matrix described previously. In this new development which also used OIN thermostats, a weighting function was introduced for each discretized space. The new algorithm also takes into account that the nearest neighbors have maximum effect on mean solvation forces for any given atom. The efficiency of this new algorithm was shown by the MTS-MD/OIN/GFSE/3D-RISM-KH simulation of a miniprotein (PDB: 1L2Y) and protein G with their reported folded forms [67–69]. The miniprotein folding was achieved via the MTS-MD formalism, starting from a fully extend denatured state, at about 60 ns simulation in comparison to the average physical folding time in the order of µs observed via experiment [68].

4. Binding Site Mapping

Receptor-ligand binding is in the heart of early-stage drug discovery. A correct mapping helps to stop waste of resources, both financially and computationally. Traditionally, lead-like molecules are used to find potential binding site(s) on a receptor surface. An alternative option to this is fragment-based mapping. These processes will lead to a set of fragments/probe molecules that are potential binders on a receptor surface with defined binding sites [70,71]. Chemical linking based on available linker databases and knowledge of chemical space yields potential leads. The success of this process depends on correctly finding a binding site, usually using empirical scoring algorithms. The 3D-RISM-KH theory essentially provides a 3D-distribution of solvent sites around a solute of arbitrary shape. Thus, one can replace the solvent with a small molecule fragment and even a mixture of fragments, and develop a distribution of unique sites from a mixture of fragments, around a solute of interest. The concept behind this process is easy to visualize, but requires specialized algorithms that can reduce computational burden and help in finding a physically meaningful solution. The 3D-distribution function for ligand site γ is given as:

$$g_\gamma(r_\gamma(R,\Omega)) = \int g_\gamma(r)\rho_\gamma(r - r_\gamma(R,\Omega))dr \tag{15}$$

The ligand sites spatial position is defined with three cartesian coordinates (R, translational) and three Euler angles (Ω, rotational). In practice, the ligand site density distribution is described using a Gaussian-type function. The so-called "site-integrated" potential of mean force (W, SI-PMF) is used to find the most probable binding site(s) of a ligand probe on a receptor [72]:

$$W\left(\Delta\vec{R},\Omega\right) = -k_B T \sum_{\gamma} ln\, g_\gamma\left(r_\gamma\left(\vec{R},\Omega\right)\right) \tag{16}$$

The latest development of this concept used spatial distribution function of a ligand around a protein and thus explored all possible binding modes of the ligand, and final filtering was done based on a scoring function [73]. This scoring function is based on estimated free energy terms and is written as:

$$W_{SP}(\{r\};\Omega) \approx -RT\, ln\left[\prod_i g_i(r_i;\Omega)\right] \tag{17}$$

These methodologies were validated against several small molecule binders and protein-ligand datasets [72–74].

Another important aspect of protein-ligand interactions is the role played by binding-site water molecules in ligand recognition [75–77]. For a regular molecular simulation with explicit solvent molecules, it is cumbersome to look for such binding site water molecules. This search of binding site waters can be eased with the help of the 3D-RISM-KH water distribution function around a solute molecule. Water distribution in the Lysozyme cavity was first successfully explored to locate binding site water using the 3D-RISM-KH theory [78]. The most updated protocol was reported by Sindhikara and Hirata [79,80]. In their method (placevent), a new successive orthogonal image (SOI) technique for sampling was employed to analyze the distribution function (Figure 3). The SOI method calculates the rotational space of three orthogonal vectors for a spherical search space by using a heavy atom of the solvent as anchor of rotation (e.g., oxygen atoms of water molecules). The solvation site volume ($\tilde{V}_n$) is calculated by applying the Kirkwood-Buff equation in a 3D-RISM-KH calculation as:

$$\tilde{V}_n = k_B T \chi_T \left(1 - \rho_0 \sum_{\gamma} \int_{V_n} c_\gamma(r)\, dr\right) \tag{18}$$

The success of this applications was reported against experimentally determined binding site and poses of ligands in biologically relevant targets. The 3D-RISM-KH based water site prediction is implemented in the MOE© suite [81]. Some examples of successful applications of the 3D-RISM-KH theory in exploiting the explicit role of water maps are reported for in-drug design and protein aggregation studies [82–84]. A very recent modification of the 3D-RISM theory in mapping solvation sites in enzyme active site was reported by Nguyen et al. [85]. This new development extended the GIST (grid inhomogeneous solvation theory) based mapping technique in to the 3D-RISM grids. Briefly, an approximated distribution of oxygen site α (from water) around a site of interest is related to thermochemical property of interest (A) as position r as:

$$A(r) \approx A_\alpha(r) + g_\alpha(r) \sum_{\gamma \neq \alpha} \omega_{\alpha\gamma}(r) * A_\gamma(r) \tag{19}$$

The number density distribution $g_\alpha(r)$ is used to weight the convolution (*) in the right-hand side of Equation (19). This formalism did not consider the non-local effect in the distribution, although the authors reported negligible errors in the final distribution resulting from this issue in comparison to molecular simulations and experimental data.

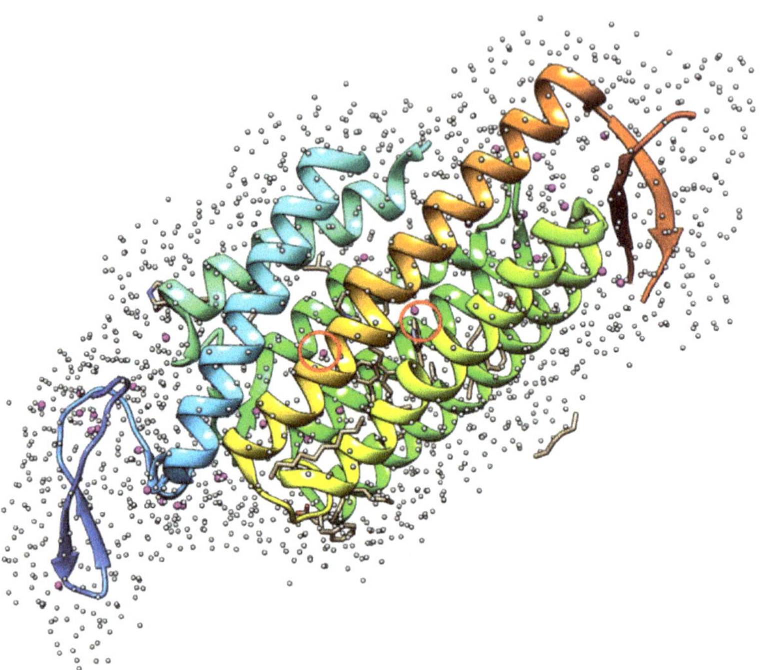

Figure 3. Distribution of water oxygen atoms from 3D-RISM-KH calculations on protein 3UG9 (white spheres). The crystallographic waters are represented with magenta spheres. The catalytic binding site waters were marked in red circle.

Among other reports of biomolecular simulations using the 3D-RISM-KH theory, the effect of (micro-) solvent environments on amyloid structure and potential of mean force calculations of solute permeation across UT-B and AQP1 proteins provided further extension of the applications of the 3D-RISM-KH theory based molecular solvation [86,87].

5. Protein-Ligand Binding Energy

Heart to drug development is correct prediction or binding affinity of small molecules toward target receptor, if not quantitatively accurate then a trend of binding affinity. Methodologies developed for calculating binding energy for the process, Protein + Ligand → Complex, are the linear-response approximation (LRA), protein-dipole Langevin-dipole approach (PDLD), linear interaction energy (LIE) approach, and MM/PB(GB)SA approach [88–93]. The MM/PBSA method is favored over the others, as it avoids empirical parameterizations. In this method, the free energy of binding (G_{bind}) is expressed via the molecular mechanics (MM) based energy (E_{MM}, gas phase, for reactants covering internal, electrostatic, and van der Wall's terms), the solvation energy term (G_{solv}), and the entropy term ($-TS_{\text{MM}}$), computed at temperature T.

$$G_{\text{bind}} = E_{\text{MM}} + G_{\text{solv}} - TS_{\text{MM}} = E_{\text{int}} + E_{\text{el}} + E_{\text{vdw}} + G_{\text{solv, polar}} + g_{\text{solv, non-polar}} - TS_{\text{MM}} \quad (20)$$

The solvation terms are calculated by solving the Poisson-Boltzmann (PB) equation (or via generalized Born, GB, model). While this method is fast enough to estimate binding energy in complex, the detailed inter- and intra-species interactions are not transferred properly due to the use of implicit solvation method(s). In the 3D-RISM-KH based binding energy calculation method, the MM/PBSA part is replaced with the 3D-RISM-KH calculations using solvent distribution functions around solutes in the MD simulation trajectory. The PB/GB polar and solvent accessible surface area (SASA) nonpolar solvation terms are replaced with the solvation free energy term from Equation (10). This modification was shown to be equally effective as of traditional MM-PBSA or MM-GBSA methods. The most notable difference was reported between the 3D-RISM-KH and SASA computed non-bonded terms. The later was found to be always favoring binding, whereas the former was not [91,94]. The 3D-RISM-KH based binding calculations were also reported for other protein-ligand complexes and host-guest complexes [95–97].

6. Molecular Solvation Energy Calculations

The excess chemical potential obtained from 3D-RISM-KH calculations are theoretically a direct measure of solvation energy. However, as mentioned previously, due to erroneous calculations of internal pressure, the computed solvation energy (Gaussian fluctuation excess chemical potential) showed large deviation from actual experimental solvation energy. Other possible reasons for deviations in calculated solvation energy in the 3D-RISM are approximate nature of the closure relations, absence of explicit cavitation energy terms, and inadequacy in force field terms, etc. Incomplete sampling of solutes and short simulation times are also responsible in errors in solvation energy calculations, which reflects in physical property calculations. A correction scheme was developed in order to address this shortcoming, initially for solvation free energy [98]. This so-termed universal correction has the form:

$$\Delta G_{\text{hydration}} = \Delta G^{\text{GF}}{}_{\text{hydration, RISM}} + a \times \text{PMV} + b \quad (21)$$

The partial molar volume (PMV) of a solute in water is an output of a RISM calculation. The coefficients a and b were obtained from regression analysis against the experimental solvation energy database of Mobley and co-workers [99]. For hydration energy calculations with the 3D-RISM-KH formalism, a modified correction scheme was reported by Truchon et al. which aimed to account for the cavitation energy term [100]. Further development of solvation energy calculations in various solvents were reported from the lab of the authors, both in the context of exploring liquid state of pure solvents as well as in calculating molecular partitioning and permeability properties. Effect of atomic charge assignment schemes in hydration free energy calculation was reported by Roy et al. for an extended database of compounds with experimental hydration free energy [37]. Literature reports on hydration free energy calculations with the 3D-RISM-KH theory obtained excellent results with the GAFF force field parameters of the solute with the modified point charge models of water. While water is one of the most polar solvents used in biochemical simulations, the RISM-KH theory is extended to non-polar solvents too. Non-polar solvents that are modeled using the 3D-RISM-KH theory are hydrocarbons (hexadecane, cyclohexane), haloalkane (chloroform), and alcohol (n-octanol, t-butanol) [41,47,101,102]. Other solvents like nitro-compounds (nitromethane, nitroethane, and nitrobenzene), acetonitrile, and dimethyl sulfoxide (DMSO) were also used in RISM-KH calculations for both liquid structure and solvation energetic studies [42,103,104]. The performance of the 3D-RISM-KH calculations is summarized in Table 1. The performance of the 3D-RISM-KH theory in solvation free energy calculation, in comparison to the performance of MD and/or quantum chemical models, together with computation speed made this theory ideal for such applications.

Table 1. Performance of the 3D-RISM-KH theory in predicting solvation free energy of solutes in various solvents reported in the literature. Performance of different computational method in solvation free energy calculation is provided in parentheses.

Solvent	Dielectric Constant	No. of Solutes	Accuracy (Kcal/Mol)	Reference
Water	78.5	504	0.91–0.95 [a] (1.51) [c]	[99]
			0.89 [a]	[100]
n-Octanol	9.86	205	0.94 [b]	[41]
		158	1.03 [b]	[102]
Cyclohexane	2.0165	91	1.12 [a]	[37,101]
Hexadecane	2.0402	189	0.88 [a]	[37,101]
Chloroform	4.7113	105	0.75 [a]	[37,101]
Acetonitrile	35.688	7	2.2 [a] (1.9) [d]	[47]
Nitromethane	36.562	7	1.32 [a] (1.83) [d]	[103]
Nitroethane	28.29	7	0.38 [a] (2.00) [d]	[103]
Nitrobenzene	34.809	15	0.88 [a] (2.91) [d]	[103]
DMSO	46.826	8	2.09 [a]	[42]

[a] Mean absolute error. [b] Relative mean square error. [c] RMSE computed from MD simulation in ref. [99]. [d] RMSE computed using CPCM continuum solvation model on Minnesota solvation database [105].

7. Conclusions

The 3D-RISM theory is under continuous development, and the range of the application of this theory is ever expanding. For biomolecular simulations, MTS-MD provides a platform to combine fast dynamics of the solute with slow solute-solvent dynamics, and is proven to be able to avoid the local minima problem in molecular dynamics. Several important modifications covering the algorithm of the 3D-RISM code for application with massive parallel computer architectures were reported [106–108]. The algorithm was also coupled with density functional theory based electronic grids for electronic structure calculations [33,109]. It is important to understand that 3D-RISM-KH molecular solvation theory deals with liquid state. Thus, a direct comparison of the simulation results obtained from 3D-RISM calculations with structures determined from solid state experiments (e.g., solid-state X-Ray, neutron diffraction, etc.) may not result in a great match. For a better comparison, data obtained from experiments with liquid state should be used. Further, the Gaussian fluctuation excess chemical potentials from a 3D-RISM calculation should not be taken as an absolute measure of solvation free energy. For solvation energy calculations, the results should be compared against experimental datasets and should be fitted for use against a test set, should such a need arise. The 3D-RISM calculations provide a unique machinery to represent liquid medium with specific concentrations of cosolvent(s), additives, etc., and thus providing an opportunity to model a more realistic environment. The theory is extendable to multiphasic systems with inhomogeneous version of molecular solvation theory. However, one should keep in mind that the 3D-RIM-KH theory is not one a "size fits all" theory. It requires detailed benchmarking of every aspects of a specific problem before using it for predictive modeling.

Author Contributions: The manuscript was written through equal contributions of all the authors. Both authors have read and agreed to the published version of the manuscript.

Funding: This work was financially supported by the NSERC Discovery Grant (RES0029477), and the Alberta Prion Research Institute Explorations VII Research Grant (RES0039402).

Informed Consent Statement: Not applicable.

Data Availability Statement: All data are available in the original research articles referenced in this review.

Acknowledgments: Generous computing time provided by WestGrid (www.westgrid.ca) and Compute Canada/Calcul Canada (www.computecanada.ca) is acknowledged in completing different sub-projects resulting in developments of the 3D-RISM-KH simulation protocols in AK's laboratory.

Conflicts of Interest: The authors declare no conflict of interest.

References

1. Alder, B.J.; Wainwright, T.E. Studies in Molecular Dynamics. I. General Method. *J. Chem. Phys.* **1959**, *31*, 459–466. [CrossRef]
2. Schlick, T. *Molecular Modeling and Simulation: An Interdisciplinary Guide*, 2nd ed.; Springer: Berlin/Heidelberg, Germany, 2010.
3. Tuckerman, M. *Statistical Mechanics: Theory and Molecular Simulation*; Oxford University Press: Oxford, UK, 2010.
4. Frenkel, D.; Smit, B. *Understanding Molecular Simulation: From Algorithms to Applications*, 2nd ed.; Elsevier: Amsterdam, The Netherlands, 2002.
5. Harrison, J.A.; Schall, J.D.; Maskey, S.; Mikulski, P.T.; Knippenberg, M.T.; Morrow, B.H. Review of force fields and intermolecular potentials used in atomistic computational materials research. *Appl. Phys. Rev.* **2018**, *5*, 031104. [CrossRef]
6. Lopes, P.E.M.; Guvench, O.; MacKerell, A.D., Jr. Current Status of Protein Force Fields for Molecular Dynamics. *Methods Mol. Biol.* **2015**, *1215*, 47–71. [PubMed]
7. Martín-García, F.; Papaleo, E.; Gomez-Puertas, P.; Boomsma, W.; Lindorff-Larsen, K. Comparing Molecular Dynamics Force Fields in the Essential Subspace. *PLoS ONE* **2015**, *10*, e0121114. [CrossRef] [PubMed]
8. Ploetz, E.A.; Karunaweera, S.; Bentenitis, N.; Chen, F.; Dai, S.; Gee, M.B.; Jiao, Y.; Kang, M.; Kariyawasam, N.L.; Naleem, N.; et al. Kirkwood–Buff-Derived Force Field for Peptides and Proteins: Philosophy and Development of KBFF20. *J. Chem. Theory Comput.* **2021**. [CrossRef]
9. Ploetz, E.A.; Smith, P.E. A Kirkwood–Buff force field for the aromatic amino acids. *Phys. Chem. Chem. Phys.* **2011**, *13*, 18154–18167. [CrossRef] [PubMed]
10. Weerasinghe, S.; Smith, P.E. A Kirkwood-Buff derived force field for the simulation of aqueous guanidinium chloride solutions. *J. Chem. Phys.* **2004**, *121*, 2180–2186. [CrossRef] [PubMed]
11. Bernardi, R.C.; Melo, M.C.R.; Schulten, K. Enhanced sampling techniques in molecular dynamics simulations of biological systems. *Biochim. Biophys. Acta* **2015**, *1850*, 872–877. [CrossRef]
12. Liao, A.; Parrinello, M. Escaping free-energy minima. *Proc. Natl. Acad. Sci. USA* **2002**, *99*, 12562–12566. [CrossRef] [PubMed]
13. Li, Z.; Schheraga, H. Monte Carlo-minimization approach to the multiple-minima problem in protein folding. *Proc. Natl. Acad. Sci. USA* **1987**, *84*, 6611–6615. [CrossRef]
14. Sicher, M.; Mohr, S.; Goedecker, S. Efficient moves for global geometry optimization methods and their application to binary systems. *J. Chem. Phys.* **2011**, *134*, 044106. [CrossRef]
15. Chmiela, S.; Sauceda, H.E.; Müller, K.-R.; Takatchenko, A. Towards exact molecular dynamics simulations with machine-learned force fields. *Nat. Commun.* **2018**, *9*, 3887. [CrossRef] [PubMed]
16. Ramabhadran, R.O.; Raghavachari, K. Extrapolation to the Gold-Standard in Quantum Chemistry: Computationally Efficient and Accurate CCSD(T) Energies for Large Molecules Using an Automated Thermochemical Hierarchy. *J. Chem. Theory Comput.* **2013**, *9*, 3986–3994. [CrossRef] [PubMed]
17. Bartlett, R.J. How and Why Coupled-Cluster Theory Became the Pre-eminent Method in ab initio Quantum Chemistry. In *Theory and Applications of Computational Chemistry: The First Fifty Years*; Dykstra, C.E., Frenking, G., Kim, K.S., Scuseria, G.E., Eds.; Elsevier: Amsterdam, The Netherlands, 2005; pp. 1191–1221.
18. Smith, J.S.; Nabgen, B.T.; Zubatyuk, R.; Lubbers, N.; Devereux, C.; Barros, K.; Tretiak, S.; Isayev, O.; Roitberg, A.E. Approaching coupled cluster accuracy with a general-purpose neural network potential through transfer learning. *Nat. Commun.* **2019**, *10*, 2903. [CrossRef] [PubMed]
19. Bartlett, R.J.; Lotrich, V.F.; Schweigert, I.V. Ab initio density functional theory: The best of both worlds? *J. Chem. Phys.* **2005**, *123*, 062205. [CrossRef] [PubMed]
20. Cossi, M.; Barone, V.; Cammi, R.; Tomasi, J. Ab initio study of solvated molecules: A new implementation of the polarizable continuum model. *Chem. Phys. Lett.* **1996**, *255*, 327–335. [CrossRef]
21. Mennucci, B.; Tomasi, J. Continuum solvation models: A new approach to the problem of solute's charge distribution and cavity boundaries. *J. Chem. Phys.* **1997**, *106*, 5151–5158. [CrossRef]
22. Tomasi, J.; Mennucci, B.; Cammi, R. Quantum mechanical continuum solvation models. *Chem. Rev.* **2005**, *105*, 2999–3093. [CrossRef]
23. Marenich, A.V.; Cramer, C.J.; Truhlar, D.G. Universal solvation model based on solute electron density and a continuum model of the solvent defined by the bulk dielectric constant and atomic surface tensions. *J. Phys. Chem. B* **2009**, *113*, 6378–6396. [CrossRef]
24. Klamt, A. The COSMO and COSMO-RS solvation models. *WIREs Comput. Mol. Sci.* **2011**, *1*, 699–709. [CrossRef]
25. Svensson, M.; Humbel, S.; Froese Robert, D.J.; Matsubara, T.; Sieber, S.; Morokuma, K. ONIOM: A Multilayered Integrated MO + MM Method for Geometry Optimizations and Single Point Energy Predictions. A Test for Diels−Alder Reactions and Pt(P(t-Bu)3)2+ H_2Oxidative Addition. *J. Phys. Chem.* **1996**, *100*, 19357. [CrossRef]
26. Senn, H.; Thiel, W. QM/MM studies of enzymes. *Curr. Opin. Chem. Biol.* **2007**, *11*, 182–187. [CrossRef]
27. Vreven, T.; Byun, K.S.; Komáromi, I.; Dapprich, S.; Montgomery, J.A., Jr.; Morokuma, K.; Frisch, M.J. Combining quantum mechanics methods with molecular mechanics methods in ONIOM. *J. Chem. Theory Comput.* **2006**, *2*, 815–826. [CrossRef] [PubMed]
28. Ahmadi, S.; Herrera, L.B.; Chehelamirani, M.; Hostaš, J.; Jalife, S.; Salahub, D.R. Multiscale modeling of enzymes: QM-cluster, QM/MM, and QM/MM/MD: A tutorial review. *Int. J. Quant. Chem.* **2018**, *118*, e25558. [CrossRef]
29. Hansen, J.-P.; McDonald, I.R. Chapter 11—Molecular Liquids. In *Theory of Simple Liquids*, 4th ed.; Hansen, J.-P., McDonald, I.R., Eds.; Academic Press: Oxford, UK, 2013; pp. 455–510.

30. Kovalenko, A. *Molecular Theory of Solvation*; Chapter 4; Hirata, F., Ed.; Kluwer Academic Publishers: New York, NY, USA, 2003; Volume 24, pp. 169–276.
31. Ratkova, E.L.; Palmer, D.S.; Fedorov, M.V. Solvation Thermodynamics of Organic Molecules by the Molecular Integral Equation Theory: Approaching Chemical Accuracy. *Chem. Rev.* **2015**, *115*, 6312. [CrossRef]
32. Luchko, T.; Gusarov, S.; Roe, D.R.; Simmerling, C.; Case, D.A.; Tuszynski, J.; Kovalenko, A. Three-Dimensional Molecular Theory of Solvation Coupled with Molecular Dynamics in Amber. *J. Chem. Theory Comput.* **2010**, *6*, 607. [CrossRef] [PubMed]
33. Gusarov, S.; Ziegler, T.; Kovalenko, A. Self-Consistent Combination of the Three-Dimensional RISM Theory of Molecular Solvation with Analytical Gradients and the Amsterdam Density Functional Package. *J. Phys. Chem. A* **2006**, *110*, 6083. [CrossRef]
34. Casanova, D.; Gusarov, S.; Kovalenko, A.; Ziegler, T. Evaluation of the SCF Combination of KS-DFT and 3D-RISM-KH; Solvation Effect on Conformational Equilibria, Tautomerization Energies, and Activation Barriers. *J. Chem. Theory Comput.* **2007**, *3*, 458. [CrossRef]
35. Aono, S.; Mori, T.; Sakaki, S. 3D-RISM-MP2 Approach to Hydration Structure of Pt(II) and Pd(II) Complexes: Unusual H-Ahead Mode vs Usual O-Ahead One. *J. Chem. Theory Comput.* **2016**, *12*, 1189–1206. [CrossRef]
36. Sosnis, S.; Misin, M.; Palmer, D.S.; Fedorov, M.V. 3D matters! 3D-RISM and 3D convolutional neural network for accurate bioaccumulation prediction. *J. Phys. Condens. Matter* **2018**, *30*, 32LT03. [CrossRef]
37. Roy, D.; Kovalenko, A. Performance of 3D-RISM-KH in Predicting Hydration Free Energy: Effect of Solute Parameters. *J. Phys. Chem. A* **2019**, *123*, 4087–4093. [CrossRef]
38. Nikolić, D.; Blinov, N.; Wishart, D.; Kovalenko, A. 3D-RISM-Dock: A New Fragment-Based Drug Design Protocol. *J. Chem. Theory Comput.* **2012**, *8*, 3356–3372. [CrossRef] [PubMed]
39. Kiyota, Y.; Yoshida, N.; Hirata, F. A New Approach for Investigating the Molecular Recognition of Protein: Toward Structure-Based Drug Design Based on the 3D-RISM Theory. *J. Chem. Theory Comput.* **2011**, *7*, 3803–3815. [CrossRef]
40. Palmer, D.S.; Mišin, M.; Fedorov, M.V.; Llinas, A. Fast and General Method To Predict the Physicochemical Properties of Druglike Molecules Using the Integral Equation Theory of Molecular Liquids. *Mol. Pharm.* **2015**, *12*, 3420–3432. [CrossRef] [PubMed]
41. Roy, D.; Hinge, V.K.; Kovalenko, A. To Pass or Not To Pass: Predicting the Blood–Brain Barrier Permeability with the 3D-RISM-KH Molecular Solvation Theory. *ACS Omega* **2019**, *4*, 16774–16780. [CrossRef]
42. Roy, D.; Kovalenko, A. Application of the Approximate 3D-Reference Interaction Site Model (RISM) Molecular Solvation Theory to Acetonitrile as Solvent. *J. Phys. Chem. B* **2020**, *124*, 4590–4597. [CrossRef] [PubMed]
43. Subramanian, V.; Ratkova, E.; Palmer, D.; Engkvist, E.; Fedorov, M.; Llinas, A. Multisolvent Models for Solvation Free Energy Predictions Using 3D-RISM Hydration Thermodynamic Descriptors. *J. Chem. Inf. Model.* **2020**, *60*, 2977–2988. [CrossRef] [PubMed]
44. Roy, D.; Dutta, D.; Wishsart, D.S.; Kovalenko, A. Predicting PAMPA permeability using the 3D-RISM-KH theory: Are we there yet? *J. Comput. Aided Mol. Des.* **2021**, *35*, 261–269. [CrossRef]
45. Hinge, V.K.; Roy, D.; Kovalenko, A. Predicting skin permeability using the 3D-RISM-KH theory based solvation energy descriptors for a diverse class of compounds. *J. Comput. Aided Mol. Des.* **2019**, *33*, 605–611. [CrossRef]
46. Hinge, V.K.; Roy, D.; Kovalenko, A. Prediction of P-glycoprotein inhibitors with machine learning classification models and 3D-RISM-KH theory based solvation energy descriptors. *J. Comput. Aided Mol. Des.* **2019**, *33*, 965–971. [CrossRef]
47. Huang, W.J.; Blinov, N.; Kovalenko, A. Octanol-Water Partition Coefficient from 3D-RISM-KH Molecular Theory of Solvation with Partial Molar Volume Correction. *J. Phys. Chem. B* **2015**, *119*, 5588–5597. [CrossRef] [PubMed]
48. Luchko, T.; Blinov, N.; Limon, G.C.; Joyce, K.P.; Kovalenko, A. SAMPL5: 3D-RISM partition coefficient calculations with partial molar volume corrections and solute conformational sampling. *J. Comput. Aided Mol. Des.* **2016**, *30*, 1115–1127. [CrossRef]
49. Chandler, D. Equilibrium structure and molecular motion in liquids. *Acc. Chem. Res.* **1974**, *7*, 246. [CrossRef]
50. Lowden, L.J.; Chandler, D. Solution of a new integral equation for pair correlation functions in molecular liquids. *J. Chem. Phys.* **1973**, *59*, 6587. [CrossRef]
51. Lowden, L.J.; Chandler, D. Theory of intermolecular pair correlations for molecular liquids. Applications to the liquids carbon tetrachloride, carbon disulfide, carbon diselenide, and benzene. *J. Chem. Phys.* **1974**, *61*, 5228. [CrossRef]
52. Chandler, D. Derivation of an integral equation for pair correlation functions in molecular fluids. *J. Chem. Phys.* **1973**, *59*, 2742. [CrossRef]
53. Chandler, D.; Hsu, C.S.; Street, W.B. Comparisons of Monte Carlo and RISM calculations of pair correlation functions. *J. Chem. Phys.* **1977**, *66*, 5231. [CrossRef]
54. Singer, S.J.; Chandler, D. Free energy functions in the extended RISM approximation. *Mol. Phys.* **1985**, *55*, 621. [CrossRef]
55. Chandler, D.; Silbey, R.; Ladanyi, B.M. New and proper integral equations for site-site equilibrium correlations in molecular fluids. *Mol. Phys.* **1982**, *46*, 1335. [CrossRef]
56. Richardson, D.M.; Chandler, D. Calculation of orientational pair correlation factors with the interaction site formalism. *J. Chem. Phys.* **1984**, *80*, 4484. [CrossRef]
57. Perkyns, J.; Pettitt, B.M. A site–site theory for finite concentration saline solutions. *J. Chem. Phys.* **1992**, *97*, 7656. [CrossRef]
58. Kovalenko, A.; Hirata, F. Potentials of mean force of simple ions in ambient aqueous solution. I. Three-dimensional reference interaction site model approach. *J. Chem. Phys.* **2000**, *112*, 10391. [CrossRef]
59. Kovalenko, A. Multiscale modeling of solvation in chemical and biological nanosystems and in nanoporous materials. *Pure Appl. Chem.* **2013**, *85*, 159–199. [CrossRef]

60. Sergiievskyi, V.; Jeanmairet, G.; Levesque, M.; Borgis, D. Solvation free-energy pressure corrections in the three dimensional reference interaction site model. *J. Chem. Phys.* **2015**, *143*, 184116. [CrossRef] [PubMed]
61. Misin, M.; Vainikka, P.A.; Fedorov, M.V.; Palmer, D.S. Salting-out effects by pressure-corrected 3D-RISM. *J. Chem. Phys.* **2016**, *145*, 194501. [CrossRef]
62. Pettitt, B.M.; Rossky, P.J. Integral equation predictions of liquid state structure for waterlike intermolecular potentials. *J. Chem. Phys.* **1982**, *77*, 1451–1457. [CrossRef]
63. Hirata, F.; Levy, R.M. A new RISM integral equation for solvated polymers. *Chem. Phys. Lett.* **1987**, *136*, 267–273. [CrossRef]
64. Noé, F. Beating the Millisecond Barrier in Molecular Dynamics Simulations. *Biophys. J.* **2015**, *108*, 228. [CrossRef]
65. Shaw, D.E. Millisecond-long molecular dynamics simulations of proteins on a special-purpose machine. *Biophys. J.* **2013**, *104*, 45A. [CrossRef]
66. Voelz, V.A.; Bowman, G.R.; Beauchamp, K.; Pande, V.S. Molecular Simulation of ab Initio Protein Folding for a Millisecond Folder NTL9(1−39). *J. Am. Chem. Soc.* **2010**, *132*, 1526. [CrossRef]
67. Miyata, T.; Hirata, F. Combination of molecular dynamics method and 3D-RISM theory for conformational sampling of large flexible molecules in solution. *J. Comput. Chem.* **2008**, *29*, 871–882. [CrossRef]
68. Omelyan, I.; Kovalenko, A. MTS-MD of Biomolecules Steered with 3D-RISM-KH Mean Solvation Forces Accelerated with Generalized Solvation Force Extrapolation. *J. Chem. Theory Comput.* **2015**, *11*, 1875–1895. [CrossRef]
69. Omelyan, I.; Kovalenko, A. Enhanced solvation force extrapolation for speeding up molecular dynamics simulations of complex biochemical liquids. *J. Chem. Phys.* **2019**, *151*, 214102. [CrossRef]
70. Murray, C.W.; Rees, D.C. The rise of fragment-based drug discovery. *Nat. Chem.* **2009**, *1*, 187–192. [CrossRef]
71. Tounge, B.A.; Parker, M.H. Chapter one—Designing a Diverse High-Quality Library for Crystallography-Based FBDD Screening. *Methods Enzymol.* **2011**, *493*, 3–20.
72. Imai, T.; Oda, K.; Kovalenko, A.; Hirata, F.; Kidera, A. Ligand Mapping on Protein Surfaces by the 3D-RISM Theory: Toward Computational Fragment-Based Drug Design. *J. Am. Chem. Soc.* **2009**, *131*, 12430. [CrossRef]
73. Sugita, M.; Hamano, M.; Kasahara, K.; Kikuchi, T.; Hirata, F. New Protocol for Predicting the Ligand-Binding Site and Mode Based on the 3D-RISM/KH Theory. *J. Chem. Theory Comput.* **2020**, *16*, 2864. [CrossRef] [PubMed]
74. Imai, T. A Novel Ligand-Mapping Method Based on Molecular Liquid Theory. *Curr. Pharm. Des.* **2011**, *17*, 1685–1694. [CrossRef] [PubMed]
75. Lemmon, G.; Meiler, J. Towards Ligand Docking Including Explicit Interface Water Molecules. *PLoS ONE* **2013**, *8*, e67536. [CrossRef] [PubMed]
76. Ross, G.A.; Morris, G.M.; Biggin, P.C. Rapid and Accurate Prediction and Scoring of Water Molecules in Protein Binding Sites. *PLoS ONE* **2012**, *7*, e32036. [CrossRef]
77. Rudling, A.; Orro, A.; Carlsson, J. Prediction of Ordered Water Molecules in Protein Binding Sites from Molecular Dynamics Simulations: The Impact of Ligand Binding on Hydration Networks. *J. Chem. Inf. Model.* **2018**, *58*, 350–361. [CrossRef]
78. Imai, T.; Hiraoka, R.; Kovalenko, A.; Hirta, F. Water Molecules in a Protein Cavity Detected by a Statistical−Mechanical Theory. *J. Am. Chem. Soc.* **2005**, *127*, 15334. [CrossRef] [PubMed]
79. Sindhikara, D.J.; Yoshida, N.; Hirata, F. Placevent: An algorithm for prediction of explicit solvent atom distribution-application to HIV-1 protease and F-ATP synthase. *J. Comput. Chem.* **2012**, *33*, 1536. [CrossRef] [PubMed]
80. Hinge, V.K.; Blinov, N.; Roy, D.; Wishart, D.S.; Kovalenko, A. The role of hydration effects in 5-fluorouridine binding to SOD1: Insight from a new 3D-RISM-KH based protocol for including structural water in docking simulations. *J. Comput. Aided Mol. Des.* **2019**, *33*, 913. [CrossRef]
81. *Molecular Operating Environment (MOE), 2019.01*; Chemical Computing Group ULC: Montreal, QC, Canada, 2021.
82. Nukaga, M.; Yoon, M.J.; Taracilia, M.A.; Hoshino, T.; Becka, S.A.; Zeiser, E.T.; Johnson, J.R.; Papp-Wallace, K.M. Assessing the Potency of β-Lactamase Inhibitors with Diverse Inactivation Mechanisms against the PenA1 Carbapenemase from *Burkholderia multivorans*. *ACS Infect. Dis.* **2021**, *7*, 826–837. [CrossRef]
83. Hüfner-Wulsdorf, T.; Klebe, G. Mapping Water Thermodynamics on Drug Candidates via Molecular Building Blocks: A Strategy to Improve Ligand Design and Rationalize SAR. *J. Med. Chem.* **2021**. [CrossRef]
84. Aggarwal, L.; Biswas, P. Hydration Thermodynamics of Familial Parkinson's Disease-Linked Mutants of α-Synuclein. *J. Chem. Inf. Model.* **2021**. [CrossRef]
85. Nguyen, C.; Yamazaki, T.; Kovalenko, A.; Case, D.A.; Gilson, M.K.; Kurtzman, T.; Luchko, T. A molecular reconstruction approach to site-based 3D-RISM and comparison to GIST hydration thermodynamic maps in an enzyme active site. *PLoS ONE* **2019**, *14*, e0219473. [CrossRef]
86. Blinov, N.; Wishart, D.S.; Kovalenko, A. Solvent Composition Effects on the Structural Properties of the Aβ42 Monomer from the 3D-RISM-KH Molecular Theory of Solvation. *J. Phys. Chem. B* **2019**, *123*, 2491–2506. [CrossRef] [PubMed]
87. Ariz-Extreme, I.; Hub, J.S. Potential of Mean Force Calculations of Solute Permeation across UT-B and AQP1: A Comparison between Molecular Dynamics and 3D-RISM. *J. Phys. Chem. B* **2017**, *121*, 1506–1519. [CrossRef] [PubMed]
88. Lee, F.S.; Chu, Z.-T.; Bolger, M.B.; Warshel, A. Calculations of antibody-antigen interactions: Microscopic and semi-microscopic evaluation of the free energies of binding of phosphorylcholine analogs to McPC603. *Protein Eng.* **1992**, *5*, 215–228. [CrossRef]
89. Sham, Y.Y.; Chu, Z.T.; Tao, H.; Warshel, A. Examining methods for calculations of binding free energies: LRA, LIE, PDLD-LRA, and PDLD/S-LRA calculations of ligands binding to an HIV protease. *Proteins: Struct. Funct. Genet.* **2000**, *39*, 393–407. [CrossRef]

90. Warshel, A.; Sharma, P.K.; Kato, M.; Parson, W.W. Modeling electrostatic effects in proteins. *Biochim. Biophys. Acta* **2006**, *1764*, 1647–1676. [CrossRef]
91. Åqvist, J.; Medina, C.; Samuelsson, J.E. A new method for predicting binding affinity in computer-aided drug design. *Protein Eng.* **1994**, *7*, 385–391. [CrossRef]
92. Hansson, T.; Marelius, J.; Åqvist, J. Ligand binding affinity prediction by linear interaction energy methods. *J. Comput. Aided Mol. Des.* **1998**, *12*, 27–35. [CrossRef]
93. Kollman, P.A.; Massova, I.; Reyes, C.; Kuhn, B.; Huo, S.; Chong, L.; Lee, M.; Lee, T.; Duan, Y.; Wang, W.; et al. Calculating Structures and Free Energies of Complex Molecules: Combining Molecular Mechanics and Continuum Models. *Acc. Chem. Res.* **2000**, *33*, 889–897. [CrossRef] [PubMed]
94. Genheden, S.; Luchko, T.; Gusarov, S.; Kovalenko, A.; Ryde, U. An MM/3D-RISM Approach for Ligand Binding Affinities. *J. Phys. Chem. B* **2010**, *114*, 8505–8516. [CrossRef] [PubMed]
95. Sugita, M.; Kuwano, I.; Higashi, T.; Motoyama, K.; Arima, H.; Hirata, F. Computational Screening of a Functional Cyclodextrin Derivative for Suppressing a Side Effect of Doxorubicin. *J. Phys. Chem. B* **2021**, *125*, 2308–2316. [CrossRef]
96. Suárez, D.; Díaz, N. Affinity Calculations of Cyclodextrin Host–Guest Complexes: Assessment of Strengths and Weaknesses of End-Point Free Energy Methods. *J. Chem. Inf. Model.* **2019**, *59*, 421–440. [CrossRef]
97. Miller, B.R., III; McGee, T.D., Jr.; Swails, J.M.; Homeyer, N.; Gohlke, H.; Roitberg, A.E. MMPBSA.py: An Efficient Program for End-State Free Energy Calculations. *J. Chem. Theory Comput.* **2012**, *8*, 3314–3321. [CrossRef] [PubMed]
98. Palmer, D.S.; Frolov, A.I.; Ratkova, E.L.; Fedorov, M.V. Towards a universal method for calculating hydration free energies: A 3D reference interaction site model with partial molar volume correction. *J. Phys. Condens. Matter* **2010**, *22*, 492101. [CrossRef]
99. Mobley, D.L.; Guthrie, J.P. FreeSolv: A database of experimental and calculated hydration free energies, with input files. *J. Comput. Aided Mol. Des.* **2014**, *28*, 711–720. [CrossRef] [PubMed]
100. Truchon, J.-F.; Pettit, B.M.; Labute, O. A Cavity Corrected 3D-RISM Functional for Accurate Solvation Free Energies. *J. Chem. Theory Comput.* **2014**, *10*, 934. [CrossRef]
101. Roy, D.; Hinge, V.K.; Kovalenko, A. Predicting Blood–Brain Partitioning of Small Molecules Using a Novel Minimalistic Descriptor-Based Approach via the 3D-RISM-KH Molecular Solvation Theory. *ACS Omega* **2019**, *4*, 3055–3060. [CrossRef]
102. Roy, D.; Blinov, N.; Kovalenko, A. Predicting Accurate Solvation Free Energy in n-Octanol Using 3D-RISM-KH Molecular Theory of Solvation: Making Right Choices. *J. Phys. Chem. B* **2017**, *121*, 9268–9273. [CrossRef] [PubMed]
103. Roy, D.; Kovalenko, A. Application of the 3D-RISM-KH molecular solvation theory for DMSO as solvent. *J. Comput. Aided Mol. Des.* **2019**, *33*, 905–912. [CrossRef] [PubMed]
104. Roy, D.; Kovalenko, A. A 3D-RISM-KH study of liquid nitromethane, nitroethane, and nitrobenzene as solvents. *J. Mol. Liq.* **2021**, *332*, 115857. [CrossRef]
105. Marenich, A.V.; Kelly, C.P.; Thompson, J.D.; Hawkins, G.D.; Chambers, C.C.; Giesen, D.J.; Winget, P.; Cramer, C.J.; Truhlar, D.G. *Minnesota Solvation Database—Version 2012*; University of Minnesota: Minneapolis, MN, USA, 2012.
106. Maruyama, Y.; Yoshida, N.; Tadano, H.; Takahashi, D.; Sato, M.; Hirata, F. Massively parallel implementation of 3D-RISM calculation with volumetric 3D-FFT. *J. Comput. Chem.* **2014**, *35*, 1347–1355. [CrossRef]
107. Maruyama, Y.; Hirata, F. Modified Anderson Method for Accelerating 3D-RISM Calculations Using Graphics Processing Unit. *J. Chem. Theory Comput.* **2012**, *8*, 3015–3021. [CrossRef] [PubMed]
108. Onishi, I.; Tsuji, H.; Irisa, M. A tool written in Scala for preparation and analysis in MD simulation and 3D-RISM calculation of biomolecules. *Biophys. Physicobiol.* **2019**, *16*, 485–489. [CrossRef]
109. Reimann, M.; Kaupp, M. Evaluation of an Efficient 3D-RISM-SCF Implementation as a Tool for Computational Spectroscopy in Solution. *J. Phys. Chem. A* **2020**, *124*, 7439–7452. [CrossRef] [PubMed]

International Journal of
Molecular Sciences

MDPI

Review

"Dividing and Conquering" and "Caching" in Molecular Modeling

Xiaoyong Cao [1] and Pu Tian [1,2,*]

1 School of Life Sciences, Jilin University, Changchun 130012, China
2 School of Artificial Intelligence, Jilin University, Changchun 130012, China
* Correspondence: tianpu@jlu.edu.cn; Tel.: +86-(0)431-85155287

Abstract: Molecular modeling is widely utilized in subjects including but not limited to physics, chemistry, biology, materials science and engineering. Impressive progress has been made in development of theories, algorithms and software packages. To divide and conquer, and to cache intermediate results have been long standing principles in development of algorithms. Not surprisingly, most important methodological advancements in more than half century of molecular modeling are various implementations of these two fundamental principles. In the mainstream classical computational molecular science, tremendous efforts have been invested on two lines of algorithm development. The first is coarse graining, which is to represent multiple basic particles in higher resolution modeling as a single larger and softer particle in lower resolution counterpart, with resulting force fields of partial transferability at the expense of some information loss. The second is enhanced sampling, which realizes "dividing and conquering" and/or "caching" in configurational space with focus either on reaction coordinates and collective variables as in metadynamics and related algorithms, or on the transition matrix and state discretization as in Markov state models. For this line of algorithms, spatial resolution is maintained but results are not transferable. Deep learning has been utilized to realize more efficient and accurate ways of "dividing and conquering" and "caching" along these two lines of algorithmic research. We proposed and demonstrated the local free energy landscape approach, a new framework for classical computational molecular science. This framework is based on a third class of algorithm that facilitates molecular modeling through partially transferable in resolution "caching" of distributions for local clusters of molecular degrees of freedom. Differences, connections and potential interactions among these three algorithmic directions are discussed, with the hope to stimulate development of more elegant, efficient and reliable formulations and algorithms for "dividing and conquering" and "caching" in complex molecular systems.

Keywords: molecular modeling; multiscale; coarse graining; molecular dynamics simulation; Monte Carlo simulation; force fields; neural network; many body interactions; sampling; local sampling; local free energy landscape; generalized solvation free energy

Citation: Cao, X.; Tian, P. "Dividing and Conquering" and "Caching" in Molecular Modeling. *Int. J. Mol. Sci.* **2021**, *22*, 5053. https://doi.org/10.3390/ijms22095053

Academic Editor: Małgorzata Borówko

Received: 3 March 2021
Accepted: 27 April 2021
Published: 10 May 2021

1. Introduction

The impact of molecular modeling in scientific research is clearly embodied by the number of publications. Statistics from a Web of Science (www.webofknowledge.com (accessed on 15 February 2021)) search with various relevant key words is listed in Table 1. However, despite widespread applications of molecular modeling, we remain far from accurately predicting and designing molecular systems in general. Further methodological development is highly desired to tap its full potential. Historically, molecular modeling has been approached from a physical or application point of view, and numerous excellent reviews are available in this regard [1–16]. From an algorithmic perspective, "dividing and conquering" (DC) and "caching" intermediate results that need to be computed repetitively are two fundamental principles in development of many important algorithms (e.g., dynamic programming [17]). As a matter of fact, the major focus of modern statistical machine learning is to learn ("caching" relevant information) and then carry out inference on top of which [18]. In this review, we provide a brief discussion of important methodological development in molecular modeling as specific applications of these two principles. The content will be organized as follows. Part 2 describes fundamental challenges in molecular modeling; Part 3 summarizes application of these two fundamental algorithmic principles in two lines of methodological research, coarse graining (CG) [18–27] and enhanced sampling (ES) [28–31]; Part 4 covers how machine learning, particularly deep learning, facilitates DC and "caching" in CG and ES [29,30,32–35], Part 5 introduces local free energy landscape (LFEL) approach, a new framework for computational molecular science based on partially transferable in resolution "caching" of local sampling. The first implementation of this new framework in protein structural refinement based on generalized solvation free energy (GSFE) theory [36] is briefly discussed; and Part 6 discusses connections among these three lines of algorithmic development, their specific advantages and prospective explorations. Due to the large body of literature and limited space, we apologize to authors whose excellent works are not cited here.

Table 1. Number of publications from Web of Science search on 8 September 2020.

Key Words	Number of Publications
Molecular dynamics simulation	241,748
Monte Carlo simulation	189,550
QM-MM (quantum mechanical—molecular mechanical) simulation	9907
Dissipative particle dynamics simulation	3693
Langevin dynamics simulation	3893
Molecular modeling	2,072,091
All of the above	2,243,182

2. Challenges in Molecular Modeling

2.1. Accurate Description of Molecular Interactions

Molecular interactions may be accurately described with high level molecular orbital theories (e.g., coupled cluster theory [37,38]) or sophisticated density functionals combined with large basis sets [39–43]. However, such quantum mechanically detailed computation is prohibitively expensive for any realistic complex molecular systems. Molecular interactions are traditionally represented by explicit functions and pairwise approximations as exemplified by typical physics based atomistic molecular mechanical (MM) force fields (FFs) [44–47]:

$$\begin{aligned} U(\vec{R}) = & \sum_{bonds} K_b(b-b_0)^2 + \sum_{angles} K_\theta(\theta-\theta_0)^2 \\ & + \sum_{dihedral} K_\chi(1+cos(n\chi-\delta)) \\ & + \sum_{impropers} K_{imp}(\phi-\phi_0)^2 \\ & + \sum_{nonbonded} \left(\epsilon_{ij} \left[\left(\frac{Rmin_{ij}}{r_{ij}} \right)^{12} - \left(\frac{Rmin_{ij}}{r_{ij}} \right)^{6} \right] \right) + \frac{q_i q_j}{\epsilon r_{ij}} \end{aligned} \quad (1)$$

or knowledge based potential functions [48–50]: these simple functions, while being amenable to rapid computation and are physically sound grounded near local energy minima (e.g., harmonic behavior of bonding, bending near equilibrium bond lengths and bend angles), are potentially problematic for anharmonic interactions, which are very common in some molecular systems [51]. It is well understood that properly parameterized Lennard–Jones potentials are accurate only near the bottom of its potential well. Frustration are ubiquitous in biomolecular systems and are likely fundamental driving force for conformational fluctuations [52–54]. One may imagine that a molecular system with all its comprising particles at their respective "happy" energy minima positions would likely be a stable "dead" molecule, which may be a good structural support but is likely not able to provide dynamic functional behavior.

Pairwise approximations are usually adopted for its computational convenience, both in terms of dramatically reduced computational cost and tremendously smaller (when compared with possible many body potentials) number of necessary parameters to be fit in FF parameterization. It is widely acknowledged that construction of traditional FF (e.g., Equation (1)) is a laborious process. Development of polarizable [9,55,56] and more complex FF with larger parameter set [57] alleviates some shortcomings of earlier counterpart. Expansion based treatments were incorporated to address anharmonicity [58]. However, to tackle limitation of explicit simple functional form and pairwise approximation for better description of molecular interactions remains challenge to be met for molecular modeling community. Additionally, even atomistic simulations are prohibitively expensive for large biomolecular complexes at long time scales (e.g., milliseconds and beyond) [59–61].

2.2. Inherent Low Efficiency in Sampling of Configurational Space

Complexity of molecular systems is rooted in their molecular interactions, which engender complex and non-linear correlations among molecular degrees of freedom (DOFs). Consequently, effective number of DOFs are greatly reduced. Therefore, complex molecular systems are confined to manifolds [62] of much lower dimensionality with near zero measure in corresponding nominal high dimensional space (NHDS). Consequently, sufficient brute force random sampling in NHDS of interested molecular systems is hopeless.

In stochastic trajectory generation by Monte Carlo (MC) simulations or candidate structural model proposal in protein structure prediction and refinement (or other similar scenarios), new configuration proposal is carried out in NHDS. A lot of effort is inevitably wasted due to sampling outside the actual manifold occupied by the target molecular system. Such wasting may be avoided if we understood all correlations. However, understanding all correlations implicates accurate description of global free energy landscape (FEL) and there is no need to investigate it further. Due to preference of lower energy configurations by typical importance sampling strategies (e.g., Metropolis MC), stochastic trajectories tend to be trapped in local minima of FEL. This is especially true for complex molecular (e.g., biomolecular) systems which have hierarchical rugged FEL with many local minima [63,64]. In trajectory generation by molecular dynamics (MD) simulations, configurational space is explored by laws of classical mechanics and no wasting due to random moves exists. However, molecular systems may well drift away from their true manifolds due to insufficient accuracy of FF, resulting incorrect and wasted sampling. Similar to stochastic trajectory generation, it takes long simulations to map FEL since molecular system tend to staying at any local minimum, achieve equilibrium among many local minima is just as challenging as in the case of stochastic counterpart.

3. DC and "Caching" in Traditional Molecular Modeling

To cope with fundamental difficulties in molecular modeling, two distinct lines of methodological development (CG and ES) based on DC and "caching" strategies have been conducted and tremendous progress has been made in understanding of molecular systems. As summarized below:

3.1. Coarse Graining, a Partially Transferable "Caching" Strategy

Atomistic FF parameterization is the most well established coarse graining based on time scale separation between each nuclei and its surrounding electrons. Theoretically, MMFFs are potential of mean force (PMF) obtained by averaging over many electronic DOFs for given atomic configurations. In practice, due to the fact that ab initio calculations are expensive and may have significant error when level of theory (and/or basis set) is not sufficient, reference data usually include results from both quantum mechanical (QM) calculations and well-established experimental data [65–67]. The DC strategy is utilized by selecting atomic clusters of various size to facilitate generation of QM reference data. The essential information learned from reference data is then permanently and approximately "cached" in FF parameters through the parameterization process. Time scale separation ensures that elimination of electronic DOFs is straight forward but comes with the price of incapability in describing chemical reactions. To harvest benefits of both quantum and atomistic simulations, a well-established DC strategy is to treat a small region involved in interested chemical reaction at QM detail and its surrounding with MMFF [68–73]. This series of pioneering work was awarded Nobel prize in 2013, and QM-MM treatment continues to be the mainstream methodology for computational description of chemical reactions [74,75].

The united atom model (UAM) is the next step in coarse graining [76], where hydrogen atoms are merged into bonded heavy atoms. This is quite intuitive since hydrogens have much smaller mass on the one hand, and are difficult to see by experimental detection techniques utilizing electron diffraction (e.g., X-ray crystallography) on the other hand. For both polymeric and biomolecular systems, UAM remains to be expensive for many interested spatial and temporal scales. Therefore, further coarse graining in various forms have been constructed. As a matter of fact, CG is usually used to denote modeling with particles that representing multiple atoms in contrast to atomistic simulations, and the same convention will be adopted in the remaining part of this review unless stated otherwise. Both "Top-down" (that based on reproducing experimental data) and "bottom-up" (that based on reproducing certain properties of atomistic simulations) approaches are utilized [21,77]. For polymeric materials, beads are either utilized to represent monomers or defined on consideration of persistent length [78], and dissipative particle dynamics (DPD) were proposed to deal with complexities arise from much larger particles [79]. For biomolecular systems, a wide variety of coarse grained models have been developed [20,21,23,24,80,81]. Another important subject of CG methodology development is materials science [82,83]. Earlier definition of CG particles are rather ad hoc [20]. More formulations with improved statistical mechanical rigor appeared later on [22], with radial distribution function based inversion [78,84–86], entropy divergence [19] and force matching algorithm [87–89] being outstanding examples of systematic development. Present CG is essentially to realize the following mapping as disclosed by Equation (4) in ref. [22] :

$$exp[-\beta V_{CG}(\mathbf{R}_{CG})] \equiv \int d\mathbf{r}\delta(M_R(\mathbf{r}) - \mathbf{R}_{CG})exp[-\beta V(\mathbf{r})] \tag{2}$$

with $\mathbf{r}$ and $\mathbf{R}$ being coordinates in higher resolution and CG coordinates, $M_R(\mathbf{r})$ being the map operator from $\mathbf{r}$ to $\mathbf{R}$, V and V_{CG} being relevant potential of mean force in higher resolution and CG representation respectively. Due to lack of time scale separation (see Figure 1) for essentially all CG mapping, strict realization of this equation/mapping with exact transferability is not rigorously possible. A naive treatment of CG particles as basic units (with no internal degrees of freedom) would result in wrong thermodynamics [22]. Due to corresponding significant loss of information, it is not possible to develop a definition of CG and corresponding FF parameterization for comprehensive reproduction of atomistic description of corresponding molecular systems. Different coarse graining have distinct advantages and disadvantages, so choosing proper CG strategy is highly dependent upon specific goal in mind. CG particles are usually isotropic larger and softer particles with pairwise interactions, or simple convex anisotropic object (e.g., soft spheroids) that may be treated analytically [24,90–92]. Such simplifications provide both convenience of computation and certain deficiency for capturing physics of target molecular systems. CG may be carried out iteratively to address increasingly larger spatial scales by "caching" lower resolution CG distributions with ultra CG (UCG) FF [22,93–97]. Pairwise approximation remains to be limitation of interaction description for traditional CG FF, which may use either explicit simple function form or tables. When compared with atomistic FF, pairwise approximation deteriorate further due to lack of time scale separation (Figure 1).

Another simple and powerful type of CG model for biomolecular systems is Gō model [98,99] and elastic network models (ENM) [100–102] or gaussian network models (GNM) [103,104] with native structure being defined as the equilibrium state, and with quadratic/harmonic interactions between all residues within given cutoff. Only a few parameters (e.g., cutoff distance, spring constant) are needed. Such models "caching" the experimental structures and are proved to be useful in understanding major conformational transitions and slow dynamics of many biomolecular systems [105,106].

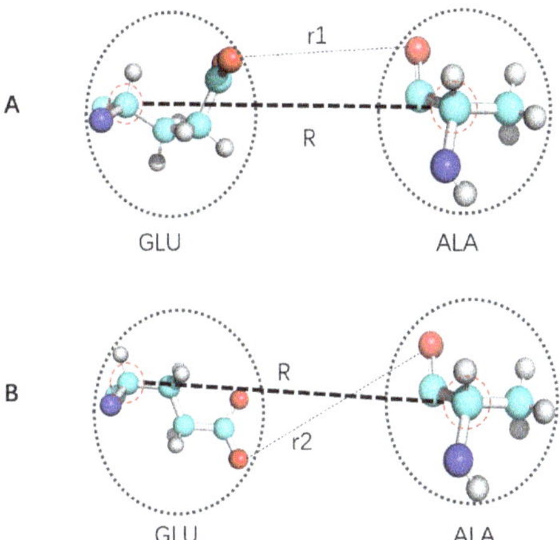

Figure 1. Schematic illustration of time scale separation issue in CG. (**A**,**B**) show two situations with C_α distances between two amino acids GLU and ALA being R, but with GLU have different conformations. If C_α atoms were defined as CG site, then these two relative conformation with distinct interactions would be treated as the same. In (**A**,**B**), CG site distance in both (**A**,**B**) are R, but many other pairs of atoms have distinct distances as exemplified by r_1 and r_2. Such treatment would only be true if for any small amount of displacement of C_α, side chains accomplished many rotations and thus may be accurately represented by averaging (i.e., with good time scale separation). This issue is apparently not limited to the specific definition of C_α being CG site, but rather general for essentially all CG development.

3.2. Enhanced Sampling, a Nontransferable in Resolution DC and "Caching" Strategy

Umbrella sampling (US) [107] is probably the first combination of DC and "caching" strategy for better sampling of molecular system along a given reaction coordinate (RC) (or order parameter) s. DC strategy is first applied by dividing s into windows, information for each window is then partially "cached" by corresponding bias potentials and local statistics. Later on, adaptive US (AUS) [108,109] and weighted histogram analysis method (WHAM) [110] were developed to improve both efficiency and accuracy. MBAR [111,112] was developed to achieve error bound analysis which was not available in WHAM. Further development including adaptive bias force (ABF) [113–115] and metadynamics [116–118]. Details of these methodologies were well explained by excellent reviews [119–122]. The common trick to all of these algorithms (and their variants) is to "cache" visited configurational space with bias potentials/force and local statistics, thus reduce time spent in local minima and dramatically accelerate sampling of interested rare events. Denote CV as $\mathbf{s}(\mathbf{r})$ ($\mathbf{r}$ being physical coordinates of atoms/particles in the target molecular system), equilibrium distribution and free energy on the CV may be expressed as [30]:

$$p_0(\mathbf{s}) = \int d\mathbf{r}\delta[\mathbf{s} - \mathbf{s}(\mathbf{r})]p_0(\mathbf{r}) = \langle\delta[\mathbf{s} - \mathbf{s}(\mathbf{r})]\rangle \tag{3}$$

$$p_0(\mathbf{r}) = \frac{e^{-\beta U(\mathbf{r})}}{\int dre^{-\beta U(\mathbf{r})}} \tag{4}$$

$$F(\mathbf{s}) = -\frac{1}{\beta}log[p_0(\mathbf{s})] \tag{5}$$

$$F(\mathbf{s}) = -\frac{1}{\beta}log[p(\mathbf{s})] - V(\mathbf{s}) \tag{6}$$

with $p(\mathbf{s})$ being the sampled distribution in simulations with corresponding bias potential $V(\mathbf{s})$ for "caching" of visited configurational space.

The starting point of these "caching" algorithms is specification of reaction coordinates (RC) or collective variables (CVs), which is a very challenging task for complex molecular systems in many cases. Traditionally, principle component analysis (PCA) [123] is the most widely utilized and a robust way for disclosing DOFs associated with the largest variations. To deal with ubiquitous nonlinear correlations, kernels are often used albeit with the difficulty of choosing proper kernels [124]. Additional methodologies, include multidimensional scaling (MDS) [125], isomap [126], locally linear embedding (LLE) [127], diffusion map [128,129] and sketch map [130] have been developed to map out manifold for high dimensional data. However, each has it own limitations. For example, LLE [127] is sensitive to noise and therefore has difficulty with molecular simulation trajectories which are quite noisy; Isomap [126] requires relatively homogeneously sampled manifold to be accurate. Both LLE and Isomap do not provide explicit mapping between molecular coordinates and CVs; diffusion and sketch maps are likely to be more suitable to analyze molecular simulation trajectories. Nonetheless, their successful application for large and complex molecular systems remains to be tested. All of above non-linear mapping algorithm are mainly suitable for manifold on a single scale, and capturing manifold on multiple scales simultaneously in molecular simulations has not been reported yet. When we are interested in finding paths for transitions among known metastable states, transition path sampling (TPS) [131–133] methodology maybe utilized to establish CV.

Apparently, RC and/or CV based ES is a different path for facilitate simulation of complex molecular systems on longer time scales from coarse graining. One apparent plus side is that these algorithms are "in resolution" as no systematic discarding of molecular DOFs occur. With specification of RC and/or CVs, computational resource is presumably directed toward the most interesting dynamics of the target molecular system, and RC and/or CV maybe repetitively refined to obtain mechanistic understanding of interested molecular processes. However, the down side is that "cached" information on local configurational space is not transferable to other similar molecular systems. While rigorous transferability may not be easily established for any CG FF, practical utility of CG FF for molecular systems with similar composition and thermodynamic conditions have been quite common and useful [24]. Therefore, CG FF may be deemed as partially transferable.

An important recent development of DC strategy for enhanced sampling is Markov state models (MSM) [134–137], one great advantage of which is that no RC or CV is needed. Instead, it extracts long-time dynamics from independent short trajectories distributed in configurational space. Many important biomolecular functional processes have been characterized with this great technique [138–140]. The most fundamental assumption is that all states for a target molecular system form an ergodic Markov chain:

$$\pi(t+\tau) = \pi(t)\mathbf{P} \tag{7}$$

with $\pi(t)$ and $\pi(t+\tau)$ being a vector of probabilities for all states at time t and $t+\tau$ respectively. $\mathbf{P}$ is the transition matrix with its element P_{ij} being probability of the molecular system being found in state j after an implied lag time (τ) from the previous state i. Apparently as t goes to infinity for an equilibrium molecular system, a stationary distribution π will arise as defined below:

$$\pi = \pi P \tag{8}$$

The advantage of not needing RC/CV does not come for free but with accompanying difficulties. Firstly, one has to distribute start points of trajectories to statistically important and different part of configurational space, then select proper (usually hierarchical, with each level of hierarchy corresponds to a specific lag time) partition of configurational space into discrete states. This is the key step of DC strategy in MSM. No formal rule is available and experience is important. In many cases some try and error is necessary. Secondly, within each discrete state at a given level of hierarchy, equilibration is assumed to be achieved instantly and this assumption causes systematic discretization error, which fortunately may be controlled with proper partition and sufficiently long lag time [141]. Apparently, metastable states obtained from MSM analysis is molecular system specific and thus not transferable.

Another important class of enhanced sampling is to facilitate sampling with non-Boltzmann distributions and restore property at targeted thermodynamic condition through proper reweight [142]. Most outstanding examples are Tsallis statistics [143,144], parallel tempering [145,146], replica exchange molecular dynamics [147,148], Landau-Wang algorithm [149] and integrated tempering sampling (ITS) [150–152]. These algorithms are not direct applications of DC and "caching" strategies and are not discussed further here.

4. Machine Learning Improves "Caching"

4.1. Toward Ab Initio Accuracy of Molecular Simulation Potentials

Fixed functional form and pairwise approximation of non-bonded interactions are two major factors limiting the accuracy of molecular interaction description in both atomistic and some CG FF. Neural network (NN) has capability of approximating arbitrary functions and therefore has potential to address these two issues. Not surprisingly, significant progress has been made in this regard as summarized by recent excellent reviews [27,153–157]. Cutoff and attention to local interactions remains the DC strategy for development of machine learning potentials. The major improvement over traditional FF is better "caching" that overcomes pairwise approximation and fixed functional form limitations. NN FF naturally tackle both issues as explicit functions are not necessary since NNs are universal approximators. The significance of many-body potentials [158] and extent of pairwise contributions were analyzed [159,160]. It is important to note that despite the fact that pairwise interactions account for the majority of energy contributions, high ordered interactions are likely to be significant in shaping differences of subtly distinct molecular systems. There are also efforts to search for different and proper simple functional forms, which are expected to be more accurate than present functional forms in traditional FF on the one hand, and alleviate overfitting/generalization difficulty and reduce computational cost of complex NN FF on the other hand [161,162], especially when training dataset is small. While most machine learning FF are trained by energy data [153,157], gradient-domain machine learning (GDML) approach [163] directly learns from forces and realizes great savings of data generation.

Just as in the case of traditional FF, transferability and accuracy is always a tradeoff. More transferability implicates less attention is paid to "cache" detailed differences among different molecular systems, hence less accuracy. Exploration in this regard, however, remains not as much as necessary [164–166]. Unlike manual fitting of traditional FF, systematic investigation of tradeoff strategies is potentially feasible for machine learning fitting [167], and yet to be done for many interesting molecular systems. With expediency of NN training, development of a NN FF hierarchy with increasing transferability/accuracy and decreasing accuracy/transferability is likely to become a pleasing reality in the near future. Rapid further development of machine learning potentials, particularly NN potentials, are expected. However, significant challenges for NN potentials remain on better generalization capability, description/treatment of long range interactions [168,169], wide range of transferability, [170] faster computation [171] and proper characterization of their error bounds. Should further significant progress be made on these issues, it is promising we may have routine molecular simulations with both classical efficiency and ab initio accuracy in the near future.

4.2. Machine Learning and Coarse Graining

As in the case of constructing atomic level potentials, machine learning has been applied to address two outstanding pending issues in coarse graining, which are definition of CG sites/particles and parameterization of corresponding interactions between/among these sites/particles. Traditional CG FF, suffers from both pairwise approximation and, for some, accuracy ceiling of simple fixed functional forms which are easy to fit. By using more complex (but fixed functional form) potentials with a machine learning fitting process, Chan et al. [172] developed ML-BOP CG water model with great success. Deep neural network (DNN) was utilized to facilitate parameterization of CG potentials when given radial distribution functions (RDF) from atomistic simulations [173]. CGnet demonstrated great success with simple model systems (alanine dipeptide) [174]. DeePCG model was developed to overcome pair approximation and fixed functional form and demonstrated with water [175]. Using oxygen site to represent water is rather intuitive. However, for more complex biomolecules such as proteins, possibility for selection of CG site explodes. To improve over intuitive or manual try and error definition of CG sites, a number of studies have been carried out [176–179] to provide better and faster options for choosing CG sites. However, no consensus strategy is available up to date and more investigations are desired. The fundamental difficulty is that there is no sufficient time scale separation between explicit CG DOFs and discarded implicit DOFs, regardless of specific selection scheme being utilized. Intuitively, one would expect CG FF parameters to be dependent upon definition of CG sites/particles. In this regard, auto-encoders were utilized to construct a generative framework that accomplishes CG representation and parameterization in a unified way [33]. The spirit of generative adversarial networks was utilized to facilitate CG construction and parameterization, particularly with virtual site representation [180]. It was found that description of off-target property by CG exhibit strong correlation with CG resolution, to which on-target property being much less sensitive [181]. Such observation suggests that adjust CG for specific target properties might be a better strategy than searching for a single best CG representation. Despite potentially more severe impact of pairwise approximation for CG FF than in atomistic FF, quantitative analysis in this regard remain to be done to the best of our knowledge.

4.3. Machine Learning in Searching for RC/CVs and Construction of MSM

To overcome difficulties of earlier nonlinear CV construction algorithms [126–128,130] and to reduce reliance on human experience, auto-encoders, which is well-established for trainable (non-linear) dimensionality reduction, are utilized in a few studies [182–185]. Chen and Ferguson [183] first utilized autoencoders to learn nonlinear CVs that are explicit and differentiable functions of molecular coordinates, thus enabling direct utility in molecular simulations for more effective exploration of configurational space. Further improvement [182] was achieved through circular network nodes and hierarchical network architectures to rank-order CVs. Wehmeyer and Noé [184] developed time-lagged auto-encoder to search for low dimensional embeddings that capture slow dynamics. Ribeiro et al. [185] proposed the reweighted autoencoded variational Bayes to iteratively refine RC and demonstrated in computation of the binding free energy profile for a hydrophobic ligand-substrate system. Building a MSM for any specific molecular system requires tremendous experience and many steps in process are error prone. To overcome these pitfalls, VAMPnet that based on variational approach for Markov process was developed to realize the complete mapping steps from molecular trajectories to Markov states [186]. As physical understanding of interested molecular systems is essential and the ultimate goal, application of these methods as black boxes are not encouraged.

5. The Local Free Energy Landscape Approach

Both CG and ES methodologies facilitate molecular simulation by effectively reducing local sampling. In CG, it is realized through "caching" (integration) of distributions for faster/discarded DOFs with proper CG FF, and thus has the inevitable cost of losing resolution (information), accompanied by the desired attribute of (partial) transferability to various extent. ES reduces lingering time of molecular systems in local minima through "caching" visited local configurational space, which is usually defined by relevant DC strategies, with biasing potentials. When compared with CG, there is no resolution loss. However, "cached" manifold of configurational space is molecular process specific and thus not transferable at all. In molecular modeling community, these two lines of methodologies are developed quite independently. Nevertheless, one might want to ask why not have both advantages in one method, that is to reduce repetitive local sampling without loss of resolution and with "cached" results being partially transferable. The local free energy landscape (LFEL) approach [187] is proposed with this intention in mind. Historically, parameterization of FF by coarse graining has been the only viable framework due to two fundamental constraints. Firstly, in earlier days of molecular modeling, typical computers have memory space of megabytes or less, render it impossible to accommodate millions or more parameters needed to fit complex LFEL; secondly, while both neural network and autodifferentiation were invented decades ago, the computational molecular science community did not master these techniques for fitting large number of parameters efficiently until recently. With these two constraints removed, possibility for alternative path arise to break monopoly of classical molecular modeling by FF parameterization via coarse graining. Specifically, one may carry out direct fitting of LFEL and all important information on local distributions of molecular DOFs obtained from expensive local sampling may be "cached". This is in strong contrast to coarse graining based parameterization, in which local distributions are substituted by averaging in relevant lower dimensional space projection (e.g., pairwise distances among CG sites). However, it is essential to assemble LFEL and construct FEL of the interested molecular system, and this is the core of the LFEL approach. For a molecular system with N DOFs, this LFEL approach may be expressed as:

$$P(r_1, r_2, \cdots, r_N) = P(R_1, R_2, \cdots, R_M)(M \leq N) \tag{9}$$

$$R_i = (r_{i_1}, r_{i_2}, \cdots, r_{i_l}) \tag{10}$$

$$\begin{aligned} &P(R_1, R_2, \cdots, R_M) \\ &= \prod_{i=1}^{M} P(R_i) \frac{P(R_1, R_2, \cdots, R_M)}{\prod_{i=1}^{M} P(R_i)} && (11) \\ &\approx \prod_{i=1}^{M} P(R_i), \text{and sampling all } Rs \text{ with mediated GCF} && (12) \end{aligned}$$

$$\begin{aligned} G &= -k_B T lnP(r_1, r_2, \cdots, r_N) \\ &= -k_B T lnP(R_1, R_2, \cdots, R_M) \\ &\approx -k_B T \sum_{i}^{M} lnP(R_i), \text{and sampling all } Rs \text{ with mediated GCF} && (13) \end{aligned}$$

an N-DOF molecular system is reorganized into M overlapping regions (Equation (9)), each region has some number of DOFs (Equation (10)). The key step of LFEL approach is expressed in Equation (11), in which the first product term (addressed as "local term(s)" hereafter) treat M regions as if they were independent, and all correlations among different regions are incorporated by the fraction term, which is termed global correlation fraction (GCF) and is extremely difficult, if ever possible, to be calculated directly. However, GCF is a unnormalized probability distribution, when all molecular DOFs in local terms are (approximately) sampled according to GCF, then we do not need GCF explicitly anymore (Equations (12) and (13)). GCF represents two types of global correlations. The first type is mediated correlations among different regions by the fact that they overlap, and relevant molecular DOFs in such overlapping space shared by different regions should have exact same state for all concerning regions. The second type is direct global correlations among molecular DOFs in different regions caused by genuine long-range interactions (e.g., electrostatic interactions). Satisfying the first type with sampling is trivial, and ensuring all overlapping regions share the exact same state is sufficient (Equations (12) and (13)). The second type of global correlations need more involved treatment. These equations are apparently of general utility for any multiple-variable (high-dimensional) problem. In the specific case of a complex molecular system, using one set of coordinates realizes the mediated contribution of GCF. The approximation in Equation (12) is made by ignoring the second type of global correlations. Free energy minimization of a molecular system in thermodynamic equilibrium may be treated as maximization of joint probability (Equation (13)). For molecular systems (or biological systems) off equilibrium, the joint distribution remains our focus despite free energy is not well defined anymore. A schematic representation of the LFEL approach in contrast to FF framework is shown in (Figure 2). While we only demonstrated GSFE implementation of LFEL at residue level for protein structural refinement. LFEL approach may be utilized to "cache" local distributions at any spatial scales. Just as there are many methodological developments in the mainstream FF framework, there are certainly many possible ways to develop algorithms in the LFEL approach. We explored a first step toward this direction through a neural network implementation of the generalized solvation free energy (GSFE) theory [36]. In GSFE theory, each comprising unit in a complex molecular system is solvated by its neighboring units. Therefore, each unit is both a solute itself and a comprising solvent unit of its solvent units. Let $(x_i, y_i) = R_i$ denote a region i defined by a solute x_i and its solvent y_i, a molecular system of N units has N overlapping regions. Each local term may be further expanded:

$$\begin{aligned} P(R_i) &= P(x_i, y_i) \\ &= P(x_i|y_i)P(y_i) \end{aligned} \tag{14}$$

Both terms may be learned from either experimental or computational datasets, as long as they are sufficiently representative and reliable. The first term in Equation (14) is the likelihood term when x_i is the given, it quantifies the extent of match between the solute x_i and its solvent y_i. The second term is the local prior term, it quantifies the stability of the solvent environment y_i. Computation of the prior term is more difficult than the likelihood term, but certainly learnable when sufficient data is available. A local maximum likelihood approximation of GSFE (LMLA-GSFE) is to simply ignore local prior terms.

A particular implementation of the LMLA-GSFE for protein structure refinement with residues defined as comprising unit was conducted [187]. In this scheme, GSFE is integrated with autodifferentiation and coordinate transformation to construct a computational graph for free energy optimization. With fully trainable LFEL derived from backbone and C_β atom coordinates of selected experimental protein structures, we achieved superb efficiency and competitive accuracy when compared with state of the art atomistic protein refinement refinement methodologies. With our newly developed pipeline, refinement of typical protein structure decoys (within 300 amino acids) takes a few seconds on a single CPU core, in contrast to a few hours by typical efficient sampling/minimization based algorithms (e.g., FastRelax [188]) and thousands of hours for MD based refinement [189]. In the latest CASP14 refinement contest (predictioncenter.org/casp14/index.cgi (accessed on 15 February 2021)), our method ranked the the first for the 13 targets with start GDT-TS score larger than 60. We expect incorporation of complete heavy atom information and local prior terms to further improve this method in the future. GSFE theory in particular and the LFEL approach in general, are certainly extendable to modeling of other soft matter molecular systems.

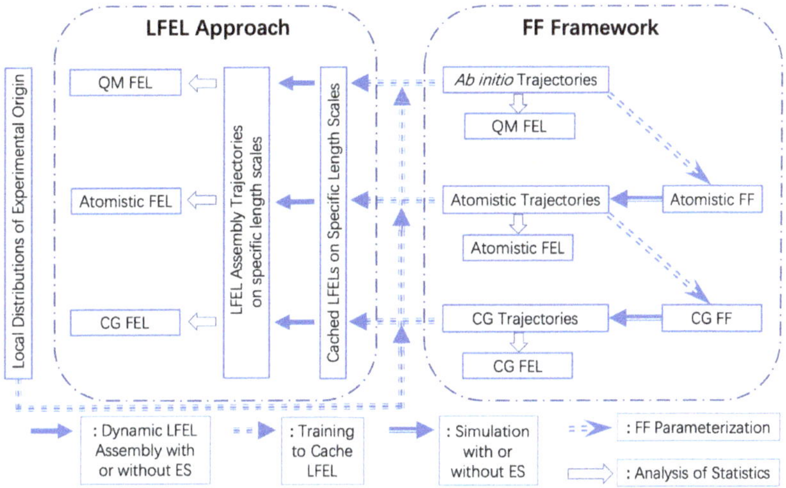

Figure 2. Schematic illustration of the LFEL approach in contrast to present mainstream FF framework. FF parameterization is the foundation for present classical computational molecular science. Training of neural network for "caching" LFEL is the foundation for LFEL approach, the source data can be either of experimental or computational origin. In FF framework, simulation (with or without ES) is driven by FF, in LFEL approach, propagation of molecular systems to minimize free energy (or maximize joint probability) is driven by compromise among many LFELs. Expensive repetitive local sampling in FF framework is substituted by differentiation w.r.t. LFELs.

6. More on Connections among CG, ES and LFEL Approach

All of these algorithms have a common goal of accelerating computation of a joint distribution for a given molecular system at some target resolution, albeit from distinct perspectives. The fundamental underpinning is the fact that molecular correlations among its various DOFs limit a molecular system to a manifold of significantly lower dimension. Both ES and CG in the FF framework and the LFEL approach are distinct strategies to "cache" manifolds from either configurational space (Figure 3) or physical space perspective (Figure 4). Commonality and differences of these strategies are summarized in Table 2 and discussed below.

Table 2. Commonality and difference among three types of algorithms.

Algorithm	Coarse Graining	Enhanced Sampling	LFEL Approach
Resolution	Lower	In	In
Transferable?	Partial	No	Partial
Dividing space	Physical	Configurational	Physical
Free energy unit	Partially Specified	Specified	Arbitrary

Both MSM and RC/CV based ES are designed to first describe local parts of the approximate manifold in the configurational space formed by all molecular DOFs of the target molecular system. Information for such local configurational space is partially "cached" either as bias potentials or transition counts, which are further processed to map FEL and dynamics of interested molecular processes. Computational process (or educated guess) for establishment of RC/CVs is essentially "caching" results from sampling/guessing local parts of the configurational as approximate relevant manifold (Figure 3B). Subsequent sampling along RC/CV is hoped to disclose our interested molecular processes (e.g., biomolecular conformational transitions, substrate binding/release in catalysis). Involved molecular DOFs for RC/CVs are not necessarily spatially adjacent on the one hand, and may be different for different molecular processes of the same molecular system. Apparently, RC/CVs are molecular process specific and not transferable, even among different molecular process of the same molecular system. Nonetheless, the methodology for searching CVs may be applied to many different molecular processes/systems.

In contrast, both CG in the FF framework and the LFEL approach are motivated to "cache" relevant information on the complete configurational distribution for local clusters of molecular DOFs. Such local clusters are building blocks for many similar molecular systems (e.g., AAs in protein molecular systems) and consequently have limited and approximate transferability. In CG, strongly correlated local clusters of molecular DOFs are represented as a single particle, complex many body correlations/interactions of CG particles within selected cutoff distances are represented by simplified CG FF in a lower resolution and longer range correlations/interactions are incorporated either through more coarser CG models or by separate long-range interaction computation. In LMLA-GSFE implementation of LFEL, all complex many body correlations within selected regions (i.e., each solute and its specific solvent) are decomposed into two terms in Equation (14), local likelihoods and local priors in the same resolution, with local priors and direct genuine long-range interactions simply ignored, and LFEL being approximated by local likelihood terms. More and better ways for implementing LFEL are expected in the future.

The first step of CG is to partition atoms/particles of high resolution representation into highly correlated local clusters that will be represented by corresponding single CG particles, and moderately correlated regions define interaction cutoff for CG particles; The second step is to select a site (usually one of the comprising high resolution particles) to represent the corresponding highly correlated cluster; The third step is to select functional forms to describe molecular interactions among newly defined CG particles, and parameters are optimized by selected loss functions (e.g., differences of average force in force matching [87,88]) based on sampling in the whole configurational space of molecular systems and hopefully to be transferable to some extent. One may imagine that both best clustering and optimal representation sites of clusters may vary with different functional forms used to describe CG particle interactions and in different part of configurational space. Neural network based CG potentials do not have limitation of fixed functional form and pairwise approximations. However, the need to partition molecular systems into transferable clusters and to specify representation site/particle remain. For all different forms of CG, the fundamental essence is to "cache" many body potential of mean force (PMF) in simplified CG FF at a lower resolution. In contrast, LFEL approach is to first using a DC strategy to divide molecular systems into local regions, then directly "cache" many body PMF (or LFEL) of such local regions in the original resolution. The cached complex local multivariate distributions in NN are subsequently utilized to construct FEL of target molecular system through dynamic puzzle assembly based on sampling with GCF as expressed in Equations (12) and (13). In language of statistical machine learning. Training of LFEL is the learning step, while construction of global FEL is the inference step. The advantage of CG is a simpler resulting physical model, but is inflexible due to fixed clustering and representation on the one hand, and lost resolution/information on the other hand. Properly implemented LFEL while has selected spatial regions comprising many molecular DOFs, composition of such regions are fully dynamic. For example, in GSFE implementation of LFEL, a region is defined by a solute unit and all of its solvent units, and comprising units for the solvent is dynamically updated in each iteration of free energy optimization. Additionally, no loss of resolution is involved for LFEL approach. Hence all difficulties and uncertainties associated with molecular DOF partition, CG site selection and time scale separation, all of which apparently limit transferability of CG FF, disappear. Correspondingly, the extent of transferability of a LFEL model is in principle at least no worse than CG FF. Differences of CG and specific implementation of LFEL by GSFE theory is schematically illustrated in Figure 4. The superior efficiency of LFEL approach comes with a price. The assembled global FEL has arbitrary unit for two reasons. Firstly, it is extremely difficult to obtain the partition function (normalization constant) for local regions directly during the training/caching stage, therefore we effectively obtain the LFEL up to an unknown constant. Secondly, for two different molecular systems, the number of local regions are usually different and so is the corresponding normalization constant.

These three lines of algorithms may be combined to facilitate molecular modeling. For example, one might first utilize deep learning based near quantum accuracy many body FF to perform atomistic simulations for protein molecular systems, and then extracting local distributions properly with some form of LFEL, which may potentially be utilized to simulate protein molecular systems with near-quantum accuracy and at regular amino-acid based CG or even much faster speed! Similarly, one may extract and "cache" large body of information from residue level CG simulations with proper LFEL implementation, which may be utilized to achieve ultra CG (UCG) efficiency with residue resolution. Application of CV and MSM based ES algorithm for CG models is straight forward. Combination of LFEL with CV or MSM based ES is more subtle and yet to be investigated.

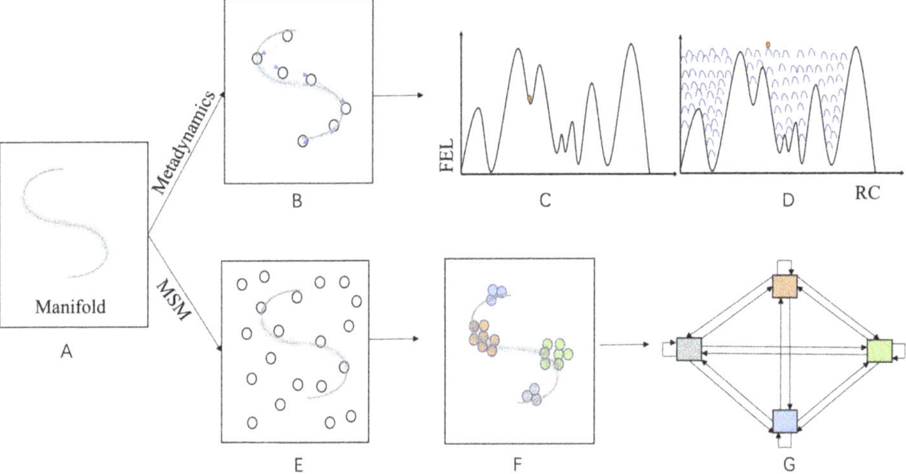

Figure 3. Schematic illustration of essential features for enhanced sampling by metadynamics and MSM. (**A**) The "S" shape grey line represents the unknown manifold in the configurational space (represented by the square) of a molecular system. (**B**) Small circles connected by blue arrows represent computed (guessed) RC/CVs for the molecular system, which is utilized to conduct metadynamics simulations. (**C**) The FEL of the molecular system along the computed/selected RC/CV in (**B**,**D**) "Caching" of the LFEL by bias potentials (gaussians represented by blue bell shaped lines) accumulated in the course of metadynamics simulations. (**E**) Distribution of the molecular system to the whole configurational space at the start of a MSM simulation, small circles represent initial start points for short MSM trajectories. (**F**) Sampling results of short MSM trajectories fall mainly near the manifold, distinct "states" are represented by different colors. (**G**) Establishment of transition matrix by transition counts between "states" obtained from short trajectories.

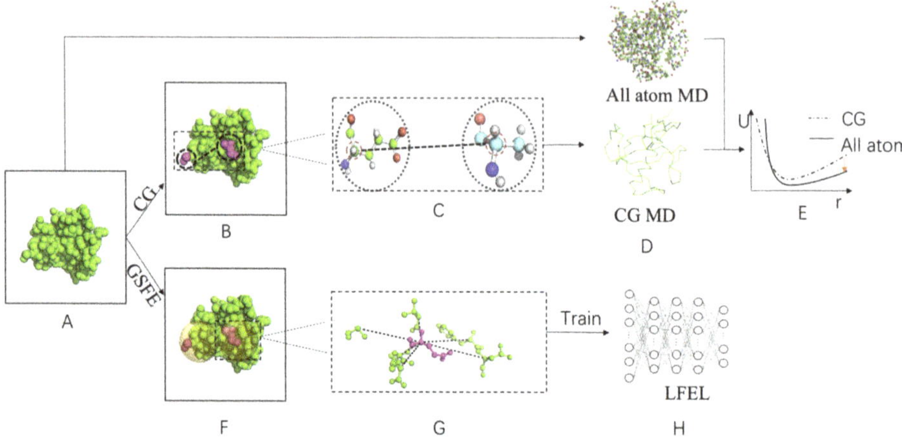

Figure 4. Schematic illustration of difference between CG and GSFE implementation of LFEL using protein as an example. (**A**) Target molecular systems in physical space. Due to the goal of constructing partially transferable models and/or force fields, usually many different but similar molecular systems are considered. (**B**) Selection of local atom/particle clusters to be represented as one particle in CG model. (**C**) Selection of CG sites. (**D**) Comparison between atomistic (or higher resolution) simulation results and CG (lower resolution) results. (**E**) Adjust of CG FF parameter according to comparison from (**D**). (**F**) Definition of solvent region for each solute unit. (**G**) Feature extraction for each solute. (**H**) "Caching" of LFEL with neural network by training with prepared data sets.

7. Conclusions and Prospect

The application of "dividing and conquering" and "caching" principle in development of molecular modeling algorithms is briefed. Historically, coarse graining and enhanced sampling have been two independent lines of methodological development in the mainstream FF framework. While they share the common goal of reducing local sampling, the formulations are completely different with distinct (dis)advantages. Coarse graining obtains partial transferable FFs but loses resolution, enhanced sampling retains resolution but results are not transferable. The LFEL approach suggests a third strategy to directly approximate global joint distribution by superposition of LFEL, which may be learned from available dataset of either experimental or computational origin. Through integration of coordinate transformation, autodifferentiation and neural network implementation of GSFE, our recent work of protein structure refinement demonstrated that simultaneous realization of transferable in-resolution "caching" of local sampling is not only feasible, but also highly efficient due to replacement of local sampling by differentiation. It is hoped that this review stimulates further development of better "dividing and conquering" strategies for complex molecular systems through more elegant, efficient and accurate ways of "caching" potentially repetitive computations in molecular modeling at various spatial and temporal scales. With diverse molecular systems (e.g., nanomaterials, biomolecular systems), specialization of methodology is essential to take advantage of distinct constraints and characteristics.

Author Contributions: Conceptualization, P.T.; Resources, P.T.; Data Curation, X.C.; Writing—Original Draft Preparation, P.T.; Writing—Review & Editing, P.T. and X.C.; Visualization, X.C.; Funding Acquisition, P.T. All authors have read and agreed to the published version of the manuscript.

Funding: This work has been supported by the National Key Research and Development Program of China under grant number 2017YFB0702500.

Institutional Review Board Statement: Not applicable.

Informed Consent Statement: Not applicable.

Data Availability Statement: Data sharing not applicable.

Conflicts of Interest: The authors declare no conflict of interest.

References

1. Chodera, J.D.; Mobley, D.L.; Shirts, M.R.; Dixon, R.W.; Branson, K.; Pande, V.S. Alchemical free energy methods for drug discovery: Progress and challenges. *Curr. Opin. Struct. Biol.* **2011**, *21*, 150–160. [CrossRef]
2. Dror, R.O.; Dirks, R.M.; Grossman, J.; Xu, H.; Shaw, D.E. Biomolecular Simulation: A Computational Microscope for Molecular Biology. *Annu. Rev. Biophys.* **2012**, *41*, 429–452. [CrossRef] [PubMed]
3. Van der Kamp, M.W.; Mulholland, A.J. Combined Quantum Mechanics/Molecular Mechanics (QM/MM) Methods in Computational Enzymology. *Biochemistry* **2013**, *52*, 2708–2728. [CrossRef]
4. Canchi, D.R.; Garcia, A.E. Cosolvent Effects on Protein Stability. *Annu. Rev. Phys. Chem.* **2013**, *64*, 273–293. [CrossRef]
5. Hansen, N.; van Gunsteren, W.F. Practical Aspects of Free-Energy Calculations: A Review. *J. Chem. Theory Comput.* **2014**, *10*, 2632–2647. [CrossRef] [PubMed]
6. Mobley, D.L.; Gilson, M.K. Predicting Binding Free Energies: Frontiers and Benchmarks. *Annu. Rev. Biophys.* **2017**, *46*, 531–558. [CrossRef]
7. Wang, E.; Sun, H.; Wang, J.; Wang, Z.; Liu, H.; Zhang, J.Z.H.; Hou, T. End-Point Binding Free Energy Calculation with MM/PBSA and MM/GBSA: Strategies and Applications in Drug Design. *Chem. Rev.* **2019**, *119*, 9478–9508. [CrossRef]
8. Enkavi, G.; Javanainen, M.; Kulig, W.; Rog, T.; Vattulainen, I. Multiscale Simulations of Biological Membranes: The Challenge To Understand Biological Phenomena in a Living Substance. *Chem. Rev.* **2019**, *119*, 5607–5774. [CrossRef]
9. Bedrov, D.; Piquemal, J.-P.; Borodin, O.; MacKerell, A.D., Jr.; Roux, B.; Schroeder, C. Molecular Dynamics Simulations of Ionic Liquids and Electrolytes Using Polarizable Force Fields. *Chem. Rev.* **2019**, *119*, 7940–7995. [CrossRef] [PubMed]
10. Marrink, S.J.; Corradi, V.; Souza, P.C.T.; Ingolfsson, H.I.; Tieleman, D.P.; Sansom, M.S.P. Computational Modeling of Realistic Cell Membranes. *Chem. Rev.* **2019**, *119*, 6184–6226. [CrossRef] [PubMed]
11. Lee, A.C.-L.; Harris, J.L.; Khanna, K.K.; Hong, J.-H. A Comprehensive Review on Current Advances in Peptide Drug Development and Design. *Int. J. Mol. Sci.* **2019**, *20*, 2383. [CrossRef] [PubMed]
12. Kang, L.; Liang, F.; Jiang, X.; Lin, Z.; Chen, C. First-Principles Design and Simulations Promote the Development of Nonlinear Optical Crystals. *Accounts Chem. Res.* **2020**, *53*, 209–217. [CrossRef] [PubMed]
13. Nelson, T.R.; White, A.J.; Bjorgaard, J.A.; Sifain, A.E.; Zhang, Y.; Nebgen, B.; Fernandez-Alberti, S.; Mozyrsky, D.; Roitberg, A.E.; Tretiak, S. Non-adiabatic Excited-State Molecular Dynamics: Theory and Applications for Modeling Photophysics in Extended Molecular Materials. *Chem. Rev.* **2020**, *120*, 2215–2287. [CrossRef]
14. Shaebani, M.R.; Wysocki, A.; Winkler, R.G.; Gompper, G.; Rieger, H. Computational models for active matter. *Nat. Rev. Phys.* **2020**, *2*, 181–199. [CrossRef]
15. Spaggiari, G.; Di Pizio, A.; Cozzini, P. Sweet, umami and bitter taste receptors: State of the art of in silico molecular modeling approaches. *Trends Food Sci. Technol.* **2020**, *96*, 21–29. [CrossRef]
16. Duncan, A.L.; Song, W.; Sansom, M.S.P. Lipid-Dependent Regulation of Ion Channels and G Protein-Coupled Receptors: Insights from Structures and Simulations *Annu. Rev. Pharmacol. Toxicol.* **2020**, *60*, 31–50. [CrossRef] [PubMed]
17. Rust, J. *The New Palgrave Dictionary of Economics*; Palgrave Macmillan UK: London, UK, 2016; pp. 1–26.
18. Sambasivan, R.; Das, S.; Sahu, S.K. A Bayesian perspective of statistical machine learning for big data. *Comput. Stat.* **2020**, *35*, 893–930. [CrossRef]
19. Rudzinski, J.F.; Noid, W.G. Coarse-graining entropy, forces, and structures. *J. Chem. Phys.* **2011**, *135*, 214101. [CrossRef]
20. Marrink, S.J.; Tieleman, D.P. Perspective on the martini model. *Chem. Soc. Rev.* **2013**, *42*, 6801–6822. [CrossRef]
21. Noid, W.G. Perspective: Coarse-grained models for biomolecular systems. *J. Chem. Phys.* **2013**, *139*, 090901. [CrossRef] [PubMed]
22. Saunders, M.G.; Voth, G.A. Coarse-graining methods for computational biology. *Annu. Rev. Biophys.* **2013**, *42*, 73–93. [CrossRef] [PubMed]
23. Ruff, K.M.; Harmon, T.S.; Pappu, R.V. CAMELOT: A machine learning approach for Coarse-grained simulations of aggregation of block-copolymeric protein sequences. *J. Chem. Phys.* **2015**, *143*, 1–19. [CrossRef]
24. Kmiecik, S.; Gront, D.; Kolinski, M.; Wieteska, L.; Dawid, A.E.; Kolinski, A. Coarse-Grained Protein Models and Their Applications. *Chem. Rev.* **2016**, *116*, 7898–7936. [CrossRef]
25. Hafner, A.E.; Krausser, J.; Saric, A. Minimal coarse-grained models for molecular self-organisation in biology. *Curr. Opin. Struct. Biol.* **2019**, *58*, 43–52. [CrossRef] [PubMed]
26. Joshi, S.Y.; Deshmukh, S.A. A review of advancements in coarse-grained molecular dynamics simulations. *Mol. Simul.* **2020**, *47*, 1–18.
27. Gkeka, P.; Stoltz, G.; Barati Farimani, A.; Belkacemi, Z.; Ceriotti, M.; Chodera, J.D.; Lelièvre, T. Machine Learning Force Fields and Coarse-Grained Variables in Molecular Dynamics: Application to Materials and Biological Systems. *J. Chem. Theory Comput.* **2020**, *16*, 4757–4775. [CrossRef]
28. Bernardi, R.C.; Melo, M. C.R.; Schulten, K. Enhanced sampling techniques in molecular dynamics simulations of biological systems. *Biochim. Biophys. Acta Gen. Subj.* **2015**, *1850*, 872–877. [CrossRef]
29. Mlynsky, V.; Bussi, G. Exploring RNA structure and dynamics through enhanced sampling simulations. *Curr. Opin. Struct. Biol.* **2018**, *49*, 63–71. [CrossRef]

30. Yang, Y.I.; Shao, Q.; Zhang, J.; Yang, L.; Gao, Y.Q. Enhanced sampling in molecular dynamics. *J. Chem. Phys.* **2019**, *151*, 070902. [CrossRef] [PubMed]
31. Wang, A.-H.; Zhang, Z.-C.; Li, G.-H. Advances in Enhanced Sampling Molecular Dynamics Simulations for Biomolecules. *Chin. J. Chem. Phys.* **2019**, *32*, 277–286. [CrossRef]
32. Okamoto, Y. Protein structure predictions by enhanced conformational sampling methods. *Biophys. Phys.* **2019**, *16*, 344–366. [CrossRef] [PubMed]
33. Wang, W.; Gómez-Bombarelli, R. Coarse-graining auto-encoders for molecular dynamics. *NPJ Comput. Mater.* **2019**, *5*, 1–9. [CrossRef]
34. Lazim, R.; Suh, D.; Choi, S. Advances in Molecular Dynamics Simulations and Enhanced Sampling Methods for the Study of Protein Systems. *Int. J. Mol. Sci.* **2020**, *21*, 6339. [CrossRef]
35. Liao, Q. *Computational Approaches for Understanding Dynamical Systems: Protein Folding and Assembly*; Progress in Molecular Biology and Translational Science; Strodel, B., Barz, B., Eds.; Academic Press Ltd.: London, UK; Elsevier Science Ltd.: London, UK, 2020; Volume 170; pp. 177–213.
36. Long, S.; Tian, P. A simple neural network implementation of generalized solvation free energy for assessment of protein structural models. *RSC Adv.* **2019**, *9*, 36227–36233. [CrossRef]
37. Čížek, J. On the Correlation Problem in Atomic and Molecular Systems. Calculation of Wavefunction Components in Ursell-Type Expansion Using Quantum-Field Theoretical Methods. *J. Chem. Phys.* **1966**, *45*, 4256–4266. [CrossRef]
38. Reyes, A.; Moncada, F.; Charry, J. The any particle molecular orbital approach: A short review of the theory and applications. *Int. J. Quantum Chem.* **2019**, *119*, e25705
39. Van Houten, J. A Century of Chemical Dynamics Traced through the Nobel Prizes. 1998: Walter Kohn and John Pople. *J. Chem. Educ.* **2002**, *79*, 1297. [CrossRef]
40. Bensberg, M.; Neugebauer, J. Density functional theory based embedding approaches for transition-metal complexes. *Phys. Chem. Chem. Phys.* **2020**, *22*, 26093–26103. [CrossRef] [PubMed]
41. Kraus, P. Basis Set Extrapolations for Density Functional Theory. *J. Chem. Theory Comput.* **2020**, *16*, 5712–5722. [CrossRef]
42. Morgante, P.; Peverati, R. The devil in the details: A tutorial review on some undervalued aspects of density functional theory calculations. *Int. J. Quantum Chem.* **2020**, *120*, e26332. [CrossRef]
43. Zhang, I.Y.; Xu, X. On the top rung of Jacob's ladder of density functional theory: Toward resolving the dilemma ofSIEandNCE. *Wiley Interdiscip. Rev. Comput. Mol. Sci.* **2021**, *11*, e1490. [CrossRef]
44. Mackerell, A.D., Jr. Empirical force fields for biological macromolecules: Overview and issues. *J. Comput. Chem.* **2004**, *25*, 1584–1604.
45. Kumar, A.; Yoluk, O.; Mackerell, A.D., Jr. FFParam: Standalone package for CHARMM additive and Drude polarizable force field parametrization of small molecules. *J. Comput. Chem.* **2020**, *41*, 958–970. [CrossRef]
46. Oweida, T.J.; Kim, H.S.; Donald, J.M.; Singh, A.; Yingling, Y.G. Assessment of AMBER Force Fields for Simulations of ssDNA. *J. Chem. Theory Comput.* **2021**, *17*, 1208–1217. [CrossRef]
47. Huai, Z.; Shen, Z.; Sun, Z. Binding Thermodynamics and Interaction Patterns of Inhibitor-Major Urinary Protein-I Binding from Extensive Free-Energy Calculations: Benchmarking AMBER Force Fields. *J. Chem. Inf. Model.* **2021**, *61*, 284–297. [CrossRef]
48. Alford, R.F.; Leaver-Fay, A.; Jeliazkov, J.R.; O'Meara, M.J.; DiMaio, F.P.; Park, H.; Gray, J.J. The Rosetta All-Atom Energy Function for Macromolecular Modeling and Design. *J. Chem. Theory Comput.* **2017**, *13*, 3031–3048. [CrossRef] [PubMed]
49. Sasse, A.; de Vries, S.J.; Schindler, C.E.M.; de Beauchêne, I.C.; Zacharias, M. Rapid Design of Knowledge-Based Scoring Potentials for Enrichment of Near-Native Geometries in Protein-Protein Docking. *PLoS ONE* **2017**, *12*, e0170625. [CrossRef] [PubMed]
50. Narykov, O.; Bogatov, D.; Korkin, D. DISPOT: A simple knowledge-based protein domain interaction statistical potential. *Bioinformatics* **2019**, *35*, 5374–5378. [CrossRef] [PubMed]
51. Wang, S. Efficiently Calculating Anharmonic Frequencies of Molecular Vibration by Molecular Dynamics Trajectory Analysis. *ACS Omega* **2019**, *4*, 9271–9283. [CrossRef]
52. Ferreiro, D.U.; Komives, E.A.; Wolynes, P.G. Frustration in biomolecules. *Q. Rev. Biophys.* **2014**, *47*, 285–363. [CrossRef] [PubMed]
53. Sulkowska, J.I. On folding of entangled proteins: Knots, lassos, links and theta-curves. *Curr. Opin. Struct. Biol.* **2020**, *60*, 131–141. [CrossRef] [PubMed]
54. Chen, M.; Chen, X.; Schafer, N.P.; Clementi, C.; Komives, E.A.; Ferreiro, D.U.; Wolynes, P.G. Surveying biomolecular frustration at atomic resolution. *Nat. Commun.* **2020**, *11*, 5944. [CrossRef] [PubMed]
55. Warshel, A.; Kato, M.; Pisliakov, A.V. Polarizable force fields: History, test cases, and prospects. *J. Chem. Theory Comput.* **2007**, *3*, 2034–2045. [CrossRef]
56. Jing, Z.; Liu, C.; Cheng, S.Y.; Qi, R.; Walker, B.D.; Piquemal, J.-P.; Ren, P. Polarizable Force Fields for Biomolecular Simulations: Recent Advances and Applications. *Annu. Rev. Biophys.* **2019**, *48*, 371–394. [CrossRef] [PubMed]
57. Wang, L.P.; Chen, J.; Van Voorhis, T. Systematic parametrization of oolarizable force fields from quantum chemistry data. *J. Chem. Theory Comput.* **2013**, *9*, 452–460. [CrossRef] [PubMed]
58. Császár, A.G. Anharmonic molecular force fields. *WIREs Comput. Mol. Sci.* **2012**, *2*, 273–289. [CrossRef]
59. Ulmschneider, J.P.; Ulmschneider, M.B. Molecular Dynamics Simulations Are Redefining Our View of Peptides Interacting with Biological Membranes. *Accounts Chem. Res.* **2018**, *51*, 1106–1116. [CrossRef]

60. Jung, J.; Nishima, W.; Daniels, M.; Bascom, G.; Kobayashi, C.; Adedoyin, A.; Wall, M.; Lappala, A.; Phillips, D.; Fischer, W.; et al. Scaling molecular dynamics beyond 100,000 processor cores for large-scale biophysical simulations. *J. Comput. Chem.* **2019**, *40*, 1919–1930. [CrossRef]
61. Wolf, S.; Lickert, B.; Bray, S.; Stock, G. Multisecond ligand dissociation dynamics from atomistic simulations. *Nat. Commun.* **2020**, *11*, 2918. [CrossRef]
62. Ferguson, A.L.; Panagiotopoulos, A.Z.; Kevrekidis, I.G.; Debenedetti, P.G. Nonlinear dimensionality reduction in molecular simulation: The diffusion map approach. *Chem. Phys. Lett.* **2011**, *509*, 1–11. [CrossRef]
63. Springer. *Rugged Free Energy Landscapes*; Springer: Berlin/Heidelberg, Germany, 2008.
64. Westerlund, A.M.; Delemotte, L. InfleCS: Clustering Free Energy Landscapes with Gaussian Mixtures. *J. Chem. Theory Comput.* **2019**, *15*, 6752–6759. [CrossRef] [PubMed]
65. Dick, T.J.; Madura, J.D. *Chapter 5 A Review of the TIP4P, TIP4P-Ew, TIP5P, and TIP5P-E Water Models*; Annual Reports in Computational Chemistry; Elsevier: Amsterdam, The Netherlands, 2005; Volume 1, pp. 59–74.
66. Vanommeslaeghe, K.; Hatcher, E.; Acharya, C.; Kundu, S.; Zhong, S.; Shim, J.; Darian, E.; Guvench, O.; Lopes, P.; Vorobyov, I.; et al. CHARMM general force field: A force field for drug-like molecules compatible with the CHARMM all-atom additive biological force fields. *J. Comput. Chem.* **2010**, *31*, 671–690. [CrossRef] [PubMed]
67. Van Der Spoel, D. Systematic Design of Biomolecular Force Fields. *Curr. Opin. Struct. Biol.* **2021**, *67*, 18–24. [CrossRef] [PubMed]
68. Michael, L.; Warshel, A. Computer simulation of protein folding. *Nature* **1975**, *253*, 694–698.
69. Levitt, M. A simplified representation of protein conformations for rapid simulation of protein folding. *J. Mol. Biol.* **1976**, *104*, 59–107. [CrossRef]
70. Field, M.J.; Bash, P.A.; Karplus, M. A combined quantum mechanical and molecular mechanical potential for molecular dynamics simulations. *J. Comput. Chem.* **1990**, *11*, 700–733. [CrossRef]
71. Gao, J. *Reviews in Computational Chemistry*; John Wiley & Sons, Ltd.: Hoboken, NJ, USA, 2007; pp. 119–185.
72. Messer, B.M.; Roca, M.; Chu, Z.T.; Vicatos, S.; Kilshtain, A.V.; Warshel, A. Multiscale simulations of protein landscapes: Using coarse-grained models as reference potentials to full explicit models. *Proteins Struct. Funct. Bioinform.* **2010**, *78*, 1212–1227. [CrossRef]
73. Mukherjee, S.; Warshel, A. Realistic simulations of the coupling between the protomotive force and the mechanical rotation of the F0-ATPase. *Proc. Natl. Acad. Sci. USA* **2012**, *109*, 14876–14881. [CrossRef] [PubMed]
74. Jones, L.O.; Mosquera, M.A.; Schatz, G.C.; Ratner, M.A. Embedding Methods for Quantum Chemistry: Applications from Materials to Life Sciences. *J. Am. Chem. Soc.* **2020**, *142*, 3281–3295. [CrossRef]
75. Nochebuena, J.; Naseem-Khan, S.; Cisneros, G.A. Development and application of quantum mechanics/molecular mechanics methods with advanced polarizable potentials. *WIREs Comput. Mol. Sci.* **2021**, e1515.
76. Chen, C.; Depa, P.; Sakai, V.G.; Maranas, J.K.; Lynn, J.W.; Peral, I.; Copley, J. R.D. A comparison of united atom, explicit atom, and coarse-grained simulation models for poly(ethylene oxide). *J. Chem. Phys.* **2006**, *124*, 234901. [CrossRef] [PubMed]
77. Potter, T.D.; Walker, M.; Wilson, M.R. Self-assembly and mesophase formation in a non-ionic chromonic liquid crystal: Insights from bottom-up and top-down coarse-grained simulation models. *Soft Matter* **2020**, *16*, 9488–9498. [CrossRef] [PubMed]
78. Tschöp, W.; Kremer, K.; Batoulis, J.; Bürger, T.; Hahn, O. Simulation of polymer melts. I. Coarse-graining procedure for polycarbonates. *Acta Polym.* **1998**, *49*, 61–74. [CrossRef]
79. Español, P.; Warren, P.B. Perspective: Dissipative particle dynamics. *J. Chem. Phys.* **2017**, *146*, 150901. [CrossRef] [PubMed]
80. Wu, K.; Xu, S.; Wan, B.; Xiu, P.; Zhou, X. A novel multiscale scheme to accelerate atomistic simulations of bio-macromolecules by adaptively driving coarse-grained coordinates. *J. Chem. Phys.* **2020**, *152*, 114115. [CrossRef] [PubMed]
81. Perdikari, T.M.; Jovic, N.; Dignon, G.L.; Kim, Y.C.; Fawzi, N.L.; Mittal, J. A coarse-grained model for position-specific effects of post-translational modifications on disordered protein phase separation. *Biophys. J.* **2021**, *120*, 1187–1197. [CrossRef]
82. Elliott, J.A. Novel approaches to multiscale modelling in materials science. *Int. Mater. Rev.* **2011**, *56*, 207–225. [CrossRef]
83. Jankowski, E.; Ellyson, N.; Fothergill, J.W.; Henry, M.M.; Leibowitz, M.H.; Miller, E.D.; Alberts, M.; Chesser, S.; Guevara, J.D.; Jones, C.D.; et al. Perspective on coarse-graining, cognitive load, and materials simulation. *Comput. Mater. Sci.* **2020**, *171*, 109129. [CrossRef]
84. Lyubartsev, A.P.; Laaksonen, A. Calculation of effective interaction potentials from radial distribution functions: A reverse Monte Carlo approach. *Phys. Rev. E* **1995**, *52*, 3730–3737. [CrossRef]
85. Chiba, S.; Okuno, Y.; Honma, T.; Ikeguchi, M. Force-field parametrization based on radial and energy distribution functions. *J. Comput. Chem.* **2019**, *40*, 2577–2585. [CrossRef]
86. Mironenko, A.V.; Voth, G.A. Density Functional Theory-Based Quantum Mechanics/Coarse-Grained Molecular Mechanics: Theory and Implementation. *J. Chem. Theory Comput.* **2020**, *16*, 6329–6342. PMID: 32877176. [CrossRef] [PubMed]
87. Noid, W.G.; Chu, J.-W.; Ayton, G.S.; Krishna, V.; Izvekov, S.; Voth, G.A.; Das, A.; Andersen, H.C. The multiscale coarse-graining method. I. A rigorous bridge between atomistic and coarse-grained models. *J. Chem. Phys.* **2008**, *128*, 244114. [CrossRef]
88. Noid, W.G.; Liu, P.; Wang, Y.; Chu, J.-W.; Ayton, G.S.; Izvekov, S.; Andersen, H.C.; Voth, G.A. The multiscale coarse-graining method. II. Numerical implementation for coarse-grained molecular models. *J. Chem. Phys.* **2008**, *128*, 244115. [CrossRef]
89. Jin, J.; Han, Y.; Pak, A.J.; Voth, G.A. A new one-site coarse-grained model for water: Bottom-up many-body projected water (BUMPer). I. General theory and model. *J. Chem. Phys.* **2021**, *154*, 044104. [CrossRef] [PubMed]

90. Shen, H.; Li, Y.; Ren, P.; Zhang, D.; Li, G. Anisotropic Coarse-Grained Model for Proteins Based On Gay–Berne and Electric Multipole Potentials. *J. Chem. Theory Comput.* **2014**, *10*, 731–750. [CrossRef] [PubMed]
91. Li, G.; Shen, H.; Zhang, D.; Li, Y.; Wang, H. Coarse-Grained Modeling of Nucleic Acids Using Anisotropic Gay–Berne and Electric Multipole Potentials. *J. Chem. Theory Comput.* **2016**, *12*, 676–693. [CrossRef]
92. Tanis, I.; Rousseau, B.; Soulard, L.; Lemarchand, C.A. Assessment of an anisotropic coarse-grained model for cis-1,4-polybutadiene: A bottom-up approach. *Soft Matter* **2021**, *17*, 621–636. [CrossRef] [PubMed]
93. Dama, J.F.; Sinitskiy, A.V.; McCullagh, M.; Weare, J.; Roux, B.; Dinner, A.R.; Voth, G.A. The Theory of Ultra-Coarse-Graining. 1. General Principles. *J. Chem. Theory Comput.* **2013**, *9*, 2466–2480. [CrossRef] [PubMed]
94. Davtyan, A.; Dama, J.F.; Sinitskiy, A.V.; Voth, G.A. The Theory of Ultra-Coarse-Graining. 2. Numerical Implementation. *J. Chem. Theory Comput.* **2014**, *10*, 5265–5275. [CrossRef]
95. Jin, J.; Voth, G.A. Ultra-Coarse-Grained Models Allow for an Accurate and Transferable Treatment of Interfacial Systems. *J. Chem. Theory Comput.* **2018**, *14*, 2180–2197. [CrossRef]
96. Zhang, Y.; Cao, Z.; Zhang, J.Z.; Xia, F. Double-Well Ultra-Coarse-Grained Model to Describe Protein Conformational Transitions. *J. Chem. Theory Comput.* **2020**, *16*, 6678–6689. [CrossRef]
97. Jin, J.; Yu, A.; Voth, G.A. Temperature and Phase Transferable Bottom-up Coarse-Grained Models. *J. Chem. Theory Comput.* **2020**, *16*, 6823–6842. [CrossRef]
98. Ueda, Y.; Taketomi, H.; Gō, N. Studies on protein folding, unfolding, and fluctuations by computer simulation. II. A. Three-dimensional lattice model of lysozyme. *Biopolymers* **1978**, *17*, 1531–1548. [CrossRef]
99. Nymeyer, H.; García, A.E.; Onuchic, J.N. Folding funnels and frustration in off-lattice minimalist protein landscapes. *Proc. Natl. Acad. Sci. USA* **1998**, *95*, 5921–5928. [CrossRef] [PubMed]
100. Tirion, M.M. Large Amplitude Elastic Motions in Proteins from a Single-Parameter, Atomic Analysis. *Phys. Rev. Lett.* **1996**, *77*, 1905–1908. [CrossRef]
101. Atilgan, A.; Durell, S.; Jernigan, R.; Demirel, M.; Keskin, O.; Bahar, I. Anisotropy of fluctuation dynamics of proteins with an elastic network model. *Biophys. J.* **2001**, *80*, 505–515. [CrossRef]
102. Togashi, Y.; Flechsig, H. Coarse-Grained Protein Dynamics Studies Using Elastic Network Models. *Int. J. Mol. Sci.* **2018**, *19*, 3899. [CrossRef] [PubMed]
103. Haliloglu, T.; Bahar, I.; Erman, B. Gaussian Dynamics of Folded Proteins. *Phys. Rev. Lett.* **1997**, *79*, 3090–3093. [CrossRef]
104. Wang, S.; Gong, W.; Deng, X.; Liu, Y.; Li, C. Exploring the dynamics of RNA molecules with multiscale Gaussian network model. *Chem. Phys.* **2020**, *538*, 110820. [CrossRef]
105. Yang, L.-W.; Chng, C.-P. Coarse-Grained Models Reveal Functional Dynamics—I. Elastic Network Models—Theories, Comparisons and Perspectives. *Bioinform. Biol. Insights* **2008**, *2*, BBI-S460. [CrossRef]
106. Chng, C.-P.; Yang, L.-W. Coarse-Grained Models Reveal Functional Dynamics—II. Molecular Dynamics Simulation at the Coarse-Grained Level—Theories and Biological Applications. *Bioinform. Biol. Insights* **2008**, *2*, BBI-S459. [CrossRef] [PubMed]
107. Torrie, G.; Valleau, J. Nonphysical sampling distributions in Monte Carlo free-energy estimation: Umbrella sampling. *J. Comput. Phys.* **1977**, *23*, 187–199. [CrossRef]
108. Mezei, M. Adaptive umbrella sampling: Self-consistent determination of the non-Boltzmann bias. *J. Comput. Phys.* **1987**, *68*, 237–248. [CrossRef]
109. Park, S.; Im, W. Theory of adaptive optimization for umbrella sampling. *J. Chem. Theory Comput.* **2014**, *10*, 2719–2728. [CrossRef]
110. Kumar, S.; Rosenberg, J.M.; Bouzida, D.; Swendsen, R.H.; Kollman, P.A. THE weighted histogram analysis method for free-energy calculations on biomolecules. I. The method. *J. Comput. Chem.* **1992**, *13*, 1011–1021. [CrossRef]
111. Shirts, M.R.; Chodera, J.D. Statistically optimal analysis of samples from multiple equilibrium states. *J. Chem. Phys.* **2008**, *129*, 124105. [CrossRef] [PubMed]
112. Paliwal, H.; Shirts, M.R. Using Multistate Reweighting to Rapidly and E ffi ciently Explore Molecular Simulation Parameters Space for Nonbonded Interactions. *J. Chem. Theory Comput.* **2013**, *9*, 4700–4717. [CrossRef] [PubMed]
113. Darve, E.; Pohorille, A. Calculating free energies using average force. *J. Chem. Phys.* **2001**, *115*, 9169–9183. [CrossRef]
114. Zhao, T.; Fu, H.; Lehievre, T.; Shao, X.; Chipot, C.; Cai, W. The Extended Generalized Adaptive Biasing Force Algorithm for Multidimensional Free-Energy Calculations. *J. Chem. Theory Comput.* **2017**, *13*, 1566–1576. [CrossRef]
115. Miao, M.; Fu, H.; Zhang, H.; Shao, X.; Chipot, C.; Cai, W. Avoiding non-equilibrium effects in adaptive biasing force calculations. *Mol. Simul.* **2020**, *46*, 1–5. [CrossRef]
116. Laio, A.; Parrinello, M. Escaping free-energy minima. *Proc. Natl. Acad. Sci. USA* **2002**, *99*, 12562–12566. [CrossRef] [PubMed]
117. Raniolo, S.; Limongelli, V. Ligand binding free-energy calculations with funnel metadynamics. *Nat. Protoc.* **2020**, *15*, 2837–2866. [CrossRef]
118. Kondo, T.; Sasaki, T.; Ruiz-Barragan, S.; Ribas-Ariño, J.; Shiga, M. Refined metadynamics through canonical sampling using time-invariant bias potential: A study of polyalcohol dehydration in hot acidic solutions. *J. Comput. Chem.* **2021**, *42*, 156–165. [CrossRef]
119. Comer, J.; Gumbart, J.C.; Hénin, J.; Lelièvre, T.; Pohorille, A.; Chipot, C. The Adaptive Biasing Force Method: Everything You Always Wanted To Know but Were Afraid To Ask. *J. Phys. Chem. B* **2015**, *119*, 1129–1151. [CrossRef]
120. Fu, H.; Shao, X.; Cai, W.; Chipot, C. Taming Rugged Free Energy Landscapes Using an Average Force. *Accounts Chem. Res.* **2019**, *52*, 3254–3264. [CrossRef]

121. Valsson, O.; Tiwary, P.; Parrinello, M. Enhancing Important Fluctuations: Rare Events and Metadynamics from a Conceptual Viewpoint. *Annu. Rev. Phys. Chem.* **2016**, *67*, 159–184. [CrossRef]
122. Bussi, G.; Laio, A. Using metadynamics to explore complex free-energy landscapes. *Nat. Rev. Phys.* **2020**, *2*, 200–212. [CrossRef]
123. Jolliffe, I.T. *Principal Component Analysis*; Springer: New York, NY, USA; Berlin/Heidelberg, Germany; Hong Kong, China; Milan, Italy; Paris, France; Tokyo, Japan, 2002.
124. Shan, P.; Zhao, Y.; Wang, Q.; Ying, Y.; Peng, S. Principal component analysis or kernel principal component analysis based joint spectral subspace method for calibration transfer. *Spectrochim. Acta Part Mol. Biomol. Spectrosc.* **2020**, *227*, 117653. [CrossRef] [PubMed]
125. Cox, T.F.; Cox, M.A.A. *Multidimensional SCALING*; Chapman & HALL/CRC: London, UK 2000
126. Tenenbaum, J.B.; Silva, V.D.; Langford, J.C. A Global Geometric Framework for Nonlinear Dimensionality Reduction. *Science* **2000**, *290*, 2319–2323. [CrossRef] [PubMed]
127. Roweis, S.T.; Saul, L.K. Nonlinear Dimensionality Reduction by Locally Linear Embedding. *Science* **2000**, *290*, 2323–2326. [CrossRef]
128. Coifman, R.R.; Lafon, S.; Lee, A.B.; Maggioni, M.; Nadler, B.; Warner, F.; Zucker, S.W. Geometric diffusions as a tool for harmonic analysis and structure definition of data: Diffusion maps. *Proc. Natl. Acad. Sci. USA* **2005**, *102*, 7426–7431. [CrossRef]
129. Coifman, R.R.; Lafon, S.; Lee, A.B.; Maggioni, M.; Nadler, B.; Warner, F.; Zucker, S.W. Geometric diffusions as a tool for harmonic analysis and structure definition of data: Multiscale methods. *Proc. Natl. Acad. Sci. USA* **2005**, *102*, 7432–7437. [CrossRef]
130. Ceriotti, M.; Tribello, G.A.; Parrinello, M. Simplifying the representation of complex free-energy landscapes using sketch-map. *Proc. Natl. Acad. Sci. USA* **2011**, *108*, 13023–13028. [CrossRef]
131. Dellago, C.; Bolhuis, P.G.; Geissler, P.L. *Advances in Chemical Physics*; John Wiley & Sons, Ltd.: Hoboken, NJ, USA, 2003; pp. 1–78. Chapter 1.
132. Rogal, J.; Bolhuis, P.G. Multiple state transition path sampling. *J. Chem. Phys.* **2008**, *129*, 224107. [CrossRef] [PubMed]
133. Buijsman, P.; Bolhuis, P.G. Transition path sampling for non-equilibrium dynamics without predefined reaction coordinates. *J. Chem. Phys.* **2020**, *152*, 044108. [CrossRef]
134. Yao, Y.; Cui, R.Z.; Bowman, G.R.; Silva, D.-A.; Sun, J.; Huang, X. Hierarchical Nyström methods for constructing Markov state models for conformational dynamics. *J. Chem. Phys.* **2013**, *138*, 174106. [CrossRef] [PubMed]
135. Chodera, J.D.; Noé, F. Markov state models of biomolecular conformational dynamics. *Curr. Opin. Struct. Biol.* **2014**, *25*, 135–144. [CrossRef]
136. Husic, B.E.; Pande, V.S. Markov State Models: From an Art to a Science. *J. Am. Chem. Soc.* **2018**, *140*, 2386–2396. [CrossRef] [PubMed]
137. Nagel, D.; Weber, A.; Stock, G. MSMPathfinder: Identification of Pathways in Markov State Models. *J. Chem. Theory Comput.* **2020**, *16*, 7874–7882. [CrossRef]
138. Sadiq, S.K.; Noé, F.; De Fabritiis, G. Kinetic characterization of the critical step in HIV-1 protease maturation. *Proc. Natl. Acad. Sci. USA* **2012**, *109*, 20449–20454. [CrossRef]
139. Kohlhoff, K.J.; Shukla, D.; Lawrenz, M.; Bowman, G.R.; Konerding, D.E.; Belov, D.; Altman, R.B.; Pande, V.S. Cloud-based simulations on Google Exacycle reveal ligand modulation of GPCR activation pathways. *Nat. Chem.* **2014**, *6*, 15–21. [CrossRef]
140. Yang, M.; Tang, Y.; Weng, J.; Liu, Z.; Wang, W. The Role of Calcium in Regulating the Conformational Dynamics of d-Galactose/d-Glucose-Binding Protein Revealed by Markov State Model Analysis. *J. Chem. Inf. Model.* **2021**, *61*, 891–900. [CrossRef]
141. Wu, H.; Noé, F. Variational Approach for Learning Markov Processes from Time Series Data. *J. Nonlinear Sci.* **2020**, *30*, 23–66. [CrossRef]
142. Zuckerman, D.M.; Chong, L.T. Weighted Ensemble Simulation: Review of Methodology, Applications, and Software. *Annu. Rev. Biophys.* **2017**, *46*, 43–57. [CrossRef] [PubMed]
143. Tsallis, C. Some comments on Boltzmann-Gibbs statistical mechanics. *CHaos Solitons Fractals* **1995**, *6*, 539–559. [CrossRef]
144. Plastino, A. Why Tsallis' statistics? *Phys. A Stat. Mech. Its Appl.* **2004**, *344*, 608–613. [CrossRef]
145. Swendsen, R.H.; Wang, J.-S. Replica Monte Carlo Simulation of Spin-Glasses. *Phys. Rev. Lett.* **1986**, *57*, 2607–2609. [CrossRef]
146. Appadurai, R.; Nagesh, J.; Srivastava, A. High resolution ensemble description of metamorphic and intrinsically disordered proteins using an efficient hybrid parallel tempering scheme. *Nat. Commun.* **2021**, *12*, 958. [CrossRef]
147. Sugita, Y.; Okamoto, Y. Replica-exchange molecular dynamics method for protein folding. *Chem. Phys. Lett.* **1999**, *314*, 141–151. [CrossRef]
148. Peng, C.; Wang, J.; Shi, Y.; Xu, Z.; Zhu, W. Increasing the Sampling Efficiency of Protein Conformational Change by Combining a Modified Replica Exchange Molecular Dynamics and Normal Mode Analysis. *J. Chem. Theory Comput.* **2021**, *17*, 13–28. [CrossRef] [PubMed]
149. Wang, F.; Landau, D.P. Efficient, Multiple-Range Random Walk Algorithm to Calculate the Density of States. *Phys. Rev. Lett.* **2001**, *86*, 2050–2053. [CrossRef]
150. Gao, Y.Q. An integrate-over-temperature approach for enhanced sampling. *J. Chem. Phys.* **2008**, *128*, 064105. [CrossRef]
151. Yang, L.; Liu, C.-W.; Shao, Q.; Zhang, J.; Gao, Y.Q. From Thermodynamics to Kinetics: Enhanced Sampling of Rare Events. *Accounts Chem. Res.* **2015**, *48*, 947–955. [CrossRef]
152. Yang, Y.I.; Niu, H.; Parrinello, M. Combining Metadynamics and Integrated Tempering Sampling. *J. Phys. Chem. Lett.* **2018**, *9*, 6426–6430. [CrossRef]

153. Behler, J. Perspective: Machine learning potentials for atomistic simulations. *J. Chem. Phys.* **2016**, *145*, 170901. [CrossRef]
154. Ceriotti, M. Unsupervised machine learning in atomistic simulations, between predictions and understanding. *J. Chem. Phys.* **2019**, *150*, 150901. [CrossRef] [PubMed]
155. Lunghi, A.; Sanvito, S. A unified picture of the covalent bond within quantum-accurate force fields: From organic molecules to metallic complexes' reactivity. *Sci. Adv.* **2019**, *5*, eaaw2210. [CrossRef]
156. Mueller, T.; Hernandez, A.; Wang, C. Machine learning for interatomic potential models. *J. Chem. Phys.* **2020**, *152*, 050902. [CrossRef] [PubMed]
157. Noé, F.; Tkatchenko, A.; Müller, K.-R.; Clementi, C. Machine Learning for Molecular Simulation. *Annu. Rev. Phys. Chem.* **2020**, *71*, 361–390. [CrossRef] [PubMed]
158. Takahashi, A.; Seko, A.; Tanaka, I. Linearized machine-learning interatomic potentials for non-magnetic elemental metals: Limitation of pairwise descriptors and trend of predictive power. *J. Chem. Phys.* **2018**, *148*, 234106. [CrossRef]
159. Deringer, V.L.; Csányi, G. Machine learning based interatomic potential for amorphous carbon. *Phys. Rev. B* **2017**, *95*, 094203. [CrossRef]
160. Bartók, A.P.; Kermode, J.; Bernstein, N.; Csányi, G. Machine Learning a General-Purpose Interatomic Potential for Silicon. *Phys. Rev. X* **2018**, *8*, 041048. [CrossRef]
161. Slepoy, A.; Peters, M.D.; Thompson, A.P. Searching for globally optimal functional forms for interatomic potentials using genetic programming with parallel tempering. *J. Comput. Chem.* **2007**, *28*, 2465–2471. [CrossRef]
162. Qu, C.; Bowman, J.M. A fragmented, permutationally invariant polynomial approach for potential energy surfaces of large molecules: Application to N-methyl acetamide. *J. Chem. Phys.* **2019**, *150*, 141101. [CrossRef]
163. Chmiela, S.; Sauceda, H.E.; Müller, K.-R. Towards exact molecular dynamics simulations with machine-learned force fields. *Nat. Commun.* **2018**, *9*, 3887. [CrossRef] [PubMed]
164. Artrith, N.; Behler, J. High-dimensional neural network potentials for metal surfaces: A prototype study for copper. *Phys. Rev. B* **2012**, *85*, 045439. [CrossRef]
165. Podryabinkin, E.V.; Tikhonov, E.V.; Shapeev, A.V.; Oganov, A.R. Accelerating crystal structure prediction by machine-learning interatomic potentials with active learning. *Phys. Rev. B* **2019**, *99*, 064114. [CrossRef]
166. Jinnouchi, R.; Karsai, F.; Kresse, G. On-the-fly machine learning force field generation: Application to melting points. *Phys. Rev. B* **2019**, *100*, 014105. [CrossRef]
167. Huan, T.D.; Batra, R.; Chapman, J.; Kim, C.; Chandrasekaran, A.; Ramprasad, R. Iterative-Learning Strategy for the Development of Application-Specific Atomistic Force Fields. *J. Phys. Chem. C* **2019**, *123*, 20715–20722. [CrossRef]
168. Ghasemi, S.A.; Hofstetter, A.; Saha, S.; Goedecker, S. Interatomic potentials for ionic systems with density functional accuracy based on charge densities obtained by a neural network. *Phys. Rev. B* **2015**, *92*, 045131. [CrossRef]
169. Ko, T.W.; Finkler, J.A.; Goedecker, S.; Behler, J. A fourth-generation high-dimensional neural network potential with accurate electrostatics including non-local charge transfer. *Nat. Commun.* **2021**, *12*, 398. [CrossRef]
170. Harris, W.H. Machine Learning Transferable Physics-Based Force Fields using Graph Convolutional Neural Networks. Ph.D. Thesis, Massachusetts Institute of Technology, Cambridge, MA, USA, 2020.
171. Zhang, Y.; Hu, C.; Jiang, B. Embedded Atom Neural Network Potentials: Efficient and Accurate Machine Learning with a Physically Inspired Representation. *J. Phys. Chem. Lett.* **2019**, *10*, 4962–4967. [CrossRef]
172. Chan, H.; Cherukara, M.J.; Narayanan, B.; Loeffler, T.D.; Benmore, C.; Gray, S.K.; Sankaranarayanan, S.K. Machine learning coarse grained models for water. *Nat. Commun.* **2019**, *10*, 1–14. [CrossRef]
173. Moradzadeh, A.; Aluru, N.R. Transfer-Learning-Based Coarse-Graining Method for Simple Fluids: Toward Deep Inverse Liquid-State Theory. *J. Phys. Chem. Lett.* **2019**, *10*, 1242–1250. [CrossRef]
174. Wang, J.; Olsson, S.; Wehmeyer, C.; Pérez, A.; Charron, N.E.; De Fabritiis, G.; Noé, F.; Clementi, C. Machine Learning of Coarse-Grained Molecular Dynamics Force Fields. *ACS Cent. Sci.* **2019**, *5*, 755–767. [CrossRef]
175. Zhang, L.; Han, J.; Wang, H.; Car, R.; Weinan, W.E. DeePCG: Constructing coarse-grained models via deep neural networks. *J. Chem. Phys.* **2018**, *149*, 034101. [CrossRef]
176. Chakraborty, M.; Xu, C.; White, A.D. Encoding and selecting coarse-grain mapping operators with hierarchical graphs. *J. Chem. Phys.* **2018**, *149*, 134106. [CrossRef]
177. Webb, M.A.; Delannoy, J.Y.; De Pablo, J.J. Graph-Based Approach to Systematic Molecular Coarse-Graining. *J. Chem. Theory Comput.* **2019**, *15*, 1199–1208. [CrossRef] [PubMed]
178. Giulini, M.; Menichetti, R.; Shell, M.S.; Potestio, R. An information theory-based approach for optimal model reduction of biomolecules. **2020**, *16*, 6795–6813. [CrossRef]
179. Li, Z.; Wellawatte, G.P.; Chakraborty, M.; Gandhi, H.A.; Xu, C.; White, A.D. Graph neural network based coarse-grained mapping prediction. *Chem. Sci.* **2020**, *11*, 9524–9531. [CrossRef]
180. Durumeric, A.E.; Voth, G.A. Adversarial-residual-coarse-graining: Applying machine learning theory to systematic molecular coarse-graining. *J. Chem. Phys.* **2019**, *151*, 124110. [CrossRef] [PubMed]
181. Khot, A.; Shiring, S.B.; Savoie, B.M. Evidence of information limitations in coarse-grained models. *J. Chem. Phys.* **2019**, *151*, 244105. [CrossRef]
182. Chen, W.; Tan, A.R.; Ferguson, A.L. Collective variable discovery and enhanced sampling using autoencoders: Innovations in network architecture and error function design. *J. Chem. Phys.* **2018**, *149*, 072312. [CrossRef] [PubMed]

183. Chen, W.; Ferguson, A.L. Molecular enhanced sampling with autoencoders: On-the-fly collective variable discovery and accelerated free energy landscape exploration. *J. Comput. Chem.* **2018**, *39*, 2079–2102. [CrossRef]
184. Wehmeyer, C.; Noé, F. Time-lagged autoencoders: Deep learning of slow collective variables for molecular kinetics. *J. Chem. Phys.* **2018**, *148*, 241703. [CrossRef]
185. Ribeiro, J. M.L.; Bravo, P.; Wang, Y.; Tiwary, P. Reweighted autoencoded variational Bayes for enhanced sampling (RAVE). *J. Chem. Phys.* **2018**, *149*, 241703. [CrossRef]
186. Mardt, A.; Pasquali, L.; Wu, H.; Noé, F. VAMPnets for deep learning of molecular kinetics. *Nat. Commun.* **2018**, *9*, 1–11. [CrossRef] [PubMed]
187. Cao, X.; Tian, P. Molecular free energy optimization on a computational graph. *RSC Adv.* **2020**, *11*, 12929. [CrossRef]
188. Khatib, F.; Cooper, S.; Tyka, M.D.; Xu, K.; Makedon, I.; Popović, Z.; Baker, D.; Players, F. Algorithm discovery by protein folding game players. *Proc. Natl. Acad. Sci. USA* **2011**, *108*, 18949–18953. [CrossRef]
189. Feig, M. Computational protein structure refinement: Almost there, yet still so far to go. *Wiley Interdiplinary Rev. Comput. Mol. Sci.* **2017**, *7*, e1307. [CrossRef]

International Journal of
Molecular Sciences

MDPI

Article

Time-Dependent Unitary Transformation Method in the Strong-Field-Ionization Regime with the Kramers-Henneberger Picture

Je-Hoi Mun [1,2,*], Hirofumi Sakai [3,4] and Dong-Eon Kim [1,2]

1 Department of Physics and Center for Attosecond Science and Technology, POSTECH, Pohang 37673, Korea; kimd@postech.ac.kr
2 Max Planck POSTECH/KOREA Research Initiative, Pohang 37673, Korea
3 Department of Physics, Graduate School of Science, The University of Tokyo, 7-3-1 Hongo, Bunkyo-ku, Tokyo 113-0033, Japan; hsakai@phys.s.u-tokyo.ac.jp
4 Institute for Photon Science and Technology, Graduate School of Science, The University of Tokyo, 7-3-1 Hongo, Bunkyo-ku, Tokyo 113-0033, Japan
* Correspondence: mun1219@mpk.or.kr

Abstract: Time evolution operators of a strongly ionizing medium are calculated by a time-dependent unitary transformation (TDUT) method. The TDUT method has been employed in a quantum mechanical system composed of discrete states. This method is especially helpful for solving molecular rotational dynamics in quasi-adiabatic regimes because the strict unitary nature of the propagation operator allows us to set the temporal step size to large; a tight limitation on the temporal step size ($\delta t << 1$) can be circumvented by the strict unitary nature. On the other hand, in a strongly ionizing system where the Hamiltonian is not Hermitian, the same approach cannot be directly applied because it is demanding to define a set of field-dressed eigenstates. In this study, the TDUT method was applied to the ionizing regime using the Kramers-Henneberger frame, in which the strong-field-dressed discrete eigenstates are given by the field-free discrete eigenstates in a moving frame. Although the present work verifies the method for a one-dimensional atom as a prototype, the method can be applied to three-dimensional atoms, and molecules exposed to strong laser fields.

Keywords: numerical method; laser-matter interaction; time-dependent Schrödinger equation; time-dependent unitary transformation method; strong-field ionization; Kramers-Henneberger frame

Citation: Mun, J.H.; Sakai, H.; Kim, D.E. Time-Dependent Unitary Transformation Method in the Strong-Field-Ionization Regime with the Kramers-Henneberger Picture. *Int. J. Mol. Sci.* **2021**, *22*, 8514. https://doi.org/10.3390/ijms22168514

Academic Editor: Małgorzata Borówko

Received: 25 June 2021
Accepted: 4 August 2021
Published: 7 August 2021

1. Introduction

Over the last few decades, the ionization of atoms and molecules by ultrafast strong infrared laser fields has attracted considerable interest because of the availability of high-intensity lasers. Field-induced ionization can be divided into two regimes, according to the Keldysh parameter $\gamma \equiv \sqrt{I_P/(2U_P)}$ [1], where I_P and U_P are the ionization and the ponderomotive potentials, respectively. The ionization dynamics is considered to be governed by tunneling when $\gamma << 1$, while for $\gamma >> 1$ the process is mostly affected by multiphoton ionization. For a typical 800-nm infrared laser field, the dynamics of the atoms and molecules can be characterized by the laser intensity. In the low-intensity laser field ($\gamma >> 1$), the dynamics can be studied with the aid of the perturbation theory. When the laser field strength is comparable to or even higher than the Coulomb field strength in atoms and molecules ($\gamma << 1$), the ionization dynamics can be described by tunneling ionization, which can be solved analytically by using a strong-field approximation (SFA) model [1,2]. The tunneling picture based on the SFA model explains many qualitative features of strong-field phenomena, such as high-harmonic generation (HHG) [3–6] and above-threshold ionization (ATI) [4,7,8].

However, the conventional SFA model is not capable to render a description of phenomena mediated by other bound states in addition to the ground state. For example, some tunneling-ionized electrons that receive relatively low drift energies from the laser fields are observed to still stay in the excited bound states after the laser pulse has passed [9]. A simple man model, which is based on the solutions of the Newtonian equations of motion, has been used in many studies on this phenomenon known as the frustrated tunneling ionization (FTI) [9–14]. Moreover, coherent EUV generation via FTI [15,16] and resonantly enhanced HHG [17–21] require the high-lying electronic bound states to be considered. In solving the full TDSE to understand these phenomena, discrete-level-based calculations and/or analysis are essential. The discrete-level-based analysis of the strong-field ionization can provide fresh insights into the underlying physical mechanisms. In Ref. [22], by calculating a strong-field-dressed discrete adiabatic basis set, it has been revealed that tunneling ionization is diabatic rather than adiabatic in a language based on the so-called adiabatic representation. Tunneling ionization is often regarded as an adiabatic process, which is not true in terms of the adiabatic representation [22]. Furthermore, a series of discrete-level-based numerical calculations have shown that atomic ionization passages can be manipulated by chirp control of an incident laser pulse [23].

Analytical and numerical studies with discrete basis sets are commonly used in various branches of atomic and molecular physics, such as the Rydberg atoms [24,25], ultracold gases and trapped ions [26–28], and molecular rotational dynamics [29–31]. In molecular rotational dynamics, we have developed a time-dependent unitary transformation (TDUT), which has been particularly useful in quasi-adiabatic regimes. In this method, the field-dressed eigenstates and eigenenergies are calculated in every temporal step to obtain strict unitary propagation operators. The TDUT method is free from a tight limitation on the temporal step size ($\delta t << 1$), existing in conventional numerical methods (e.g., Crank–Nicolson method and Runge–Kutta method), so that rapid numerical calculations are possible. In the case of strong-field-induced ionization dynamics, however, specific efforts are needed to apply this approach using discrete field-dressed states as a basis set. The eigenstates of an atomic potential tilted by a strong electric field form continuum states [22], and the ground state is localized at the edge of a spatial boundary position. Thus, the direct formulation of the TDUT method itself [29] cannot solve the dynamics in the presence of ionization events.

In this study, this problem was solved using the Kramers-Henneberger (KH) frame [32], in which the strong-field-dressed discrete eigenstates are given by the field-free discrete eigenstates in a moving frame. Once a unitary translation operator is calculated correctly, the TDSE can be solved numerically using the TDUT method in the KH frame. Using a one-dimensional atom as a prototype, the numerical method is validated by comparing it with the Crank–Nicolson method. The final electronic states excited by a few-optical cycled near-infrared (800 nm) laser are calculated by changing the field strength and initial state. This study numerically clarifies that the number of discrete states for the TDUT calculation depends on the field strength. Below the tunneling intensity regime, only a few bound states can be used, while in the stronger field regime, much higher-lying continuum states need to be included. The dynamics of three-dimensional atoms and molecules can be calculated using the TDUT method.

The paper is organized as follows. Section 2, revisits the general TDUT method in the ionization-free regime, which was developed for the molecular rotational dynamics [29]. The method is then implemented to a strong-field ionization regime within the KH frame. Section 3 presents the numerical results of a one-dimensional atom exposed to an intense laser pulse to test the numerical method. A summary and outlook are provided in Section 4. Atomic units are used throughout the paper unless otherwise stated.

2. Numerical Method

This section provides a detailed explanation of the numerical method.

2.1. Time-Dependent Unitary Transformation Method (TDUT) in Discretized Systems

In conventional methods for solving the TDSE, a wave function at $t+\delta t$, $\psi(t+\delta t)$, is approximated by $[\exp(-i\hat{H}(t)\delta t)]\psi(t) \sim [1-i\hat{H}(t)\delta t]\psi(t)$ under the assumption of $\delta t \ll 1$. Here, $\hat{H}(t)$ is the time-dependent Hamiltonian. In this simple sum formulation, the operator $[1-i\hat{H}(t)\delta t]$ is not a unitary one in principle. The non-unitary nature is accumulated over a number of evolution steps, and the simulation can catastrophically fail. The popular Runge–Kutta method and Crank–Nicolson method can reduce the non-unitary nature. On the other hand, the constraint $\delta t \ll 1$ must be satisfied to a reasonable degree.

Several numerical methods, such as methods based on the Chebyshev propagator [33] and Lanczos propagator [34], the split-operator method [35], and a huge number of variations, are available. The time-dependent unitary transformation (TDUT) method is one of the effective and intuitive methods for the TDSE [29], where every temporal evolution operator is strictly unitary. To describe the method, we need to consider two arbitrary frames labeled by (a) and (b), respectively. When we consider a single stepwise temporal variation δt, the time-dependent Hamiltonian evolves from $\hat{H}(t)$ to $\hat{H}(t+\delta t)$. The two Hamiltonians can be labeled as $\hat{H}(t) \equiv \hat{H}^{\mathrm{a}}$ and $\hat{H}(t+\delta t) \equiv \hat{H}^{\mathrm{b}}$. Because the time-independent Hamiltonian operator $\hat{H}^{\mathrm{a}}$ or $\hat{H}^{\mathrm{b}}$ is a Hermite operator in general, there are a set of real-valued eigenvalues ϵ_i^{a} (ϵ_i^{b}) and the corresponding set of eigenstates ϕ_i^{a} (ϕ_i^{b}), obtained by solving the TISE as follows.

$$\begin{aligned}\hat{H}^{\mathrm{a}}\phi_i^{\mathrm{a}} &= \epsilon_i^{\mathrm{a}}\phi_i^{\mathrm{a}},\\ \hat{H}^{\mathrm{b}}\phi_i^{\mathrm{b}} &= \epsilon_i^{\mathrm{b}}\phi_i^{\mathrm{b}}.\end{aligned} \tag{1}$$

Here the integer i represents a state number.

Let us set the moment of the step-wise change of the Hamiltonian from (a) to (b) at t_0. A wave function $\psi(t_0)$ can be expressed as a superposition of eigenstates ϕ_i^{a} as follows.

$$\psi(t_0) = \sum_m c_m^{\mathrm{a}}(t_0)\phi_m^{\mathrm{a}}(t_0), \tag{2}$$

where $c_m^{\mathrm{a}}(t_0)$ corresponds to the probability amplitude $\langle\phi_m^{\mathrm{a}}|\psi(t_0)\rangle$ at time t_0. The same eigenstate expansion is also possible in the (b) frame. The probability amplitude $c_m^{\mathrm{b}}(t_0) \equiv \langle\phi_m^{\mathrm{b}}|\psi(t_0)\rangle$ in the (b) representation, can be expressed using the (a) frame eigenstates as $c_n^{\mathrm{b}}(t_0) = \sum_m\langle\phi_n^{\mathrm{b}}|\phi_m^{\mathrm{a}}\rangle c_m^{\mathrm{a}}(t_0)$. In the (b) frame, for $t \geq t_0$ until the next step-wise change of external field intensity occurs, the time evolution of the wave function is given by adding a phase shift to each eigenstate of ϕ_n^{b}, which is operated by

$$\begin{aligned}\psi(t) = \sum_n\Big[\sum_m\langle\phi_n^{\mathrm{b}}|\phi_m^{\mathrm{a}}\rangle c_m^{\mathrm{a}}(t_0)\Big]\phi_n^{\mathrm{b}}\times\\ \exp(-i\epsilon_n^{\mathrm{b}}(t-t_0)).\end{aligned} \tag{3}$$

Here, the matrix $\langle\phi_n^{\mathrm{b}}|\phi_m^{\mathrm{a}}\rangle$ is equivalent to $\sum_l\langle\phi_n^{\mathrm{b}}|\chi_l\rangle\langle\chi_l|\phi_m^{\mathrm{a}}\rangle$, with χ_l being a field-free eigenstate, i.e., the spherical harmonics in the case of the linear molecules. The unitary matrices $\langle\phi_m^{\mathrm{a}}|\chi_l\rangle$ and $\langle\phi_m^{\mathrm{b}}|\chi_l\rangle$ are obtained by solving the TISE given in Equation (1). In a matrix-based expression, the temporal evolution can be rewritten by an evolution matrix, $\hat{U}_{t_0}$, satisfying $\hat{\psi}(t_0+\delta t) \equiv \hat{U}_{t_0}\hat{\psi}(t_0)$, as follows:

$$\hat{U}_{t_0} = \exp(-i\hat{\epsilon}^{\mathrm{b}}\delta t)\hat{U}^{\mathrm{b}}\hat{U}^{\dagger\mathrm{a}}. \tag{4}$$

Here $\hat{U}^{\dagger\mathrm{a}} \equiv \langle\chi_l|\phi_m^{\mathrm{a}}\rangle$ transforms the (a)-frame eigenstate back to the one in the field-free frame. Next, $\hat{U}^{\mathrm{b}} \equiv \langle\phi_m^{\mathrm{b}}|\chi_l\rangle$ transforms the state defined in the field-free frame to the one in the (b) frame. The time-evolution is operated easily in the (b) frame by adding the phase-shift to the eigenstates defined in the (b) frame, which is performed by multiplying a diagonal matrix $\exp(-i\hat{\epsilon}^{\mathrm{b}}\delta t)$. Figure 1 presents a schematic diagram of the numerical method.

We emphasize again that all the operations are strictly unitary in this time evolution if the eigenvalue problems are solved correctly. Hence, it is free from any numerical errors generated in the temporal propagation. Instead, the temporally-discretized Hamiltonian is solely responsible for some possible deviations from the correct solution. The discretized Hamiltonian approaches a real Hamiltonian by reducing the step size. The reliable maximum step size depends on the temporal shape of the Hamiltonian rather than the wave function condition. In this method, most of the numerical tasks involve calculating the eigenstates and eigenvalues of the Hamiltonians in every discretized step, which can be conducted using the available built-in functions in various programming languages. A similar numerical approach is used in the Lanczos propagator [34], but with a set of quasi-eigenvalues and quasi-eigenstates defined in a reduced subspace rather than exact eigenvalues and eigenstates in the full Hilbert space. For example, if the exact TISE calculation is numerically demanding, the Lanczos approach can be used with a trade-off between the numerical accuracy and the computing cost.

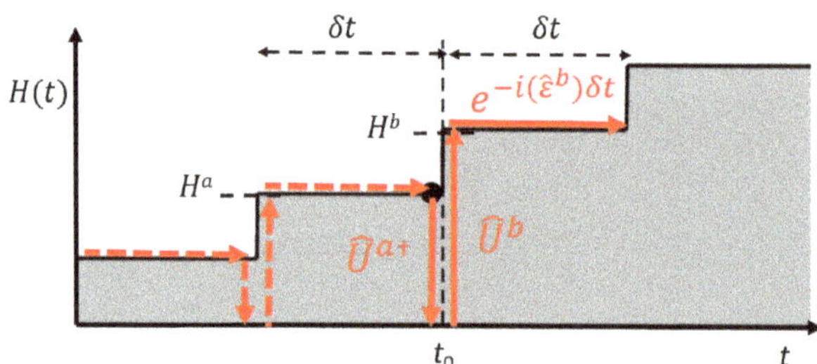

Figure 1. Schematic diagram of time-dependent unitary transformation method for wave packet evolution. In the presence of the time-dependent Hamiltonian $\hat{H}(t)$, temporal evolution of a wave function from $t = t_0$ to $t = t_0 + \delta t$ is operated by multiplying three strict unitary operators $\hat{U}^{\dagger \mathrm{a}}$, $\hat{U}^{\mathrm{b}}$, and $\exp(-i\hat{\varepsilon}^{\mathrm{b}}\delta t)$. See the main texts.

2.2. TDUT in the Strong-Field Ionization Regime

When a target material ionized by an external laser field is considered, a set of time-independent field-dressed eigenstates of the material at a sepecific time will form a continuum state that is not localized in real space [22]. As a result, the TDUT approach introduced in Section 2.1 is not directly applicable in the strong-field regime because the method requires a set of field-dressed eigenstates in every temporal step. On the other hand, a set of strong-field-dressed eigenstates and eigenenergies can be obtained in the moving frame, so-called the Kramers-Henneberger frame.

In the KH frame [32], the laser-field-dressed Hamiltonian of an atom or a molecule is given by

$$H = \frac{\vec{p}^2}{2} + V(\vec{r} + \vec{\alpha}(t)), \tag{5}$$

where $V(\vec{r})$ is the binding potential and $\vec{\alpha}(t) \equiv \int_{-\infty}^{t} \vec{A}(t)dt$ indicates the classical trajectory of a free electron exposed to the laser field $\vec{E}(t) \equiv -\frac{d\vec{A}(t)}{dt}$. In this moving frame, the eigenstates of a field-dressed Hamiltonian are equivalent to the field-free eigenstates except that they are displaced uniformly from the origin by $-\vec{\alpha}(t)$. Once the field-free eigenstates of an atom or a molecule are accurately obtained, we can apply the method given in Equation (4).

The wave function $\psi(t)$ can be described in terms of the discrete level expansion as Equation (2). Hence, the wave function $\psi(t)$ can be represented by the vector $\hat{\psi}(t) = (a_1, a_2, ..., a_{n_{\max}})^T$, e.g., the initial ground state is $(1, 0, ..., 0)^T$. In the moving (laser-field-dressed) frame, $\hat{\psi}(t)$ is replaced with $\hat{U}(\vec{\alpha}(t)) \cdot \hat{\psi}(t)$, where the matrix $\hat{U}(\vec{\alpha}(t))$ is a translation operator (Figure 2):

$$\hat{U}(\vec{\alpha}(t))_{m,n} \equiv \int_{-\infty}^{\infty} \phi_m^*(\vec{r} + \vec{\alpha}(t))\phi_n(\vec{r})d\boldsymbol{r}. \tag{6}$$

The time evolution of the wave function for a time step dt can be expressed as

$$\hat{U}(\vec{\alpha}(t)) \cdot \hat{\psi}(t + dt) = \exp(-i\hat{e}dt) \cdot \hat{U}(\vec{\alpha}(t)) \cdot \hat{\psi}(t). \tag{7}$$

Here, $\hat{e}$ is the matrix whose diagonal elements are eigenenergies of the field-free eigenstates. To rewrite the wave function in Equation (7) in the field-free frame, the transpose of $\hat{U}(\vec{\alpha}(t))$ needs to be multiplied, resulting in

$$\hat{\psi}(t + dt) = \hat{U}^T(\vec{\alpha}(t)) \cdot \exp(-i\hat{e}dt) \cdot \hat{U}(\vec{\alpha}(t)) \cdot \hat{\psi}(t). \tag{8}$$

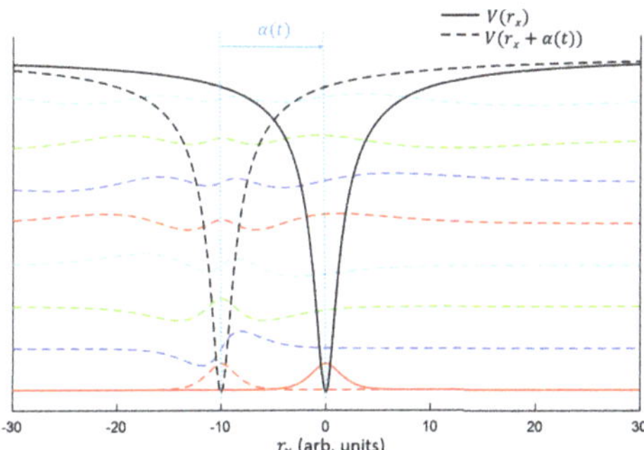

Figure 2. Representation of two frames converted by a translation operator $\hat{U}(\vec{\alpha}(t))$. The black-solid and black-dashed lines are the atomic potentials in the field-free and the field-dressed KH frames, respectively, while the solid red line is the ground state of the atom, and the other colored-dashed lines are eigenstates of the field-dressed KH Hamiltonian Equation (5). The initial ground state is expanded by the eigenstates of the field-dressed Hamiltonian by multiplying $\hat{U}(\vec{\alpha}(t))$ to the initial ground state.

Although the translation operator $\hat{U}(\vec{\alpha}(t))$ is a unitary matrix in principle, in case of the numerical calculation, the unitarity of $\hat{U}(\vec{\alpha}(t))$ is not ensured. Furthermore, the matrix becomes less unitary as the spatial displacement $\vec{\alpha}(t)$ increases. The matrix $\hat{U}(\vec{\alpha}(t))$ corresponds to the unity matrix when $\vec{\alpha}(t)$ is zero. For other cases, because $\hat{U}(\vec{\alpha}(t))$ is responsible for transforming a wave function in the $\vec{r}$ frame into that in the $\vec{r} + \vec{\alpha}(t)$ frame, if $\vec{\alpha}(t)$ is too large, the initial wave function will be placed outside of the spatial boundary so that $\hat{U}(\vec{\alpha}(t))$ is no longer unitary.

We have identified that a spatial displacement of 0.2, which is the size of spatial grid used in the numerical calculations, does not ensure that the matrix is unitary, i.e., $det|\hat{U}(\alpha_x(t) = 0.2)| < 1$. To obtain a unitary translation operator $\hat{U}(\delta x)$, the matrix elements were calculated in the momentum domain by

$$\hat{U}(\delta x)_{m,n} \equiv \int_{-\infty}^{\infty} \exp(ip_x\delta x)\tilde{\phi}_m^*(\vec{p})\tilde{\phi}_n(\vec{p})d\boldsymbol{p}, \tag{9}$$

where $\tilde{\phi}_n(\vec{p})$ is the Fourier-transformed, normalized wave function.

In the momentum domain, we have calculated the unit translation operator $\hat{U}(\delta x)$ with $\delta x = 10^{-4}$. This operator is quasi-unitary, satisfying $|\hat{U}(\delta x)| \sim 1$ with a possible deviation of only 10^{-15}. When the laser field is linearly polarized in the x direction and $\vec{\alpha}(t) \equiv \alpha_x(t) \equiv \alpha(t)$, an arbitrary translation operator $\hat{U}(\alpha(t))$ can be obtained from the multiplication of the unit displacement operator, i.e., $\hat{U}(\alpha(t)) \equiv \hat{U}(\delta x)^{\alpha(t)/\delta x}$. By applying this operator $\hat{U}(\delta x)$, the TDSE can be solved by the following procedure.

For the time evolution of the wave function, first we expand the wave function $\hat{\psi}(t)$ with respect to the field-dressed eigenstates, which are the field-free eigenstates spatially displaced by $\alpha(t)$ from the origin. This expansion is conducted by multiplying the quasi-unitary matrix $\hat{U}(\delta x)^{\alpha(t)/\delta x}$ to the wave function $\hat{\psi}(t)$. The negative unit translation operator $\hat{U}(\delta x)^{-1}$ is, therefore, defined as $\hat{U}(\delta x)^T$. Afterwards, the expanded wave function $\hat{U}(\delta x)^{\alpha(t)/\delta x} \cdot \hat{\psi}(t)$ is multiplied by the diagonal matrix $\exp(-i\hat{e}dt)$, resulting in phase shifts of the expanded eigenstates. Thereafter, we recover the wave function described from the spatially displaced (field-dressed) frame to the field-free frame. This is conducted by multiplying the transpose of $\hat{U}(\delta x)^{\alpha(t)/\delta x}$. By repeating the procedure for the overall time evolution, we can express the full time-evolution operator $\hat{U}_{\text{total}}$ given by

$$\hat{U}_{\text{total}} = [\prod_{l=0}^{l_f} \hat{U}(\delta x)^{-\alpha(t_l)/\delta x} \cdot \exp(-i\hat{e}dt) \cdot \hat{U}(\delta x)^{\alpha(t_l)/\delta x}], \tag{10}$$

where $t_l \equiv t_0 + ldt$, with l being an integer. l_f is defined as $t_f = t_0 + l_f dt$.

By combining the relation $\alpha(t + dt) - \alpha(t) = A(t)dt$ with a temporal boundary condition $A(t_0) = A(t_f) = 0$, the expression for $\hat{U}_{\text{total}}$ can be further simplified to

$$\hat{U}_{\text{total}} = [\prod_{l=0}^{l_f} \exp(-i\hat{e}dt) \cdot \vec{U}(\delta x)^{A(t_l)dt/\delta x}]. \tag{11}$$

In Equation (11), the number of multiplications $A(t_l)dt/\delta x$ for each time step l is rounded. The numerical error caused by the rounding is ignorable by setting δx at $\sim 10^{-4}$, far smaller than the spatial grid size of 0.2. We use a set of eigenstates of the field-free Hamiltonian to calculate the time evolution operator. To obtain the field-free eigenstates by solving a Hermitian TISE, we considered a reflection boundary rather than an absorbing or transparent one. In the absorbing and transparent boundaries, the Hamiltonians are non-Hermitian, so that the numerical solutions of the TISE require additional techniques [22]. The resulting high-lying continuum states in the reflection boundary are, therefore, well confined inside the boundary, indicating that the full propagator Equation (11) intrinsically includes unphysical reflections of the wave function at the boundary. To avoid such unphysical reflections, an absorbing boundary matrix $\hat{W}$ is multiplied to the wave function at every time step. The matrix elements are described as

$$\hat{W}_{m,n} \equiv \int_{-\infty}^{\infty} \phi_m^*(x) W(x) \phi_n(x) dx, \tag{12}$$

where $W(x)$ is a unity function that smoothly decays to zero near the boundary. We can express the full time-evolution propagator, including the absorbing boundary, as

$$\hat{U}_{\text{total}} = [\prod_{l=0}^{l_f} \hat{W} \cdot \exp(-i\hat{e}dt) \cdot \hat{U}(\delta x)^{A(t_l)dt/\delta x}]. \tag{13}$$

An initial state converts to the corresponding final state by multiplying the operator Equation (13).

3. Results of the Simulation

The numerical method shown in Section 2 was tested with a one-dimensional soft-core potential $V(x) = -\frac{1}{\sqrt{x^2+1}}$. The atom was irradiated with a pulsed laser field,

$$E(t) = E_0 \exp[-2\ln 2(t^2/\tau^2)] \sin(\omega t), \tag{14}$$

where E_0 is the peak strength of the laser field, and τ is the FWHM (fixed at 5 fs). The central frequency of the laser ω was set to $\lambda \equiv 2\pi c/\omega = 800$ nm, where c is the speed of light. Figure 3 presents the temporal profiles of the vector potential, $A(t) \equiv -\int_{-\infty}^{t} E(t)dt$, and the function, $A(t)dt/\delta x$, used to define the translation operator, $\hat{U}(\delta x)^{A(t_l)dt/\delta x}$ as included in Equation (11). dt was set to 0.2 in the calculations. We tested the dt dependence of the numerical solution by varying its values from 0.5 to 0.01 (not shown). The numerical solution converged well for all the values.

To obtain the eigenstates and eigenvalues of the atom, the TISE was solved. The one-dimensional x space, whose reflection boundaries were set at -614.4 and 614.2, was employed. The space was discretized by 6144 grids, each with a size of 0.2. There were 43 bound states with negative energies and 6101 continuum states with positive energies. The ionization energy of the atom, i.e., the eigenenergy of the ground state, was obtained as -18.23 eV. We set n_{max}, the number of the lowest-lying states, to 50$\sim$2000. The number of the necessary states depends on the peak laser intensity.

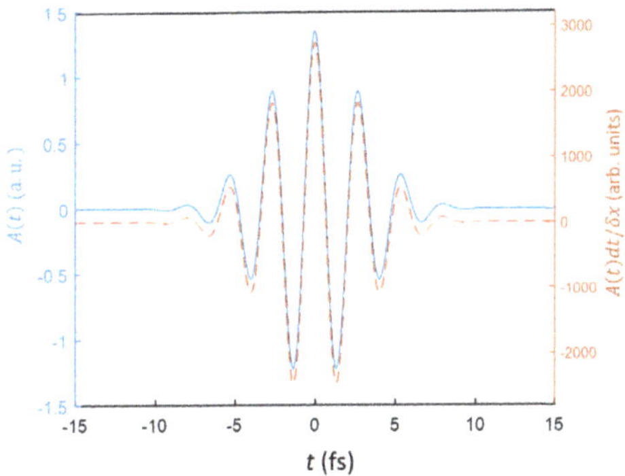

Figure 3. Temporal profiles of the vector potential $A(t)$ and the function $A(t)dt/\delta x$, which is rounded.

Figure 4 shows the final wave function after a laser pulse, described by Equation (14), has passed. The peak intensity of the pulse is 2.0×10^{14} W/cm^2. We show the results obtained when the numbers of discrete states n_{max} are 400, 1000, and 1600. For the above laser condition, the results obtained using $n_{\text{max}} = 1000$ and 1600 do not show any noticeable variations in the all represented domains: space (Figure 4a), momentum (Figure 4b), and quantum number (Figure 4c). The agreement between the results for n_{max} = 1000 and 1600 indicates that the TDUT method converges by using $n_{\text{max}} > 1000$ discrete states, as the basis set. When n_{max} is set to 400, the results from the TDUT show clear deviations from the other results. The deviations are observed not only in the continuum states ($n > 44$), but also in the bound states ($n \leq 43$), meaning that a noticeable amount of field-ionized electrons can be recaptured to the bound states.

When the applied laser field is stronger and the excitation of the higher-lying states becomes essential, more discrete states are required for the TDUT calculation. We have evaluated, therefore, this quantity as a function of the peak laser intensity. For a given peak intensity condition and from the initial ground state, we have recalculated the final states until a converged result is obtained by increasing n_{max} by every 50. The convergence is evaluated from the parameter,

$$\delta O(n_{\text{max}}) \equiv |U_{\text{total}}(n_{\text{max}})|1> - U_{\text{total}}(n_{\text{max}} + 50)|1> |^2, \tag{15}$$

where $|1>$ represents the ground state. By increasing n_{max}, the final wave function $U_{\text{total}}(n_{\text{max}})|1>$ must converge to a state so that $\delta O(n_{\text{max}})$ becomes negligibly small. When $\delta O(n_{\text{max}})$ is smaller than 2.5^{-7}, n_{max} is selected as n_{cutoff} and the convergence test is terminated.

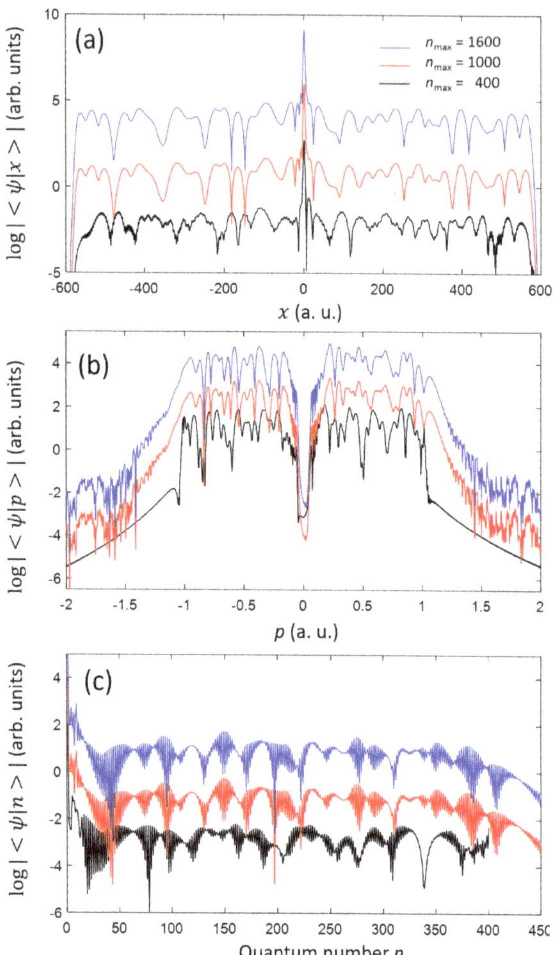

Figure 4. Atomic wave function irradiated with a short infrared laser pulse described by Equation (14). The ground state is used as an initial state. The wave function at $t = t_f = 15$ fs is shown in the space (**a**) and momentum (**b**) domains, and also by the quantum number n representation (**c**). For the plots in the momentum domain (**b**), only the continuum states are considered. The lines are displaced vertically for visual convenience.

In Figure 5a, n_{cutoff} is shown as a function of the peak laser intensity. For lower intensities, such as 0.125–0.25 $\times$ 10^{14} W/cm^2, the TDUT method provides converged results by using less than 50 discrete states. In this case, the atomic response can be described by perturbation theory, considering only a few discrete low-lying bound states. When the peak laser intensity is high, the dynamics is governed by tunneling ionization. In Figure 5a, the dramatic increase in n_{cutoff} at an intensity of 0.375 $\times$ 10^{14} W/cm^2 is observed, showing the transition from the perturbative to the tunneling regimes. After this transition point, n_{cutoff} continues to increase with the increasing peak laser intensity. In Figure 5b, the eigenenergy of a discrete state, whose quantum number corresponds to n_{cutoff} is plotted as a function of the peak laser intensity. Additionally, 10 U_P, where U_P is the laser-intensity-dependent ponderomotive energy, is represented by the blue dashed line, and 10 U_P is a maximum possible energy of an electron, which is generated after being elastically rescattered by the atomic core in the above-threshold ionization (ATI) dynamics [7]. The number of states needed for the TDUT method can be approximately determined by the maximum electron energy, because much higher-energy states cannot physically exist. The necessary number of states in the calculation will depend on the dynamics to be investigated, because such high-energy states have ignorable influence on the bound electron dynamics. In fact, it has been clarified that some of strong-field-ionized electrons having low kinetic energy can survive as Rydberg states [9,15,16]. To study only the bound-state dynamics, which is possibly coupled with the strong-field ionization, the number of included states can be further reduced.

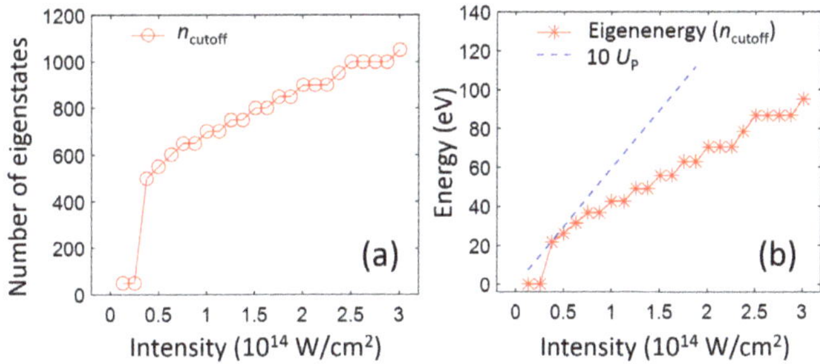

Figure 5. n_{cutoff} (**a**) and the eigenenergy at n_{cutoff} (**b**) as functions of the peak laser intensity. See the main text.

We are able to apply the full time-evolution operator Equation (13) to any initial wave function. Figure 6 shows the final states obtained by multiplying a full-time evolution operator to several different initial bound states ($n_0 = 1, 2, 3, 4, 11, 12$, and 13), where n_0 is the quantum number of the initial state. The temporal shape of the laser is the same as Equation (14), and the peak laser intensity is set at 1.0×10^{14} W/cm^2. The time evolution operator is calculated with $n_{\text{max}} = 2000$. The results obtained by the Crank–Nicolson method are also shown in Figure 6.

For the initial ground state $n_0 = 1$, the ground state population remains almost equal to 1, indicating that the depletion by tunneling ionization is not so significant. Other initial low-lying bound states, i.e., when $n_0 = 2, 3$, and 4, whose eigenenergy values are -7.49, -4.13, and -2.53 eV, respectively, result in significant depletion by tunneling. However, for the Rydberg bound states $n_0 = 11, 12$, and 13, with the eigenenergy values of -0.40, -0.34, and -0.29 eV, respectively, the depletion of the initial states is less dominant than that in the lower-lying bound states. This is due to the stabilization of Rydberg states [36,37]. For the defined initial conditions, the final quantum state distributions obtained by both methods exhibit some visible deviations, due to the difference in the absorbing boundary conditions, in the continuum states ($n > 43$). The absorbing boundary function $W(x)$, defined in the space domain for the Crank–Nicolson method, has been transformed by the matrix $\hat{W}_{m,n}$ of the discrete state representation given by Equation (12). For the bound states ($n \leq 43$), which are not directly affected by the absorbing boundary conditions, the numerical results from both the methods are consistent.

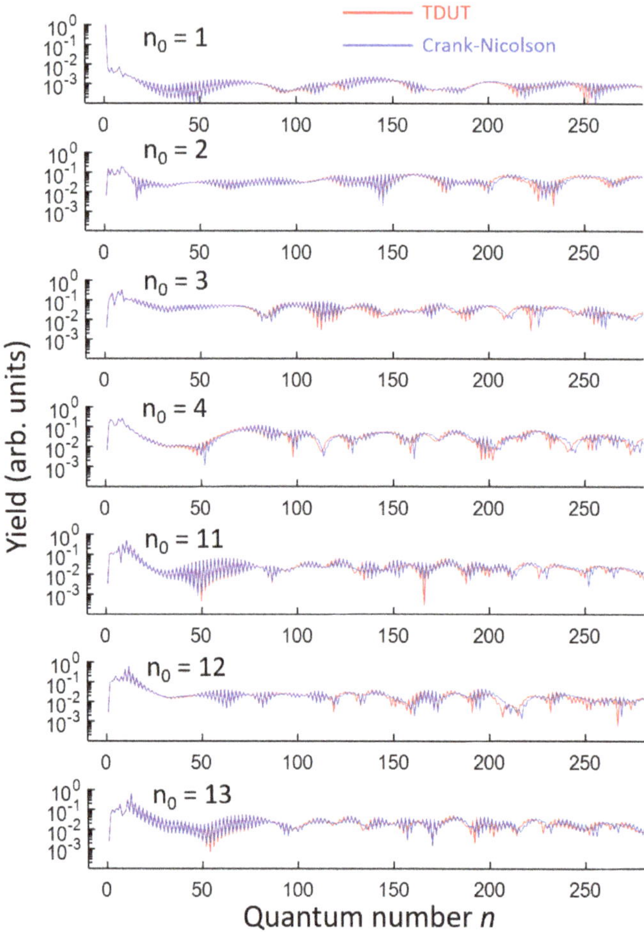

Figure 6. Final states in the quantum number representation for several different initial wave functions, numbered by $n_0 = 1, 2, 3, 4, 11, 12,$ and 13, shown in the discrete state representation.

4. Summary and Outlook

In this paper, we have introduced a numerical method based on the time-dependent unitary transformation (TDUT). This approach, first demonstrated for molecular rotational dynamics in Refs. [29,30], is implemented to the strong-field-ionization regime of an atom or a molecule. In the Kramers-Henneberger frame, the field-dressed eigenstates are identical to the field-free eigenstates excluding their spatial displacement, which becomes a useful advantage to calculate the unitary operators to propagate the electronic wave function in every temporal step. In the TDUT method, matrices and vectors associated with an atom, such as eigenstates, eigenenergies, and a unit translation operator $\hat{U}(\delta x)$ (Equation (9)), need to be calculated in advance (the calculation took 5 min for $n_{\max} = 2000$). The matrices and vectors can be reapplied under different laser conditions. For the same reason, the method will be even more beneficial in a very long-pulsed case. A final bound state population after pico- to nanosecond laser pulse irradiation can be calculated. In the long pulsed regime, it becomes more important to remove any unphysical reflections at the boundaries, which has been done by multiplying an absorbing boundary matrix (Equation (12)) at every temporal step. In the present work, after the matrix elements were prepared, it took approximately 4 s and 20 s with the TDUT and Crank-Nicolson methods, respectively, by using a personal computer (Intel(R) Core(TM) i9-9900K CPU, 128 GB RAM, Windows 10). This method can be a useful tool in calculating and analyzing bound electron dynamics coupled with strong-field ionization, not only in the one-dimensional system, but also in three-dimensional molecular systems.

Author Contributions: Conceptualization, J.-H.M., H.S. and D.-E.K.; Data curation, J.-H.M.; Funding acquisition, J.-H.M. and D.-E.K.; Investigation, J.-H.M., H.S. and D.-E.K.; Software, J.-H.M.; Visualization, J.-H.M.; Writing—original draft, J.-H.M., H.S. and D.-E.K.; Writing—review and editing, J.-H.M., H.S., and D.-E.K. All authors have read and agreed to the published version of the manuscript.

Funding: This work was supported by the National Research Foundation of Korea (NRF) grant funded by the Korea government (MSIT) [Grant No 2020R1C1C1012953]. This research was also supported in part by the Max Planck POSTECH/KOREA Research Initiative Program through the National Research Foundation of Korea (NRF) funded by the Ministry of Science and ICT [Grant No 2016K1A4A4A01922028].

Institutional Review Board Statement: Not applicable.

Informed Consent Statement: Not applicable.

Data Availability Statement: The data presented in this study are available on request from the corresponding author. The data are not publicly available due to ethical restriction.

Conflicts of Interest: The authors declare no conflict of interest.

Abbreviations

The following abbreviations are used in this manuscript:

TISE	Time-independent Schrödinger equation
TDSE	Time-dependent Schrödinger equation
KH	Kramers-Henneberger
SFA	Strong-field approximation
FTI	Frustrated tunneling ionization
ATI	Above-threshold ionization
HHG	High-harmonic generation
TDUT	Time-dependent unitary transformation
FWHM	Full width at half maximum

References

1. Popruzhenko, S.V. Keldysh Theory of Strong Field Ionization: History, Applications, Difficulties and Perspectives. *J. Phys. B At. Mol. Opt. Phys.* **2014**, *47*, 204001. [CrossRef]
2. Amini, K.; Biegert, J.; Calegari, F.; Chacón, A.; Ciappina, M.F.; Dauphin, A.; Efimov, D.K.; Faria, C.F.d.M.; Giergiel, K.; Gniewek, P.; et al. Symphony on Strong Field Approximation. *Rep. Prog. Phys.* **2019**, *82*, 116001. [CrossRef]
3. Lewenstein, M.; Balcou, P.; Ivanov, M.Y.; L'Huillier, A.; Corkum, P.B. Theory of High-Harmonic Generation by Low-Frequency Laser Fields. *Phys. Rev. A* **1994**, *49*, 2117–2132. [CrossRef]
4. Corkum, P.B. Plasma Perspective on Strong Field Multiphoton Ionization. *Phys. Rev. Lett.* **1993**, *71*, 1994–1997. [CrossRef]
5. Krause, J.L.; Schafer, K.J.; Kulander, K.C. High-Order Harmonic Generation from Atoms and Ions in the High Intensity Regime. *Phys. Rev. Lett.* **1992**, *68*, 3535–3538. [CrossRef]
6. Macklin, J.J.; Kmetec, J.D.; Gordon, C.L. High-Order Harmonic Generation Using Intense Femtosecond Pulses. *Phys. Rev. Lett.* **1993**, *70*, 766–769. [CrossRef]
7. Suárez, N.; Chacón, A.; Ciappina, M.F.; Biegert, J.; Lewenstein, M. Above-Threshold Ionization and Photoelectron Spectra in Atomic Systems Driven by Strong Laser Fields. *Phys. Rev. A* **2015**, *92*, 063421. [CrossRef]
8. Mohideen, U.; Sher, M.H.; Tom, H.W.K.; Aumiller, G.D.; Wood, O.R.; Freeman, R.R.; Boker, J.; Bucksbaum, P.H. High Intensity Above-Threshold Ionization of He. *Phys. Rev. Lett.* **1993**, *71*, 509–512. [CrossRef]
9. Nubbemeyer, T.; Gorling, K.; Saenz, A.; Eichmann, U.; Sandner, W. Strong-Field Tunneling without Ionization. *Phys. Rev. Lett.* **2008**, *101*, 233001. [CrossRef]
10. Eichmann, U.; Nubbemeyer, T.; Rottke, H.; Sandner, W. Acceleration of Neutral Atoms in Strong Short-Pulse Laser Fields. *Nature* **2009**, *461*, 1261–1264. [CrossRef]

11. Xiong, W.H.; Xiao, X.R.; Peng, L.Y.; Gong, Q. Correspondence of Below-Threshold High-Order-Harmonic Generation and Frustrated Tunneling Ionization. *Phys. Rev. A* **2016**, *94*, 013417. [CrossRef]
12. Popruzhenko, S.V. Quantum Theory of Strong-Field Frustrated Tunneling. *J. Phys. B At. Mol. Opt. Phys.* **2017**, *51*, 014002. [CrossRef]
13. Popruzhenko, S.V.; Lomonosova, T.A. Frustrated Ionization of Atoms in the Multiphoton Regime. *Laser Phys. Lett.* **2020**, *18*, 015301. [CrossRef]
14. Cao, C.; Li, M.; Liang, J.; Guo, K.; Zhou, Y.; Lu, P. Frustrated Tunneling Ionization in Strong Circularly Polarized Two-Color Laser Fields. *J. Phys. B At. Mol. Opt. Phys.* **2021**, *54*, 035601. [CrossRef]
15. Mun, J.H.; Ivanov, I.A.; Yun, H.; Kim, K.T. Strong-Field-Approximation Model for Coherent Extreme-Ultraviolet Emission Generated through Frustrated Tunneling Ionization. *Phys. Rev. A* **2018**, *98*, 063429. [CrossRef]
16. Yun, H.; Mun, J.H.; Hwang, S.I.; Park, S.B.; Ivanov, I.A.; Nam, C.H.; Kim, K.T. Coherent Extreme-Ultraviolet Emission Generated through Frustrated Tunnelling Ionization. *Nat. Photonics* **2018**, *12*, 620–624. [CrossRef]
17. Strelkov, V. Role of Autoionizing State in Resonant High-Order Harmonic Generation and Attosecond Pulse Production. *Phys. Rev. Lett.* **2010**, *104*, 123901. [CrossRef]
18. Chini, M.; Wang, X.; Cheng, Y.; Wang, H.; Wu, Y.; Cunningham, E.; Li, P.C.; Heslar, J.; Telnov, D.A.; Chu, S.I.; et al. Coherent Phase-Matched VUV Generation by Field-Controlled Bound States. *Nat. Photonics* **2014**, *8*, 437–441. [CrossRef]
19. Li, P.C.; Sheu, Y.L.; Laughlin, C.; Chu, S.I. Role of Laser-Driven Electron-Multirescattering in Resonance-Enhanced below-Threshold Harmonic Generation in He Atoms. *Phys. Rev. A* **2014**, *90*, 041401. [CrossRef]
20. Camp, S.; Schafer, K.J.; Gaarde, M.B. Interplay between Resonant Enhancement and Quantum Path Dynamics in Harmonic Generation in Helium. *Phys. Rev. A* **2015**, *92*, 013404. [CrossRef]
21. Beaulieu, S.; Camp, S.; Descamps, D.; Comby, A.; Wanie, V.; Petit, S.; Légaré, F.; Schafer, K.J.; Gaarde, M.B.; Catoire, F.; et al. Role of Excited States In High-Order Harmonic Generation. *Phys. Rev. Lett.* **2016**, *117*, 203001. [CrossRef]
22. Karamatskou, A.; Pabst, S.; Santra, R. Adiabaticity and Diabaticity in Strong-Field Ionization. *Phys. Rev. A* **2013**, *87*, 043422. [CrossRef]
23. Saalmann, U.; Giri, S.K.; Rost, J.M. Adiabatic Passage to the Continuum: Controlling Ionization with Chirped Laser Pulses. *Phys. Rev. Lett.* **2018**, *121*, 153203. [CrossRef]
24. Rubbmark, J.R.; Kash, M.M.; Littman, M.G.; Kleppner, D. Dynamical Effects at Avoided Level Crossings: A Study of the Landau-Zener Effect Using Rydberg Atoms. *Phys. Rev. A* **1981**, *23*, 3107–3117. [CrossRef]
25. Clark, W.; Greene, C.H. Adventures of a Rydberg Electron in an Anisotropic World. *Rev. Mod. Phys.* **1999**, *71*, 821–833. [CrossRef]
26. von Stecher, J.; Greene, C.H. Spectrum and Dynamics of the BCS-BEC Crossover from a Few-Body Perspective. *Phys. Rev. Lett.* **2007**, *99*, 090402. [CrossRef]
27. Bloch, I.; Dalibard, J.; Zwerger, W. Many-Body Physics with Ultracold Gases. *Rev. Mod. Phys.* **2008**, *80*, 885–964. [CrossRef]
28. Dürr, S.; Volz, T.; Marte, A.; Rempe, G. Observation of Molecules Produced from a Bose-Einstein Condensate. *Phys. Rev. Lett.* **2004**, *92*, 020406. [CrossRef]
29. Mun, J.H.; Sakai, H. Improving Molecular Orientation by Optimizing Relative Delay and Intensities of Two-Color Laser Pulses. *Phys. Rev. A* **2018**, *98*, 013404. [CrossRef]
30. Mun, J.H.; Sakai, H.; González-Férez, R. Orientation of Linear Molecules in Two-Color Laser Fields with Perpendicularly Crossed Polarizations. *Phys. Rev. A* **2019**, *99*, 053424. [CrossRef]
31. Mun, J.H.; Kim, D.E. Field-Free Molecular Orientation by Delay- and Polarization-Optimized Two Fs Pulses. *Sci. Rep.* **2020**, *10*, 18875. [CrossRef] [PubMed]
32. Henneberger, W.C. Perturbation Method for Atoms in Intense Light Beams. *Phys. Rev. Lett.* **1968**, *21*, 838–841. [CrossRef]
33. Chen, R.; Guo, H. The Chebyshev Propagator for Quantum Systems. *Comput. Phys. Commun.* **1999**, *119*, 19–31. [CrossRef]
34. Guiang, C.S.; Wyatt, R.E. Quantum dynamics with Lanczos subspace propagation: Application to a laser-driven molecular system. *Int. J. Quantum Chem.* **1998**, *67*, 273–285. [CrossRef]
35. Bandrauk, A.D.; Shen, H. Exponential Split Operator Methods for Solving Coupled Time-dependent Schrödinger Equations. *J. Chem. Phys.* **1993**, *99*, 1185–1193. [CrossRef]
36. Parzyński, R.; Wieczorek, S. Interference Stabilization of Rydberg Atoms Enhanced by Multiple V-Type Resonances. *Phys. Rev. A* **1998**, *58*, 3051–3057. [CrossRef]
37. Scrinzi, A.; Elander, N.; Piraux, B. Stabilization of Rydberg Atoms in Superintense Laser Fields. *Phys. Rev. A* **1993**, *48*, R2527–R2530. [CrossRef]

International Journal of
Molecular Sciences

Article

Simple Models to Study Spectral Properties of Microbial and Animal Rhodopsins: Evaluation of the Electrostatic Effect of Charged and Polar Residues on the First Absorption Band Maxima

Andrey A. Shtyrov [1], Dmitrii M. Nikolaev [1], Vladimir N. Mironov [1,2], Andrey V. Vasin [3], Maxim S. Panov [2], Yuri S. Tveryanovich [2] and Mikhail N. Ryazantsev [1,2,*]

1 Nanotechnology Research and Education Centre RAS, Saint Petersburg Academic University, 8/3 Khlopina Street, 194021 St. Petersburg, Russia; andriei.shtyrov@gmail.com (A.A.S.); dmitrii.m.nikolaev@gmail.com (D.M.N.); vova_mironov_97@mail.ru (V.N.M.);
2 Institute of Chemistry, Saint Petersburg State University, 7/9 Universitetskaya Nab., 199034 St. Petersburg, Russia; m.s.panov@spbu.ru (M.S.P.); tys@bk.ru (Y.S.T.)
3 Institute of Biomedical Systems and Botechnologies, Peter the Great St. Petersburg Polytechnic University, 29 Polytechnicheskaya Street, 195251 St. Petersburg, Russia; vasin_av@spbstu.ru
* Correspondence: mikhail.n.ryazantsev@gmail.com

Citation: Shtyrov, A.A.; Nikolaev, D.M.; Mironov, V.N.; Vasin, A.V.; Panov, M.S.; Tveryanovich, Y.S.; Ryazantsev, M.N. Simple Models to Study Spectral Properties of Microbial and Animal Rhodopsins: Evaluation of the Electrostatic Effect of Charged and Polar Residues on the First Absorption Band Maxima. *Int. J. Mol. Sci.* **2021**, *22*, 3029. https://doi.org/10.3390/ijms22063029

Academic Editor: Małgorzata Borówko

Received: 24 February 2021
Accepted: 5 March 2021
Published: 16 March 2021

Abstract: A typical feature of proteins from the rhodopsin family is the sensitivity of their absorption band maximum to protein amino acid composition. For this reason, studies of these proteins often require methodologies that determine spectral shift caused by amino acid substitutions. Generally, quantum mechanics/molecular mechanics models allow for the calculation of a substitution-induced spectral shift with high accuracy, but their application is not always easy and requires special knowledge. In the present study, we propose simple models that allow us to estimate the direct effect of a charged or polar residue substitution without extensive calculations using only rhodopsin three-dimensional structure and plots or tables that are provided in this article. The models are based on absorption maximum values calculated at the SORCI+Q level of theory for cis- and trans-forms of retinal protonated Schiff base in an external electrostatic field of charges and dipoles. Each value corresponds to a certain position of a charged or polar residue relative to the retinal chromophore. The proposed approach was evaluated against an example set consisting of twelve bovine rhodopsin and sodium pumping rhodopsin mutants. The limits of the applicability of the models are also discussed. The results of our study can be useful for the interpretation of experimental data and for the rational design of rhodopsins with required spectral properties.

Keywords: rhodopsins; spectral properties of rhodopsins; spectral tuning in rhodopsins; engineering of red-shifted rhodopsins; photobiology; biological photosensors

1. Introduction

Rhodopsins are photosensitive membrane proteins that have been discovered in many species across all three life domains. The natural diversity of the rhodopsins' first absorption band maxima (λ_{max}) is achieved via the variation of the proteins' amino acid compositions during evolutionary processes [1,2]. The same strategy is used in modern technologies to obtain rhodopsin variants with an optimal λ_{max} [3–6]. In this context, it is desirable to develop methodologies for prediction of the λ_{max} change caused by the modifications of the primary protein structure, e.g., single or multiple amino acid substitutions ($\Delta\lambda_{max}$).

Site-directed mutagenesis is a common experimental technique that allows for $\Delta\lambda_{max}$ evaluation. In many studies, amino acid substitutions are introduced into rhodopsins to measure $\Delta\lambda_{max}$ and, consequently, to estimate the substituted residue contribution to the absorption maximum [7–9]. The objectives of these studies are to evaluate the correlations between the type/position of a rhodopsin residue and the contribution of this residue to λ_{max} and, ultimately, to establish general rules that allow for controlling λ_{max}. However, the interpretation of the measured $\Delta\lambda_{max}$ is not straightforward. Generally, a substitution of a single residue can lead to reorganization of the protein internal H-bond network changing positions of other residues and, therefore, their impact on the spectral properties. For such substitutions, a measured $\Delta\lambda_{max}$ cannot be attributed exclusively to the mutated residue, and the aforementioned effects must be taken into account. Apparently, this indirect spectral tuning due to reorganization of the internal H-bond network is common for rhodopsins. For example, such H-bond network reorganization is responsible for the origin of the spectral shift between anion-free and chloride-ion-bound forms of halorhodopsin from *Natronomonas pharaonis* [10], evolutionary switch between ultraviolet and violet vision in vertebrates [11], and between visual rhodopsins from *Alloteuthis subulata* and *Loligo forbesii* squids [12].

In addition to experimental studies, computational modeling can be involved. In general, computational models enable not only the calculation of $\Delta\lambda_{max}$ but also the evaluation of its direct and indirect parts. Currently, hybrid quantum mechanics/molecular mechanics (QM/MM) models are able to reproduce experimental λ_{max} and $\Delta\lambda_{max}$ values with good accuracy (within 20–30 nm and just a few nm from experiment, respectively), assuming that a high-quality three-dimensional protein structure is provided [11,13–18]. However, evaluation of QM/MM $\Delta\lambda_{max}$ values is computationally expensive and not always easy. Thus, alternative less-demanding models are desirable for the interpretation of experimental data and initial rational design. These models can be less general and rigorous than QM/MM models, but they should allow for fast and simple $\Delta\lambda_{max}$ prediction.

Here, for visual and microbial rhodopsins, we proposed such simple models that allow us to estimate the direct electrostatic part of $\Delta\lambda_{max}$ for charged/polar amino acid substitution. The models are based on the precalculated high-level ab initio data. Application of these models requires only three-dimensional structures of rhodopsins, i.e., either X-ray structures or structures generated by comparative modeling. As a test, these models were applied to estimate the direct part of $\Delta\lambda_{max}$ for charged and polar residue substitutions in bovine rhodopsin and sodium pumping rhodopsin KR2. The obtained data were validated both against more sophisticated ab initio QM/MM calculations and against experiment.

2. Results

The results are described in the following sequence. First, the major principles of spectral tuning that make the proposed models possible are described. Then, models are introduced and applied to mutants of bovine rhodopsin and sodium pumping rhodopsin as an example. Finally, the limits of applicability of these models are discussed

2.1. Steric and Electrostatic Factors in Rhodopsin Spectral Tuning

The tuning of the rhodopsins' first spectral band has been widely investigated [13,16,17,19,20]. Two factors are found to be the most important: the steric and electrostatic interactions of PSBs with surrounding amino acids of the opsin. Several other factors, such as the polarization of the retinal environment or inter-residual charge transfer within the binding pocket, have also been studied but were found to be less significant [21,22].

A substantial modification of the steric interaction between the protein pocket and the chromophore can lead to the distortion of the chromophore and, consequently, to a change in λ_{max}. If this is the case, $\Delta\lambda_{max}$ evaluation requires detailed information about the chromophore geometrical parameters. Generally, the resolution of available X-ray structures is not good enough to obtain these parameters with required precision. To date, the geometrical parameters of sufficient quality can be obtained only from high-level computational models, and simpler approaches are hardly possible. To cause prominent change in the steric interaction of PSB with opsin, one has to either introduce/remove a bulky residue in the protein binding pocket or to introduce/remove a distant residue that causes a substantial deformation of the binding pocket. Although modification of steric interactions for λ_{max} tuning occurs in nature [23], rational design of rhodopsins with specific λ_{max} by adjustment of steric interactions is not straightforward.

On the contrary, the electrostatic tuning mechanism is not only quite common in nature but also can be more easily utilized for rational design. Due to the charge transfer character of PSB $S_0 \rightarrow S_1$ transition [24,25], λ_{max} is very sensitive to the electrostatic field generated by the amino acids constituting an opsin. The contribution of any residue to this electrostatic field is primarily determined by its charge or dipole moment and its position relative to the chromophore. The contributions of quadrupoles and the multipoles of higher order can be neglected [26]. Generally, amino acid substitution can change $\Delta\lambda_{max}$ in two ways: either directly or indirectly. In the first case, $\Delta\lambda_{max}$ is obtained via substitution of the original amino acid by an amino acid with different charge/dipole moment. In the second case, $\Delta\lambda_{max}$ is caused by a substitution that induces reorganization of the whole protein, including charged and polar residues and changing the electrostatic field in the chromophore region.

Unlike the steric part of $\Delta\lambda_{max}$, the electrostatic part can be treated by a simple, practical model. To make it possible, the following assumptions must be done:

1. λ_{max} can be evaluated as the vertical excitation energy of one characteristic snapshot that is close to the Gibbs free energy minimum of the whole protein. Protein dynamics can modify absorption band counters, but it does not affect the position of its maximum significantly.
2. The impact of a charged residue on λ_{max} is equal to the impact of a unit negative/positive charge located at the center of the charged group of this residue;
3. The impact of a polar residue to λ_{max} is equal to the impact of a dipole located at the center of the polar group of the residue.
4. If substantial reorganization of the H-bond network does not occur, the impact of each residue on λ_{max} can be treated independently from the rest of the residues; i.e., we assume that all impacts are additive.
5. The impact of a charged/polar residue on λ_{max} depends only on its charge/dipole moment and its distance to/orientation along the chromophore axis (see Figure 1). For charges, this "cylindrical" symmetry allows for reducing a four-dimensional function $\Delta\lambda_{max} = f$ (three Cartesian coordinates for a charge location) to a simpler three-dimensional function $\Delta\lambda_{max} = f$(two Cartesian coordinates for a charge location). For polar residues, an additional argument, which describes the orientation of the dipole moment relative to the chromophore axis, should be added.
6. Although the electrostatic field always modifies a chromophore geometry by alternate changing of the length of double and single bonds, the effect of this geometry change on $\Delta\lambda_{max}$ can be neglected.

The first point is the widely used approximation for rhodopsin λ_{max} modeling. Although more molecular dynamics studies are necessary to understand the limits of applicability for this assumption, inhomogeneous broadening of the absorption band is not yet reported for rhodopsins in contrast to some other photosensitive proteins [27,28]. Moreover, static QM/MM models have already been tested intensively and proven to be able to reproduce experimental λ_{max} for dozens of rhodopsins [10,12,13,16,29,30]. It is worth mentioning that a system of interest can consist of a mixture of two or more stable forms, such as 13-cis and all-trans retinal containing the ground state of bacteriorhodopsin or different protonation states of titratable residues in Anabaena sensory rhodopsin [31]. If this is the case, several representative snapshots should be taken into account. Points 2 and 3 are the common coarse-grained approximation with well-known limitations [32,33]. For the last three points, additional justifications and discussion is given in the section "Limitations of the proposed models" and in the Supplementary Materials to avoid readers' distraction from the main subject.

If the statements above are assumed, a two-dimensional grid for charges and a three-dimensional grid for dipoles can be calculated only once with a robust quantum mechanical method, and, then, one can use graphical representations, tables, or fitting by suitable functions for fast data acquisition. An approximate evaluation of $\Delta\lambda_{max}$ using these tabulated data can be performed only based on the knowledge of charges/dipoles of altered residues and their positions relative to the chromophore. In other words, all required information can be obtained from an X-ray structure or a structure predicted by comparative modeling.

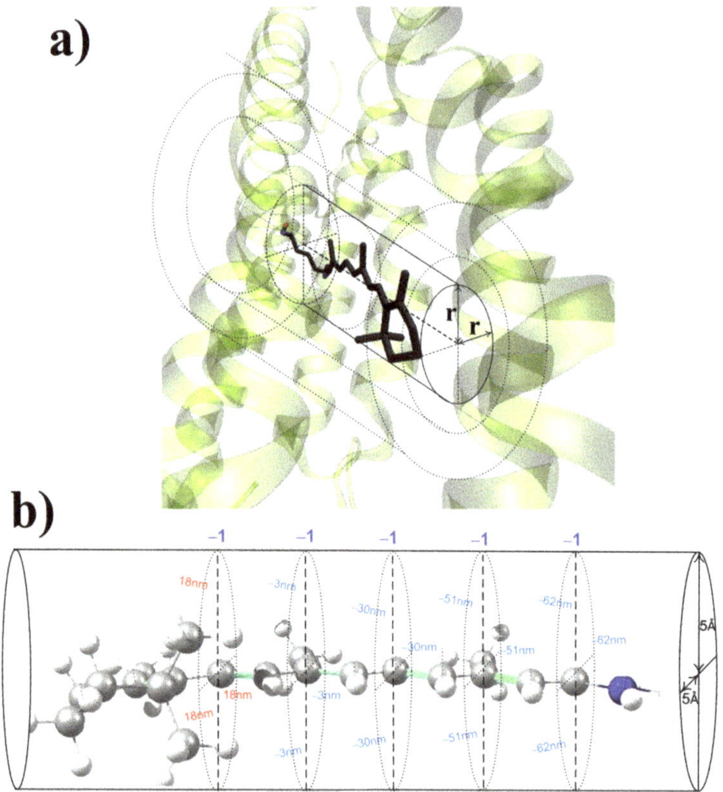

Figure 1. (**a**) Illustration of the "cylindrical" symmetry assumption for the protonated Schiff base (PSB). The impact of a charged/polar residue on λ_{max} depends only on its charge/dipole moment and its distance to/orientation along the chromophore axis. (**b**) Spectral shift values caused by a unit negative charge (denoted as '−1') located at 5 Å from different reference atoms of all-trans PSB. Negative spectral shift values (reference atoms C15, C13, C11, C9) are presented in blue; positive spectral shift value (reference atom C7) is presented in red.

2.2. Models to Evaluate the Direct Electrostatic Effect of Charged Residues

For each chromophore and negative/positive charges, we derived the numerical function that relates the position of a charged residue to the chromophore and its impact on λ_{max} (Figure 2).

The geometries of the chromophores were kept as calculated in the gas phase, i.e., without external charges. We constructed a grid for the placement of unit charges as follows: We plotted thirteen grid lines from thirteen PSB reference atoms perpendicular to the chromophore axis (see Figure 2). Along each grid line, we placed unit charges at the fixed distances from the retinal chromophore, from 3 Å to 18 Å with the 1 Å interval

(16 points along each grid line). In total, we performed 13 × 16 = 208 calculations for each charge and for each chromophore.

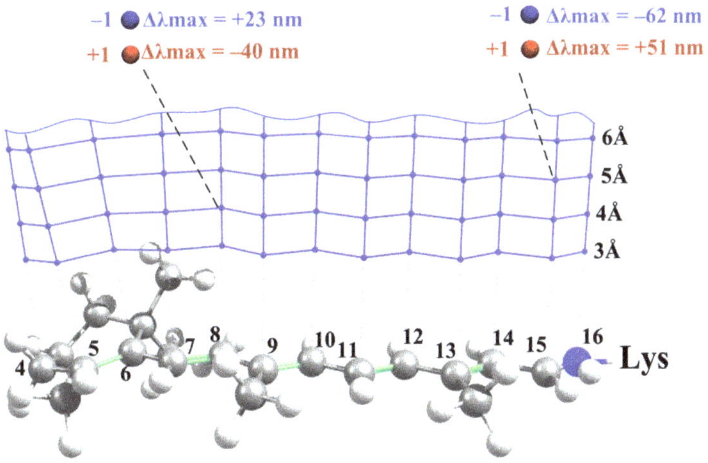

Figure 2. Grid representing the positions of a unit charge relative to the all-trans PSB. At each grid point we performed an ab initio calculation of $\Delta\lambda_{max}$ values for the positive and negative unit charges.

The calculated reference absorption maxima for the 11-cis and all-trans PSB without external charges λ_{max}^{ref} were found to be 595 nm and 596 nm, respectively. For each chromophore and charge type and position, we performed the SORCI+Q calculation of the absorption maximum value (λ_{max}^{i}). Then, we derived the effect of the charged residue placed at a certain position relative to the retinal chromophore as $\Delta\lambda_{max}^{i} = \lambda_{max}^{i} - \lambda_{max}^{ref}$. The results of these calculations are illustrated as 2D functions $\Delta\lambda_{max}^{i}$ (reference atom, distance) in Figure 3, Figure S1 and Tables S1–S4.

The four panels in Figure 3 correspond to the four considered systems: (a) a negative charge, 11-cis chromophore; (b) a positive charge, 11-cis chromophore; (c) a negative charge, all-trans chromophore; (d) a positive charge, all-trans chromophore. These 2D functions allow us to estimate $\Delta\lambda_{max}$ caused by the charged residue placed at a certain position in the rhodopsin. For example, a positively charged residue (lysine, arginine, or protonated histidine) placed at 7 Å from the C_{14} PSB atom would cause an approximately +40 nm red shift for the 11-cis chromophore and a +36 nm red shift for the all-trans chromophore. On the other hand, a negatively charged residue (glutamic or aspartic acid) at the same position would cause an approximately −40 nm blue shift for 11-cis chromophore and a −35 nm red shift for the all-trans chromophore.

Several well-known rules/patterns can also be clearly seen from the plots in Figure 3:

1. The effect of a charged residue on λ_{max} depends on (a) the sign of the charge and (b) the distance from the charge to the closest atom of the retinal.
2. A negative charge located in the NH region causes a blue shift; a negative charge located in the β-ionone ring region causes a red shift.
3. On the contrary, a positive charge located in the NH region causes a red shift; a positive charge located in the β-ionone ring region causes a blue shift.
4. The charges that are closer to the ends of the chromophore (atoms N_{16}, C_{15}, C_6, and C_5) cause larger shifts, while the charges that are close to the middle of the chromophore (atoms C_{12}, C_{11}, and C_{10}) cause smaller shifts.

The effect of charged residues on λ_{max} is slightly larger for 11-cis PSB than for all-trans PSB (Figure 3). To rationalize this fact, we calculated the charge distributions in the ground and the first excited states of 11-cis and all-trans PSB. The calculations were performed at the CASSCF/6-31G* level of theory. We found that the portion of positive charge, which is translocated from the NH region to the beta-ionone ring region upon photoexcitation, is larger for 11-cis PSB (0.30) than for all-trans PSB (0.21).

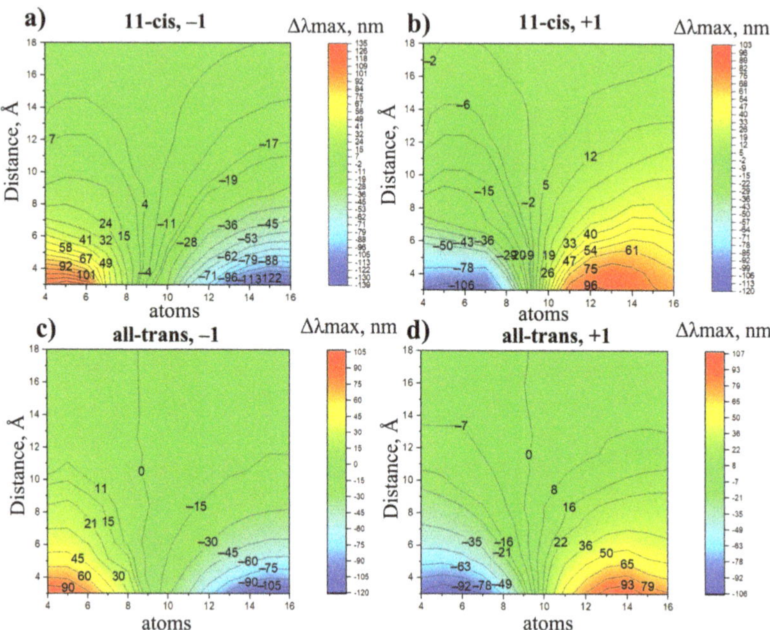

Figure 3. The impact of a unit charge on PSB λ_{max} as a 2D function of its position relative to the chromophore. (**a**) 11-cis PSB, negative charge; (**b**) 11-cis PSB, positive charge; (**c**) all-trans PSB, negative charge; (**d**) all-trans PSB, positive charge

2.3. *Models to Evaluate the Direct Electrostatic Effect of Polar Residues*

To derive models for polar residues, we followed the strategy similar to the approach used in the previous section. $\Delta\lambda^{i}_{max}$ caused by a polar residue depends not only on the distance of this residue to the chromophore but also on the orientation of this residue polar group (see Figure 4). Therefore, we calculated a numerical function that relates $\Delta\lambda^{i}_{max}$ and both the distances to the chromophore and the angle between a dipole and the chromophore axis. Each dipole was represented by two charges of a different sign (−0.43 and +0.43) that are situated 1.0 Å from each other (see Figure 4). The magnitudes of the charges and the distance between them were chosen in order to represent the Amber parameters for the -OH group of polar residues, such as Ser or Thr. The geometries of 11-cis PSB and all-trans PSB were kept as calculated in the gas phase, i.e., without an external electrostatic field. Four atoms of the 11-cis and all-trans chromophore (N_{16}, C_{12}, C_8, C_4) were taken as the reference atoms. We plotted four grid lines from these four reference atoms perpendicular to the chromophore (see Figure 4). Along each grid line we placed the center of the dipole at the fixed distances from the retinal chromophore, from 3.5 Å to 6.5 Å with the 1 Å interval (4 points along each grid line in total). The orientation of the dipole was varied by changing the angle γ that is defined as the angle between the chromophore axis and the line connecting oxygen and hydrogen atoms. γ varied from 0° to 360° with the step of 30°. The results of our calculations are presented in Figures 5 and 6, Tables S5 and S6. We also performed spline interpolation of the calculated data to plot the largest possible negative and positive contributions of polar residues to λ_{max} as functions of a dipole moment position along the chromophore axis (Figure 7).

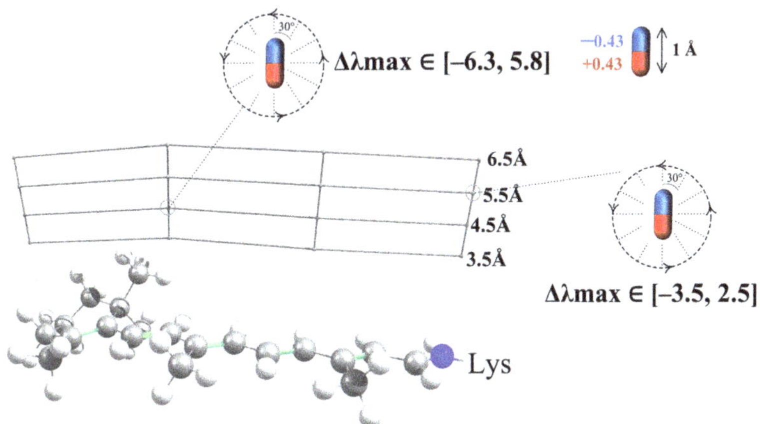

Figure 4. Grid representing the positions of a dipole moment relative to the all-trans PSB. At each grid point, we performed the ab initio calculation of $\Delta\lambda_{max}$ values for different orientations of the dipole moment relative to the chromophore axis.

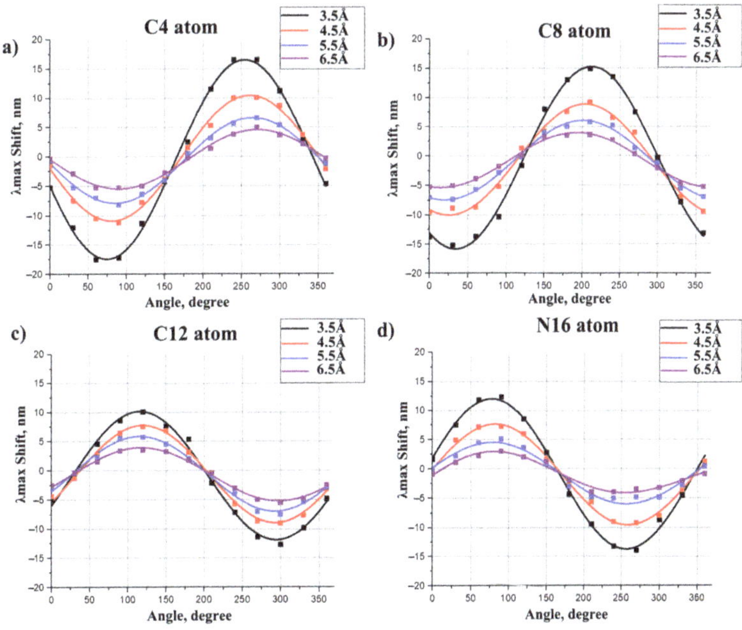

Figure 5. Impact of the dipole moment on λ_{max} of the 11-cis PSB as a function of the angle between the dipole and the chromophore axis. Functions were calculated at different positions of the dipole relative to the chromophore. Dipoles were located at the distances 3.5 Å, 4.5 Å, 5.5 Å, 6.5 Å from the C_4 (**a**), C_8 (**b**), C_{12} (**c**), and N_{16} (**d**) chromophore atoms along the grid line perpendicular to the chromophore axis (see Figure 4). Dots represent the ab initio calculated values. Functions were derived as the spline interpolation of the calculated data.

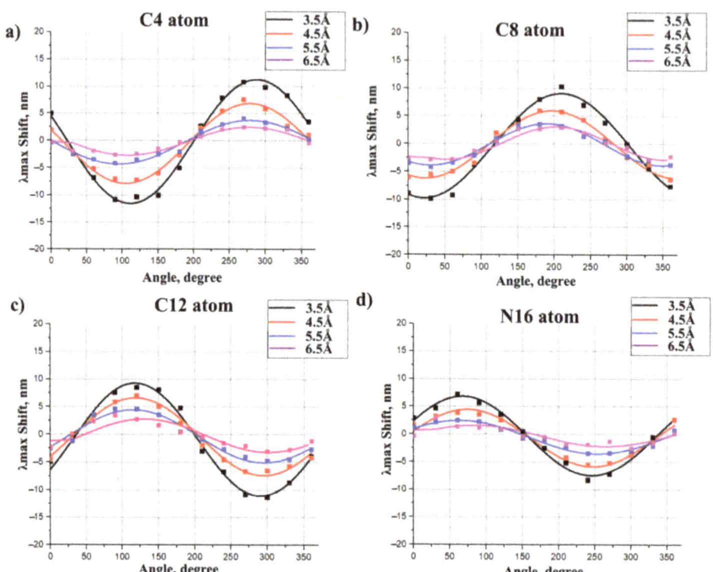

Figure 6. Impact of the dipole moment on λ_{max} of the all-trans PSB as a function of the angle between the dipole and the chromophore axis. Functions were calculated for different positions of the dipole relative to the chromophore. Dipoles were located at the distances 3.5 Å, 4.5 Å, 5.5 Å, 6.5 Å from the C_4 (**a**), C_8 (**b**), C_{12} (**c**), and N_{16} (**d**) chromophore atoms along the grid line perpendicular to the chromophore axis (see Figure 4). Dots represent the ab initio calculated values. Functions were derived as the spline interpolation of the calculated data.

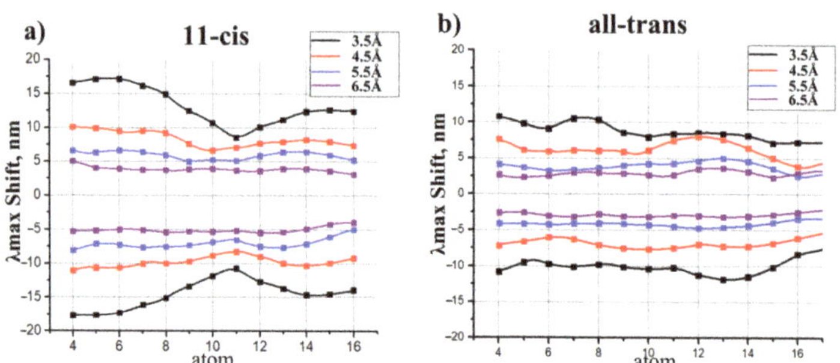

Figure 7. Largest possible negative and positive contributions of dipole moments to λ_{max} of 11-cis PSB (**a**) and all-trans PSB (**b**) as functions of a dipole moment position along the chromophore axis. Functions were calculated at different distances of the dipole moment from the chromophore. Dots represent the ab initio calculated values. Functions were derived as the spline interpolation of the calculated data.

Several rules/patterns can be derived from the plots presented in Figures 5 and 6.

1. The effect of polar residues located further than 6–7 Å from the PSB can be neglected.
2. The impact of a polar residue on λ_{max} is determined not only by the distance from the polar group of the residue to a given atom of the chromophore, as it is for charged residues, but also by the orientation of a polar group relative to the chromophore.
3. $\Delta\lambda_{max}$ ranges from a negative value (for example, −9 nm for a dipole situated at 4.5 Å from an atom of 11-cis PSB) to zero and then to a positive value (+7.5 nm for

this dipole). Therefore, to estimate the effect of a polar residue on the rhodopsin absorption maximum, accurate structural information is required.

The effect of polar residues on λ_{max} is slightly larger for 11-cis PSB than for all-trans PSB. This fact can be rationalized by the difference in the portion of positive charge translocated from the NH region to the beta-ionone ring region upon photoexcitation, as discussed in the Section 2.2.

Only the rotation of a dipole moment in the grid plane (Figure 4) was considered. Due to the symmetry of the system, the dipole moment component that is perpendicular to the grid surface should have a negligible effect on λ_{max}. To prove this, we located the dipole moment at 3.5 Å from the N16 PSB atom and rotated it perpendicular to the grid surface; $\Delta\lambda_{max}$ values were calculated with a step of 30°. The calculated spectral shift values did not exceed 1.5 nm, which is within the limits of QM calculation error.

2.4. Application of the Proposed Models to Evaluate the Direct Effect of Amino Acid Substitutions

Using the protocol described in the Methods section, we applied the proposed models to evaluate the direct effect of twelve amino acid substitutions in bovine rhodopsin (Rh) and sodium pumping rhodopsin (KR2). For charged residues, the data presented in Figure 3 were used to obtain the correspondence between their position and the possible spectral shift. For polar residues, the data presented in Figures 5–7 were used to evaluate the range of possible spectral shift values due to the lack of information about the orientation of a polar group relative to PSB. The exact orientation of a polar group could be defined only by constructing the corresponding protein QM/MM model. The estimated direct effect values were compared with the experimentally observed and QM/MM calculated spectral shift values.

According to the proposed models, for seven mutants, the direct effect of amino acid substitution completely explains the experimentally observed spectral shift (Figure 8 and green color-coding in Table 1). These estimations were also confirmed by the analysis of the corresponding QM/MM models. For five mutants, the direct effect of amino acid substitution cannot completely explain the experimentally observed spectral shift, and the indirect effect has to be taken into account (Figure 9 and brown color-coding in Table 1). The analysis of the corresponding QM/MM models confirmed that these five substitutions cause structural reorganization of the protein (Figure 10). Reorganization can include three components: (1) Reorientation of charged/polar residues in the protein due to the substitution; (2) Addition/deletion of water molecules. Water molecules possess a dipole moment and for this reason can impact on λ_{max}; (3) Change in the protonation state of titratable residues. Below, we describe the indirect effect of considered amino acid substitutions in more detail.

(1) The structural reorganization caused by E122A replacement in Rh (3.8 Å from C5 PSB atom) involves the reorientation of C167 residue and the addition of two water molecules located in the increased cavity at the substitution site (Figure 10a).

(2) The structural reorganization caused by P219T replacement in KR2 (3.9 Å from C5 PSB atom) involves the reorientation of the polar Y247 residue and the addition of two water molecules at the substitution site (Figure 10b).

(3) The structural reorganization caused by S254A replacement in KR2 (3.5 Å from C15 PSB atom) involves the reorientation of the polar N112 residue located in the vicinity of the N16 PSB atom (Figure 10c). The distance from the NH2 group of N112 to the N16 PSB atom decreases from 4.5 Å to 3.6 Å.

(4) The structural reorganization caused by G171S replacement in KR2 (4.7 Å from C4 PSB atom) involves the reorientation of the positively charged R246 residue located at 12 Å from C6 PSB in the wild-type protein. The charged center of R246 comes closer to beta-ionone part of PSB, leading to an additional slight blue shift. The water molecule located between G171 and the beta-ionone ring of PSB in wild-type KR2 moves away in the KR2 G171S mutant.

(5) The structural reorganization caused by W265F replacement in Rh (4.9 Å from C4 PSB atom) involves the reorientation of the polar Y191 residue and the addition of three water molecules in the increased cavity at the substitution site. According to our QM/MM model, W265F replacement has a non-negligible effect on retinal geometry. The spectral shift related to retinal geometry modification is −8 nm, while the experimentally observed spectral shift is −18 nm.

Overall, the proposed models can be applied not only to estimate the direct effect of amino acid substitution but also to determine if the indirect effect of amino acid substitution occurs.

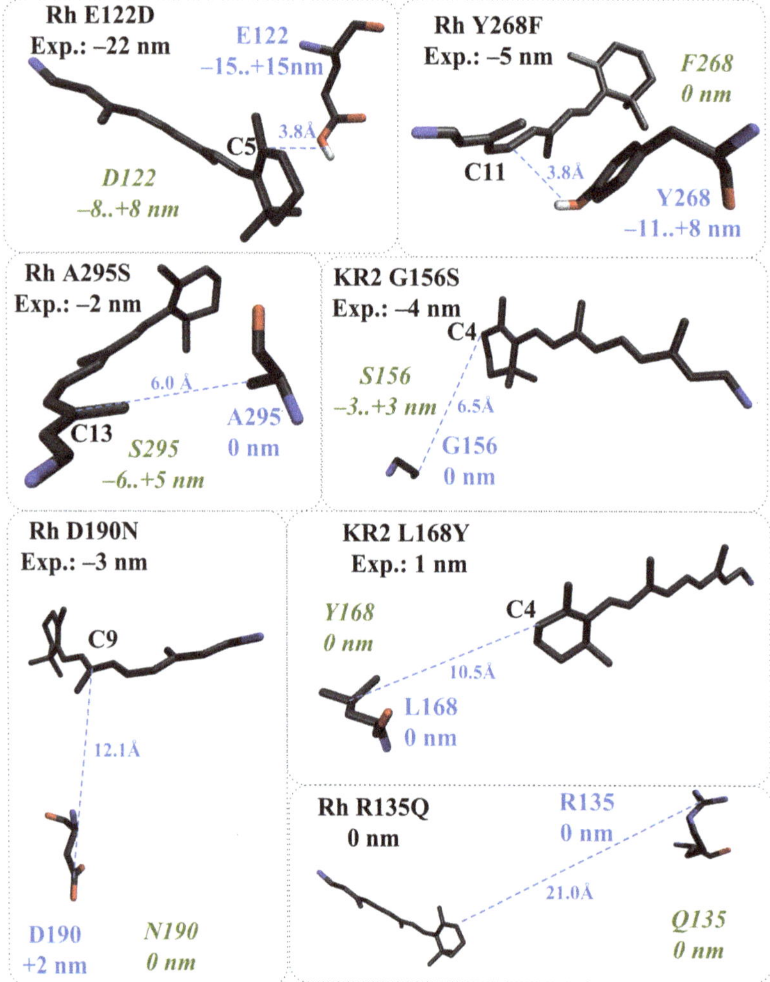

Figure 8. Bovine rhodopsin (Rh) and sodium pumping rhodopsin (KR2) mutants, for which the direct effect of amino acid substitution completely explains the experimentally observed spectral shift. Experimental spectral shift values are shown in black. The distances from the wild-type residues to the PSB and corresponding contributions to λ_{max} are shown in blue. The evaluated contributions of new residues to λ_{max} are shown in green. The distances are given in Å.

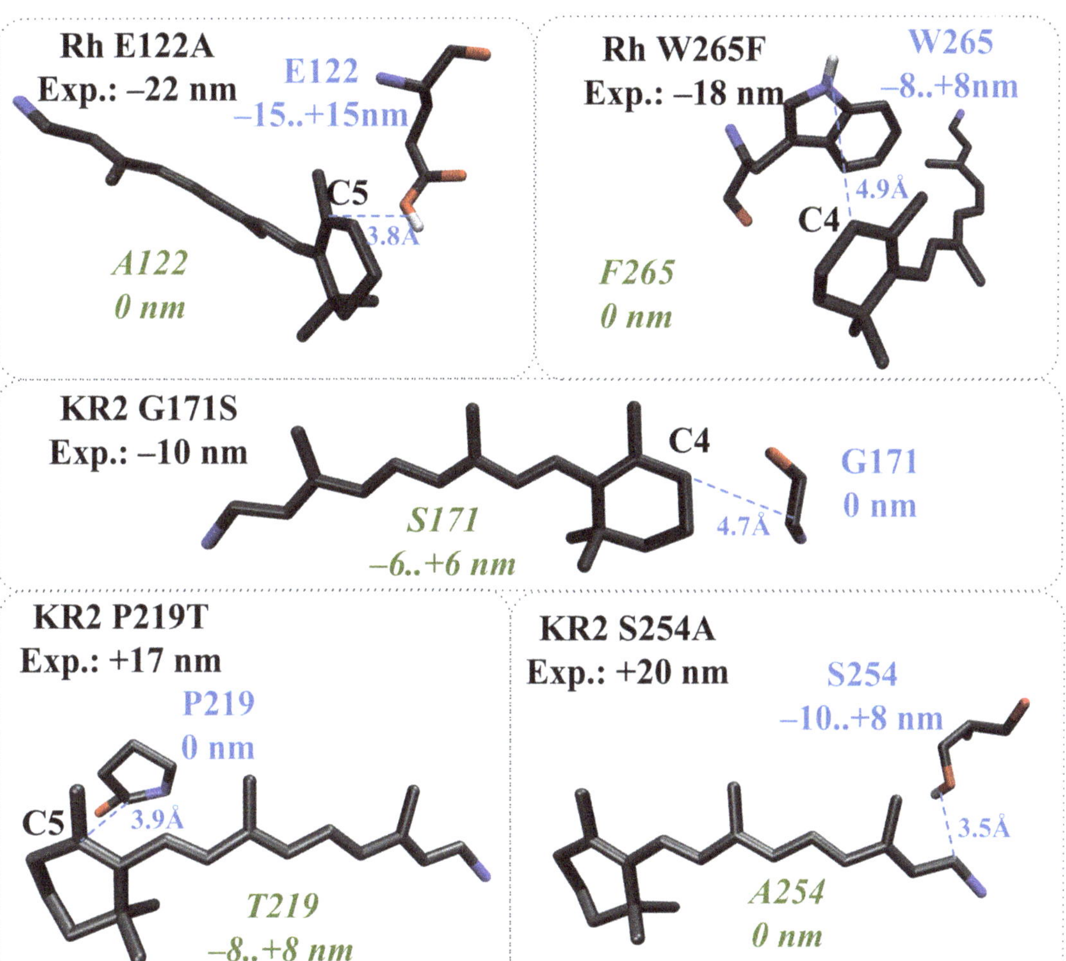

Figure 9. Bovine rhodopsin (Rh) and sodium pumping rhodopsin (KR2) mutants, for which the direct effect of amino acid substitution cannot completely explain the experimentally observed spectral shift. Experimental spectral shifts are shown in black. The distances from the wild-type residues to the PSB and corresponding contributions to λ_{max} are shown in blue. The evaluated contributions of new residues to λ_{max} are shown in green. The distances are given in Å.

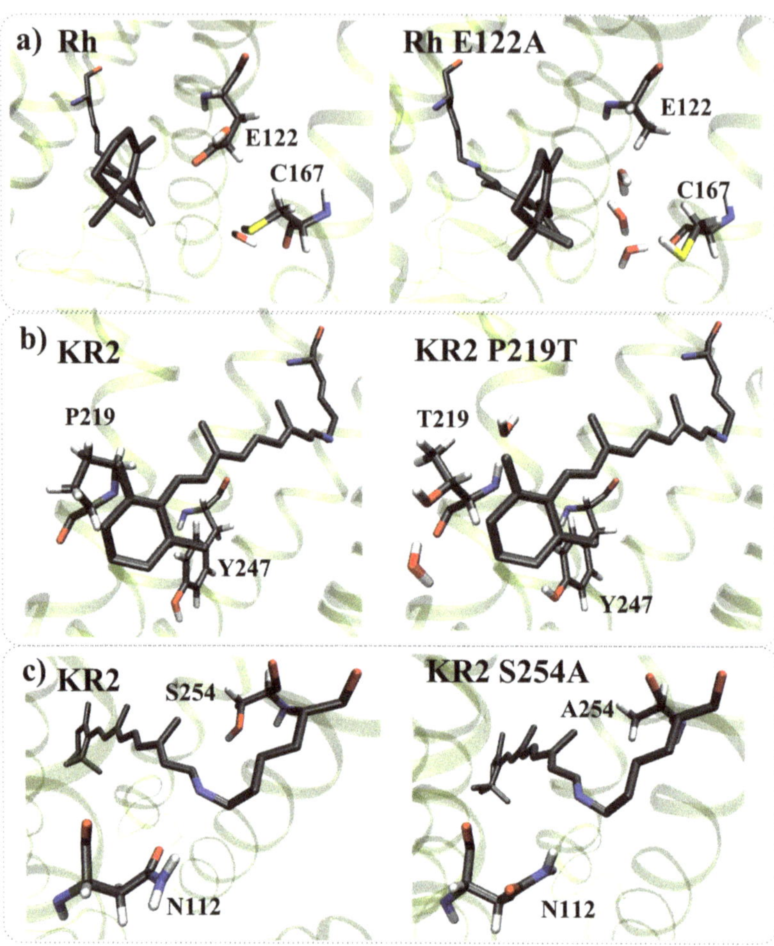

Figure 10. Structural reorganization caused by amino acid replacements. (**a**) E122A substitution in bovine rhodopsin (Rh); (**b**) P219T substitution in sodium pumping rhodopsin (KR2); (**c**) S254A substitution in sodium pumping rhodopsin (KR2).

Table 1. Spectral shifts in bovine rhodopsin (Rh) and sodium pumping rhodopsin (KR2) caused by amino acid substitutions. $\Delta\lambda_{max}^{exp}$—experimental spectral shift. $\Delta\lambda_{max}^{direct}$—the magnitude of the spectral shift estimated by the proposed models (Figures 3, 5–7 and Tables S1–S6). $\Delta\lambda_{max}^{QM/MM}$—the spectral shift calculated with the quantum mechanics/molecular mechanics models of rhodopsin mutants. Color coding: green—substitutions that do not induce structural reorganization (direct spectral tuning), brown—substitutions that cause a substantial structural reorganization (indirect spectral tuning).

Mutant	Type	$\Delta\lambda_{max}^{exp}$	$\Delta\lambda_{max}^{direct}$	$\Delta\lambda_{max}^{QM/MM}$
Rh Y268F	polar/ nonpolar	−5 nm [34]	Y268: −11 to +8 nm 3.8 Å from C_{11} F268: 0 nm	−7 nm
Rh A295S	nonpolar/ polar	−2 nm [35]	A295: 0 nm S295: −6 to +5 nm 6.0 Å from C_{13}	−7 nm
Rh D190N	charged/ polar	−3 nm[36]	D190: +2 nm 12.1 Å from C_9 N190: 0 nm	+1 nm
Rh R135Q	charged/ polar	0 nm [37]	R295: 0 nm 21.0 Å from C_4 Q135: 0 nm	−2 nm
KR2 L168Y	nonpolar/ polar	1 nm [38]	L168: 0 nm Y168: 0 nm 10.5 Å from C_4	4 nm
Rh E122D	charged / charged	−22 nm[34]	E122: +62 nm 4.8 Å from C_5 D122: +38 nm E122D: −27 nm	–
Rh E122D	charged (protonated)/ charged (protonated)	−22 nm [34]	E122$^+$: −15 to +15 nm 3.8 Å from C_5 D122$^+$: −8 to +8 nm E122D: −23 nm to 23 nm	−24 nm
KR2 G156S	nonpolar/ polar	−4 nm [38]	G156: 0 nm S156: −3 to +3 nm 6.5 Å from C_4	−8 nm
KR2 G171S	nonpolar/ polar	−10 nm[38]	G171: 0 nm S171: −6 to +6 nm 4.7 Å from C_4	−9 nm
Rh E122A	charged (protonated)/ neutral	−22 nm [34]	E122$^+$: −15 to +15 nm 3.8 Å from C_5 A122: 0 nm	−23 nm
Rh W265F	polar/ neutral	−18 nm [34]	W265: −8 to +8 nm 4.9 Å from C_4 F265: 0 nm	−16 nm
KR2 S254A	polar neutral	+20 nm [38]	S254: -10 to +8 nm 3.5 Å from C_{15} A254: 0 nm	+24 nm
KR2 P219T	neutral/ polar	+17 nm [38]	P219: 0 nm T265: −8 to +8 nm 3.9 Å from C_5	+12 nm

2.5. Limitations of the Proposed Models

As demonstrated in a number of experimental and computational studies [24,39–42], the positive charge located at the NH moiety of the chromophore after excitation partially relocates to the β-ionone ring moiety, making the NH part less positive and, accordingly, the β-ionone ring part more positive. This difference in the charge distribution between S_0 and S_1 states leads to different electrostatic interactions of the chromophore with external charges (see Figure 11).

The interaction between the charge density of the chromophore and the negative charge located in the NH region stabilizes the ground state more than the first excited state and, therefore, leads to a blue shift in the S_0 to S_1 band. On the other hand, a negative charge in the β-ionone region stabilizes the excited state more than the ground state, leading to a red shift. A positive charge in the NH region destabilizes the ground state more than the excited state, leading to a red shift. Finally, a positive charge in the β-ionone region destabilizes the excited state more than the ground state, leading to a blue shift (Figure 11).

If $\Delta\lambda_{max}$ was determined only by this "charge transfer" factor, and both S_0-to-S_1 charge redistribution and the geometry of the chromophore did not depend on an external electrostatic field, the impact of each residue on $\Delta\lambda_{max}$ would be independent of the rest of the residues; i.e., $\Delta\lambda_{max}$ of each residue would be additive. In fact, this additivity is broken due to the polarization effect caused by any charged or polar residue. The additional electrostatic field modifies both the magnitude of the $S_0 \rightarrow S_1$ charge transfer due to different polarization of the ground and the excited states and the ground state geometry of the chromophore by changing the so-called bond length alternation (BLA), i.e., averaged difference between single- and double-bond lengths of the chromophore (Figure 12). However, as demonstrated, for example, for *N. Pharaonis* halorhodopsin [10], the contribution of these polarization effects to $\Delta\lambda_{max}$ is much smaller than the contribution of the "charge transfer" effect, and, in the absence of other protein residues reorganization, the $\Delta\lambda_{max}$ additivity can be considered as a good approximation.

The proposed models assume that the impact of a charged/polar residue on λ_{max} depends only on its charge/dipole moment and its distance to/orientation along the chromophore axis but not a radial angle. To confirm this "cylindrical symmetry" assumption, we performed an additional set of calculations rotating negative unit charges at 4 Å around the 11-cis chromophore axis. The results confirm that calculated $\Delta\lambda_{max}$ only slightly depend on radial angles (See Supplementary Table S7 for details).

Finally, the proposed models do not take into account the possible distortion of the chromophore due to the steric interactions caused by amino acid substitution [23]. This effect can be accurately taken into account only by QM/MM models.

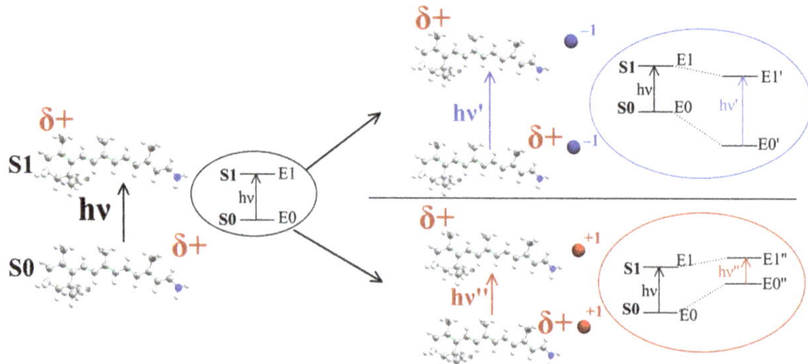

Figure 11. Difference in the charge distribution between S_0 and S_1 states of PSB leads to different electrostatic interactions of the chromophore with external charges.

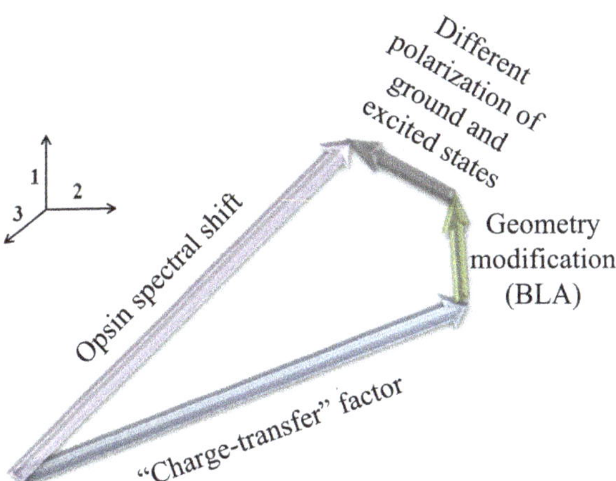

Figure 12. The electrostatic spectral tuning mechanism in rhodopsins involves three main factors: 1. The "charge transfer" factor related to difference in charge distributions of ground and excited states of the PSB. The positive charge partially translocates from the NH region to the β-ionone ring region upon photoexcitation. Therefore, ground and excited states of the chromophore possess different interactions with external charges. 2. The modification of the bond length alternation (BLA) of the chromophore by the external electrostatic field. 3. Differences in polarization of the ground and excited states of the PSB by the external electrostatic field.

3. Materials and Methods

3.1. Ab Initio-Based Models

Geometries of 11 -cis PSB (protonated Schiff base) and all-trans PSB were optimized at the B3LYP/6-31G* level of theory. Absorption maxima values were calculated at the SORCI(6,6)+Q/6-31G* level of theory. Electrostatic embedding scheme was used to include the effect of external charges. The calculations were performed with the ORCA program, version 3.0.3 [43].

3.2. Evaluation of Spectral Shifts Caused by Amino Acid Replacements

To evaluate $\Delta\lambda_{max}$ values caused by amino acid substitutions, the corresponding three-dimensional structures of the wild-type proteins were used. For bovine rhodopsin (Rh), the 2.2 Å X-ray structure was used, RCSB code 1U19 [44]; for sodium pumping *Krokinobacter eikastus* rhodopsin 2 (KR2), the 1.8 Å structure was used, RCSB code 6RF6 [45]. The distance from the substituted amino acid to the closest atom of the retinal chromophore was measured using visualizing software (VMD program, v.1.9.3) [46]. The pdb file of the X-ray structure was used without any preliminary modifications. When the position of the residue was defined, we used the figures and tables presented in the Results section and the Supporing Information to determine the correspondence between the position of the residue and the possible spectral shift.

3.3. QM/MM Models Construction

To generate QM/MM models of rhodopsin mutants, we started from the corresponding wild-type X-ray structures. The amino acid substitutions were inserted into the wild-type X-ray structures using the Mutate Model algorithm implemented in Modeller v.9.15 program package [47]. The algorithm replaces the indicated amino acid in the protein X-ray structure and optimizes its position, leaving other protein residues intact. The retinal chromophore was inserted into the models and bound to the proper lysine residue (11-cis PSB, Lys296 for Rh; all-trans PSB, Lys255 for KR2). Afterward, models were hydrated with the Dowser++ algorithm [48]; the configuration parameters for running Dowser++ and the parameter set for the PSB were described in our previous work [29]. The PROPKA program, version 3.1 [49], was used to calculate the pKa values of titratable residues (pH = 7.0) and assign their protonation states; hydrogen atoms were added with the pdb2pqr program, version 2.1.1 [50]. The obtained models were optimized gradually first at the MM level (Amber96 force field [51], TIP3P for water) then at the QM/MM level utilizing the hybrid two-layer ONIOM (QM:MM-EE) scheme (QM = B3LYP/6-31G*; MM = AMBER96 for amino acids and ions, TIP3P for water, EE = electronic embedding). The ONIOM calculations were performed with Gaussian09 [52]. Fifty atoms of the retinal chromophore were included in the QM part; the link atom was placed at the N_Z-C_ϵ bond of Lys296. The SORCI+Q/6-31G* method was used to calculate the PSB absorption maxima values in the opsin environment represented as Amber96 point charges. The absorption maxima calculations were performed with the ORCA program, version 3.0.3 [43]. The reliability of the applied methodology for rhodopsins was tested in several previous studies [10,12,13,29,30].

4. Conclusions

The main goal of this article was to present a simple ab initio-based approach to evaluate $\Delta\lambda_{max}$ caused by substitution of charged or polar amino acids in visual and microbial rhodopsins. If the rhodopsin three-dimensional structure is available, $\Delta\lambda_{max}$ can be obtained from the plots and tables given in the article and Supplementary Materials. The performance of the proposed models is evaluated against a test set consisting of ten mutants of bovine and sodium pumping rhodopsins.

Additional, general conclusions of this study can be summarized as follows:

1. The contribution of charged residues to λ_{max} strongly depends on their positions and varies from over 100 nm for counterions at the distance of around 3.5 Å from the nitrogen atom of the chromophore to several nm for the residues located at 18 Å.
2. The contribution of polar residues outside the binding pocket, i.e., more than 6–7 Å from the chromophore, is negligible.
3. The distance from a charged/polar residue to the closest atom of the chromophore is the main parameter that is required to estimate the contribution of this residue to λ_{max}. In addition, the information about the dipole moment orientation relative to the chromophore is important for the evaluation of contributions of polar residues.
4. An adequate model to evaluate λ_{max} of a rhodopsin must take into account the effect of polar/charged residues in the binding pocket, i.e., within 6–7 Å, and the charged residues at least up to 16–18 Å. On one hand, these findings explain the success of "binding pocket models" [14,53], in which the main difference in λ_{max} between two rhodopsins is attributed to the amino acid compositions of their binding pockets. On the other hand, these findings also reveal the limitations of the "binding pocket models" models, such as neglecting the charged residues beyond the binding pocket and the reorganization of polar/charged residues within the binding pocket due to distant amino acid substitutions.

The models proposed in this study can be used to estimate the direct part of $\Delta\lambda_{max}$ caused by residue substitution and, therefore, can be utilized both for the interpretation of experimental data and for the rational design of rhodopsins with specific spectral properties.

Supplementary Materials: Supplementary materials can be found at www.mdpi.com/xxx/s1.

Author Contributions: Conceptualization, M.N.R., A.V.V., Y.S.T.; investigation, data curation, A.A.S., D.M.N., V.N.M., M.S.P.; formal analysis, M.N.R., A.V.V., Y.S.T.; writing—original draft preparation, D.M.N.; writing—review and editing, M.N.R., A.V.V., Y.S.T.; supervision, project administration, M.N.R. All authors have read and agreed to the published version of the manuscript.

Funding: The work was funded by the Russian Science Foundation (RSF), grant No. 20-13-00303.

Institutional Review Board Statement: Not applicable.

Informed Consent Statement: Not applicable.

Data Availability Statement: The data that support the findings of this study are available from the corresponding author upon reasonable request.

Acknowledgments: Authors acknowledge HPC computing resources at the Resource Center "Computer Center of SPbU".

Conflicts of Interest: The authors declare no conflict of interest. The funders had no role in the design of the study; in the collection, analyses, or interpretation of data; in the writing of the manuscript, or in the decision to publish the results.

References

1. Man, D.; Wang, W.; Sabehi, G.; Aravind, L.; Post, A.F.; Massana, R.; Spudich, E.N.; Spudich, J.L.; Béja, O. Diversification and spectral tuning in marine proteorhodopsins. *EMBO Rep.* **2003**, *22*, 1725–1731. [CrossRef]
2. Spudich, J.L.; Jung, K.H. Microbial Rhodopsins. In *Protein Science Encyclopedia: Online*; Wiley-VCH: Weinheim, Germany, 2008.
3. Engqvist, M.K.; McIsaac, R.S.; Dollinger, P.; Flytzanis, N.C.; Abrams, M.; Schor, S.; Arnold, F.H. Directed evolution of Gloeobacter violaceus rhodopsin spectral properties. *J. Mol. Biol.* **2015**, *427*, 205–220. [CrossRef]
4. Lin, J.Y.; Knutsen, P.M.; Muller, A.; Kleinfeld, D.; Tsien, R.Y. ReaChR: A red-shifted variant of channelrhodopsin enables deep transcranial optogenetic excitation. *Nat. Neurosci.* **2013**, *16*, 1499. [CrossRef]
5. McIsaac, R.S.; Engqvist, M.K.; Wannier, T.; Rosenthal, A.Z.; Herwig, L.; Flytzanis, N.C.; Imasheva, E.S.; Lanyi, J.K.; Balashov, S.P.; Gradinaru, V.; et al. Directed evolution of a far-red fluorescent rhodopsin. *Proc. Natl. Acad. Sci. USA* **2014**, *111*, 13034–13039. [CrossRef] [PubMed]
6. Nikolaev, D.M.; Panov, M.S.; Shtyrov, A.A.; Boitsov, V.M.; Vyazmin, S.Y.; Chakchir, O.B.; Yakovlev, I.P.; Ryazantsev, M.N. Perspective tools for optogenetics and photopharmacology: From design to implementation. In *Progress in Photon Science*; Springer: Berlin/Heidelberg, Germany, 2019; pp. 139–172.
7. Shimono, K.; Ikeura, Y.; Sudo, Y.; Iwamoto, M.; Kamo, N. Environment around the chromophore in pharaonis phoborhodopsin: mutation analysis of the retinal binding site. *Biochim. Biophys. Acta. Biomembr.* **2001**, *1515*, 92–100. [CrossRef]
8. Fasick, J.I.; Lee, N.; Oprian, D.D. Spectral tuning in the human blue cone pigment. *Biochemistry* **1999**, *38*, 11593–11596. [CrossRef] [PubMed]
9. Kim, S.Y.; Waschuk, S.A.; Brown, L.S.; Jung, K.H. Screening and characterization of proteorhodopsin color-tuning mutations in Escherichia coli with endogenous retinal synthesis. *Biochim. Biophys. Acta. Bioenerg.* **2008**, *1777*, 504–513. [CrossRef] [PubMed]
10. Ryazantsev, M.N.; Altun, A.; Morokuma, K. Color tuning in rhodopsins: The origin of the spectral shift between the chloride-bound and anion-free forms of halorhodopsin. *J. Am. Chem. Soc.* **2012**, *134*, 5520–5523. [CrossRef] [PubMed]
11. Altun, A.; Morokuma, K.; Yokoyama, S. H-bond network around retinal regulates the evolution of ultraviolet and violet vision. *ACS Chem. Biol.* **2011**, *6*, 775–780. [CrossRef] [PubMed]
12. Nikolaev, D.M.; Osipov, D.E.; Strashkov, D.M.; Vyazmin, S.Y.; Akulov, V.E.; Kravtcov, D.V.; Chakchir, O.B.; Panov, M.S.; Ryazantsev, M.N. Molecular mechanisms of adaptation to the habitat depth in visual pigments of A. subulata and L. forbesi squids: On the role of the S270F substitution. *J. Integr. OMICS* **2019**, *9*, 44–50.
13. Altun, A.; Yokoyama, S.; Morokuma, K. Color tuning in short wavelength-sensitive human and mouse visual pigments: Ab initio quantum mechanics/molecular mechanics studies. *J. Phys. Chem. A* **2009**, *113*, 11685–11692. [CrossRef]
14. Frahmcke, J.S.; Wanko, M.; Elstner, M. Building a model of the blue cone pigment based on the wild type rhodopsin structure with QM/MM methods. *J. Phys. Chem. B* **2012**, *116*, 3313–3321. [CrossRef]
15. Ryazantsev, M.N.; Jamal, A.; Maeda, S.; Morokuma, K. Global investigation of potential energy surfaces for the pyrolysis of C1-C3 hydrocarbons: Toward the development of detailed kinetic models from first principles. *Phys. Chem. Chem. Phys.* **2015**, *17*, 27789–27805. [CrossRef] [PubMed]
16. Hoffmann, M.; Wanko, M.; Strodel, P.; König, P.H.; Frauenheim, T.; Schulten, K.; Thiel, W.; Tajkhorshid, E.; Elstner, M. Color tuning in rhodopsins: The mechanism for the spectral shift between bacteriorhodopsin and sensory rhodopsin II. *J. Am. Chem. Soc.* **2006**, *128*, 10808–10818. [CrossRef] [PubMed]
17. Ryazantsev, M.N.; Nikolaev, D.M.; Struts, A.V.; Brown, M.F. Quantum mechanical and molecular mechanics modeling of membrane-embedded rhodopsins. *J. Membr. Biol.* **2019**, *252*, 425–449. [CrossRef] [PubMed]
18. Parker, D.S.; Dangi, B.B.; Kaiser, R.I.; Jamal, A.; Ryazantsev, M.; Morokuma, K. Formation of 6-Methyl-1, 4-dihydronaphthalene in the Reaction of the p-Tolyl Radical with 1, 3-Butadiene under Single-Collision Conditions. *J. Phys. Chem. A* **2014**, *118*, 12111–12119. [CrossRef]
19. Fujimoto, K.; Hayashi, S.; Hasegawa, J.Y.; Nakatsuji, H. Theoretical studies on the color-tuning mechanism in retinal proteins. *J. Chem. Theory Comput.* **2007**, *3*, 605–618. [CrossRef]
20. Tsujimura, M.; Ishikita, H. Insights into the Protein Functions and Absorption Wavelengths of Microbial Rhodopsins. *J. Phys. Chem. B* **2020**, *124*, 11819–11826. [CrossRef]

21. Wanko, M.; Hoffmann, M.; Frauenheim, T.; Elstner, M. Effect of polarization on the opsin shift in rhodopsins. 1. A combined QM/QM/MM model for bacteriorhodopsin and pharaonis sensory rhodopsin II. *J. Phys. Chem. B* **2008**, *112*, 11462–11467. [CrossRef]
22. Wanko, M.; Hoffmann, M.; Frahmcke, J.; Frauenheim, T.; Elstner, M. Effect of polarization on the opsin shift in rhodopsins. 2. Empirical polarization models for proteins. *J. Phys. Chem. B* **2008**, *112*, 11468–11478. [CrossRef]
23. Mao, J.; Do, N.N.; Scholz, F.; Reggie, L.; Mehler, M.; Lakatos, A.; Ong, Y.S.; Ullrich, S.J.; Brown, L.J.; Brown, R.C.; et al. Structural basis of the green–blue color switching in proteorhodopsin as determined by NMR spectroscopy. *J. Am. Chem. Soc.* **2014**, *136*, 17578–17590. [CrossRef] [PubMed]
24. Mathies, R.; Stryer, L. Retinal has a highly dipolar vertically excited singlet state: implications for vision. *Proc. Natl. Acad. Sci. USA* **1976**, *73*, 2169–2173. [CrossRef]
25. Schenkl, S.V.; Van Mourik, F.; Van der Zwan, G.; Haacke, S.; Chergui, M. Probing the ultrafast charge translocation of photoexcited retinal in bacteriorhodopsin. *Science* **2005**, *309*, 917–920. [CrossRef] [PubMed]
26. Soderhjelm, P.; Husberg, C.; Strambi, A.; Olivucci, M.; Ryde, U. Protein influence on electronic spectra modeled by multipoles and polarizabilities. *J. Chem. Theory Comput.* **2009**, *5*, 649–658. [CrossRef]
27. Wu, M.; Eriksson, L.A. Absorption Spectra of Riboflavin: A Difficult Case for Computational Chemistry. *J. Phys. Chem. A* **2010**, *114*, 10234–10242. [CrossRef]
28. Schwinn, K.; Ferré, N.; Huix-Rotllant, M. Efficient analytic second derivative of electrostatic embedding QM/MM energy: Normal mode analysis of plant cryptochrome. *J. Chem. Theory Comput.* **2020**, *16*, 3816–3824. [CrossRef] [PubMed]
29. Nikolaev, D.M.; Shtyrov, A.A.; Mereshchenko, A.S.; Panov, M.S.; Tveryanovich, Y.S.; Ryazantsev, M.N. An assessment of water placement algorithms in quantum mechanics/molecular mechanics modeling: The case of rhodopsins' first spectral absorption band maxima. *Phys. Chem. Chem. Phys.* **2020**, *22*, 18114–18123. [CrossRef] [PubMed]
30. Nikolaev, D.M.; Emelyanov, A.; Boitsov, V.M.; Panov, M.S.; Ryazantsev, M.N. A voltage-dependent fluorescent indicator for optogenetic applications, archaerhodopsin-3: Structure and optical properties from in silico modeling. *F1000Research* **2017**, *6*, 33. [CrossRef]
31. Pieri, E.; Ledentu, V.; Sahlin, M.; Dehez, F.; Olivucci, M.; Ferré, N. CpHMD-Then-QM/MM identification of the amino acids responsible for the anabaena sensory rhodopsin pH-dependent electronic absorption spectrum. *J. Chem. Theory Comput.* **2019**, *15*, 4535–4546. [CrossRef]
32. Marrink, S.J.; Risselada, H.J.; Yefimov, S.; Tieleman, D.P.; De Vries, A.H. The MARTINI force field: Coarse grained model for biomolecular simulations. *J. Phys. Chem. B* **2007**, *111*, 7812–7824. [CrossRef]
33. Kmiecik, S.; Gront, D.; Kolinski, M.; Wieteska, L.; Dawid, A.E.; Kolinski, A. Coarse-grained protein models and their applications. *Chem. Rev.* **2016**, *116*, 7898–7936. [CrossRef]
34. Nakayama, T.; Khorana, H.G. Mapping of the amino acids in membrane-embedded helices that interact with the retinal chromophore in bovine rhodopsin. *J. Biol. Chem.* **1991**, *266*, 4269–4275. [CrossRef]
35. Janz, J.M.; Farrens, D.L. Engineering a functional blue-wavelength-shifted rhodopsin mutant. *Biochemistry* **2001**, *40*, 7219–7227. [CrossRef]
36. Nathans, J. Determinants of visual pigment absorbance: Identification of the retinylidene Schiff's base counterion in bovine rhodopsin. *Biochemistry* **1990**, *29*, 9746–9752. [CrossRef]
37. Sakmar, T.P.; Franke, R.R.; Khorana, H.G. Glutamic acid-113 serves as the retinylidene Schiff base counterion in bovine rhodopsin. *Proc. Natl. Acad. Sci. USA* **1989**, *86*, 8309–8313. [CrossRef]
38. Inoue, K.; del Carmen Marín, M.; Tomida, S.; Nakamura, R.; Nakajima, Y.; Olivucci, M.; Kandori, H. Red-shifting mutation of light-driven sodium-pump rhodopsin. *Nat. Commun.* **2019**, *10*, 1–11. [CrossRef] [PubMed]
39. Coto, P.B.; Strambi, A.; Ferré, N.; Olivucci, M. The color of rhodopsins at the ab initio multiconfigurational perturbation theory resolution. *Proc. Natl. Acad. Sci. USA* **2006**, *103*, 17154–17159. [CrossRef] [PubMed]
40. Tomasello, G.; Olaso-Gonzalez, G.; Altoe, P.; Stenta, M.; Serrano-Andres, L.; Merchan, M.; Orlandi, G.; Bottoni, A.; Garavelli, M. Electrostatic control of the photoisomerization efficiency and optical properties in visual pigments: On the role of counterion quenching. *J. Am. Chem. Soc.* **2009**, *131*, 5172–5186. [CrossRef] [PubMed]
41. Ferré, N.; Olivucci, M. Probing the rhodopsin cavity with reduced retinal models at the CASPT2//CASSCF/AMBER level of theory. *J. Am. Chem. Soc.* **2003**, *125*, 6868–6869. [CrossRef] [PubMed]
42. Sumita, M.; Ryazantsev, M.N.; Saito, K. Acceleration of the Z to E photoisomerization of penta-2,4-dieniminium by hydrogen out-of-plane motion: Theoretical study on a model system of retinal protonated Schiff base. *Phys. Chem. Chem. Phys.* **2009**, *11*, 6406–6414. [CrossRef]
43. Neese, F. The ORCA program system. *Wiley Interdiscip. Rev. Comput. Mol. Sci.* **2012**, *2*, 73–78. [CrossRef]
44. Okada, T.; Sugihara, M.; Bondar, A.N.; Elstner, M.; Entel, P.; Buss, V. The retinal conformation and its environment in rhodopsin in light of a new 2.2 Å crystal structure. *J. Mol. Biol.* **2004**, *342*, 571–583. [CrossRef]
45. Kovalev, K.; Polovinkin, V.; Gushchin, I.; Alekseev, A.; Shevchenko, V.; Borshchevskiy, V.; Astashkin, R.; Balandin, T.; Bratanov, D.; Vaganova, S.; et al. Structure and mechanisms of sodium-pumping KR2 rhodopsin. *Sci. Adv.* **2019**, *5*, eaav2671. [CrossRef]
46. Humphrey, W.; Dalke, A.; Schulten, K.; et al. VMD: Visual molecular dynamics. *J. Mol. Graph.* **1996**, *14*, 33–38. [CrossRef]
47. Webb, B.; Sali, A. Comparative protein structure modeling using MODELLER. *Curr. Protoc. Bioinform.* **2016**, *54*, 5–6. [CrossRef] [PubMed]

48. Morozenko, A.; Stuchebrukhov, A. Dowser++, a new method of hydrating protein structures. *Proteins* **2016**, *84*, 1347–1357. [CrossRef]
49. Olsson, M.H.; Søndergaard, C.R.; Rostkowski, M.; Jensen, J.H. PROPKA3: Consistent treatment of internal and surface residues in empirical pKa predictions. *J. Chem. Theory Comput.* **2011**, *7*, 525–537. [CrossRef] [PubMed]
50. Dolinsky, T.J.; Czodrowski, P.; Li, H.; Nielsen, J.E.; Jensen, J.H.; Klebe, G.; Baker, N.A. PDB2PQR: Expanding and upgrading automated preparation of biomolecular structures for molecular simulations. *Nucleic Acids Res.* **2007**, *35*, W522–W525. [CrossRef] [PubMed]
51. Cornell, W.D.; Cieplak, P.; Bayly, C.I.; Gould, I.R.; Merz, K.M.; Ferguson, D.M.; Spellmeyer, D.C.; Fox, T.; Caldwell, J.W.; Kollman, P.A. A second generation force field for the simulation of proteins, nucleic acids, and organic molecules. *J. Am. Chem. Soc.* **1996**, *118*, 2309–2309. [CrossRef]
52. Frisch, M.J.; Trucks, G.; Schlegel, H.; Scuseria, G.; Robb, M.; Cheeseman, J.; Scalmani, G.; Barone, V.; Mennucci, B.; Petersson, G.; et al. *Gaussian 09, Revision D. 01*; Gaussian Inc.: Wallingford, CT, USA, 2009.
53. Welke, K.; Frahmcke, J.S.; Watanabe, H.C.; Hegemann, P.; Elstner, M. Color tuning in binding pocket models of the chlamydomonas-type channelrhodopsins. *J. Phys. Chem. B* **2011**, *115*, 15119–15128. [CrossRef] [PubMed]

International Journal of
Molecular Sciences

MDPI

Article

Recent Developments on gMicroMC: Transport Simulations of Proton and Heavy Ions and Concurrent Transport of Radicals and DNA

Youfang Lai [1,2], Xun Jia [2,*] and Yujie Chi [1,*]

1 Department of Physics, University of Texas at Arlington, Arlington, TX 76019, USA; youfang.lai@mavs.uta.edu
2 Innovative Technology of Radiotherapy Computation and Hardware (iTORCH) Laboratory, Department of Radiation Oncology, University of Texas Southwestern Medical Center, Dallas, TX 75287, USA
* Correspondence: xun.jia@utsouthwestern.edu (X.J.); yujie.chi@uta.edu (Y.C.)

Abstract: Mechanistic Monte Carlo (MC) simulation of radiation interaction with water and DNA is important for the understanding of biological responses induced by ionizing radiation. In our previous work, we employed the Graphical Processing Unit (GPU)-based parallel computing technique to develop a novel, highly efficient, and open-source MC simulation tool, gMicroMC, for simulating electron-induced DNA damages. In this work, we reported two new developments in gMicroMC: the transport simulation of protons and heavy ions and the concurrent transport of radicals in the presence of DNA. We modeled these transports based on electromagnetic interactions between charged particles and water molecules and the chemical reactions between radicals and DNA molecules. Various physical properties, such as Linear Energy Transfer (LET) and particle range, from our simulation agreed with data published by NIST or simulation results from other CPU-based MC packages. The simulation results of DNA damage under the concurrent transport of radicals and DNA agreed with those from nBio-Topas simulation in a comprehensive testing case. GPU parallel computing enabled high computational efficiency. It took 41 s to simultaneously transport 100 protons with an initial kinetic energy of 10 MeV in water and 470 s to transport 10^5 radicals up to 1 μs in the presence of DNA.

Keywords: Monte Carlo simulation; GPU programming; DNA damage; proton transport

Citation: Lai, Y.; Jia, X.; Chi, Y. Recent Developments on gMicroMC: Transport Simulations of Proton and Heavy Ions and Concurrent Transport of Radicals and DNA. *Int. J. Mol. Sci.* **2021**, *22*, 6615. https://doi.org/10.3390/ijms22126615

Academic Editor: Małgorzata Borówko

Received: 18 May 2021
Accepted: 16 June 2021
Published: 21 June 2021

1. Introduction

Understanding biological responses to ionizing radiation is of crucial importance for cancer treatment using radiotherapy. Mechanistic Monte Carlo (MC) simulation of radiation's effect on DNA in a water medium is a promising tool for relevant studies after decades of development [1]. The central idea of such an approach is to obtain the initial DNA damage spectrum via mechanistic modeling of the radio-biological interactions at the atomic or molecular levels. This includes the development of track-structure codes [2–15] and the subsequent computation of DNA damages by incorporating DNA models [5,7,16–18]. The track-structure simulation can be divided into the simulations of the physical stage and the chemical stage. The physical-stage simulation deals with the ionization, excitation, and elastic scattering processes between the ionizing radiation particles and the water media and records the 3D coordinates of energy deposition events. The chemical-stage simulation computes how the chemical radicals, produced after the physical-stage simulation, diffuse and react mutually with the recording of the residual radicals' positions. The positions of these energy deposition events and radicals are then utilized to compute the initial DNA damage sites, followed by an analysis to characterize DNA strand break patterns.

Although many developments have been performed to generate state-of-the-art mechanistic MC simulation tools, it is still necessary to further improve the simulation methods to accommodate different scenarios [8]. For instance, to make the code versatile for studies on the Oxygen Enhancement Ratio (OER) [19] and the Fenton reaction effect [20], it is desired to include more types of molecules other than free radicals generated by the initial radiation into the chemical-stage simulation. However, due to the computational complexity of the "many-body" problem and the long temporal duration of the chemical stage, a step-by-step simulation of these relevant processes on conventional CPU computational platforms can be extremely time consuming [21]. Under the constraint of computational resources, studies typically suffer from a restricted simulation region or a shortened temporal duration [22–24], limiting their broad applications.

To overcome these obstacles, Graphical Processing Unit (GPU)-based parallel computing can be a cost-effective option [25,26]. We developed an open-source, GPU-based microscopic MC simulation toolkit, gMicroMC [9], with the first version available on GitHub (https://github.com/utaresearch/gMicroMC (accessed on 17 June 2021)). We initially focused on boosting the chemical-stage simulation for radicals produced from water radiolysis, achieving a speedup of several hundred folds compared to CPU-based packages [27]. Later, we supported the physical track simulation for energetic electrons and implemented a DNA model of a lymphocyte cell nucleus at the base-pair resolution for the computation of electron-induced DNA damages [9]. Recently, we also included oxygen molecules in the chemical-stage simulation in a step-by-step manner, which enabled the study of the radiolytic depletion effect of dissolved oxygen molecules [28]. With these efforts, we were able to quantitatively study multiple critical problems that are computationally demanding. For example, we performed comprehensive simulations with gMicroMC to answer how uncertainties from the simulation parameters affect the accuracy of the final DNA damage computations [29]. We also studied the radiolytic depletion of oxygen under ultra-high dose rate radiation (FLASH) to investigate the fundamental mechanism behind FLASH radiotherapy with the developed oxygen module in gMicroMC [28].

In this work, we reported our recent progress on two new and important features that we recently introduced to gMicroMC, namely: (1) enabling the physical-stage simulations of protons and heavy ions; and (2) considering the presence of the DNA structure and its chemical reactions with radicals in the chemical-stage simulation. It was expected that the first feature would contribute greatly to the mechanistic study of particle irradiation, such as particle radiotherapy [30,31]. The presence of the DNA structure in the chemical-stage simulation will allow us to realistically describe the indirect DNA damage process. With the GPU acceleration, we were able to afford computationally challenging simulations that included detailed physics modeling and chemical reactions that spanned over a large temporal duration, enabling more realistic simulations of the relevant processes.

2. Materials and Methods

2.1. Cross-Sections for the Transport Simulation of Protons and Heavy Ions

When a proton or heavy ion moves through a medium, it interacts with the atomic electrons inside the medium [32]. Considering that there have been various models developed and implemented to describe this process, in this work, we focused on a novel implementation of existing models on GPU parallel computing platforms. Specifically, we only considered the interactions between particles and water molecules because this is representative of modeling the cell environment. We employed the Rudd model [33] to compute the ionization of a water molecule by a proton in the energy range from 10 eV to 1 TeV. We implemented the Plante model [34] and Dingfelder's model [35] to simulate the excitation of a water molecule for protons with an energy above and below 500 keV, respectively. We also applied Booth's empirical formula [36] to include the charge effect on the cross-section computation. Lastly, we used the charge scaling rule [37] to obtain the cross-sections for heavy ions based on those for a proton. To make the manuscript easy to follow, we briefly introduce these models in the following subsections.

2.1.1. Ionization for Protons

An energetic proton could eject a secondary electron from different atomic subshells when it ionizes a water molecule. In the Rudd model [33,38], the partial Singly Differential Cross-Section (SDCS) can be described as:

$$\frac{d\sigma_i^{ion}}{dw} = \frac{S_i}{B_i}\frac{(F_1(\nu) + wF_2(\nu))}{(1+w)^3}\frac{1}{1+\exp\left[\frac{\alpha(w-w_i)}{\nu}\right]} \tag{1}$$

Here, i refers to the subshells of the water molecule, namely $^1b_1$, $^3a_1$, $^1b_2$, $^2a_1$, and $^1a_1$. B_i is the binding energy for electrons on shell i. $w = E_e/B_i$, and E_e is the energy of the secondary electron. $S_i = 4\pi a_0^2 * N_i * (E_R/B_i)^2$, where $a_0 = 5.3 \times 10^{-11}$ m is the Bohr radius, $E_R = 13.6$ eV is the Rydberg energy, and N_i is the number of electrons on shell i. $\nu = \sqrt{T/B_i}$ denotes the scaled velocity of the projectile, with $T = \frac{m}{M} * E_k$. m and M are the masses of the electron and proton, while E_k is the kinetic energy of the proton. $w_i = 4\nu^2 - 2\nu - E_R/4B_i$ is the scaled cutoff energy, and α is a numerical parameter related to the relative size of the target molecule. The specific values for B_i, N_i and α are listed in Table 1. $F_1(\nu)$ and $F_2(\nu)$ are fitting functions, defined as:

$$F_1(\nu) = L_1 + H_1 \tag{2}$$

$$F_2(\nu) = \frac{L_2H_2}{L_2 + H_2} \tag{3}$$

where:

$$L_1 = \frac{C_1\nu^{D_1}}{1 + E_1\nu^{D_1+4}}, \tag{4}$$

$$H_1 = \frac{A_1 \ln(1+\nu^2)}{\nu^2 + B_1/\nu^2}, \tag{5}$$

$$L_2 = C_2\nu^{D_2}, \tag{6}$$

$$H_2 = \frac{A_2}{\nu^2} + \frac{B_2}{\nu^4}. \tag{7}$$

The values for the nine basic parameters $A_1, \ldots, E_1$ and $A_2, \ldots, D_2$ used in Equations (4)–(7) can be seen in Table 1. These values differed for inner shell orbitals and external orbitals, and an orbital was regarded as an inner one when its binding energy exceeded twice the binding energy of the least-tightly bound orbital [33].

Table 1. Parameters used in this work for Equations (1)–(7). Data were extracted from [35,38].

Parameter	Inner Orbitals		External Orbitals		
	$^1a_1$	$^2a_1$	$^1b_2$	$^3a_1$	$^1b_1$
A_1	1.25	1.25	1.02	1.02	1.02
B_1	0.5	0.5	82	82	82
C_1	1	1	0.45	0.45	0.45
D_1	1	1	−0.80	−0.80	−0.80
E_1	3	3	0.38	0.38	0.38
A_2	1.1	1.1	1.07	1.07	1.07
B_2	1.3	1.3	14.6	14.6	14.6
C_2	1	1	0.6	0.6	0.6
D_2	0	0	0.04	0.04	0.04
α	0.66	0.66	0.64	0.64	0.64
N_i	2	2	2	2	2
B_i	539.7	32.2	18.55	14.73	12.61

From Equation (1), we can calculate the total cross-section for subshell i as:

$$\sigma_i^{ion} = \int_0^{w_m} \frac{d\sigma_i^{ion}}{dw} dw, \tag{8}$$

where $w_m = \frac{E_m}{B_i}$ and E_m is the scaled maximum transferable energy from the proton to the ejected electron. The relativistic expression of E_m was given by Plante et al. [34] as:

$$E_m = \frac{2mc^2(\gamma^2 - 1)}{1 + 2\gamma(\frac{m}{M}) + (\frac{m}{M})^2}, \tag{9}$$

with:

$$\gamma = 1 + \frac{E_k}{Mc^2}, \tag{10}$$

and c is the speed of light. The relativistic format for the scaled velocity ν is then written as:

$$\nu^2 = \frac{mc^2}{2B_i}[1 - \frac{1}{\gamma^2}]. \tag{11}$$

With the ionization model and parameters determined under both the relativistic and nonrelativistic formalism, we could integrate Equation (8) numerically to obtain the ionization cross-section table for different subshells of a water molecule in a broad proton energy range. In our implementation, we computed the table for proton energies ranging from 10 eV to 1 TeV with a 0.01 increment along the logarithmic scale. It is worth mentioning that the computation of the cross-section table only needed to be computed once in an offline manner and was stored in a data file that could be loaded to GPU memory for the query of the cross-sections of any incident energy. More details of this usage are given in Section 2.3.1.

2.1.2. Excitation for Proton

Due to the lack of experimental data, different models could have differential configurations of the excitation pathways. In our implementation, we adopted the three-pathway model [39,40] containing $\widetilde{A}^1B_1$ and $\widetilde{B}^1A_1$ and plasma mode for protons with energy >500 keV and the model with the excitation channels of $\widetilde{A}^1B_1$ and $\widetilde{B}^1A_1$, Ryd A + B and Ryd C + D, and the diffusion band [35] for a proton energy < 500 keV. Specifically, in the three-pathway model, the differential cross-section for the excitation channel j is expressed as:

$$\frac{d\sigma_j^{exc}}{dW} = \rho(W)Wf_j(W)\ln\left(\frac{W}{Q_{min}}\right), \tag{12}$$

where:

$$\rho(W) = \frac{8\pi Z^2 a_0^2}{mu^2}\frac{Ry^2}{W^2}, \tag{13}$$

$$Q_{min} = 2T\left(\frac{M}{m}\right)^2\left(1 - \frac{1}{2}\frac{m}{M}\frac{W}{T} - \sqrt{1 - \frac{m}{M}\frac{W}{T}}\right). \tag{14}$$

Here, j denotes different excitation channels, namely $\widetilde{A}^1B_1$ and $\widetilde{B}^1A_1$, and plasma mode. u, Z, and W are the velocity, charge, and energy loss of the incident proton. Other parameters were the same as those used in Section 2.1.1. When $\frac{m}{M}\frac{W}{T} = \frac{W}{E_k} \ll 1$, Equation (12) can be simplified as:

$$\frac{d\sigma_j^{exc}}{dW} = \rho(W)Wf_j(W)\ln\left(\frac{4T}{W}\right). \tag{15}$$

In the relativistic situation, mu^2 in Equation (13) can be expressed as:

$$mu^2 = mc^2\left[1 - \gamma^{-2}\right]. \tag{16}$$

$f_j(W)$, as a function of the excitation pathway j, has the form of:

$$f_j(W) = \begin{cases} f_j^0\sqrt{\alpha_j/\pi}e^{\left[-\alpha_j(W-w_j)^2\right]}, & \text{if } j = \widetilde{A}^1B_1,\ \widetilde{B}^1A_1 \\ f_j^0\alpha_j e^x/(1+e^x)^2, & \text{otherwise} \end{cases} \tag{17}$$

where $x = \alpha_j(W - w_j)$ and f_j^0, α_j, and w_j are parameters with their values summarized in Table 2. Under the assumption that the proton only loses a small portion of its kinetic energy to excite a water molecule, i.e., $\frac{W}{E_k} \ll 1$, Equation (15) can be used to calculate the total cross-section for excitation channel j as:

$$\sigma_j^{exc} = \int_{W_{min}}^{W_{max}} \frac{d\sigma_j^{exc}}{dW}dW. \tag{18}$$

In principle, the upper and lower boundaries of the integration can be E_k and zero. However, in practical usage, it is common to set W_{max} =50 eV and W_{min} = 2 eV. The reason is that $\frac{d\sigma_j^{exc}}{dW}(W)$ drops to a negligible value when $W \notin [W_{min}, W_{max}]$, and the boundary cutoffs also ensure a positive and convergent integrated total cross-section.

Table 2. Parameters used in Equations (17).

j	$\widetilde{A}^1B_1$	$\widetilde{B}^1A_1$	Plasma Mode
f_j^0	0.0187	0.0157	0.7843
α_j	3 (eV^{-2})	1 (eV^{-2})	0.6 (eV^{-1})
w_j	8.4	10.1	21.3

When a proton's energy is smaller than 500 keV, the Born approximation is no longer a good approximation [35], and Equation (15) may have problem in evaluating the cross-sections. We then applied the semi-empirical model [35] to obtain the excitation cross-section for a low-energy proton as:

$$\sigma_j^{exc}(E_k) = \frac{\sigma_0(Za)^{\Omega}(E_k - E_j)^{v}}{J^{\Omega+v} + E_k^{\Omega+v}}. \tag{19}$$

Here, j denotes the excitation channels $\widetilde{A}^1B_1$ and $\widetilde{B}^1A_1$, Ryd A+B and Ryd C+D, and the diffusion band. The corresponding energy loss E_j of the proton is discrete instead of continuous. Further details of the model can be found in [35].

With the excitation cross-section given in Equation (18) and the relevant parameters determined, we could integrate it numerically to obtain the excitation cross-section table for different subshells of a water molecule at proton energies above 500 keV. Meanwhile, we could rely on Equation (19) to handle protons with energies below 500 keV. To make the cross-section data computed from the two models smoothly connected at the proton energy of 500 keV, we adjusted the obtained cross-section data as follows. We applied coefficients of 1.23 and 3.5 to the cross-section data for the $\widetilde{A}^1B_1$ and $\widetilde{B}^1A_1$ channels obtained from Equation (18) to make them smoothly connected to that obtained from Equation (19) at 500 keV for these two modes. We then multiplied 0.339 in plasma mode to make the total cross-section also smoothly connected. Similar to the strategy applied to obtain the ionization cross-section table, we also only needed to compute the excitation cross-section table once and stored it in a data file. Its usage on a GPU is also given in Section 2.3.1.

2.1.3. Charge Effect

When charged particles travel through a water medium, except for ionizing or exciting the water molecule, they could also drag electrons from the medium to move with them, forming a reduced effective charge $Z_{\text{eff}} < Z$. This effect is found reversely proportional to the kinetic energy of the incident particle. In our simulation, we adopted the empirical Booth model [36] to obtain the effective charge Z_{eff} as:

$$Z_{\text{eff}} = Z\left(1 - \exp\left[-1.316y + 0.112y^2 - 0.065y^3\right]\right), \tag{20}$$

where $y = 100Z^{-2/3}\sqrt{1 - (1 + E_k/(AMc^2))^{-2}}$ and A is the mass number of the particle. The correction is larger than 5% ($Z_{\text{eff}} < 0.95 \cdot Z$) when $y < 2.172$, which gives $E_k \sim 18$ MeV per nucleon for the Fe ion and 0.22 MeV for the proton.

2.1.4. Cross-Section for Heavy Ions

Within the first-order plane-wave Born approximation, we could correlate the ionization and excitation cross-section for bare and sufficiently fast heavy ions to that of the proton by the scaling law. Specifically, for a heavy ion with velocity u and charge number Z, the doubly differential cross-section can be scaled from that of the proton with the same velocity u by a factor of Z^2 [37]:

$$\frac{d^2\sigma_{ion}}{dWdQ}(u) = Z^2 \times \frac{d^2\sigma_{proton}}{dWdQ}(u), \tag{21}$$

where W and Q refer to the energy transfer and the recoil energy, respectively. After integrating over Q, we could obtain the SDCS as a function of W. Considering that an ion of mass number A and kinetic energy E_k has the same velocity with a proton of kinetic energy $E_{k,p} = E_k \frac{M}{M_{ion}} \approx \frac{E_k}{A}$, we could rewrite the scaling formula as a function of E_k as:

$$\frac{d\sigma_{ion}}{dW}(E_k) = Z^2 \times \frac{d\sigma_{proton}}{dW}(E_k/A), \tag{22}$$

It holds for ions for both the nonrelativistic and relativistic formats. The electron attachment effect can be more significant for a heavy ion than for a proton of the same velocity since a heavy ion typically carries more charge than a proton. With the electron attachment effect considered, we replaced Z with Z_{eff} when scaling the cross-section from a proton to a heavy ion using Equation (22), with Z_{eff} calculated from Equation (20).

2.2. Concurrent Transport Method

Due to the computational challenge, existing MC tools compute the DNA damage formed by radical attachment typically via two successive steps. First, the radicals are diffused and mutually reacted in the chemical stage without DNA. Second, the coordinates of $OH\cdot$ radicals obtained at the end of the chemical stage are overlapped with the DNA geometry such that DNA damages caused by radicals can be computed [9,29]. We refer to this approach as the "overlay method". This approach is effective for simple applications. However, it can be problematic for those scenarios sensitive to radical evolution. In reality, DNA could be present and react with radicals during the radical diffusion. This could affect the radical yields and the damage site distribution on the DNA chain, consequently impacting the final characterization of the DNA strand breaks. To model this effect, in this work, we included the simulation of the reactions between radicals and DNA at the time of transporting the radicals in the chemical stage and refer to this approach as the "concurrent transport method".

In our previous development of gMicroMC without considering DNA in the chemical stage, we simulated the diffusion and reactions among radicals in a step-by-step fashion. The relevant parameters were the diffusion coefficient for each radical species and the reaction rates for possible radical–radical reaction types. To include DNA in this transport frame, we need to know the diffusion coefficient of DNA and the reaction rates between radicals and DNA. Considering the relatively large mass of DNA, we assumed that the whole DNA molecule was static during the chemical transport and took its diffusion coefficient as zero to simplify the simulation. As for the reaction rates between radicals and DNA, we considered two types of reactions based on the DNA model for a whole lymphocyte cell nucleus [9]. The DNA was described in a voxel-based format with each voxel of side length 55 nm. The voxel was either empty or filled with a DNA chain that connected two faces of the voxel. The DNA chain consisted of a group of spheres representing the basic structures of the DNA: the base pair, the sugar-phosphate group, and the histone protein. With it, we considered the first reaction type as the damaging effect of $OH\cdot$ and e_h radicals on the DNA bases and sugar-phosphate groups. The reaction rates are listed in Table 3. Here, although there were four types of DNA bases, that is Adenine (A), Guanine (G), Cytosine (C), and Thymine (T), associated with four different reaction rates with the $OH\cdot$ or e_h radical, we used the average reaction rate in our simulation since our DNA model had no differentiation among the base types. The second type was the scavenging reaction for all radical species by the histone protein. In this reaction, the radical was assumed to be fully absorbed once it was within the histone protein volume.

Table 3. Reaction rates ($\times 10^9\ L \cdot mol^{-1} \cdot s^{-1}$) used in gMicroMC for concurrent DNA transport [41].

Radicals	A	G	C	T	DNA Base	DNA Sugar-Phosphate Group
$OH\cdot$	6.1	9.2	6.4	6.1	6.95	1.9
e_h	9	14	18	13	13.5	−1 *

* A negative value means no reaction between the radical and the DNA substructure.

After introducing DNA into the chemical-stage simulation, two consequences required special attention. First, radicals were not supposed to be produced inside the DNA region; hence, at the beginning of the chemical stage, we needed to exclude those radicals produced inside the chromatin zone from the subsequent diffusion [42] without recording any damages on DNA. Second, as there were a huge number of DNA basic structures in our DNA model, for instance 6.2×10^9 base pairs, checking for reactions between radicals and DNA after each diffusion step would generate numerous computations. To circumvent the problem, we defined a time interval t_i to control the frequency of checking for reactions between radicals and DNA. During the simulation of the chemical stage, as the time evolved, we only checked for radical–DNA reactions every t_i. In the limit of a small t_i, the frequent inspection for reactions ensured simulation accuracy. In the other extreme of a large t_i, the DNA-related reactions would be less frequently checked, which eventually converged to the overlay picture. We study the impact of t_i in later sections.

2.3. GPU Implementation

2.3.1. Physical Transport for Protons and Heavy Ions

Before transporting protons and heavy ions on the GPU, we prepared lookup tables on the CPU host to store the tabulated cross-sections for a proton, as stated in Section 2.1. The lookup tables were then transferred to the GPU texture memory such that we could employ the GPU built-in fast interpolation technique to obtain cross-section data for particle transport. We supported two types of source particle generation: reading from a Phase Space File (PSF) or sampling from a given distribution. We sorted the source particles in descending order based on their charge number Z. If the particles were protons or heavy ions, we transported the sorted particles into groups using the GPU kernel we developed in this work dedicated to proton and heavy ion transport. If they were electrons, we transported them with our previously developed kernel for electron transport [9]. The purpose of particle sorting and grouping was to minimize the thread divergence on the GPU, and hence to improve simulation efficiency [25,26].

For the GPU kernel in charge of the transport of protons and heavy ions, each thread took care of one primary particle. For a particle with charge number Z, atomic mass number A, and incident energy E_k, the thread sampled its free travel distance s in water and its interaction with the water molecules in the iteration. Specifically, we first calculated the effective charge number Z_{eff} according to Equation (20) and the kinetic energy $E_p = \frac{E_k}{A}$ for a proton with the same velocity as that of the primary particle. Based on the logarithmic value of E_p, we interpolated the lookup tables to obtain the cross-section $\sigma_i(E_p)$ for the proton. Here, i represents all ionization and excitation channels listed in the tables. We then scaled and summed σ_i to obtain the total cross-section for the primary particle as $\sigma_t = Z_{\text{eff}}^2 \sum \sigma_i$ based on Equation (22). With σ_t, we sampled the free travel distance s in water as $s = -\frac{M_w}{\rho \cdot \sigma_t \cdot N_A} \ln \zeta$, where ρ and M_w are the density and atomic mass of water. N_A is the Avogadro constant, and ζ is a random number uniformly distributed between zero and one. We forwarded the particle position by s along the momentum direction followed by a sampling of the interaction type based on the relative cross-section distribution $\frac{\sigma_i}{\sum \sigma_j}$.

If the sampled interaction i_0 was an ionization event, we then sampled the energy E_e and the emission angle of the ejected secondary electron, along with updating the kinetic energy of the primary particle. Noticing that the partial SDCS in Equation (1) had the form of $\frac{d\sigma_i^{ion}}{dw} \propto f(w)\phi(w)$, with $f(w) = (F_1(\nu) + wF_2(\nu))/(1+w)^3$, which was relatively easy to integrate, and $\phi(w) = 1/(1 + \exp[\alpha(w - w_i)/\nu])$, we took $f(w)$ as a sampling function and $\phi(w)$ as the rejection function to effectively sample E_e. Specifically, for a given proton energy E_p, ν can be solely determined based on Equation (11), and hence, $F_1(\nu)$ and $F_2(\nu)$ are just numbers. We wrote them as F_1 and F_2 for simplicity in the following equations. Applying the direct inversion method, we could sample w_s from $f(w)$ as:

$$w_s = (-F_1 + 2N_c\zeta + \sqrt{F_1^2 + 2F_2N_c\zeta - 2F_1N_c\zeta})/(F_1 + F_2 - 2N_c\zeta). \tag{23}$$

Here, $\zeta \in [0,\ 1)$ is a randomly sampled number and N_c has the form of:

$$N_c = \int_0^{w_m} f(w)dw = (w_m(F_2w_m + 2F_1 + F_1w_m))/(2(1+w_m)^2). \tag{24}$$

We repeatedly sampled w_s with Equation (23) and updated $\phi(w_s)$ until we obtained $\phi(w_s) > \zeta'$ with ζ' a random number $\in [0,\ 1/(1 + e^{-\alpha w_i/\nu})]$. Once reaching this stopping criterion, we accepted w_s and computed E_e as:

$$E_e = w_s * B_{i_0}. \tag{25}$$

The polar scattering angle θ_e of the electron relative to the moving direction of the primary particle satisfied $\cos\theta_e = \sqrt{\frac{E_e}{4*T}}$ for $E_e > B_{i_0}$ and was uniformly distributed between zero and π otherwise [40]. The azimuth scattering angle was uniformly sampled between zero and 2π. The residual energy of the primary particle after ionization was $E_k' = E_k - E_e - B_{i_0}$, and its polar scattering angle was zero.

If the sampled interaction i_0 belonged to the excitation category, there was no secondary electron emission, and we only needed to sample the energy loss W of the primary particle. In this case, the polar scattering angle for the primary particle was zero as well. When $E_p > 500$ keV, we sampled W based on Equation (15). Noticing that $\frac{d\sigma_{i_0}^{exc}}{dW} \propto f_{i_0}(W)g(W)$, where $g(W) = \frac{1}{W} ln(\frac{4T}{W})$, we then used $f_{i_0}(W)$ for the sampling of W and $g(W)$ for rejection. For the $\widetilde{A}^1B_1$ and $\widetilde{B}^1A_1$ channels, W_s can be directly sampled as:

$$W_s = w_{i_0} + \frac{1}{\sqrt{2\alpha_{i_0}}}\zeta_n, \tag{26}$$

where ζ_n is a random number following the standard normal distribution. As for plasma mode, we applied the direct inversion method to obtain W_s as:

$$W_s = w_{i_0} + \frac{1}{\alpha_{i_0}} \ln\left(\frac{u_1}{1-\alpha_{i_0} u_1 N_c \zeta} - 1\right). \tag{27}$$

Here, ζ is a random number $\in [0,\ 1)$ and N_c has the form of:

$$N_c = \int_{W_{min}}^{W_{max}} f_{i_0}(W)dW = \frac{1}{\alpha_{i_0}}\left(\frac{1}{u_1} - \frac{1}{u_2}\right), \tag{28}$$

with $u_1 = 1 + e^{\alpha_{i_0}(W_{min} - w_{i_0})}$ and $u_2 = 1 + e^{\alpha_{i_0}(W_{max} - w_{i_0})}$. We repeated the sampling of W_s and updating $g(W_s)$ until we obtained $g(W_s) > \zeta'$ with ζ' a random number $\in [0, \frac{1}{W_{min}} \ln\left(\frac{4T}{W_{min}}\right)]$. The residual energy of the primary particle after excitation was then $E_k' = E_k - W_s$. When $E_p \leq 500$ keV, the energy loss E_{i_0} was directly obtainable from the discrete energy spectrum [37]. The residual energy of the primary particle was then $E_k' = E_k - E_{i_0}$.

After transporting the primary particle with one step and simulating its interaction with one water molecule, we updated E_k with E_k' and started the next round of transport sampling until the kinetic energy of the primary particle reached the cutoff energy or ran out in the simulation region. During the process, all secondary electrons were stored in a global stack to be further simulated using our previously developed kernel in charge of electron transport, the physics models that covered the electron spectrum as low as a few eV [9]. The ionized and excited water molecules were also tagged for further analysis.

2.3.2. Concurrent Transport

In the concurrent transport picture, we simulated the reactions among radicals and DNA and the diffusion of the radicals in a step-by-step manner. Considering the complex structure of DNA and the possibly different checking frequencies for radical–DNA interactions and radical–radical reactions depending on the value of t_i, we utilized two GPU kernels for the chemical stage transport in the presence of DNA. One GPU kernel was responsible for the interactions between radicals and DNA, and the other kernel was in charge of the radical–radical reactions and the diffusion of the radicals.

For the GPU kernel managing the reactions between radicals and DNA, each GPU thread was responsible for one radical. To obtain the possible reaction and reaction type between the radical and DNA, we needed to search the DNA geometry and compute the distances from the radical to the centers of the DNA basic structures (DNA base, sugar-phosphate group, and histone protein). The smallest distance d_{min} was then compared to the reaction range of $R + R_c$ with R the radius of that DNA structure. $R_c = k/4\pi N_A D$ for reactions between the radical and DNA base or sugar-phosphate group, where k is the reaction rate, N_A is the Avogadro constant, and D is the diffusion rate for the radical. For all considered radical species, R_c was < 1 nm. Due to the lack of experimental data for the reaction between radicals and the histone protein, we assumed that the radical was only absorbed when it hit the histone, and hence, we set $R_c = 0$ for this case. If $d_{min} < R + R_c$, a reaction was recorded. Otherwise, the Brownian bridge method [9] was applied to compensate for possible reactions between the radical and DNA during the diffusion. As our DNA model was constructed with a huge amount of basic structures, it would be too time consuming to search the entire space to obtain the smallest distance from the radical to the DNA. To reduce the searching burden, we relied on the voxelized geometry of the DNA model and only performed the search at most on two voxels. Specifically, noticing that the outer boundary of the DNA chain was > 2 nm away from all edges of the voxel it occupied [9] and $R_c < 1$ nm for all reactions between the radical and DNA, this indicated that a radical could only react with those DNA structures in two special voxels: the current voxel in which it was located and the adjacent voxel having the surface closest to it. The latter voxel was not considered unless the radical was < 2 nm from its closest surface. In this way, we reduced the searching burden significantly. Once a reaction was recorded, the radical was removed from the reactant list. If the reaction was with the

DNA base or sugar-phosphate group, the reaction site was stored in a global stack for further analysis.

For the radical–radical reactions and radical diffusion, we continued to employ the GPU kernel developed in our previous work [9]. Each thread was in charge of one radical. To reduce the searching burden for mutual reactions, the entire space was divided into small grids with the grid size twice the largest reaction radius. This ensured that each radical only reacted with other radicals in the same or adjacent grids. The distances from the radical to other radicals were then computed and compared to the reaction radii to obtain whether a reaction would occur. If a reaction happened, the new products were placed, and radical–radical reactions were checked again. Otherwise, the radical was diffused by one step followed by the check for radical–radical reactions based on the Brownian bridge method.

At the beginning of the chemical stage, the GPU kernel for the DNA–radical reactions was executed to remove the radicals within the chromatin region from the subsequent chemical-stage simulation. This was followed by the launch of the GPU kernel in charge of the radical–radical reaction and radical diffusion. After that, we compared t_{elap}, the time elapsed from the last execution of the former GPU kernel, to t_i. If $t_{\text{elap}} \geq t_i$, the former and latter kernels were called in sequence. Otherwise, only the latter kernel was executed. The process was repeated until reaching the end of the chemical stage.

2.4. Simulation Setup

2.4.1. Simulation Setup for the Transport of Protons and Heavy Ions

We performed a series of simulations to validate the physical-stage transport for protons and heavy ions. These included: (1) the computations of the cross-section, linear energy transfer (LET), and traveling range; (2) the validation of the energy spectrum of secondary electrons, the radial dose distribution, and the track structure; and (3) the evaluation of the DNA damage spectrum.

We first calculated the total cross-section for both the ionization and excitation channels according to Equations (8) and (18), under the relativistic formats. The results were compared to Plante et al.'s [34] and Dingfelder et al.'s [35] works, as shown in Figure 1. Based on this, we calculated the track-length-averaged unrestricted LET for different ion species with their energy ranging from $0.1 \sim 10^4$ MeV amu^{-1}. For an ion with energy E_k, we sampled the energy loss of primary particles ε_i and the free-fly distance s_i. We then repeated the simulation $N = 10^5$ times and computed the length-averaged unrestricted LET as:

$$LET = \sum_{i=1}^{N} \frac{\varepsilon_i}{s_i} \cdot \frac{s_i}{\sum_{j=1}^{N} s_j} = \frac{\sum_{i=1}^{N} \varepsilon_i}{\sum_{j=1}^{N} s_j}. \tag{29}$$

We compared the LETs to those reported by Plante et al. [34]. After that, we simulated the proton range by tracking its starting and ending positions for a proton energy of $0.1 \sim 100$ MeV and compared this to the NIST data. We show both results in Figure 2.

As for the validation of the energy spectrum of secondary electrons, we simulated the interactions of a 5 MeV proton and 4 MeV alpha particles with a liquid water target, recorded the energy of the secondary electrons, and compared it to those obtained with the GEANT4-DNA simulation [43] (GEANT4 Version 10.5.1). The result is plotted in Figure 3. As for the radial dose distribution, we transported 5 and 10 MeV protons within a water slab of 10 μm in thickness and infinitely long at the other two dimensions and analyzed the dose distribution within a thin slice 4 to 6 μm away from the proton starting point along the thickness direction. We accounted for the dose distributed in an annular ring with inner and outer radii of r and $r + \Delta r$ as the dose at radius r. We set Δr = 1 nm, the same as that used in Wang et al.'s work [44]. We repeated the simulation 10^5 times, averaged the obtained radial dose, and compared it with that reported in Wang et al.'s work [44]. Finally, we show a representative physical track structure for a 5 MeV proton in Figure 5, including the track for both the primary proton and secondary electrons.

We used the lymphocyte nucleus model developed in our previous work for this evaluation study [9]. We initiated a proton emission plane of 11 × 11 μm^2 and 5.5 μm away from the center of the cell nucleus for two proton energies, 0.5 and 0.9 MeV. For each energy, we randomly sampled the proton position at the emission plane and its momentum towards the positive z direction, transported the proton until reaching a cutoff energy of 1 keV, and recorded the dose inside the cell nucleus. We repeated the simulation until reaching the accumulated dose of 2 Gy. After that, we simulated the physio-chemical and chemical stage with a chemical stage duration of t_c = 1 ns. We then applied the overlay method to obtain the DNA damage sites and grouped them into DNA Single-Strand Breaks (SSBs) and Double-Strand Breaks (DSBs) [29]. The result was compared to Nikjoo et al.'s work with the KURBUC model [45] and shown in Table 4.

2.4.2. Simulation Setup for Concurrent Transport

We studied the impact of t_i on the radical yields. We simulated the cases with a chemical stage duration of $t_c = 1$, 10 ns, and 1 μs and t_i from 1 ps to 1 μs with an increment of one at the logarithmic ten scale. Again, the lymphocyte cell nucleus with a radius of 5.5 μm was used as the Region Of Interest (ROI). As for the radical yield, we transported a 4.5 keV electron with its initial position randomly sampled inside the ROI and its direction towards the ROI center. We then took the generated radicals as inputs for the chemical-stage simulation. The final G values for the e_h, $OH\cdot$, $H\cdot$, and H_2O_2 radicals were recorded. We repeated the simulation 100 times to reduce the statistical uncertainties and reported the averaged G values over all the simulations. The results are shown in Figure 6.

We also computed the DNA damage as a function of the incident proton energy and the chemical stage duration under the concurrent DNA transport frame. A proton energy E_k of 0.5, 0.6, 0.8, 1.0, 1.5, 2, 5 10, 20, and 50 MeV and a chemical stage duration t_c of 1, 2.5, and 10 ns were considered, following the parameters used in Zhu et al.'s work [42]. We initiated the proton on a spherical shell with a radius of 5.5 μm and shot it randomly towards the inner space of the sphere. We repeated the simulation until having the accumulated dose in the sphere of 1 Gy. We then simulated the chemical stage with DNA concurrent transport ($t_i = 1$ ps) and computed the total DSB yield. For each proton energy, using the DSB yield at $t_c = 1\ ns$ as a reference, we defined $R(t) = N_{DSB}(t_c = t)/N_{DSB}(t_c = 1\ ns)$ to represent the relative DSB yields at $t_c = t$. For each pair of E_k and t_c, we ran the simulation 20 times and computed the mean and standard deviation for the relative DSB yield. We then compared the data with $t_c = 2.5$ ns (R(2.5)) and 10 ns (R(10)) to Zhu et al.'s work [42] and show the results in Figure 7.

3. Results

3.1. Validation of Development for Protons and Heavy Ions

Figure 1 presents the total and partial cross-sections for ionization and excitation as a function of incident proton energy. From Figure 1a, the total cross-section for ionization from our simulation agreed well with that from Plante et al.'s work [34]. From Figure 1b, for a proton energy > 500 keV, our simulated total cross-section for excitation matched that from Plante et al.'s work [34]. As for the slow proton, it followed that from Dingfelder et al.'s work [35]. The results revealed that the ionization model and the two-stage excitation model were successfully implemented as expected.

In Figure 1b, we noticed a dramatic drop-off of the total excitation cross-section at around 10 keV for the Plante model. This is due to the cross-section formula shown in Equation (15) depending on the scaled energy $T = \frac{m}{M}E_k$. When E_k drops below 10 keV, T is too small to excite even the lowest excitation channel ($j = \widetilde{A}^1B_1$). After replacing it with Dingfelder's model (Equation (19)) at the low-energy region, the excitation cross-section drops much more slowly. Considering that the low-energy proton largely appears after the Bragg peak, a proper excitation model could be important for the distal dose computation in proton therapy.

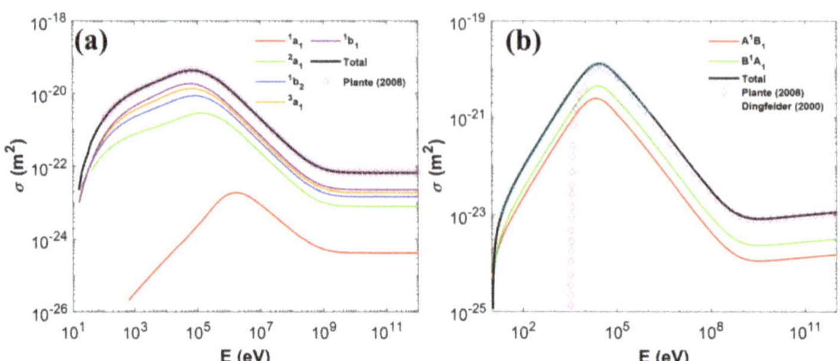

Figure 1. Total and partial cross-sections of (**a**) ionization and (**b**) excitation channels for protons with different energies.

The calculated unrestricted LETs for different ions are plotted in Figure 2a. They agreed well with Plante et al.'s work for ions with an energy greater than 0.5 MeV per nucleon. At the low energy range, LETs from our simulation were lower than those from Plante et al.'s work, which can be explained by the different excitation models used in the two simulations. As shown in Figure 1b, the excitation cross-section from our work was higher than that from Plante et al.'s work at the low energy range, resulting in a higher sampling rate of excitation interactions in our simulation. Considering that the energy loss from an excitation event was typically smaller than that from an ionization event (Table 1), a higher sampling of excitation could result in a lower LET. In Figure 2b, we show the proton range from our simulation and its comparison with the NIST data. As is shown, our simulation result matched well with the NIST PSTAR data (https://physics.nist.gov/PhysRefData/Star/Text/PSTAR.html (accessed on 17 June 2021)), with the relative difference smaller than 1%.

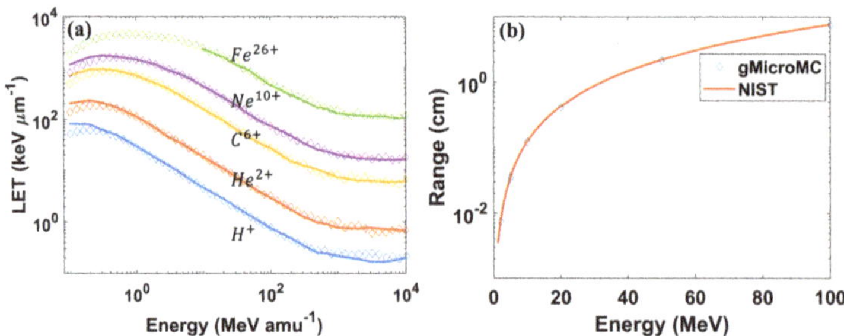

Figure 2. (**a**) The unrestricted LETs for different ions with different energies. The unit amu^{-1} means per nucleon. Solid lines represent data extracted from Plante et al.'s work, while data with diamond symbols are from our simulation with gMicroMC. (**b**) The simulated proton range for different energies.

Figure 3 shows the energy spectrum of secondary electrons generated from a 5 MeV proton and a 4 MeV alpha particle. From the figure, the yielding rates of secondary electrons dropped quickly along with the increase of the electron energy. For the entire plot, our simulated results agreed well with that from GEANT4-DNA. We did not compare the spectrum for electron energy greater than 200 eV due to a too low yielding rate and the consequent large uncertainty.

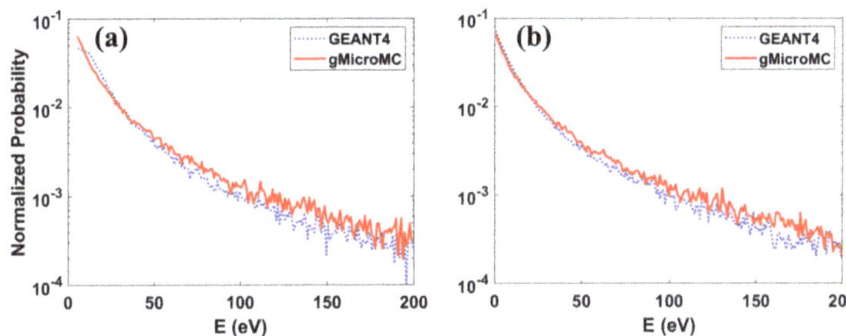

Figure 3. Secondary electron spectrum for (**a**) a 5 MeV proton and (**b**) a 4 MeV alpha particle.

In Figure 4, we see the radial dose distributions for 10 and 50 MeV protons from ours and Wang et al.'s work (Equations (1)–(7) in [44]) under the same setup. As is shown, the two curves matched quite well, although the curve from our simulation suffered a relatively large statistical fluctuation for the regions > 1000 nm from the primary track axis. Figures 3 and 4 together validated the energy spectrum and angular distribution of secondary electrons from our simulation, furthering proving the successful implementation of the transport models for protons and heavy ions.

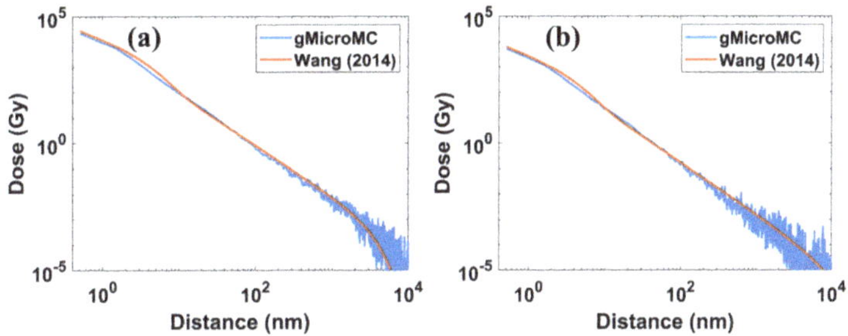

Figure 4. Radial dose distributions for (**a**) 10 MeV and (**b**) 50 MeV protons.

We then present a track structure for a 5 MeV primary proton and its produced secondary electrons in liquid water in Figure 5. For simplicity, we only present the entrance (Figure 5a) and Bragg peak (Figure 5b) regions. At the entrance region, the secondary electron tracks were quite sparse. In contrast, they were much denser in the Bragg peak region. In addition, the electron track lengths were shorter in the Bragg peak region. This was mainly due to the kinetic energy of the proton being much higher at the entrance than the Bragg peak region. This led to a smaller total cross-section and a longer interval between the production of secondary electrons. Plus, high-energy electrons (Equation (23)) would be favored when the proton energy is high. In general, most of the electrons travel a tiny distance before being locally deposited, forming the dense blue area around the central proton line, and hence a high radial dose distribution in the regions with small radii (Figure 4).

Finally, in Table 4, we report the DNA damages in the form of DSBs induced by 0.5 and 0.9 MeV protons. The results from our simulation were compared with those from Nikjoo et al.'s work with the KURBUC model [45]. The difference was found within 10%.

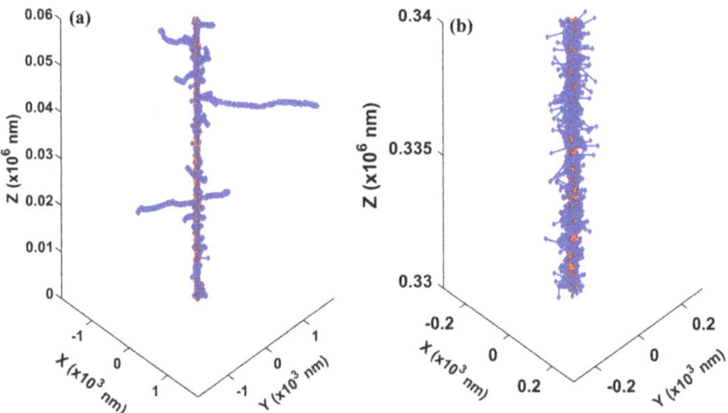

Figure 5. A representative track structure for a 5 MeV proton at the entrance part (**a**) and in the Bragg peak region (**b**). The proton was emitted along the positive Z direction. Red and blue dots represent the energy depositions by the proton and secondary electrons, respectively. Note: in the two subplots, we kept the same aspect ratio between the z and x/y axes, but plotted them with different ranges.

Table 4. The DSB yields (number per Gy per Gbp) obtained under the overlay method for two proton energies.

Energy (MeV)	from gMicroMC	from Nikjoo's Work
0.9	20.1	18.2
0.5	25.1	23.9

3.2. Validation of Concurrent Transport

As for the validation of the concurrent DNA transport module, we first studied the influence of different checking time intervals t_i and chemical stage durations t_c on the yields of different radicals. As shown in Figure 6, at a fixed t_i, the yields of the e_h and $OH\cdot$ radicals reduced when t_c increased. This was because longer t_c enabled more reactions among radicals and DNA. For the e_h and $OH\cdot$ radicals, these reactions were mainly consumption channels, resulting in a reduced yield rate when t_c increased. In contrast, although the presence of DNA also consumed $H\cdot$ and H_2O_2 radicals, reactions among radicals could contribute positively to the yields of these two radicals. Hence, the production of the $H\cdot$ and H_2O_2 radicals could be dependent on t_c in a more complex way. In addition, at a fixed t_c, varying t_i from 1 ps to t_c transformed the simulation from the concurrent method to the overlay method. All lines were connected smoothly, and the G value with $t_i = t_c$ matched with that in our previous publication under the overlay method, indicating the self-consistency of the concurrent DNA transport in gMicroMC.

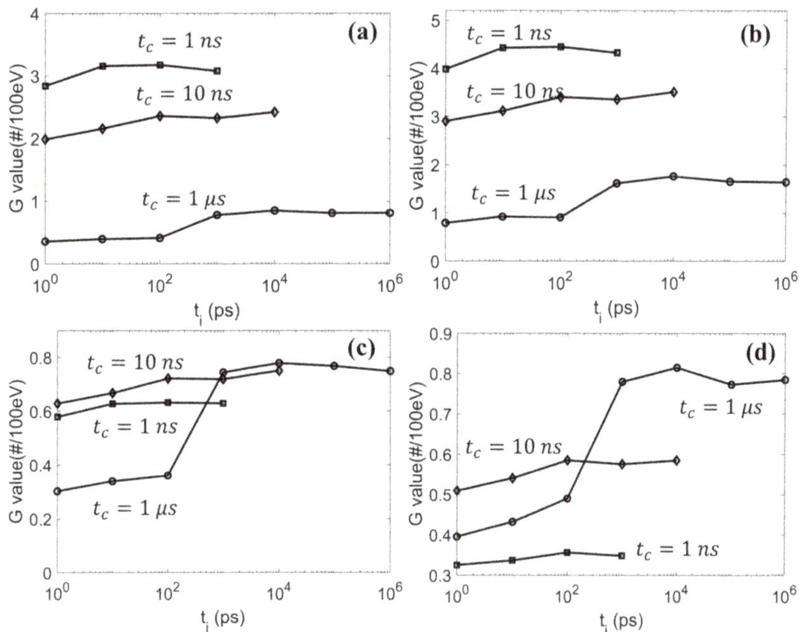

Figure 6. The yields of (**a**) e_h, (**b**) $OH\cdot$, (**c**) $H\cdot$, and (**d**) H_2O_2 chemical species at different checking time intervals t_i and chemical stage durations t_c.

After examining the self-consistency of the developed concurrent DNA transport method, we comprehensively studied the DSB yields as a function of proton energy E_k and chemical stage duration t_c. The results were compared to Zhu et al.'s work [42] and are shown in Figure 7. From the figure, all data points had relative DSB yields >1, and the R(10) values were larger than R(2.5) for the same E_k. This indicated the DSB yields increased when the chemical stage expanded from 1 ns to 10 ns under the concurrent transport frame. The reason was that the longer the chemical stage lasted, the more checks between radicals and DNA were performed, and hence, more DSBs could be formed. Along with the increase of the proton energy, the relative DSB yields exhibited a maximum in the middle energy range. Comparing the data from our simulation to that from Zhu et al.'s work, the trends generally agreed, especially for the R(2.5) data. Some larger discrepancies existed for the R(10) values, which could be explained partially by the different radical diffusion rates and different DNA geometries applied in the two works. For example, the diffusion rate of the $OH\cdot$ radical was larger in our package. This could make $OH\cdot$ diffuse longer and experience a higher scavenging rate from the histone protein within one diffusion step. In addition, a larger diffusion rate could result in a smaller reaction radius between $OH\cdot$ and the DNA sugar-phosphate moiety. Both led to a reduced DSB yield. The longer the t_c was, the more reduction effect it could create, such that we would obtain a smaller relative DSB yield than that from Zhu et al.'s work for the R(10) data than for the R(2.5) data.

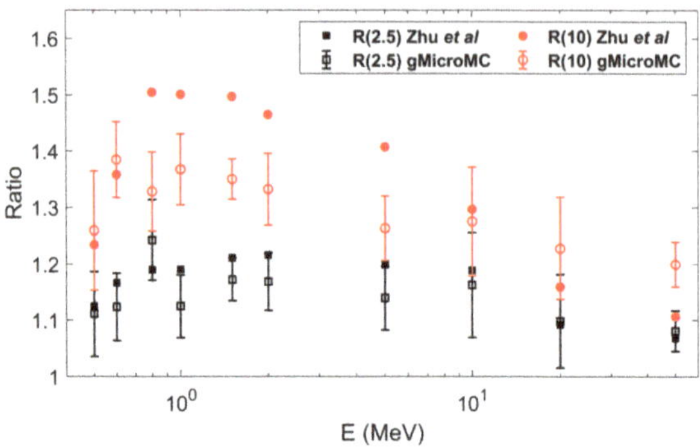

Figure 7. The relative DSB yields at different chemical stage durations t_c and different proton energies with $R(t) = \frac{DSB(t_c=t)}{DSB(t_c=1\ ns)}$. The data from gMicroMC simulation were compared to that from Zhu et al.'s work [42].

3.3. Computational Efficiency

After evaluating the two newly developed features of gMicroMC by comparing to the NIST data or simulation results from other packages, it was important to evaluate the time performance of the new modules for practical applications. In Table 5, we show the simulation time for the physical transport of 1, 10, and 100 protons at 1 and 10 MeV with gMicroMC on a single GPU card (Nvidia V100). As can be seen, it only took 2 and 4 s to transport a single proton with an initial kinetic energy of 1 and 10 MeV. In contrast, it could take multiple hundreds to thousands of seconds to perform similar simulations with single-CPU-based packages, showing the high efficiency of gMicroMC. It is also important to point out that when raising the proton numbers from one to one-hundred, the simulation time only increased by <5 and 10 folds for 1 and 10 MeV protons, respectively. This feature was against the linearly increasing behavior for typical CPU-based simulations, making gMicroMC especially suitable for large-scale simulations. Actually, when the proton number was small, the parallel computing capacity of the GPU was far from being exhausted when launching the kernel for the primary particle transport, such that increasing the proton number, the running time would not significantly increase.

Table 5. The simulation time (s) of physically transporting 1, 10, and 100 protons for proton energies at 1 and 10 MeV by gMicroMC on a single GPU card.

Energy (MeV)	Number of Primary Protons		
	1	10	100
1	1.9	3.1	9.3
10	3.9	9.8	40.5

As for the simulation time of the concurrent transport ($t_i = 1$ ps) in the chemical stage, it could be affected by many parameters, such as the number of radicals, the chemical stage duration, and the DNA complexity. Here, we reported the simulation time for a representative case. Considering that the number of radicals produced from one a 100 keV electron or a 1 MeV proton was roughly 10^5 and the longest chemical stage duration used in the overlay method was ~1 μs, we set the initial number of radicals $N_i = 10^5$ and chemical stage duration $t_c = 1$ μs in our simulation. We also included our DNA model containing of ~6.2×10^9 bps for a human lymphocyte cell nucleus in the simulation, the complexity of which is high. The simulation time was found to be 470 s with gMicroMC on a single GPU. Compared to the simulation time of 31 s under the overlay scheme for gMicroMC, the concurrent transport frame was still quite efficient since the simulation became much more sophisticated with the presence of DNA. This indicated that gMicroMC can be applied in simulations with realistic settings. In comparison, restrictions on the reaction region and time duration, etc., are typically required in other packages for memory- or time-saving purpose [22,23].

4. Discussion

This is a continuous development work on gMicroMC. In this work, we successfully implemented the physical transport for proton and heavy ions and the concurrent transport of radicals and DNA in the chemical stage. For the former implementation, we considered the ionization, excitation, and charge effect during the transport and performed a series of case studies to validate the development. The obtained results matched well with the literature reports. We then computed the DNA DSBs for two proton energies, and the results agreed with those computed with the KURBUC model in Nikjoo et al.'s work [45] within 10%. As for the latter, we considered the interaction of radicals with the DNA, histone protein, base, and sugar-phosphate groups during the chemical transport. To validate it, we first performed a self-consistency check for the evolution of chemical radicals and induced DNA DSBs under different checking frequencies for radical–DNA interactions. The result showed the high self-consistency of the developed module. We then performed a comprehensive study of the DSB yields as a function of the chemical stage duration and incident particle energies under the DNA concurrent transport frame. The results generally followed that from Zhu et al.'s work. The use of the GPU made the code have high computational efficiency. Running gMicroMC on a single GPU card, it took only 41 s to transport 100 protons with a kinetic energy of 10 MeV and around 8 min to transport 10^5 radicals up to 1 μs with the presence of a DNA model containing $\sim 6.2 \times 10^9$ base pairs. The high computational performance of gMicroMC makes it suitable to simulate complex radiation scenarios such as proton FLASH radiotherapy, which is our next work. To benefit the community, we are also working on releasing the newly developed features of gMicroMC on GitHub (https://github.com/utaresearch/gMicroMC (accessed on 17 June 2021)).

Despite the above success, there are also some limitations to our current development. In the physical transport of protons and heavy ions, we ignored the nuclear inelastic interaction and the nuclear and electromagnetic elastic scattering. Among them, the nuclear inelastic interaction can fragment the target and/or projectile nuclei, which is a main factor to remove primaries from the projectile beam. However, due to the complexity of the fragmentation process and its products, this process is typically not directly included in any mechanistic MC simulation tools at this moment [46]. Since this work mainly focused on the novel GPU implementation of existing models, it will be our next work for a possible inclusion of the nuclear inelastic scattering process. As for the elastic scattering, it could change the direction of the primaries, hence affecting the track structure and radical dose distributions. However, elastic scattering mainly dominants interactions between the proton and water medium at a very low energy range, and the scattering angle is typically small [46,47]. Hence, we expect it will not affect the accuracy of the DNA damage computation greatly. Considering the powerful computational capacity of the GPU, it is promising to consider a more complete modeling of the physical interactions between ions and water molecules, making gMicroMC versatile for advanced applications.

It is also worth pointing out that we applied a low-energy five-pathway model and a high-energy three-pathway model to simulate the proton-induced excitation of a water molecule such that both very-low-energy and relativistic situations could be covered. Yet, this could cause a concern that some excitation pathways could be discontinuous at the model switching point. A previous study showed that the low-energy model could be extended up to 80 MeV with some proper parameter fitting [34]. We hence made it an option in our package to only apply Equation (19) to model the excitation process up to 80 MeV. In addition, for the two-model picture, although we currently set the model switching point at 500 keV to distinguish the slow and fast proton behavior following the same logic as stated in [35], it deserves further study to investigate its impact on the subsequent radical yielding process.

Another important procedure that could affect the computational accuracy of the proton- and heavy ion-induced DNA damage is the modeling of the secondary electron transport, especially for the low-energy electrons (sub-keV range). Previous studies revealed their critical importance in determining the initial distribution of radicals and the consequent DNA damage patterns [3]. However, due to the lack of sufficient experimental data, uncertainties about the simulation results could be introduced by the inaccurate modeling of this process. For instance, with different models adopted in gMicroMC and Geant4-DNA, the maximal discrepancies of the track length and penetration depth for electrons below 1 keV computed by the two packages were around 10 and 4 nm, respectively [9]. In recent years, there has been much effort contributing to improving the accuracy of describing the low-energy electron transport process [11–15,48]. However, more efforts are required to fully elucidate this problem.

In addition, as discussed in our previous study [29], the cross-section used to simulate the ionization and excitation processes from electrons could significantly impact the accuracy of the finally obtained DSB yields. In the case of protons and heavy ions, due to the lack of experimental data, the cross-sections and models could also be associated with large uncertainties, causing concerns about the robustness of the simulation results. To reduce these uncertainties, there have been multiple experiments and models [13,49–56] developed in a more elaborate fashion. Yet, more studies are needed to more thoroughly understand the relevant processes.

In our previous study [29] with the overlay method, the DSB yields reduced when the chemical stage duration enlarged, which was against the trend obtained in this work with the concurrent DNA transport method (Figure 7). This was due to the fact that in the overlay method, the radical–DNA interaction was only simulated after the chemical-stage simulation. The longer the chemical stage lasted, the more the $OH\cdot$ radicals were consumed during the mutual radical reactions, and the fewer the DSBs could be formed. Nonetheless, the controversial behavior between the concurrent and overlay frames reminds us to carefully check the parameters used in our simulation. One such parameter is the scavenging probability from histone proteins to chemical radicals, the value of which has not been well regulated by current experiments. We performed a new simulation to study its impact on DSB yield by gradually reducing the scavenging probability P_s from one to zero. Here, $P_s = 1$ means the total scavenging probability once radicals hit the histone proteins. Taking the DSB yields at $P_s = 1$ and $t_c = 1\ ns$ as a reference, we computed the relative DSB yields at other P_ss and t_cs. The results are shown in Figure 8. Clearly, DSB yields increased when P_s decreased. However, even under the same P_s, when t_c differed, P_s could have different impacts on the relative DSB yield. For instance, the relative DSB yield was 1.4 for $t_c = 1\ ns$, while it was 3.1 for t_c = 1 μs when $P_s = 0$. This indicated that to make the simulation tool robust for various applications, we need to apply a proper cutoff to t_c and a detailed coordination of multiple parameters used in the chemical-stage simulation. This should be considered in future studies.

In our development of the concurrent transport module, we used a complete set of cellular DNA at the base-pair resolution to simulate the radical–DNA interactions. Previous studies have incorporated other DNA models, including the simple cylinder DNA [3], other cellular DNAs [7,42,57], and the DNA model at the atomic resolution [58,59]. Due to the different simulation setups and different DNA structures adopted, the absolute DSB yields from different studies were typically non-comparable [60]. However, there were some common trends shared among different studies. For example, the DSB yields were found increasing and then decreasing when the LETs increased. The maximal yields of DSB occurred at the LET value around 60 keV/μm in [57] and around 27.2 keV/μm (the LET value for 1 MeV protons [22]) for gMicroMC and TOPAS-nBio, respectively. These consistencies motivated further studies with the concurrent simulation scheme while the expense was the lowered simulation efficiency. Our adoption of the GPU-based acceleration could shed light on this issue based on experiences from previous studies [9,27,28,61].

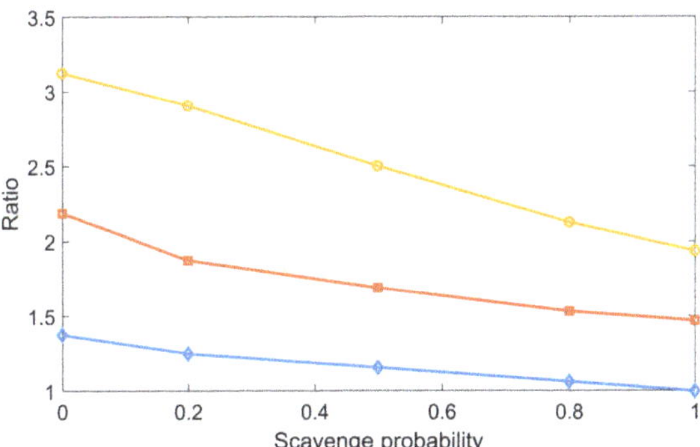

Figure 8. Ratio of DSB yields with different scavenge probabilities.

Finally, we performed a further study on the effect of t_i, which was introduced to balance the simulation efficiency and accuracy. In the *Results* Section, we showed the obvious impact of t_i on the radical evolution. It would be interesting to see how it could consequently affect the final DNA damage. As the DSB is widely accepted as the most important factor in cell death, it is thus reasonable to use the DSB as a metric to evaluate the impact of t_i. We initiated a 4.5 keV electron with its position randomly sampled inside a sphere of radius 6.1 µm and its direction following the isotropic distribution. The sphere was concentric with the cell nucleus of our DNA model. We repeated the simulation until the accumulated dose inside the cell nucleus region reached 1 Gy, equivalent to simulating ~ 2000 electrons. The generated radicals were then transported in the chemical stage along with considering the radical–DNA reactions. We calculated the resulting DNA DSBs as a function of t_i and show these in Figure 9. For the three t_c studied, the DSB lines showed a similar trend when t_i increased. The maximal DSB was obtained at $t_i = 10$ ps. The result could be interpreted as follows. At the beginning of the chemical stage, all generated radicals had a relatively dense distribution. When t_i slightly increased, the $OH\cdot$ radical could diffuse a longer distance away from its initial position before it reacted with the DNA, while its reaction probability with other radicals was not greatly affected. Hence, a sparser, but equivalent (or slightly reduced) number of DNA damage sites could be formed, which could lead to more generations of simple DSBs (composed of two damage sites) rather than the DSB+ (composed of multiple damage sites). In this way, the total DSB yields could increase. However, along with the further increase of t_i, the checking frequency for radical–radical reactions would be much higher than that for radical–DNA reactions. This could lead to a higher consumption of $OH\cdot$ radicals through radical–radical reactions than through DNA damaging reactions, which resulted in a reduced DSB yield.

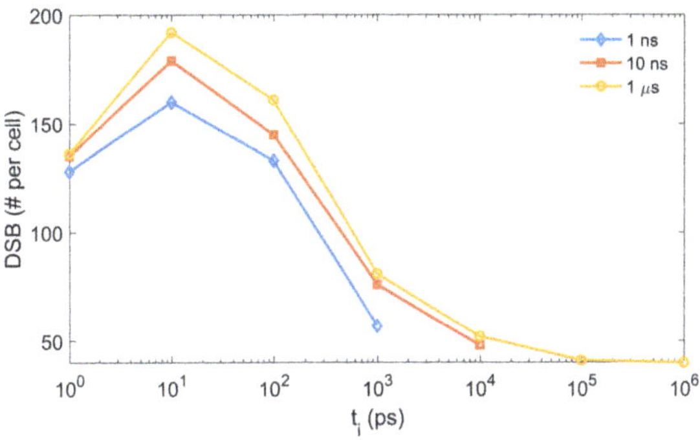

Figure 9. The yields of DSB at different t_i and t_c from the gMicroMC simulation.

5. Conclusions

We successfully developed and validated two new features in gMicroMC, the transport of protons and heavy ions in the physical stage and the concurrent transport of DNA in the chemical stage. We implemented the two features on a GPU parallel computing platform, resulting in a remarkable time performance. The physical transport of 100 protons with an initial kinetic energy of 10 MeV could be finished in s. The chemical simulation with concurrent DNA transport was far more complex, but it still only took a few minutes to run a representative case. The two newly developed features in gMicroMC that had both high accuracy and efficiency gave gMicroMC the high promise of solving large-scale problems in active radiation research areas.

Author Contributions: Conceptualization, X.J. and Y.C.; methodology, X.J. and Y.C.; software, Y.L.; validation, Y.L. and Y.C.; formal analysis, Y.L. and Y.C.; investigation, Y.L.; resources, Y.C.; data curation, Y.L., X.J., and Y.C.; writing—original draft preparation, Y.L.; writing—review and editing, Y.L., X.J., and Y.C.; visualization, Y.L.; supervision, X.J. and Y.C.; project administration, X.J. and Y.C.; funding acquisition, X.J. and Y.C. All authors read and agreed to the published version of the manuscript.

Funding: This work was supported in part by the Cancer Prevention and Research Institute of Texas (CPRIT) Grant RP160661 and by the National Institutes of Health Grants R37CA214639 and R15CA256668.

Institutional Review Board Statement: Not applicable.

Informed Consent Statement: Not applicable.

Data Availability Statement: Not applicable.

Conflicts of Interest: The authors declare no conflict of interest.

Abbreviations

The following abbreviations are used in this manuscript:

GPU	Graphical Processing Unit
LET	Linear Energy Transfer
ROI	Region Of Interest
SDCS	Singly Differential Cross-Section
OER	Oxygen Enhancement Ratio
PSF	Phase Space File

References

1. Chatzipapas, K.P.; Papadimitroulas, P.; Emfietzoglou, D.; Kalospyros, S.A.; Hada, M.; Georgakilas, A.G.; Kagadis, G.C. Ionizing radiation and complex DNA damage: Quantifying the radiobiological damage using monte carlo simulations. *Cancers* **2020**, *12*, 799. [CrossRef]
2. Paretzke, H.G. *Radiation Track Structure Theory*; Wiley: Hoboken, NJ, USA, 1987.
3. Nikjoo, H.; O'Neill, P.; Goodhead, D.; Terrissol, M. Computational modeling of low-energy electron-induced DNA damage by early physical and chemical events. *Int. J. Radiat. Biol.* **1997**, *71*, 467–483. [CrossRef]
4. Cucinotta, F.A.; Wilson, J.; Townsend, L.; Shinn, J.; Katz, R. Track structure model for damage to mammalian cell cultures during solar proton events. *Int. J. Radiat. Appl. Instrum. Part D Nucl. Tracks Radiat. Meas.* **1992**, *20*, 177–184. [CrossRef]
5. Michalik, V.; Běgusova, M. Target model of nucleosome particle for track structure calculations and DNA damage modeling. *Int. J. Radiat. Biol.* **1994**, *66*, 267–277. [CrossRef] [PubMed]
6. Nikjoo, H.; Uehara, S.; Emfietzoglou, D.; Cucinotta, F. Track-structure codes in radiation research. *Radiat. Meas.* **2006**, *41*, 1052–1074. [CrossRef]
7. Friedland, W.; Dingfelder, M.; Kundrát, P.; Jacob, P. Track structures, DNA targets and radiation effects in the biophysical Monte Carlo simulation code PARTRAC. *Mutat. Res. Mol. Mech. Mutagen.* **2011**, *711*, 28–40. [CrossRef] [PubMed]
8. Bernal, M.; Bordage, M.; Brown, J.; Davídková, M.; Delage, E.; El Bitar, Z.; Enger, S.; Francis, Z.; Guatelli, S.; Ivanchenko, V.; et al. Track structure modeling in liquid water: A review of the Geant4-DNA very low energy extension of the Geant4 Monte Carlo simulation toolkit. *Phys. Med.* **2015**, *31*, 861–874. [CrossRef]

9. Tsai, M.Y.; Tian, Z.; Qin, N.; Yan, C.; Lai, Y.; Hung, S.H.; Chi, Y.; Jia, X. A new open-source GPU-based microscopic Monte Carlo simulation tool for the calculations of DNA damages caused by ionizing radiation—Part I: Core algorithm and validation. *Med. Phys.* **2020**, *47*, 1958–1970. [CrossRef] [PubMed]
10. Bordage, M.; Bordes, J.; Edel, S.; Terrissol, M.; Franceries, X.; Bardiès, M.; Lampe, N.; Incerti, S. Implementation of new physics models for low energy electrons in liquid water in Geant4-DNA. *Phys. Med.* **2016**, *32*, 1833–1840. [CrossRef]
11. Emfietzoglou, D.; Kyriakou, I.; Garcia-Molina, R.; Abril, I.; Nikjoo, H. Inelastic cross-sections for low-energy electrons in liquid water: Exchange and correlation effects. *Radiat. Res.* **2013**, *180*, 499–513. [CrossRef]
12. Kyriakou, I.; Incerti, S.; Francis, Z. Improvements in geant4 energy-loss model and the effect on low-energy electron transport in liquid water. *Med. Phys.* **2015**, *42*, 3870–3876. [CrossRef]
13. Dingfelder, M. Updated model for dielectric response function of liquid water. *Appl. Radiat. Isot.* **2014**, *83*, 142–147. [CrossRef] [PubMed]
14. Kai, T.; Yokoya, A.; Ukai, M.; Watanabe, R. Cross sections, stopping powers, and energy loss rates for rotational and phonon excitation processes in liquid water by electron impact. *Radiat. Phys. Chem.* **2015**, *108*, 13–17. [CrossRef]
15. Shin, W.G.; Bordage, M.C.; Emfietzoglou, D.; Kyriakou, I.; Sakata, D.; Min, C.; Lee, S.B.; Guatelli, S.; Incerti, S. Development of a new Geant4-DNA electron elastic scattering model for liquid-phase water using the ELSEPA code. *J. Appl. Phys.* **2018**, *124*, 224901. [CrossRef]
16. Pomplun, E. A new DNA target model for track structure calculations and its first application to I-125 Auger electrons. *Int. J. Radiat. Biol.* **1991**, *59*, 625–642. [CrossRef] [PubMed]
17. Kreth, G.; Finsterle, J.; Von Hase, J.; Cremer, M.; Cremer, C. Radial arrangement of chromosome territories in human cell nuclei: A computer model approach based on gene density indicates a probabilistic global positioning code. *Biophys. J.* **2004**, *86*, 2803–2812. [CrossRef]
18. Schuemann, J.; McNamara, A.; Ramos-Méndez, J.; Perl, J.; Held, K.; Paganetti, H.; Incerti, S.; Faddegon, B. TOPAS-nBio: An extension to the TOPAS simulation toolkit for cellular and sub-cellular radiobiology. *Radiat. Res.* **2019**, *191*, 125–138. [CrossRef]
19. Gray, L.H.; Conger, A.; Ebert, M.; Hornsey, S.; Scott, O. The concentration of oxygen dissolved in tissues at the time of irradiation as a factor in radiotherapy. *Br. J. Radiol.* **1953**, *26*, 638–648. [CrossRef]
20. Nair, C.K.; Parida, D.K.; Nomura, T. Radioprotectors in radiotherapy. *J. Radiat. Res.* **2001**, *42*, 21–37. [CrossRef]
21. Boscolo, D.; Krämer, M.; Fuss, M.C.; Durante, M.; Scifoni, E. Impact of target oxygenation on the chemical track evolution of ion and electron radiation. *Int. J. Mol. Sci.* **2020**, *21*, 424. [CrossRef]
22. Meylan, S.; Incerti, S.; Karamitros, M.; Tang, N.; Bueno, M.; Clairand, I.; Villagrasa, C. Simulation of early DNA damage after the irradiation of a fibroblast cell nucleus using Geant4-DNA. *Sci. Rep.* **2017**, *7*, 1–15. [CrossRef]
23. Lampe, N.; Karamitros, M.; Breton, V.; Brown, J.M.; Kyriakou, I.; Sakata, D.; Sarramia, D.; Incerti, S. Mechanistic DNA damage simulations in Geant4-DNA part 1: A parameter study in a simplified geometry. *Phys. Med.* **2018**, *48*, 135–145. [CrossRef] [PubMed]
24. Sakata, D.; Incerti, S.; Bordage, M.; Lampe, N.; Okada, S.; Emfietzoglou, D.; Kyriakou, I.; Murakami, K.; Sasaki, T.; Tran, H.; et al. An implementation of discrete electron transport models for gold in the Geant4 simulation toolkit. *J. Appl. Phys.* **2016**, *120*, 244901. [CrossRef]
25. Jia, X.; Ziegenhein, P.; Jiang, S.B. GPU-based high-performance computing for radiation therapy. *Phys. Med. Biol.* **2014**, *59*, R151. [CrossRef] [PubMed]
26. Després, P.; Jia, X. A review of GPU-based medical image reconstruction. *Phys. Med.* **2017**, *42*, 76–92. [CrossRef]
27. Tian, Z.; Jiang, S.B.; Jia, X. Accelerated Monte Carlo simulation on the chemical stage in water radiolysis using GPU. *Phys. Med. Biol.* **2017**, *62*, 3081. [CrossRef]
28. Lai, Y.; Jia, X.; Chi, Y. Modeling the effect of oxygen on the chemical stage of water radiolysis using GPU-based microscopic Monte Carlo simulations, with an application in FLASH radiotherapy. *Phys. Med. Biol.* **2021**, *66*, 025004. [CrossRef] [PubMed]
29. Lai, Y.; Tsai, M.Y.; Tian, Z.; Qin, N.; Yan, C.; Hung, S.H.; Chi, Y.; Jia, X. A new open-source GPU-based microscopic Monte Carlo simulation tool for the calculations of DNA damages caused by ionizing radiation—Part II: Sensitivity and uncertainty analysis. *Med. Phys.* **2020**, *47*, 1971–1982. [CrossRef]
30. Paganetti, H.; Niemierko, A.; Ancukiewicz, M.; Gerweck, L.E.; Goitein, M.; Loeffler, J.S.; Suit, H.D. Relative biological effectiveness (RBE) values for proton beam therapy. *Int. J. Radiat. Oncol. Biol. Phys.* **2002**, *53*, 407–421. [CrossRef]
31. Vitti, E.T.; Parsons, J.L. The radiobiological effects of proton beam therapy: Impact on DNA damage and repair. *Cancers* **2019**, *11*, 946. [CrossRef] [PubMed]
32. Reeves, K.G.; Kanai, Y. Electronic excitation dynamics in liquid water under proton irradiation. *Sci. Rep.* **2017**, *7*, 1–8. [CrossRef] [PubMed]
33. Rudd, M.E.; Kim, Y.K.; Madison, D.H.; Gay, T.J. Electron production in proton collisions with atoms and molecules: Energy distributions. *Rev. Mod. Phys.* **1992**, *64*, 441. [CrossRef]
34. Plante, I.; Cucinotta, F.A. Ionization and excitation cross-sections for the interaction of HZE particles in liquid water and application to Monte Carlo simulation of radiation tracks. *New J. Phys.* **2008**, *10*, 125020. [CrossRef]
35. Dingfelder, M.; Inokuti, M.; Paretzke, H.G. Inelastic-collision cross-sections of liquid water for interactions of energetic protons. *Radiat. Phys. Chem.* **2000**, *59*, 255–275. [CrossRef]
36. Booth, W.; Grant, I. The energy loss of oxygen and chlorine ions in solids. *Nucl. Phys.* **1965**, *63*, 481–495. [CrossRef]

37. Dingfelder, M.; Jorjishvili, I.; Gersh, J.; Toburen, L. Heavy ion track structure simulations in liquid water at relativistic energies. *Radiat. Prot. Dosim.* **2006**, *122*, 26–27. [CrossRef] [PubMed]
38. Rudd, M. Cross sections for production of secondary electrons by charged particles. *Radiat. Prot. Dosim.* **1990**, *31*, 17–22. [CrossRef]
39. Kutcher, G.; Green, A. A model for energy deposition in liquid water. *Radiat. Res.* **1976**, *67*, 408–425. [CrossRef]
40. Cobut, V.; Frongillo, Y.; Patau, J.; Goulet, T.; Fraser, M.; Jay-Gerin, J. Monte Carlo simulation of fast electron and proton tracks in liquid water-I. Physical and physicochemical aspects. *Radiat. Phys. Chem.* **1998**, *51*, 229–244.
41. Buxton, G.V.; Greenstock, C.L.; Helman, W.P.; Ross, A.B. Critical review of rate constants for reactions of hydrated electrons, hydrogen atoms and hydroxyl radicals (· OH/· O- in aqueous solution. *J. Phys. Chem. Ref. Data* **1988**, *17*, 513–886. [CrossRef]
42. Zhu, H.; McNamara, A.L.; Ramos-Mendez, J.; McMahon, S.J.; Henthorn, N.T.; Faddegon, B.; Held, K.D.; Perl, J.; Li, J.; Paganetti, H.; et al. A parameter sensitivity study for simulating DNA damage after proton irradiation using TOPAS-nBio. *Phys. Med. Biol.* **2020**, *65*, 085015. [CrossRef]
43. Incerti, S.; Baldacchino, G.; Bernal, M.; Capra, R.; Champion, C.; Francis, Z.; Guèye, P.; Mantero, A.; Mascialino, B.; Moretto, P.; et al. The geant4-dna project. *Int. J. Model. Simul. Sci. Comput.* **2010**, *1*, 157–178. [CrossRef]
44. Wang, H.; Vassiliev, O.N. Radial dose distributions from protons of therapeutic energies calculated with Geant4-DNA. *Phys. Med. Biol.* **2014**, *59*, 3657. [CrossRef]
45. Nikjoo, H.; O'Neill, P.; Wilson, W.; Goodhead, D. Computational approach for determining the spectrum of DNA damage induced by ionizing radiation. *Radiat. Res.* **2001**, *156*, 577–583. [CrossRef]
46. Nikjoo, H.; Emfietzoglou, D.; Liamsuwan, T.; Taleei, R.; Liljequist, D.; Uehara, S. Radiation track, DNA damage and response—A review. *Rep. Prog. Phys.* **2016**, *79*, 116601. [CrossRef] [PubMed]
47. Nikjoo, H.; Emfietzoglou, D.; Watanabe, R.; Uehara, S. Can Monte Carlo track structure codes reveal reaction mechanism in DNA damage and improve radiation therapy? *Radiat. Phys. Chem.* **2008**, *77*, 1270–1279. [CrossRef]
48. Lemelin, V.; Bass, A.; Cloutier, P.; Sanche, L. Low energy (1–19 eV) electron scattering from condensed thymidine (dT) I: Absolute vibrational excitation cross-sections. *Phys. Chem. Chem. Phys.* **2019**, *21*, 23808–23817. [CrossRef] [PubMed]
49. Emfietzoglou, D.; Papamichael, G.; Nikjoo, H. Monte Carlo electron track structure calculations in liquid water using a new model dielectric response function. *Radiat. Res.* **2017**, *188*, 355–368. [CrossRef]
50. Ramos-Méndez, J.; Shin, W.g.; Karamitros, M.; Domínguez-Kondo, J.; Tran, N.H.; Incerti, S.; Villagrasa, C.; Perrot, Y.; Štěpán, V.; Okada, S.; et al. Independent reaction times method in Geant4-DNA: Implementation and performance. *Med. Phys.* **2020**, *47*, 5919–5930. [CrossRef] [PubMed]
51. Liamsuwan, T.; Emfietzoglou, D.; Uehara, S.; Nikjoo, H. Microdosimetry of low-energy electrons. *Int. J. Radiat. Biol.* **2012**, *88*, 899–907. [CrossRef]
52. Shin, W.G.; Ramos-Mendez, J.; Faddegon, B.; Tran, H.; Villagrasa, C.; Perrot, Y.; Okada, S.; Karamitros, M.; Emfietzoglou, D.; Kyriakou, I.; et al. Evaluation of the influence of physical and chemical parameters on water radiolysis simulations under MeV electron irradiation using Geant4-DNA. *J. Appl. Phys.* **2019**, *126*, 114301. [CrossRef]
53. Nikjoo, H.; Uehara, S.; Emfietzoglou, D.; Brahme, A. Heavy charged particles in radiation biology and biophysics. *New J. Phys.* **2008**, *10*, 075006. [CrossRef]
54. Kyriakou, I.; Šefl, M.; Nourry, V.; Incerti, S. The impact of new Geant4-DNA cross-section models on electron track structure simulations in liquid water. *J. Appl. Phys.* **2016**, *119*, 194902. [CrossRef]
55. Incerti, S.; Kyriakou, I.; Bernal, M.; Bordage, M.C.; Francis, Z.; Guatelli, S.; Ivanchenko, V.; Karamitros, M.; Lampe, N.; Lee, S.B.; et al. Geant4-DNA example applications for track structure simulations in liquid water: A report from the Geant4-DNA Project. *Med. Phys.* **2018**, *45*, e722–e739. [CrossRef] [PubMed]
56. Matsuya, Y.; Kai, T.; Yoshii, Y.; Yachi, Y.; Naijo, S.; Date, H.; Sato, T. Modeling of yield estimation for DNA strand breaks based on Monte Carlo simulations of electron track structure in liquid water. *J. Appl. Phys.* **2019**, *126*, 124701. [CrossRef]
57. Friedland, W.; Schmitt, E.; Kundrát, P.; Dingfelder, M.; Baiocco, G.; Barbieri, S.; Ottolenghi, A. Comprehensive track-structure based evaluation of DNA damage by light ions from radiotherapy-relevant energies down to stopping. *Sci. Rep.* **2017**, *7*, 1–15. [CrossRef] [PubMed]
58. Aydogan, B.; Bolch, W.E.; Swarts, S.G.; Turner, J.E.; Marshall, D.T. Monte Carlo simulations of site-specific radical attack to DNA bases. *Radiat. Res.* **2008**, *169*, 223–231. [CrossRef]
59. Verlackt, C.; Neyts, E.; Jacob, T.; Fantauzzi, D.; Golkaram, M.; Shin, Y.; Van Duin, A.; Bogaerts, A. Atomic-scale insight into the interactions between hydroxyl radicals and DNA in solution using the ReaxFF reactive force field. *New J. Phys.* **2015**, *17*, 103005. [CrossRef]
60. Schuemann, J.; McNamara, A.; Warmenhoven, J.; Henthorn, N.; Kirkby, K.J.; Merchant, M.J.; Ingram, S.; Paganetti, H.; Held, K.D.; Ramos-Mendez, J.; et al. A new standard DNA damage (SDD) data format. *Radiat. Res.* **2019**, *191*, 76–92. [CrossRef]
61. Kalantzis, G.; Emfietzoglou, D.; Hadjidoukas, P. A unified spatio-temporal parallelization framework for accelerated Monte Carlo radiobiological modeling of electron tracks and subsequent radiation chemistry. *Comput. Phys. Commun.* **2012**, *183*, 1683–1695. [CrossRef]

International Journal of
Molecular Sciences

MDPI

Article

Simu-D: A Simulator-Descriptor Suite for Polymer-Based Systems under Extreme Conditions

Miguel Herranz, Daniel Martínez-Fernández , Pablo Miguel Ramos, Katerina Foteinopoulou, Nikos Ch. Karayiannis * and Manuel Laso *

Institute for Optoelectronic Systems and Microtechnology (ISOM) and Escuela Técnica Superior de Ingenieros Industriales (ETSII), Universidad Politécnica de Madrid (UPM), José Gutierrez Abascal 2, 28006 Madrid, Spain; miguel.herranzf@upm.es (M.H.); daniel.martinez.fernandez@upm.es (D.M.-F.); pm.ramos@alumnos.upm.es (P.M.R.); k.foteinopoulou@upm.es (K.F.)

* Correspondence: nkarayiannis@etsii.upm.es (N.C.K.); manuel.laso@upm.es (M.L.); Tel.: +34-910677318 (N.C.K.); +34-910677319 (M.L.)

Abstract: We present Simu-D, a software suite for the simulation and successive identification of local structures of atomistic systems, based on polymers, under extreme conditions, in the bulk, on surfaces, and at interfaces. The protocol is built around various types of Monte Carlo algorithms, which include localized, chain-connectivity-altering, identity-exchange, and cluster-based moves. The approach focuses on alleviating one of the main disadvantages of Monte Carlo algorithms, which is the general applicability under a wide range of conditions. Present applications include polymer-based nanocomposites with nanofillers in the form of cylinders and spheres of varied concentration and size, extremely confined and maximally packed assemblies in two and three dimensions, and terminally grafted macromolecules. The main simulator is accompanied by a descriptor that identifies the similarity of computer-generated configurations with respect to reference crystals in two or three dimensions. The Simu-D simulator-descriptor can be an especially useful tool in the modeling studies of the entropy- and energy-driven phase transition, adsorption, and self-organization of polymer-based systems under a variety of conditions.

Keywords: Monte Carlo; atomistic simulation; molecular simulation; hard sphere; extreme conditions; confinement; nanocomposites; cluster; crystallization; atomic structure; packing; semi-flexible polymers; order parameter

check for updates

Citation: Herranz, M.; Martínez-Fernández, D.; Ramos, P.M.; Foteinopoulou, K.; Karayiannis, N.C.; Laso, M. Simu-D: A Simulator-Descriptor Suite for Polymer-Based Systems under Extreme Conditions. *Int. J. Mol. Sci.* **2021**, *22*, 12464. https://doi.org/10.3390/ijms222212464

Academic Editor: Małgorzata Borówko

Received: 5 October 2021
Accepted: 12 November 2021
Published: 18 November 2021

1. Introduction

The development of new materials with enhanced properties is one of the most interesting and important topics in research in materials science and engineering. To achieve this ambitious goal, one has to relate the behavior of atoms and molecules to the macroscopic properties of the end material. In this perspective, molecular simulation is of paramount importance, since it allows the study of materials at the atomistic/molecular level without needing an experimental process, which, in specific cases, can become expensive, time consuming, and environmentally hazardous. Over the years, different molecular simulation techniques and methodologies have risen to answer relevant questions of general atomic and particulate systems [1–5].

A system composed of macromolecules is a very challenging case from the perspective of molecular simulation. This stems from the fact that polymers are characterized by a wide spectrum of characteristic time and length scales. Their simulation can become prohibitively difficult when very long and well-entangled chains are involved due to the very slow dynamics. Added to this is the fact that atomistic simulations have to take into full account the chemical constitution of the repeat units and the corresponding bonded and non-bonded interactions. To address this problem, a large amount of different molecular simulation methods has been developed and constantly improved over the last decades. The choice of simulation approach/scheme depends on the system/phenomenon, its physical-chemical details, size, and properties of interest. For example, Molecular Dynamics (MD) provides dynamical information at the local level of segments and global one of chains. However, as it follows the evolution of the equations of motion in time, it can be too slow to be effective when very long chains are involved. Monte Carlo (MC), by resorting to different stochastic algorithms ("moves"), can offer rapid equilibration at all length scales. However, MC cannot provide any information about the real dynamics. Accordingly, it is not uncommon for different techniques to be combined together into powerful hierarchical modeling approaches. The MD approach is widely used when dynamical or temporal evolutions are of interest. One of the most widely used software packages for the simulation of synthetic polymers is LAMMPS [6], which has been further used in other tools such as Polymatic for the polymerization of amorphous polymers [7]. Regarding the modeling of biomolecular systems, NAMD [8] and GROMACS [9] are two of the most popular simulation software, both placing special emphasis on parallelization in order to enhance performance. Other relevant open MD software suites are ms2 [10], to extract thermodynamical properties of homogeneous fluids using hybrid parallelization on MPI and OpenMP [11]; MOLDY [12] for solids and liquids under periodic boundary conditions; or GULP [13] for solids. Commercial suites include, among others, CHARMM [14], AMBER [15], and HyperChem [16].

With respect to Monte Carlo simulations, homemade software programs usually target a specific type of polymer structure, either of its chemistry or the architecture of the chain, but most of them follow rather similar approaches. Monte Carlo simulations are applied when equilibrium structural properties, including phase transitions, constitute the main research focus. The Enhanced Monte Carlo code [17,18] is a multi-purpose modular environment for particle simulations using force fields such as COMPASS, CHARMM, or Born. This open tool has been used to study the effect of semicrystalline interphase polyethylene under different conditions of tensile deformation [19,20] or chain branching [21]. MCCCS Towhee [22] was initially developed as a simulator suitable for computing phase equilibria in the Gibbs ensemble, but later extended to different force fields and ensembles. As an example, this open tool has been used to study gas-mixture separations on clathrate hydrates [23] among many other studies. DL_MONTE [24] is a very recent MC-based open tool that can be applied to general atomistic systems under different force fields and ensembles, as well as introducing transition pathways of umbrella sampling and Wang–Landau [25]. Furthermore, it is compatible with the molecular dynamic tool DL_POLY or chain branching [21].

We should also mention other relevant open-source MC-packages, such as Cassandra [26], that can be applied to obtain thermodynamic properties of fluids and solids; RASPA [27] for simulating adsorption and diffusion phenomena; GOMP [28,29] for GPU optimized phase equilibria simulations; or DICE [30] that uses a configurational bias scheme to study flexible molecules in solute-solvent systems. Most relevant MC software packages are benchmarked in terms of computational efficiency using adsorption simulations [31]. Regarding realistic polymeric systems, Chameleon [32] is one of the latest available pieces of software. This tool employs different chain connectivity altering moves to simulate atomistically detailed polyethylene (PE), polystyrene (PS), and polyvinyl chloride (PVC) for different polymer architectures.

Usually, the development of a commercial or open code, especially when built around Monte Carlo algorithms (moves), requires a major effort and programming in order to make it user-friendly, efficient, and of general applicability. Besides, it is very common that clever MC-based or general structure-optimization algorithms have and are being developed for specific applications or general classes of physical problems in continuous or lattice cells and in systems of varied chemical detail, in the bulk and under confinement [33–52].

Equally important to the simulation itself is the post-simulation analysis. This step can include visualization, including 3-D representation and animation, of the computer-generated system configurations and calculation of relevant quantities through proper interpretation of the raw simulation data. Corresponding suites also exist for interactive visualization, description, and analysis including, among others, ParaView [53], VMD [54], disLocate [55], UCSF chimera [56], OVITO [57], and i-Rheo GT [58].

In the present manuscript, we analyze the main features of Simu-D, an MC-based simulator and structural descriptor suite for the molecular modeling of polymer-based systems under extreme conditions. The simulator, which is the central component of the present software, is effectively the accumulation of successive expansions, modifications, and improvements implemented on the MC code [59], originally built for the simulation of dense and jammed athermal polymer-based systems in the bulk. The structural descriptor is the latest version of the Characteristic Crystallographic Element (CCE) norm [60,61], a metric used to gauge the similarity of local structure with respect to reference crystals in general atomic and particulate systems. Over the last years, the MC suite has been extended to simulate athermal polymers under confinement [62] and more recently macromolecules whose monomers interact with the square well (SW) or square shoulder (SS) potential [63]. In the corresponding research studies, emphasis was placed on how the employed conditions affect the ability of chains to pack at the local and global level [64,65], the topological network of entanglements [66–68], and the entropy- or energy-driven phase behavior (crystallization) in the bulk and under extreme confinement [63,69–73]. Here, the suite is further extended to include additional factors: chain stiffness, blends of chains and monomers, spherical or cylindrical confinement, the varied potential for bonded and non-bonded interactions, nanofillers in the form of cylinders and spheres, and combinations of the above. The ongoing effort is to create a general-purpose simulator-descriptor suite that will be as efficient, general, and user-friendly as possible given the variety of simulation conditions to be considered and the stochastic nature of the underlying MC method.

The manuscript is organized as follows: Section 2 presents the molecular model, the interspecies interactions, and the systems under study. Section 3 presents the moves behind the MC simulator and briefly discusses the features of the CCE-based structural descriptor. Section 4 discusses results from representative applications of Simu-D. Finally, Section 5 summarizes the main conclusions and lists current efforts and plans.

2. Molecular Model/Systems Studied

The current version of Simu-D allows the simulation of atomistic systems composed of N_{at} spherical monomers. These monomers can be part of macromolecules and/or exist as individual particles. In the general case, the system contains N_{ch} chains with the average length of N and N_s individual particles with $N_{ch} \times N + N_s = N_{at}$. Obviously, the two limiting cases correspond to the pure polymer matrix ($N_s = 0$, $N_{at} = N_{ch} \times N$) and a system composed entirely of monomers ($N_{ch} = 0$, $N_{at} = N_s$).

Non-bonded atoms interact with a pair-wise potential, which can be discontinuous such as the hard sphere (HS) or the square well/shoulder (SW/ SS) ones or continuous such as Lennard–Jones (LJ) with the corresponding formulas being displayed in Equation (1).

$$U_{HS}(r_{ij}) = \begin{cases} 0, \ r_{ij} \geq \sigma \\ \infty, \ r_{ij} < \sigma \end{cases}, \ U_{SW/SS} = \begin{cases} 0, \ r_{ij} \geq \sigma_2 \\ -\varepsilon_{SW}, \ \sigma \leq r_{ij} < \sigma_2 \\ \infty, \ r_{ij} < \sigma \end{cases}, \ U_{LJ} = \varepsilon_{LJ}\left[\left(\frac{\sigma_{LJ}}{r_{ij}}\right)^{12} - \left(\frac{\sigma_{LJ}}{r_{ij}}\right)^{6}\right] \quad (1)$$

where r_{ij} is the distance of the centers of atoms i and j and σ is the collision diameter, which is further considered as the characteristic length of the system. σ_2 and ε_{SW} correspond, respectively, to the range and intensity of the repulsive (SS) and attractive (SW) potentials. ε_{LJ} and σ_{LJ} are the depth and zero-energy point of the LJ potential. As in any traditional molecular simulation, depending on the type of the applied non-bonded potential, the original simulation cell is split automatically into overlap cells (HS), or into overlap and cut-off cells (SW/SS, LJ) to expedite the calculation of interactions.

Polymers are modeled as linear sequences of monomers of varying chain stiffness. Bond lengths can be longer (bond gaps), equal (bond tangency), or shorter (fused spheres) than the collision distance, σ. Chain stiffness is introduced through a potential governing bending angle (supplement of bending angle, θ) formed by triplets of successive atoms along the chain backbone. The formula for the energetic calculations can be general. Configurations of semi-flexible chains have been generated in the present work with the following bending angle potential:

$$U_{bend}(\theta) = k_\theta(\theta - \theta_0)^2 \tag{2}$$

where k_θ is the bending constant and θ_0 is the equilibrium bending angle supplement (i.e., a fully extended bending angle corresponds to $\theta_0 = 0°$). For fixed bond lengths, setting $k_\theta = 0$ allows the simulation of freely jointed chains while $k_\theta \to \infty$ corresponds to the freely rotating model. In the current version of the suite, torsion angles, ϕ, can also be controlled through the implementation of a torsional potential, $U_{tor}(\phi)$. However, in all results presented below, torsion angles are allowed to fluctuate freely and thus chain stiffness is governed solely by the bending potential.

The presence and activation of specific MC moves, as will be described in the continuation, enforces dispersity in chain lengths. Such polydispersity is controlled by casting the simulation in the $N_{at}N_{ch}VT\mu^*$ ensemble where V is the total volume of the simulation cell, T is the temperature, and μ^* is the spectrum of relative chemical potentials of all chain species, as explained in detail in References [59,74]. The uniform and Flory (most probable) distributions of molecular lengths can be selected in the simulation of polydisperse systems. In the case that strictly monodisperse samples are required, then all moves that vary the chain length (sEB, x-reptation, and IdEx3, see below) are deactivated from the mix.

Depending on the system under study, initial configurations are generated under very dilute conditions and the system is brought to the desired density through compressions or simulations in the isothermal-isobaric (*NPT*) ensemble. For the latter, conventional volume fluctuation moves are attempted at regular intervals. For the former, cell compaction is achieved by a combination of volume fluctuation moves, and in the case of confined systems, the wall wrapping "MRoB" algorithm as explained in [75].

Simulations can be conducted in two or three dimensions under periodic boundary conditions or on flat surfaces. Confinement is realized through the presence of such impenetrable surfaces. The current implementation allows confinement in the form of (i) flat, parallel walls in at least one dimension, (ii) a cylinder with closed or open ends (subjected to periodic boundary conditions), and (iii) a sphere (full confinement). The intensity of confinement is controlled by the distance between the confining surfaces, i.e., the cylinder or sphere diameter or the inter-wall distance. The latter can, in general, be different in each confined dimension i, $d_{wall}(i)$. Simulation cells are always orthogonal but can be anisotropic, and the number of confined dimensions, d_{conf}, ranges from 0 (bulk cell with periodic boundary conditions) to 3 (full confinement). The cell aspect ratio, ζ, is defined as the ratio of the maximum inter-wall distance divided by the minimum one [75].

Nanocomposites can be simulated with the fillers taking the form of spherical or cylindrical particles of varied sizes and populations. In the current implementation of the suite, each nanocylinder spans the whole simulation cell and its direction is held fixed throughout the simulation. Nanospheres can, in principle, move in space, but in all computer-generated polymer nanocomposite configurations to be presented in the continuation, they are treated as immobile inclusions.

For a bulk system of pure polymer, the matrix number density, ρ, is trivially defined as $\rho = N_{at}/V$, while for non-overlapping entities (such as hard spheres), packing density, φ, is given by:

$$\varphi = \frac{V_{mon}}{V} = \frac{\pi}{6}\frac{N_{at}}{V}\sigma^3 = \frac{\pi}{6}\rho\sigma^3 \tag{3}$$

where V is the volume of the simulation cell and V_{mon} is the volume occupied by the monomers, either as individual entities ("single monomers") or by being part of polymer chains ("chain monomers").

For interfacial/confined/composite systems, the above definition provides little information on the free or accessible volume given that for very large nanofillers, the volume occupied by the nanofiller can be up to four orders of magnitude higher than the one of the monomers. Thus, we can further define an effective packing density, φ_{eff}, considering the reduction of the accessible volume due to the presence of the nanofillers as:

$$\varphi_{eff} = \frac{V_{mon}}{V_{acc}} = \frac{V_{mon}}{\left(V - V_{fill}\right)} \quad (4)$$

where V_{acc} is the volume accessible to the spherical monomers, V_{fill} (= V_{cyl} + V_{sph}) is the volume occupied by the nanofillers, being the summation of the volume occupied by N_{cyl} cylinders (V_{cyl}) and of N_{sph} spheres (V_{sph}). Additionally, in the calculations above, one could further incorporate a depletion layer as monomer centers cannot lie closer than $\sigma/2$ from the surface of nanofillers or walls. In the general case of a system under confinement and being composed of nanofillers, if d_{conf} is the number of confined dimensions, the depleted effective packing density, φ_{dep}, including the effect of all nano-entities, can be calculated as:

$$\varphi_{dep} = \frac{V_{mon}}{V_{dep}} = \frac{V_{mon}}{\left(\prod_{i=1}^{d_{conf}} (d_{wall} - \sigma) \prod_{j=d_{conf}+1}^{3} l_j - \frac{\pi}{6}\left(d_{sph} + \sigma\right)^3 N_{sph} - \frac{\pi}{4}\left(d_{cyl} + \sigma\right)^2 L_{cyl} N_{cyl}\right)} \quad (5)$$

where d_{sph} and d_{cyl} are the diameter of the nanospheres and nanocylinders, respectively, L_{cyl} is the nanocylinder length, index i runs over all confined dimensions, index j over all unrestricted ones, and l_j is the length of the simulation cell in dimension j.

3. Simulator-Descriptor Suite

3.1. Simulator

The Monte Carlo suite ("simulator") consists of four different classes of algorithms: (1) Standard localized moves that entail the displacement of a single or a sequence of atoms, (2) chain-connectivity-altering moves (CCAMs), (3) cluster-based moves, and (4) identity exchange moves, all being executed at a constant volume. When shrinkage or NPT simulations are conducted, the regular volume fluctuation moves and/or the MRoB algorithm [75] undertake the task of changing the dimensions of the orthogonal simulation cell. This size alteration can be isotropic or anisotropic.

The local moves have been described exhaustively in numerous past publications. For single monomers, the simplest possible move is that of a displacement in a random direction and length within a preset amplitude [0, $l_{disp}(i)$], which again can be different for each dimension, i. With respect to chains, the corresponding set consists of: (i) Flip (internal libration), (ii) end-mer rotation, (iii) reptation, (iv) intermolecular reptation, and (v) end-segment re-arrangement (or CCB as in [76,77]; the reason we use a different notation here is to avoid confusion with the general scheme employed in all moves is explained next). All polymer-related moves can be executed in a configurational bias (CB) pattern (as seen in Figure 1 for the reptation move), with the number of trial configurations per attempted move, n_{trials}, being an input variable in the simulator. Due to the introduction of energetic bias in the forward transition, the reverse transition must be attempted n_{trials}-1 times to guarantee microscopic reversibility. In general, the number of attempts can be different for each local move, $n_{trials}(i)$, where index i runs over all available polymer-based moves. This is because the individual MC moves are characterized by distinctly different acceptance rates, which are further heavily affected by simulation conditions, chain stiffness and especially by concentration (packing density). As intuitively expected, increasing the number of trial configurations leads to a significant increase in the computational time required per MC move. Setting n_{trials} = 1 enables the conventional execution of the local moves and eliminates the necessity to perform the reverse transition. The selection of n_{trials} is highly system dependent; for example, optimal values for hard-sphere chains in the bulk as a function of the volume fraction from dilute conditions up to the maximally random jammed (MRJ) state can be found in Table 1 of Ref. [59].

Figure 1. Schematic of the reptation move implemented through a configurational bias pattern with n_{trials} = 3. Different candidate positions could be picked by the selection of the bond length, bending, and torsion angles used for the re-construction of the monomer.

The set of chain-connectivity-altering moves consists of the simplified end-bridging (sEB), simplified intramolecular end-bridging (sIEB), and simplified double bridging (sDB) [59,75] moves. All constitute simplified versions of the original EB [74,78] and DB [79,80] algorithms, initially developed for the rapid equilibration of atomistically detailed polyethylene chains of high molecular weight. The main difference with respect to the original moves is that none of the simplified versions entails the displacement of atoms; rather they proceed by deleting and forming properly selected bonds in a pair (sEB, sDB) of chains or a single (sIEB) chain. The main advantage of the sDB algorithm is that it can be applied to strictly monodisperse systems and primarily to non-linear molecular architectures. Its main disadvantage is that it requires a bridgeable distance between two different pairs of atoms. For systems of very small bond gaps ($dl \rightarrow 0$), this condition is very rarely met except very near the jammed state where the contact network is rich as a result of the isostaticity condition [65]. Additionally, all systems to be reported in the continuation are composed of linear chains. Furthermore, it has been found that dispersity in chain lengths has no effect on the universal static scaling laws [66,67] and phase behavior [71,72] of the simulated thermal and athermal polymer packings. Based on the above, sDB is excluded from the mix of moves for all cases studied here.

The third class of MC moves is that of cluster-based ones. The two variations, implemented in Simu-D, are cluster rotation (CluRot) and cluster displacement (CluDis) as first introduced in the home-made cluster code reported in [63]. The execution of the moves proceeds according to the schematic in Figure 2. In the first step, the cluster is identified. Group similarity for cluster detection is conducted first through a Euclidean distance criterion, independently of the identity of the constituent atoms (chain versus single monomers etc.). Further linkage criteria can include additional common elements such as the same crystal similarity (as detected for example by the CCE analysis, see below). Once the clusters are identified with the corresponding members labelled accordingly, one cluster is selected randomly. That cluster, as a whole (i.e., a single object made of the corresponding sites), can be displaced by a random amount in a random direction (CluDis) or be rotated randomly with respect to its center of mass (CluRot). The cluster-related moves can be optionally and automatically de-activated when a single cluster exists in the system.

The cluster detection is a computationally demanding step, so the CluDis and CluRot moves have low attempt probabilities, as also happens with the chain-connectivity-altering ones and the algorithms that alter cell dimensions.

The fifth and final set of moves consists of algorithms that change the identity of atoms and can be applied in the case of blends of monomers and polymers but also of polymers composed of different monomers. Figure 3 presents three such identity exchange (IdEx) moves, involving a single monomer and a single chain or a pair of chains. In the top panel of Figure 3, the execution of IdEx1 is shown once a single monomer (shown in red) is within a bridgeable distance to one of the ends of the chain molecule (shown in blue). The move proceeds by connecting, via a bond, the chain end and the single monomer so that the newly incorporated atom becomes the new chain end. In parallel, the last bond connecting the other end of the chain is deleted and the end is converted to a single monomer. By construction, the move does not entail atom displacement but rather the reconstruction of properly selected bonds. Accordingly, the change in energy entering the Metropolis criterion for acceptance or rejection of the move is due to the bonded term (variation of one bond length, one bending, and one torsion angle), along with any non-bonded change due to the swap of identities. The concept of IdEx2 (middle panel) is very similar. The single monomer needs to be within a bridgeable distance from the second or penultimate atom of the chain. If the proximity condition is fulfilled, it becomes, through bond formation, the new chain end, and the corresponding chain end is converted into a single atom through bond deletion. Finally, IdEx3 (bottom panel) entails two chains. The difference with respect to the single-chain version is that the new single monomer is created by the deletion of a terminal bond of a randomly selected chain, different than the one that gains the monomer. Clearly, the implementation of IdEx3 requires dispersity in chain lengths.

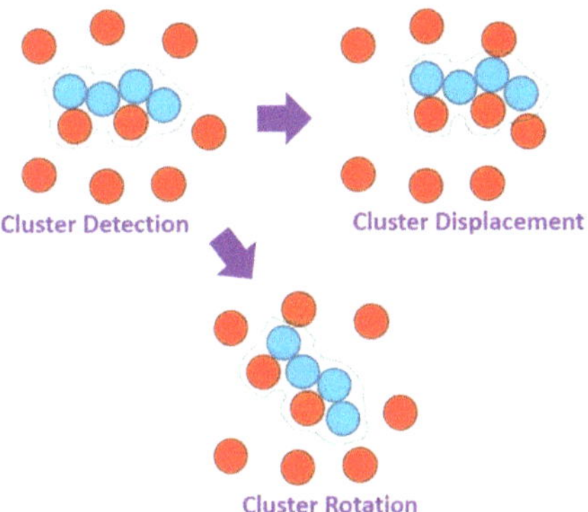

Figure 2. Schematic representation of the cluster displacement (CluDis) and cluster rotation (CluRot) moves in a mixed system of chain (**blue**) and single (**red**) monomers. The initial step of cluster detection is performed based solely on proximity criterion. The identified cluster is shown by the contour line. The cluster, as a whole unified group of monomers, can then be displaced in a random direction and length (ClusDis) or be rotated by a random amount around its center of mass (ClusRot).

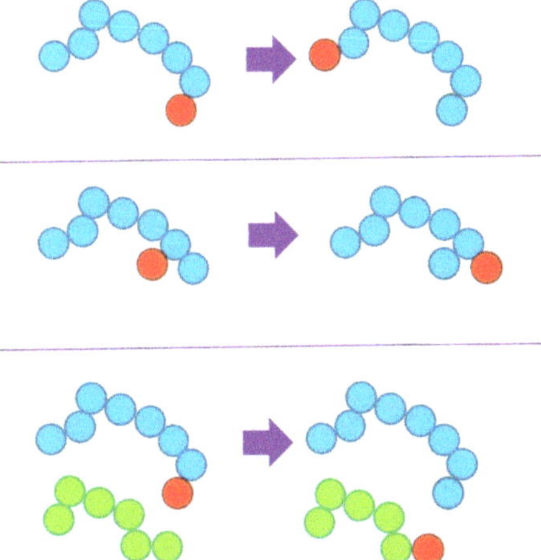

Figure 3. Schematic representation of (**top**) IdEx1 involving a chain and a single monomer within a bridgeable distance from a chain end, (**middle**) IdEx2 involving a chain and a single monomer within a bridgeable distance from the atom lying in the second of penultimate position in the chain, and (**bottom**) IdEx3, which includes a pair of chains and a single monomer. In IdEx3, the single monomer, lying within a bridgeable distance from an end of the blue chain becomes part of it, while a randomly selected chain (shown here in green) loses a randomly selected end, which becomes a single monomer. None of the moves depicted above include the displacement of atoms, rather only deletion and formation of bonds. IdEx3 requires dispersity in chain lengths in order to be applicable. In all cases, chain monomers are shown in blue (or green) and single monomers in red.

Based on the concept of identity exchange, as presented above, one can envision variations with monomers being incorporated into the inner segments of the polymer chains. However, such an approach would require the double fulfillment of the bridgeable distance and would therefore significantly reduce the pair of sites that could trigger such IdEx moves. For this reason, no further modifications have been incorporated in the present implementation of the simulator.

3.2. Descriptor

As mentioned earlier, equally important to the simulation itself is the analysis of the results, which can be "on the fly" or in a post-simulation step. Monte Carlo simulations, such as the ones presented here, provide no dynamical information, so the emphasis is placed on the study of the local and global structure, organization, topology, and phase behavior. Over recent decades, conceptually different approaches have led to the development and application of descriptors and analyzers of local structure in computer-generated configurations or of digitally processed experimental samples [81–93].

Here, since we are particularly interested in studying entropy- and energy-driven crystallization of polymer-based systems under extreme conditions, we propose a modeling scheme where the MC-based simulator is connected to a descriptor of the local structure ("descriptor") in the form of the CCE norm [60,61]. The version adopted in Simu-D is very similar to the one we presented very recently, so the concept, methodology, and technical implementation, reported in detail in [60], are all also applicable to the present context. Thus, in the continuation, we will provide a brief description on the main aspects of the CCE norm descriptor and the new features, as implemented in Simu-D. Given an atomic or particulate system in two or three dimensions, the CCE norm proceeds by comparing the local environment around each site with the ideal ones of specific reference crystals.

The main concept behind the CCE norm descriptor is that each ideal crystal is uniquely identified by a set of symmetry operations (elements of its point group) [94–97]. The identification of the totality of these crystallographic operations, or of an equally discriminating subset of them, and their application to the nearest neighbors of an atom or particle is key in the implementation of the CCE algorithm.

As explained in detail in Ref. [60], the CCE norm is defined with respect to a specific crystal X. Once the reference crystal X is selected, the point group is identified along with the generating symmetry elements. Given a site (atom or particle), i, the $N_{vor}(i)$ nearest neighbors are identified through Voronoi tessellation. The $N_{vor}(i)$ population is then compared against the coordination number of the reference crystal X, $N_{coord}(X)$. The latter is, for example, equal to 12 for the face-centered cubic (FCC) and hexagonal close-packed (HCP) crystals in 3-D and 6 for the triangular (TRI) crystal in 2-D. If the coordination number is larger than the number of nearest neighbors, $N_{coord}(X) > N_{vor}(i)$, a penalty function is applied [60]. In the opposite case, $N_{coord}(X) < N_{vor}(i)$, only the $N_{coord}(X)$ closest neighbors are kept for the successive CCE-based analysis. The characteristic crystallographic element(s) is(are) identified and the corresponding actions for each one of them are applied to the coordinates of the neighbor atoms relative to the given site. For example the HCP crystal, with $N_{coord}(\text{HCP}) = 12$, has one geometric symmetry element in the form of a sixfold roto-inversion axis, while the body centered cubic (BCC), with $N_{coord}(\text{BCC}) = 8$, has five such elements: Four three-fold roto-inversion axes and one inversion center.

One important point in the CCE norm analysis in a 3-D system is that the orientation of each symmetry axis, or at least of a sub-set of them, is not known a priory. Accordingly, we scan the orientation space SO(3) around the given site with a mesh of discretization width φ_{step}, which is the same for the azimuthal and polar angles.

For a given orientation, the actions of the symmetry element are executed. This procedure is then repeated over all symmetry elements. The goal of these crystallographic operations is to map the real coordinate system (given site i and $N_{coord}(X)$ neighbors) into the ideal one of the reference crystal X. Once this is completed, the algorithm proceeds to the next point of the discrete mesh until the whole orientation space is examined. This mapping allows to simultaneously quantify the orientational and radial similarity of the given local environment with respect to the ideal one of crystal X. This is realized through the calculation of a norm (see Equation (2) of [60]). The CCE-based norm for the given atom i with respect to reference crystal X, ε_i^X, is the one that corresponds to the global minimum of the norms as calculated over all possible orientations of all symmetry elements (axes). The same process is repeated over all particles or atoms of the systems and all reference crystals. Currently, the CCE descriptor, as implemented in Simu-D, includes the following crystals: Face-centered cubic (FCC), hexagonal close-packed (HCP), body-centered cubic (BCC), and hexagonal (HEX) for 3-D systems and honeycomb (HON), square (SQU), and triangular (TRI) for 2-D systems. Additionally, the local structure can be quantified with respect to fivefold (FIV) and pentagonal (PEN) local symmetries, in 3-D and 2-D, respectively. The lower the value of the CCE norm, the higher the similarity of the local environment to the reference crystal. A site is labelled X-type when its minimum CCE norm is lower than a critical threshold, ε^{thres}, i.e., $\varepsilon_i^X \leq \varepsilon^{thres}$. By construction, as the characteristic crystallographic elements and operations constitute a distinctive feature for a crystal, the CCE norm is highly discriminatory, so that when the CCE norm with respect to crystal X is low, the corresponding norm for other crystal types is high.

An extensive analysis of the underlying concept, the minimum distinguishable set of symmetry elements and corresponding actions for each reference crystal, the algorithmic implementation on the CCE-norm descriptor, the required computational time, and the optimal selection of parameters are all discussed in detail in Ref. [60]. The Simu-D version contains certain additional features. As an option, the "on-the-fly" implementation allows, during the scanning of the spherical space, for the CCE analysis to stop when the norm is found to be lower than the pre-set threshold and pass to the next atom so as to expedite the process and provide a preliminary structural identification. Additionally, the CCE descriptor further identifies the clusters of all atoms that bear the same similarity. For example, it detects clusters of ordered sites, calculates their size (in number of atoms), as well as their shape. The cluster-based analysis functions with the same proximity criterion as the cluster identification used for the moves of the simulator component. An additional condition for the cluster identification is that it should contain sites that have all the same similarity (with respect to a single crystal type X or to a pair of them (X or Y)). Finally, the CCE descriptor provides information on the shape, size, and statistics of the Voronoi cells, as extracted from the Voronoi tessellation.

A table with a summary of the main variables used by the Simu-D suite along with a brief description can be found in Appendix A (Table A1).

4. Simu-D: Applications

In this section, we briefly present polymer-based systems that can be simulated and successively analyzed with the Simu-D suite. Emphasis is placed on the simulation of systems under extreme conditions: These can range from a very high concentration (packing density), extreme confinement, or presence of nanofillers with dimensions significantly larger than the monomer size or any combination of the above. In the case of entropy- or energy-driven phase transitions, the corresponding analysis takes place through the CCE descriptor on the frames and trajectories generated by the simulator component.

The main point to be highlighted is that the Simu-D suite is built in a modular-based approach with the goals of general applicability and simplicity. So, all examples in the continuation have been or can be simulated and successively analyzed without any modification of the code being required from the end user. Here, it is not our intention to exhaustively analyze the physical behavior of each reported case but rather to provide evidence that such systems can be modeled and then characterized by the Simu-D software.

4.1. Packing Efficiency of Semi-Flexible Athermal Polymers (3-D)

How atoms, particles, or macroscopic objects are packed in the most efficient way is a topic of paramount importance in various fields and applications. Ordered packings of non-overlapping spheres in 3-D have an upper limit in the volume fraction, which corresponds to the one reached by the HCP or FCC crystals [98,99]. For disordered systems of the same entities, the corresponding packing density is globally accepted to be approximately 10–12% lower and corresponds to the Random Close-Packed (RCP) limit [100,101] or its equivalent Maximally Random Jammed (MRJ) state [102]. In the very first application of the MC-based code that served as the initial seed for the Simu-D suite, it was demonstrated that freely jointed chains of tangent hard spheres can be packed as efficiently as monomeric analogs [103]. However, the corresponding state of semi-flexible polymers or even of freely rotating chains is still a subject for investigation [104,105]. To this end, we used the simulator component to generate and successively equilibrate random packings of semi-flexible chains with a varied equilibrium angle, degree of stiffness, as quantified by the spring constant in Equation (2), average chain length, and volume fraction. Exploring the combined effect of the physical variables stated previously requires the conduction of numerous simulations starting from dilute systems all the way up to the RCP/MRJ limit. The range of the latter is expected to be a function of the rigidity of the chain and thus depend strongly on the bending constant and equilibrium angle [104].

Figure 4 shows bulk system configurations for semi-flexible hard-sphere chains with average length $N = 100$ ($N_{at} = 4800$) with an equilibrium (supplement) angle of $\theta_0 = 120°$ at different packing densities of $\varphi = 0.001$, 0.1, and 0.60.

Using the Simu-D generation-equilibration modules, structures of semi-flexible athermal polymers can be simulated at very high densities, which are comparable to the densest ones observed for fully flexible (freely jointed) polymers [66,103] or monomeric counterparts [102]. The acceptance rate of the employed local and chain-connectivity-altering move as a function of packing density for the 48-chain $N = 100$ system with $\theta_0 = 108°$ is shown in Figure 5 and is reminiscent of the one obtained for freely jointed chains [59]. As expected, the acceptance rate of local moves is significantly reduced as the system reaches progressively higher concentrations. Towards this, the configurational bias scheme aids in reducing this drop. The reduction for semi-flexible chains is especially apparent for the two variants of the reptation move. In sharp contrast, the acceptance rate of chain-connectivity-altering moves shows opposite trends: The higher the concentration, the higher the acceptance rate. Especially for the simplified End-Bridging at low volume fractions, acceptance is very small. This is expected as in such a dilute system there are very few or no pairs of atoms that can trigger the move. As the concentration increases, the population of such pairs also increases because chains start to feel each other and the contact network around each site becomes richer. In parallel, none of the CCAMs, as incorporated in Simu-D, requires the displacement of any atoms. Thus, their performance is enhanced at very high packing densities, and especially near the MRJ state.

According to the RCP/MRJ definition, the maximum-density state should correspond to the densest structures, which are characterized by the maximum randomness or, equivalently, the minimum order [102]. The concept of rattlers [102] and flippers [103] can be invoked to quantify the fraction of individual sites and groups of them, which are able to perform movements in their local vicinity for monomeric and polymeric packings, respectively. In both cases, it is well demonstrated that the flipper/rattler population diminishes as we approach the MRJ state. Alternatively, one could attempt to quantify the lack of order in the system through the proper definition of corresponding parameters. Towards this, we employ the CCE norm (descriptor module of Simu-D) to calculate the similarity of the local environment around each monomer site to the close-packed (HCP and FCC) crystals, which are the dominant ones in the crystallization of hard sphere packings at high volume fractions [72,106]. The absence of such crystals should correspond to a highly disordered but densely packed medium near or at the RCP/MRJ state.

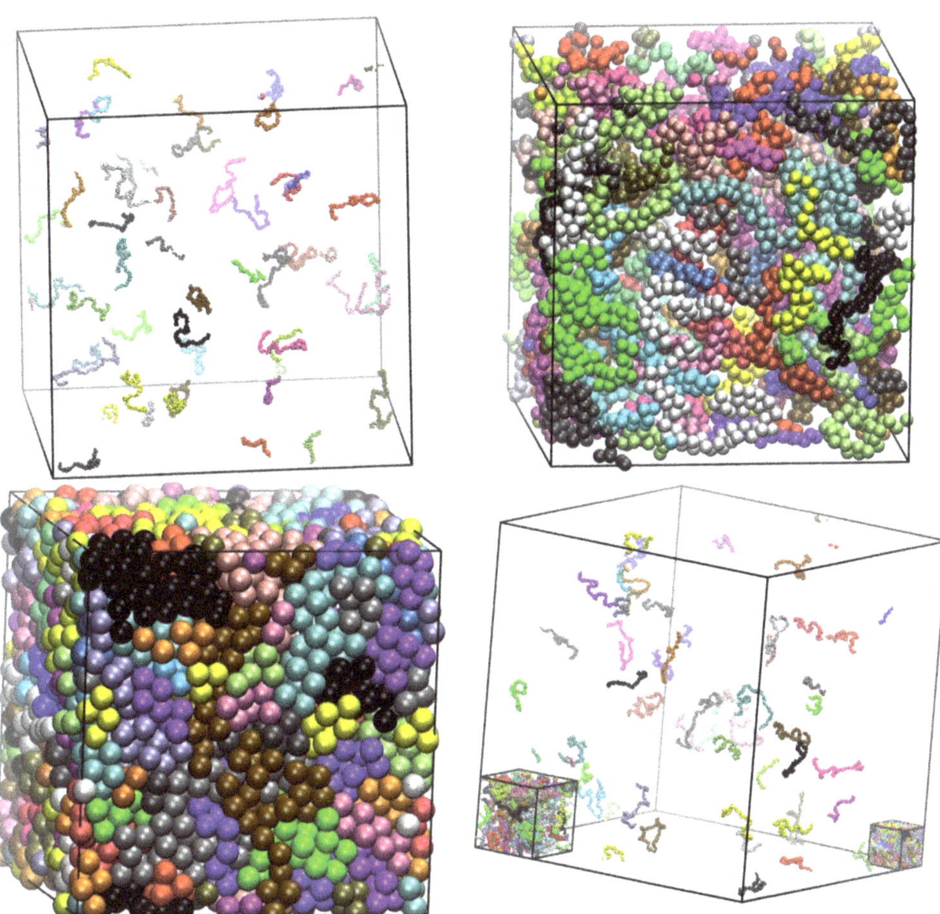

Figure 4. Bulk system configurations of semi-flexible chains of tangent hard spheres of uniform size with average length of $N = 100$ and an equilibrium angle of $\theta_0 = 120°$ at progressively higher volume fractions, φ: (**top left**) 0.001, (**top right**) 0.10 and (**bottom left**) 0.60. (**bottom right**): All three system configurations shown together allowing for a visual comparison of their dimensions. Monomers are colored according to their parent chain. Sphere monomers are shown with coordinates of their centers subjected to periodic boundary conditions. Image created with VMD visualization software [54]. Figure panels are also available as interactive, 3-D images in Supplementary Materials.

Figure 6 shows the final configuration for the 48-chain $N = 100$ system with an equilibrium bending angle of $\theta_0 = 120°$ at a density of approximately 0.64, which corresponds to the range of RCP/MRJ, as established for monomers and freely jointed chains. The left panel shows monomers colored according to the parent molecule, while the right one uses a coloring scheme according to the values of the CCE norm. More precisely, blue and red correspond to sites with HCP ($\varepsilon_i^{\mathrm{HCP}} \leq \varepsilon^{thres} = 0.245$) and FCC ($\varepsilon_i^{\mathrm{FCC}} \leq \varepsilon^{thres} = 0.245$) similarity, respectively, while green is used to represent FIV-like ($\varepsilon_i^{\mathrm{FIV}} \leq \varepsilon^{thres} = 0.245$) sites. All remaining amorphous (AMO) ones, which constitute most of the system, are shown in yellow with reduced dimensions in a 2:5 scale for clarity purposes. More accurately, amorphous (AMO) designates sites that show no similarity to any of the reference 3-D crystal (HCP, FCC, HEX, BCC) or local symmetry (FIV). This does not exclude the possibility that a specific site showing similarity to another "unknown" crystal not included in the reference list. Still, as mentioned earlier and given the very high concentration of the generated athermal packings, the presence of non-compact crystals can be excluded. This is evident as no traces of BCC or HEX crystals are detected in any of the nearly jammed polymer configurations, such as the ones visualized in Figure 6.

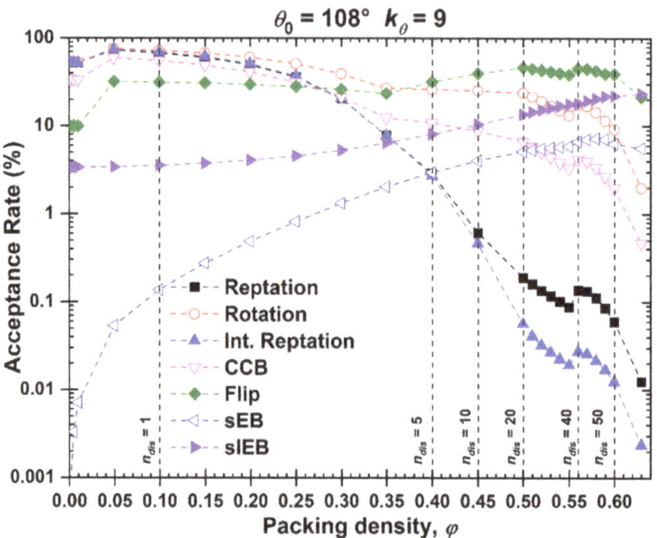

Figure 5. Percentage of acceptance as a function of packing density for the local and chain-connectivity-altering moves employed in the MC simulation of 48 chains of N = 100, θ_0 = 108° in the bulk. Vertical dash lines denote the end of the regime where n_{dis} trial configurations are attempted per local MC move.

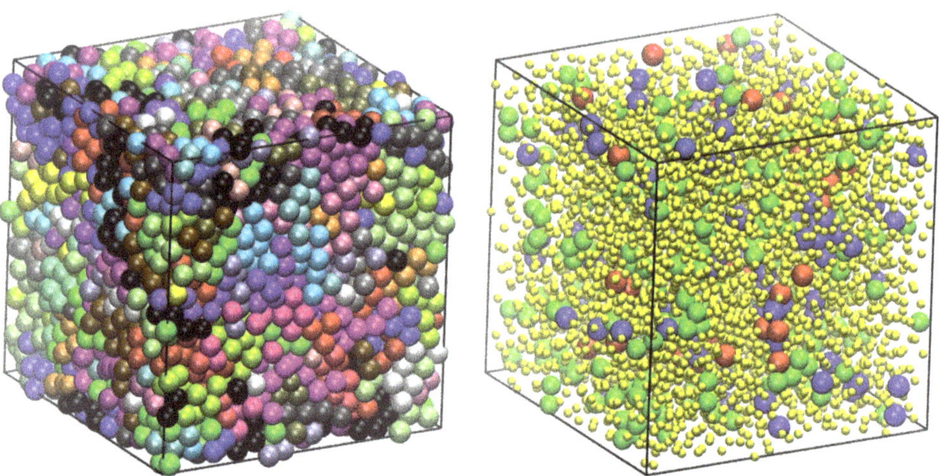

Figure 6. Jammed packing of semi-flexible chains of tangent hard spheres of uniform size with average length of N = 100 and an equilibrium angle of θ_0 = 120°at a packing density of φ = 0.637. (**Left panel**): Monomers are colored according to the parent chain; (**Right panel**): Monomers are colored according to the lowest value of the CCE norm. Blue, red, and green denote HCP, FCC, and FIV similarity, respectively. Amorphous (AMO) ones are colored yellow with reduced dimensions for clarity. Sphere monomers are shown with coordinates of their centers being subjected to periodic boundary conditions. Image created with the VMD visualization software [54]. Figure panels are also available as interactive, 3-D images in Supplementary Materials.

Visual inspection of the jammed configuration in Figure 6 suggests a predominantly amorphous structure with few ordered HCP and FCC sites randomly distributed along the whole volume of the simulation cell. In fact, one can observe that the population of FIV-like sites is higher than that of the close packed crystal ones. Moving on to quantitative analysis based on the CCE order parameter [60] for the specific structure shown in Figure 6, the HCP, FCC, and FIV fractions are 0.022, 0.021, and 0.053, respectively, further demonstrating the predominance of disorder. The random character of the maximally jammed state for semi-flexible chains is in perfect qualitative agreement with the one exhibited by freely jointed analogs; the same can be stated for the growth of fivefold local symmetry with an increasing concentration as observed for monomeric counterparts [107,108] as well as for freely jointed chains [70].

4.2. Entropy-Driven Crystallization of Semi-Flexible Athermal Polymers

The presence of fivefolds in a random particulate packing acts as an inhibitor to crystallization [107,108], especially as the concentration approaches that of the jamming state. However, after a critical volume fraction (melting point) is exceeded, and if the observation (here simulation) time is sufficiently long, hard sphere packings crystallize. Similar phase behavior is observed for freely jointed chains, albeit with differences in the critical packing density and the morphology of the established crystals, both depending strongly on the gaps between bonded atoms [69]. Using the Simu-D suite we can extend the simulations to capture the effect of chain stiffness. As an example, Figure 7 shows the phase transition as first simulated and then identified by the CCE-based analysis for the 100-chain $N = 12$ system at $\varphi = 0.58$ with an equilibrium angle of $\theta_0 = 90°$. In the left panel of Figure 7, the initial configuration is presented, as produced through the generation module, while in the right panel, the final one after the execution of 3×10^{11} MC steps is shown. In both system states, monomers are colored according to the value of the CCE norm. It can be unmistakably concluded that the specific system shows crystallization, with the final stable configuration being of defect-ridden, fivefold-free, alternating HCP and FCC layers. Given that the hard-sphere chain system is athermal, such a phase transition is dictated solely by the increase in the total entropy of the system through a mechanism similar to the one observed in freely jointed chains where the local environment around each ordered site becomes more symmetric in the crystal phase [71,72,109].

4.3. Phase Behavior of Athermal Blends (Polymers and Monomers)

Through the incorporation of MC moves involving individual monomers and polymer chains (IdEx1, IdEx2, and IdEx3), Simu-D software can tackle blends of chains and monomers of varied relative fractions and different interactions between species. Example cases include a high-density athermal blend of polymers and monomers with a varied number of chains as can be seen in the panels of Figure 8. The system consists of 54,000 interacting sites at a packing density of $\varphi = 0.56$ and an average chain length of $N = 1000$. The polymer relative concentration ranges from 0 (0 chains and 54,000 monomers), 0.185 (10 chains and 44,000 monomers), 0.741 (40 chains and 14,000 monomers) to 1 (54 chains and 0 monomers). The objective here is to study how the relative concentration of the different entities (single versus chain monomers) could affect crystallization. This is motivated by the fact that the selected volume fraction is below and above the melting point of strictly tangent chains and individual monomers, respectively.

4.4. Energy-Driven Cluster and Crystal Formation of Attractive Chains

Up to this point, all systems studied were athermal with all interactions being of the hard sphere type. In the following case, we employ the square well (SW) attractive potential. Additionally, we carry out the simulations at a constant volume (NVT) for chain systems and at a constant (and high) pressure (NPT) for monomeric ones. In both cases, the starting configuration corresponds to a low-density ($\varphi = 0.05$) hard sphere system where we activate the SW interactions between all sites. Given the attraction felt between the monomers, clusters start to form, which, depending on the applied intensity and range of interactions, could further lead to ordered morphologies or amorphous glasses [63]. From the technical point of view in NVT simulations, one should activate collective, cluster-related moves since, especially at a low concentration and a high attraction intensity, small and isolated clusters could be created, disallowing further inter-cluster aggregation and the eventual formation of a single entity. The phase diagram of SW chains can be surprisingly rich with different crystals and amorphous morphologies resulting as a function of the attraction range. As an example, Figure 9 hosts the final system configuration obtained from NVT simulations on chains (ε_{SW} = 1.2, σ_2 = 1.15, N = 12) and NPT simulations on monomers (ε_{SW} = 2.1, σ_2 = 1.15), both having N_{at} = 1200 interacting sites. For the given pairs of intensity and range of attraction, the established morphologies consist of HCP and FCC crystallites with random stacking directions. In the case of the chain cluster (left panel of Figure 9), the presence of fivefold sites in the form of twin defects is particularly evident in the meeting points of the HCP and FCC planes.

4.5. Polymers under Confinement

In a recent publication [75], we demonstrated the ability of the early version of the Simu-D suite to create polymer configurations under tube-like and plate-like confinement. Extreme conditions correspond to the case where the distance between the confining agents/surfaces is approximately equal to the diameter of the monomers. For example, for plates, this extreme condition corresponds effectively to a 2-D polymer system. The latest implementation of Simu-D allows for flexibility in the applied geometry of confinement departing potentially from orthogonal cells. Figures 10 and 11 show system configurations of freely jointed chains of tangent hard spheres (N = 12, N_{ch} = 60) being confined in cylindrical (closed ends) and spherical geometries, respectively.

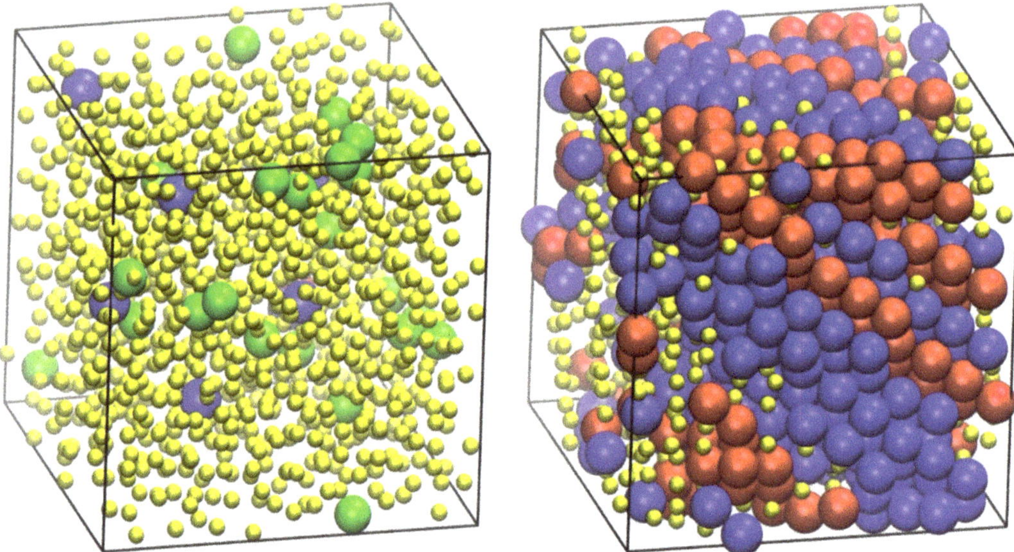

Figure 7. Snapshots of the semi-flexible N = 12 system (θ_0 = 90°) at φ = 0.58. (**Left panel**): Initial configuration as produced by the generator module of Simu-D. (**Right panel**): Final configuration of the simulation after the execution of 3×10^{11} MC steps of the simulator module. Monomers are colored according to the lowest value of the CCE norm (descriptor module). Blue, red, and green colors denote HCP, FCC, and FIV similarity, respectively. Amorphous (AMO) ones are colored yellow with reduced dimensions for clarity. Spherical monomers are shown with coordinates of their centers subjected to periodic boundary conditions. Image created with the VMD visualization software [54]. Figure panels are also available as interactive, 3-D images in Supplementary Materials.

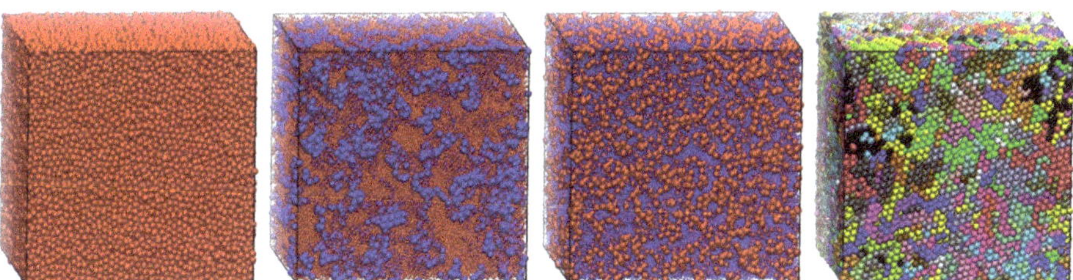

Figure 8. Bulk systems of 54,000 interacting hard spheres with varied relative concentrations of polymer content with chains having an average length of $N = 1000$ at $\varphi = 0.58$. The polymer relative concentration ranges from 0, 0.185, 0.741 to 1 (from left to right). In the pure polymer configuration (rightmost panel), sites are colored according to the parent chain. In all other cases, single and chain monomers are shown in red and blue, respectively. For chain concentrations of 0.185 and 0.741, sphere dimensions of dominant species are shown in a 2:5 scale for clarity. Image created with the VMD visualization software [54].

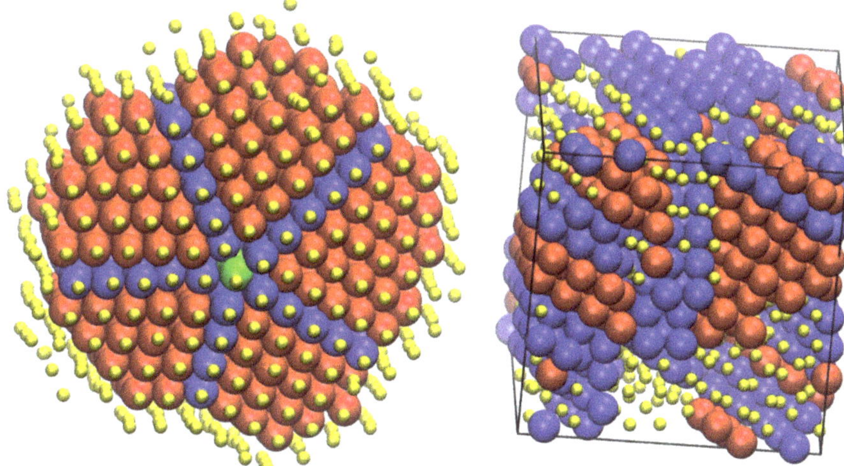

Figure 9. Final configurations of systems whose sites interact with the attractive square well potential. (**Left panel**) NVT simulations on chains ($\varepsilon_{SW} = 1.2$, $\sigma_2 = 1.15$, $N = 12$, $N_{ch} = 100$, $\varphi = 0.05$); (**Right panel**) NPT simulations on monomers ($\varepsilon_{SW} = 2.1$, $\sigma_2 = 1.15$, $N_{at} = 1200$). Sites are colored according to the CCE norm: Blue, red, green, cyan, and purple correspond to sites with HCP, FCC, FIV, BCC, and HEX similarity, respectively. Amorphous (AMO) sites are colored yellow and shown with reduced dimensions for clarity. Image created with the VMD visualization software [54]. Figure panels are also available as interactive, 3-D images in Supplementary Materials.

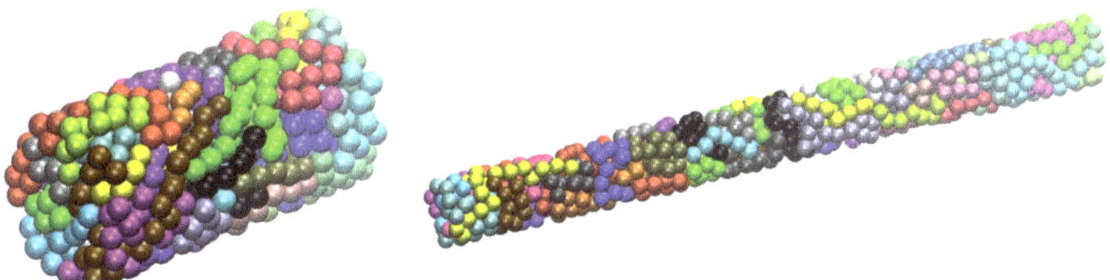

Figure 10. System snapshots of linear, fully flexible chains ($N_{ch} = 60$, $N = 12$, $\varphi = 0.40$) under cylindrical confinement with closed ends with a length-to-diameter ratio equal to 2 (**left panel**) and 10 (**right panel**). Monomers are colored according to the parent chain. Image created with the VMD visualization software [54]. Figure panels are also available as interactive, 3-D images in Supplementary Materials.

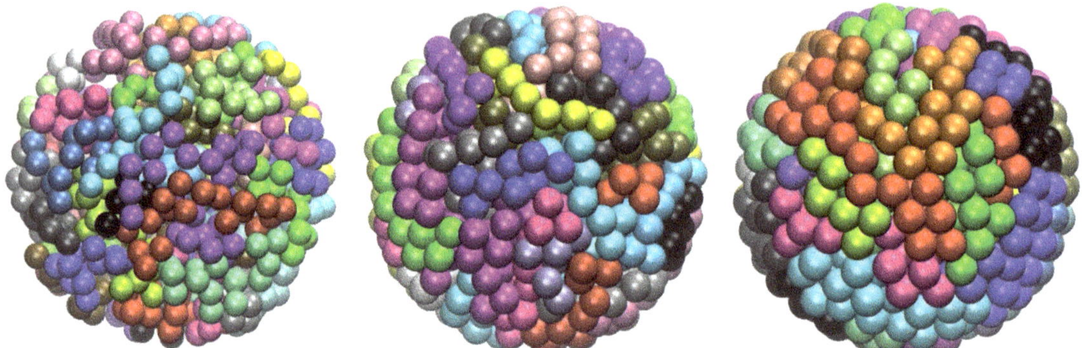

Figure 11. System snapshots of linear, fully flexible chains (N_{ch} = 60, N = 12) under spherical confinement at a packing density of (from left to right): φ = 0.30, 0.40 and 0.50. Monomers are colored according to the parent chain. Image created with the VMD visualization software [54]. Figure panels are also available as interactive, 3-D images in Supplementary Materials.

In the cylindrical confinement, the cell (length to diameter) aspect ratio increases from 2 (left panel) to 10 (right panel) while the volume fraction remains the same (φ = 0.40). In the spherical one, the packing density changes from 0.30 (leftmost panel) up to 0.50 (rightmost panel). Reaching such densities allows the investigation of crystal nucleation and the growth of chain systems and eventually the comparison with monomeric analogs, as recently reported in [110,111].

4.6. Polymer Nanocomposites

The latest implementation of Simu-D allows for the simulation of polymer-based nanocomposites (PNCs). The nanofillers can exist as compact objects of cylindrical or spherical forms and at various concentrations and interactions with the chain monomers. Figure 12 shows examples of polymer nanocomposites where all entities interact through the hard-core potential. A single nanofiller is inserted, which is a sphere of size d_{sph} (in units of monomer diameter, σ). The nanoparticle is positioned so that the coordinates of its center coincide with the center of the simulation cell. The left panel shows a PNC with d_{sph} = 5 at $\varphi = 9.9 \times 10^{-3}$ while d_{sph} = 20 at φ = 0.29 is presented in the right panel. Taking into account the presence of the nanosphere, the effective packing densities are φ_{eff} = 0.01 and 0.55 for the systems in the left and right panels of Figure 12, respectively. The minimal difference for the former case is due to the small nanoparticle size (d_{sph} = 5) compared to the large volume of the simulation cell, while in the latter case, the nanoparticle, due to its massive size (d_{sph} = 20), has a profound effect on the reduction of the available volume.

Another example from MC simulations on PNCs is provided in Figure 13. This time the nanofiller takes the form of a single, infinitely long cylinder whose direction coincides with the direction of one of the axes of the simulation cell. The diameter of the cylinder is d_{cyl} = 5 and is dispersed in a polymer matrix (N = 100, N_{ch} = 48), which consists of (left panel) freely jointed and (right panel) semi-flexible, rod-like ($\theta_0 = 0°$) chains.

4.7. Comparison with Independent Algorithms

A relevant task is to compare the results of any simulation suite with existing ones, preferably using different simulation methods. Here we should mention that with respect to jamming, our Simu-D produces very dense and nearly jammed random packings of hard spheres (chains or monomers) with volume fractions very close to the ones reported in the literature from independent sources on the RCP/MRJ state: $\varphi^{\mathrm{MRJ}} \approx 0.64$–$0.65$, with the exact value and the salient characteristics being very dependent on the employed protocol [112–114].

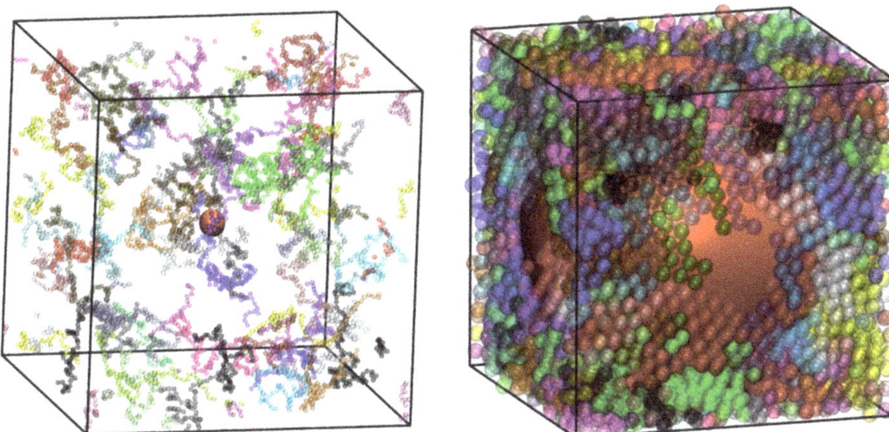

Figure 12. System snapshots of polymer nanocomposite (N = 100, N_{ch} = 48) at different effective packing density, φ_{eff}. The nanofiller, shown in red, corresponds to a single, impenetrable sphere with diameter d_{sph} (in units of σ) whose center is located at the center of the simulation cell: (**left**) φ_{eff} = 0.01, d_{sph} = 5 and (**right**) φ_{eff} = 0.55, d_{sph} = 20. Monomers are colored according to the parent chain and are shown as semitransparent spheres for clarity. Image created with the VMD visualization software [54]. Figure panels are also available as interactive, 3-D images in Supplementary Materials.

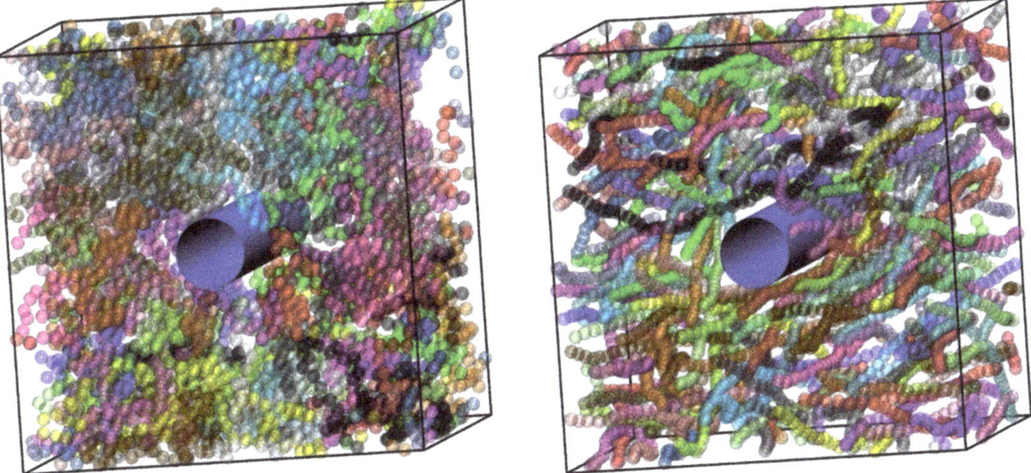

Figure 13. System snapshots of polymer nanocomposite (N = 100, N_{ch} = 48, φ_{eff} = 0.10). The nanofiller, shown in blue, corresponds to a single, impenetrable cylinder with diameter d_{cyl} = 5 (in units of σ) and infinite length. The cylinder is oriented along the direction of one of the cell axes. (**left**) Freely jointed chains and (**right**) semi-flexible, rod-like chains (θ_0 = 0°). Monomers are colored according to the parent chain and are shown as semitransparent spheres for clarity. Image created with the VMD visualization software [54]. Figure panels are also available as interactive, 3-D images in Supplementary Materials.

In parallel, the melting point of monomeric hard spheres, as determined by simulations conducted with the present protocol, coincides with the well-established value available in the literature [115]. With respect to the crystallization of athermal polymers and the effect of bond gaps (or bond tangency), the results based on the application of the Simu-D suite [69] are in excellent agreement with independent studies using event-driven Molecular Dynamics (MD) simulations [116].

Furthermore, as an additional testbed, we use the crystallization of monomeric hard spheres. Towards this, we use the same amorphous system configuration, composed of 54,000 hard spheres at a volume fraction of $\varphi = 0.56$. Using this initial structure, we embark on two different kinds of simulation: (i) Stochastic MC using the Simu-D suite and (ii) event-driven, collision-based MD. Given that the reference system consists of hard spheres at an elevated concentration, total crystallinity can be considered as the sum of the fractions of sites with HCP and FCC-like similarity, with FIV local symmetry acting as a structural competitor to compact crystals. In both cases, the local structure is quantified through the CCE metric. Even if the two simulation methods are distinctly different, one (MD) based on collision-based dynamics the other (MC) being completely stochastic, the corresponding trends on the evolution of crystallinity, as seen in Figure 14, are strikingly similar, not only in qualitative but also in quantitative terms.

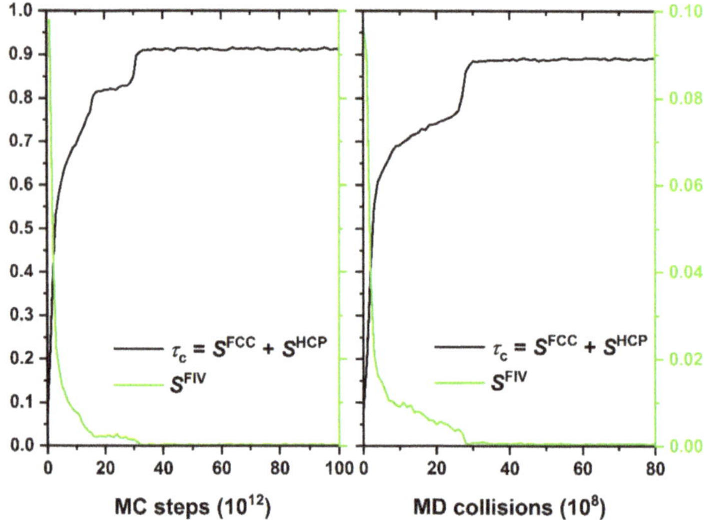

Figure 14. Evolution of crystallinity, τ_c, and of the fraction of sites with fivefold (FIV) local symmetry, S^{FIV}, as a function of MC steps (**left panel**) and MD collisions (**right panel**). Both the MC simulation, performed through the Simu-D suite, and the independent, event-driven MD simulation, are conducted on the same random initial configuration of 54,000 monomeric hard spheres of uniform size at a packing density of $\varphi = 0.56$. Total crystallinity is calculated here as the sum of fractions of sites with FCC and HCP character, as quantified by the CCE norm descriptor.

5. Conclusions

We present the latest implementation of Simu-D, a simulator-descriptor suite used to model and successively analyze/describe polymer-based systems under extreme conditions of concentration (packing density), confinement, and nanofiller content. The simulator part is based on Monte Carlo (MC) algorithms, including localized, chain-connectivity-altering, identity-exchange, and cluster moves in various statistical ensembles. The descriptor is based on the characteristic crystallographic element (CCE) norm, which is a metric to gauge the local structure by comparing it with reference crystals in two and three dimensions. The suite has a modular approach, allowing the addition of features, and is built considering efficiency, general applicability, and ease of use. Monomers/atoms/particles are presented as spheres, which interact through standard bonded and pairwise, non-bonded terms.

We have provided examples ranging from applications on bulk, pure macromolecular systems, of blends with monomers, under various conditions of confinement to polymer-based nanocomposites. Through such simulations one can study general phase transitions, packing ability, and local and global structure as a function of the aforementioned parameters. In the examples provided, emphasis is placed on the simplified hard-sphere and square-well models, but chemically realistic systems can be simulated as well.

Presently, Simu-D is further expanded to tackle more complex systems including the simulation of terminally grafted nanoparticles anchored on polymer chains as seen in Figure 15, and polymer adsorption on flat or nanostructured surfaces.

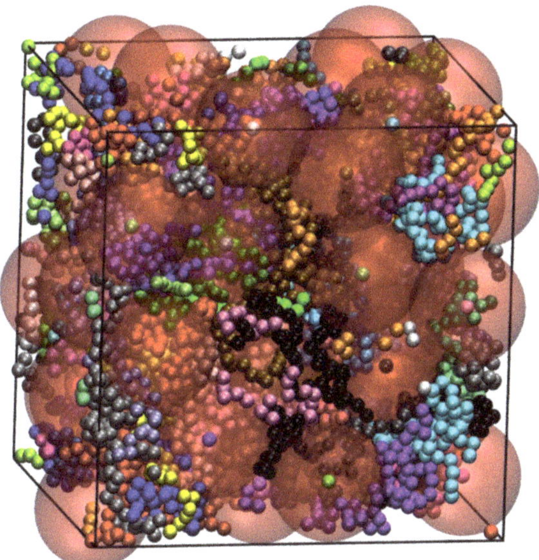

Figure 15. Terminally grafted nanoparticles on polymer chains at a volume fraction of $\varphi = 0.50$ as simulated through the Simu-D suite. Each nanoparticle, shown in red and in semitransparent format, has a size of $d_{\text{nano}} = 8$ and is anchored to a single polymer chain. Macromolecules are represented as freely jointed chains of tangent hard spheres with an average length of $N = 100$.

Our simulator-descriptor suite is rather lacking with respect to the available potentials and interactions, especially compared to latest simulators [24]. However, as explained earlier, our intention is to simulate general, but still coarse-grained, representations of atomic and particulate systems with emphasis placed on extreme conditions, such as jamming, confinement, anchoring, presence of nanoparticles or all possible combinations of the above.

Chemical reactions can also be studied by assigning reactant and product types and implementing identity-change algorithms with the MC simulations being cast in the proper reactive ensemble [117].

In a coarse-to-fine approach, the present MC suite could be further benefited by algorithms that allow the simulation of chemically complex, all-atom systems through reversible, adaptive, or bijective mapping [34–37,118]. Furthermore, the suite should be compatible with independent and efficient MC algorithms such as the event-chain ones [40,41], parallel techniques [119,120], but also with different analyzers. The latter can be in the form of geometric [121,122], stochastic [123], or energy-based [124,125] codes for the topological analysis of the primitive path network of entanglements as abundantly encountered in densely packed systems of long polymer chains.

Supplementary Materials: The following are available online at https://www.mdpi.com/article/10.3390/ijms222212464/s1. The manuscript is available in interactive, 3-D format. With the exception of Figure 8, all panels corresponding to system configurations are also available as stand-alone, interactive 3-D pdf files.

Author Contributions: Conceptualization: N.C.K. and M.L.; methodology: K.F., N.C.K. and M.L.; software (Simu-D): M.H., D.M.-F., P.M.R., K.F., N.C.K. and M.L.; data curation: M.H., D.M.-F. and P.M.R.; visualization: M.H., D.M.-F. and N.C.K.; funding acquisition: K.F., N.C.K. and M.L.; writing—original draft preparation: M.H. and N.C.K.; writing—review and editing: D.M.-F., P.M.R., K.F. and M.L. All authors have read and agreed to the published version of the manuscript.

Funding: This research was funded by MINECO/FEDER (Ministerio de Economia y Competitividad, Fondo Europeo de Desarrollo Regional), grant numbers "MAT2015-70478-P" and "RTI2018-097338-B-I00" and by UPM and Santander Bank, "Programa Propio UPM Santander".

Institutional Review Board Statement: Not applicable.

Informed Consent Statement: Not applicable.

Data Availability Statement: Presented simulation trajectories and corresponding data are fully available upon request.

Acknowledgments: Very fruitful discussions with Doros N. Theodorou, Martin Kröger, Mukta Tripathy, and Robert S. Hoy are deeply appreciated. Authors deeply thank Clara Pedrosa, Jorge Rey, Javier Benito, Olia Bouzid, and Manuel Santiago for helpful interactions. The authors acknowledge support through projects "MAT2015-70478-P" and "RTI2018-097338-B-I00" of MINECO/FEDER (Ministerio de Economia y Competitividad, Fondo Europeo de Desarrollo Regional). M.H. and D.M.F. acknowledge financial support through "Programa Propio UPM Santander" of UPM and Santander Bank. The authors gratefully acknowledge the Universidad Politécnica de Madrid for providing computing resources on the Magerit Supercomputer through projects "p208", "r553", "s341", and "t736".

Conflicts of Interest: The authors declare no conflict of interest.

Appendix A

The following Table is a summary of the main variables as used in the Simu-D suite for the simulation of different atomic systems (simulator part) and of the corresponding analysis of the local structure (descriptor part).

Table A1. Summary of the main variables as used by the Simu-D suite. Dashed line separates variables of the simulator and descriptor parts.

Name	Type	Description
D	Categorical	Number of dimensions
$dconf$	Categorical	Number of confined dimensions
N_{ch}	Categorical	Number of chains
N_{at}	Numerical	Total number of atoms
N_{high}	Numerical	Maximum number of monomers per chain
N_{low}	Numerical	Minimum number of monomers per chain
N	Categorical	Average number of monomers per chain
N_{mon}	Numerical	Number of single monomers
N_{trials}	Numerical	Number of trials per move in configurational bias scheme
Opt_{trials}	Flag	Flag to select the density-dependence of N_{trials}
ccb_{cut}	Numerical	Maximum number of monomers moved in a CCB move
$disp$	Numerical	Maximum displacement of monomer moves
φ	Numerical	Packing density
dl	Numerical	Bond gap for chains
$Nanocomp$	Flag	Inclusion of nanoparticles
N_{cyl}	Numerical	Number of nanocylinders

Table A1. *Cont.*

Name	Type	Description
N_{sph}	Numerical	Number of nanospheres
d_{cyl}	Numerical	Diameter of nanocylinders
d_{sph}	Numerical	Diameter of nanospheres
dir_{cyl}	Array	Direction of nanocylinders
σ	Numerical	Diameter designation
σ_1	Numerical	Collision diameter for Square-Well/shoulder model
σ_2	Numerical	Range of interaction for Square-Well/shoulder model
ε	Numerical	Intensity of interaction for Square-Well/shoulder model
Opt_{SW}	Flag	Creation of a second cell grid to improve SW performance.
ε_{wall}	Numerical	Intensity of interaction for Square-Well/shoulder of Walls
$\sigma_{2,wall}$	Numerical	Range of interaction for Square-Well/shoulder model of Walls
ε_{part}	Numerical	Intensity of interaction for Square-Well/shoulder of Nanoparticles
$\sigma_{2,part}$	Numerical	Range of interaction for Square-Well/shoulder model of Nanoparticles
θ_{eq}	Numerical	Supplement of the equilibrium bending angle in radians
k_{bend}	Numerical	Energy constant for bending angle potential
NPT	Flag	True: Enables NPT ensemble. False: Enables NVT ensemble
T	Numerical	Temperature
P	Numerical	Pressure
$Shrink$	Flag	True: Runs shrinkage production until a target density. False: Runs normal simulation
$fluc_{vol}$	Numerical	Maximum box length reduction when attempting shrinkage
φ_{target}	Numerical	Target density for the shrinkage production
$Isotropic$	Flag	True: Volume changes are equal in all direction. False: Volume change is anisotropic
$Cluster$	Flag	Flag to enable cluster moves when there are more than one
r_{clust}	Numerical	Radius to detect clusters
Vec	Flag	Storage of vectors for crystallographic elements
$Kiss$	Numerical	Coordination number of reference crystal
$Geom$	Flag	Check polymer geometry
$Neighs$	Numerical	Maximum number of Voronoi neighbors
HCP	Flag	CCE analysis for HCP crystal
FCC	Flag	CCE analysis for FCC crystal
BCC	Flag	CCE analysis for BCC crystal
HEX	Flag	CCE analysis for HEX crystal
FIV	Flag	CCE analysis for FIV symmetry
HON	Flag	CCE analysis for HON crystal
SQU	Flag	CCE analysis for SQU crystal
TRI	Flag	CCE analysis for TRI crystal
PEN	Flag	CCE analysis for PEN symmetry
$Thres$	Numerical	CCE threshold of similarity
$Step$	Numerical	Step of the mesh discretization (azimuthal and polar angles)
$Fast$	Flag	No full optimization if norm less than threshold

References

1. Allen, M.P.; Tildesley, D.J. *Computer Simulation of Liquids*; Oxford University Press: New York, NY, USA, 1987.
2. Frenkel, D.; Smit, B. *Understanding Molecular Simulation: From Algorithms to Applications*, 2nd ed.; Academic Press: San Diego, CA, USA, 2002.
3. Landau, D.P.; Binder, K. *A Guide to Monte Carlo Simulations in Statistical Physics*, 4th ed.; Cambridge University Press: Cambridge, UK, 2014.
4. Rapaport, D.C. *The Art of Molecular Dynamics Simulation*, 2nd ed.; Cambridge University Press: Cambridge, UK, 2004.

5. Leach, A. *Molecular Modelling: Principles and Applications*, 2nd ed.; Pearson: London, UK, 2001.
6. Plimpton, S. Fast Parallel Algorithms for Short-Range Molecular-Dynamics. *J. Comput. Phys.* **1995**, *117*, 1–19. [CrossRef]
7. Abbott, L.J.; Hart, K.E.; Colina, C.M. Polymatic: A generalized simulated polymerization algorithm for amorphous polymers. *Theor. Chem. Acc.* **2013**, *132*, 1334. [CrossRef]
8. Phillips, J.C.; Braun, R.; Wang, W.; Gumbart, J.; Tajkhorshid, E.; Villa, E.; Chipot, C.; Skeel, R.D.; Kalé, L.; Schulten, K. Scalable molecular dynamics with NAMD. *J. Comput. Chem.* **2005**, *26*, 1781–1802. [CrossRef] [PubMed]
9. Berendsen, H.J.C.; van der Spoel, D.; van Drunen, R. GROMACS: A message-passing parallel molecular dynamics implementation. *Comput. Phys. Commun.* **1995**, *91*, 43–56. [CrossRef]
10. Glass, C.W.; Reiser, S.; Rutkai, G.; Deublein, S.; Köster, A.; Guevara-Carrion, G.; Wafai, A.; Horsch, M.; Bernreuther, M.; Windmann, T.; et al. ms2: A molecular simulation tool for thermodynamic properties, new version release. *Comput. Phys. Commun.* **2014**, *185*, 3302–3306. [CrossRef]
11. Gezelter, J.D. OpenMD-Molecular Dynamics in the Open. Available online: https://openmd.org/ (accessed on 15 November 2021).
12. Refson, K. Moldy: A portable molecular dynamics simulation program for serial and parallel computers. *Comput. Phys. Commun.* **2000**, *126*, 310–329. [CrossRef]
13. Gale, J.D. GULP: A computer program for the symmetry-adapted simulation of solids. *J. Chem. Soc. Faraday Trans.* **1997**, *93*, 629–637. [CrossRef]
14. Brooks, B.R.; Bruccoleri, R.E.; Olafson, B.D.; States, D.J.; Swaminathan, S.; Karplus, M. CHARMM: A program for macromolecular energy, minimization, and dynamics calculations. *J. Comput. Chem.* **1983**, *4*, 187–217. [CrossRef]
15. Pearlman, D.A.; Case, D.A.; Caldwell, J.W.; Ross, W.S.; Cheatham, T.E.; DeBolt, S.; Ferguson, D.; Seibel, G.; Kollman, P. AMBER, a package of computer programs for applying molecular mechanics, normal mode analysis, molecular dynamics and free energy calculations to simulate the structural and energetic properties of molecules. *Comput. Phys. Commun.* **1995**, *91*, 1–41. [CrossRef]
16. Hypercube. HyperChem. Available online: http://www.hypercubeusa.com (accessed on 15 November 2021).
17. Veld, P.J. EMC: Enhanced Monte Carlo. A Multi-Purpose Modular and Easily Extendable Solution to Molecular and Mesoscale Simulations. Available online: http://montecarlo.sourceforge.net (accessed on 15 November 2021).
18. In't Veld, P.J.; Rutledge, G.C. Temperature-Dependent Elasticity of a Semicrystalline Interphase Composed of Freely Rotating Chains. *Macromolecules* **2003**, *36*, 7358–7365. [CrossRef]
19. Yeh, I.-C.; Andzelm, J.W.; Rutledge, G.C. Mechanical and Structural Characterization of Semicrystalline Polyethylene under Tensile Deformation by Molecular Dynamics Simulations. *Macromolecules* **2015**, *48*, 4228–4239. [CrossRef]
20. Kim, J.M.; Locker, R.; Rutledge, G.C. Plastic Deformation of Semicrystalline Polyethylene under Extension, Compression, and Shear Using Molecular Dynamics Simulation. *Macromolecules* **2014**, *47*, 2515–2528. [CrossRef]
21. Kumar, V.; Locker, C.R.; Veld, P.J.; Rutledge, G.C. Effect of Short Chain Branching on the Interlamellar Structure of Semicrystalline Polyethylene. *Macromolecules* **2017**, *50*, 1206–1214. [CrossRef]
22. Martin, M.G. MCCCS Towhee: A tool for Monte Carlo molecular simulation. *Mol. Simulat.* **2013**, *39*, 1212–1222. [CrossRef]
23. Tsimpanogiannis, I.N.; Costandy, J.; Kastanidis, P.; El Meragawi, S.; Michalis, V.K.; Papadimitriou, N.I.; Karozis, S.N.; Diamantonis, N.I.; Moultos, O.A.; Romanos, G.E.; et al. Using clathrate hydrates for gas storage and gas-mixture separations: Experimental and computational studies at multiple length scales. *Mol. Phys.* **2018**, *116*, 2041–2060. [CrossRef]
24. Brukhno, A.V.; Grant, J.; Underwood, T.L.; Stratford, K.; Parker, S.C.; Purton, J.A.; Wilding, N.B. DL_MONTE: A multipurpose code for Monte Carlo simulation. *Mol. Simulat.* **2021**, *47*, 131–151. [CrossRef]
25. Wang, F.; Landau, D.P. Efficient, Multiple-Range Random Walk Algorithm to Calculate the Density of States. *Phys. Rev. Lett.* **2001**, *86*, 2050–2053. [CrossRef] [PubMed]
26. Shah, J.K.; Marin-Rimoldi, E.; Mullen, R.G.; Keene, B.P.; Khan, S.; Paluch, A.S.; Rai, N.; Romanielo, L.L.; Rosch, T.W.; Yoo, B.; et al. Cassandra: An open source Monte Carlo package for molecular simulation. *J. Comput. Chem.* **2017**, *38*, 1727–1739. [CrossRef]
27. Dubbeldam, D.; Calero, S.; Ellis, D.E.; Snurr, R.Q. RASPA: Molecular simulation software for adsorption and diffusion in flexible nanoporous materials. *Mol. Simulat.* **2016**, *42*, 81–101. [CrossRef]
28. Nejahi, Y.; Barhaghi, M.S.; Mick, J.; Jackman, B.; Rushaidat, K.; Li, Y.; Schwiebert, L.; Potoff, J. GOMC: GPU Optimized Monte Carlo for the simulation of phase equilibria and physical properties of complex fluids. *SoftwareX* **2019**, *9*, 20–27. [CrossRef]
29. Mick, J.; Hailat, E.; Russo, V.; Rushaidat, K.; Schwiebert, L.; Potoff, J. GPU-accelerated Gibbs ensemble Monte Carlo simulations of Lennard-Jonesium. *Comput. Phys. Commun.* **2013**, *184*, 2662–2669. [CrossRef]
30. Cezar, H.M.; Canuto, S.; Coutinho, K. DICE: A Monte Carlo Code for Molecular Simulation Including the Configurational Bias Monte Carlo Method. *J. Chem. Inf. Model.* **2020**, *60*, 3472–3488. [CrossRef] [PubMed]
31. Gowers, R.J.; Farmahini, A.H.; Friedrich, D.; Sarkisov, L. Automated analysis and benchmarking of GCMC simulation programs in application to gas adsorption. *Mol. Simulat.* **2018**, *44*, 309–321. [CrossRef]
32. Alexiadis, O.; Cheimarios, N.; Peristeras, L.D.; Bick, A.; Mavrantzas, V.G.; Theodorou, D.N.; Hill, J.-R.; Krokidis, X. Chameleon: A generalized, connectivity altering software for tackling properties of realistic polymer systems. *WIREs Comput. Mol. Sci.* **2019**, *9*, e1414. [CrossRef]
33. Ghobadpour, E.; Kolb, M.; Ejtehadi, M.R.; Everaers, R. Monte Carlo simulation of a lattice model for the dynamics of randomly branching double-folded ring polymers. *Phys. Rev. E* **2021**, *104*, 014501. [CrossRef] [PubMed]
34. Theodorou, D.N. A reversible minimum-to-minimum mapping method for the calculation of free-energy differences. *J. Chem. Phys.* **2006**, *124*, 034109. [CrossRef] [PubMed]

35. Uhlherr, A.; Theodorou, D.N. Accelerating molecular simulations by reversible mapping between local minima. *J. Chem. Phys.* **2006**, *125*, 84107. [CrossRef]
36. Shi, W.; Maginn, E.J. Improvement in Molecule Exchange Efficiency in Gibbs Ensemble Monte Carlo: Development and Implementation of the Continuous Fractional Component Move. *J. Comput. Chem.* **2008**, *29*, 2520–2530. [CrossRef] [PubMed]
37. Shi, W.; Maginn, E.J. Continuous fractional component Monte Carlo: An adaptive biasing method for open system atomistic simulations. *J. Chem. Theory Comput.* **2007**, *3*, 1451–1463. [CrossRef]
38. Weismantel, O.; Galata, A.A.; Sadeghi, M.; Kroger, A.; Kroger, M. Efficient generation of self-avoiding, semiflexible rotational isomeric chain ensembles in bulk, confined geometries, and on surfaces. *Comput. Phys. Commun.* **2021**, *270*, 108176. [CrossRef]
39. Kroger, M.; Muller, M.; Nievergelt, J. A geometric embedding algorithm for efficiently generating semiflexible chains in the molten state. *Cmes-Comput. Modeling Eng. Sci.* **2003**, *4*, 559–569.
40. Kampmann, T.A.; Muller, D.; Weise, L.P.; Vorsmann, C.F.; Kierfeld, J. Event-Chain Monte-Carlo Simulations of Dense Soft Matter Systems. *arXiv* **2021**, arXiv:2102.05461.
41. Krauth, W. Event-Chain Monte Carlo: Foundations, Applications, and Prospects. *Front. Phys.* **2021**, *9*, 229. [CrossRef]
42. Klement, M.; Lee, S.; Anderson, J.A.; Engel, M. Newtonian Event-Chain Monte Carlo and Collision Prediction with Polyhedral Particles. *J. Chem. Theory Comput.* **2021**, *17*, 4686–4696. [CrossRef]
43. Kriuchevskyi, I.; Palyulin, V.V.; Milkus, R.; Elder, R.M.; Sirk, T.W.; Zaccone, A. Scaling up the lattice dynamics of amorphous materials by orders of magnitude. *Phys. Rev. B* **2020**, *102*, 024108. [CrossRef]
44. Auhl, R.; Everaers, R.; Grest, G.S.; Kremer, K.; Plimpton, S.J. Equilibration of long chain polymer melts in computer simulations. *J. Chem. Phys.* **2003**, *119*, 12718–12728. [CrossRef]
45. Doshi, U.; Hamelberg, D. Achieving Rigorous Accelerated Conformational Sampling in Explicit Solvent. *J. Phys. Chem. Lett.* **2014**, *5*, 1217–1224. [CrossRef]
46. Subramanian, G. A topology preserving method for generating equilibrated polymer melts in computer simulations. *J. Chem. Phys.* **2010**, *133*, 164902. [CrossRef]
47. Zhang, G.J.; Chazirakis, A.; Harmandaris, V.A.; Stuehn, T.; Daoulas, K.C.; Kremer, K. Hierarchical modelling of polystyrene melts: From soft blobs to atomistic resolution. *Soft Matter* **2019**, *15*, 289–302. [CrossRef]
48. Milchev, A.; Binder, K. Cylindrical confinement of solutions containing semiflexible macromolecules: Surface-induced nematic order versus phase separation. *Soft Matter* **2021**, *17*, 3443–3454. [CrossRef]
49. Zhou, X.L.; Wu, J.X.; Zhang, L.X. Ordered aggregation of semiflexible ring-linear blends in ellipsoidal confinement. *Polymer* **2020**, *197*, 122494. [CrossRef]
50. Milchev, A.; Nikoubashman, A.; Binder, K. The smectic phase in semiflexible polymer materials: A large scale molecular dynamics study. *Comput. Mater. Sci.* **2019**, *166*, 230–239. [CrossRef]
51. Moghimikheirabadi, A.; Kroger, M.; Karatrantos, A.V. Insights from modeling into structure, entanglements, and dynamics in attractive polymer nanocomposites. *Soft Matter* **2021**, *17*, 6362–6373. [CrossRef] [PubMed]
52. Kroger, M. Efficient hybrid algorithm for the dynamic creation of wormlike chains in solutions, brushes, melts and glasses. *Comput. Phys. Commun.* **2019**, *241*, 178–179. [CrossRef]
53. Ahrens, J.; Geveci, B.; Law, C. *ParaView: An End-User Tool for Large Data Visualization*; Elsevier: Amsterdam, The Netherlands, 2005.
54. Humphrey, W.; Dalke, A.; Schulten, K. VMD: Visual molecular dynamics. *J. Mol. Graph. Model.* **1996**, *14*, 33–38. [CrossRef]
55. Bumstead, M.; Liang, K.Y.; Hanta, G.; Hui, L.S.; Turak, A. disLocate: Tools to rapidly quantify local intermolecular structure to assess two-dimensional order in self-assembled systems. *Sci. Rep.* **2018**, *8*, 1–15.
56. Pettersen, E.F.; Goddard, T.D.; Huang, C.C.; Couch, G.S.; Greenblatt, D.M.; Meng, E.C.; Ferrin, T.E. UCSF chimera—A visualization system for exploratory research and analysis. *J. Comput. Chem.* **2004**, *25*, 1605–1612. [CrossRef] [PubMed]
57. Stukowski, A. Visualization and analysis of atomistic simulation data with OVITO-the Open Visualization Tool. *Modell. Simul. Mater. Sci. Eng.* **2010**, *18*, 015012. [CrossRef]
58. Tassieri, M.; Ramirez, J.; Karayiannis, N.C.; Sukumaran, S.K.; Masubuchi, Y. i-Rheo GT: Transforming from Time to Frequency Domain without Artifacts. *Macromolecules* **2018**, *51*, 5055–5068. [CrossRef]
59. Karayiannis, N.C.; Laso, M. Monte Carlo scheme for generation and relaxation of dense and nearly jammed random structures of freely jointed hard-sphere chains. *Macromolecules* **2008**, *41*, 1537–1551. [CrossRef]
60. Ramos, P.M.; Herranz, M.; Foteinopoulou, K.; Karayiannis, N.C.; Laso, M. Identification of Local Structure in 2-D and 3-D Atomic Systems through Crystallographic Analysis. *Crystals* **2020**, *10*, 1008. [CrossRef]
61. Karayiannis, N.C.; Foteinopoulou, K.; Laso, M. The characteristic crystallographic element norm: A descriptor of local structure in atomistic and particulate systems. *J. Chem. Phys.* **2009**, *130*, 074704. [CrossRef]
62. Foteinopoulou, K.; Karayiannis, N.C.; Laso, M. Monte Carlo simulations of densely-packed athermal polymers in the bulk and under confinement. *Chem. Eng. Sci.* **2015**, *121*, 118–132. [CrossRef]
63. Herranz, M.; Santiago, M.; Foteinopoulou, K.; Karayiannis, N.C.; Laso, M. Crystal, Fivefold and Glass Formation in Clusters of Polymers Interacting with the Square Well Potential. *Polymers* **2020**, *12*, 1111. [CrossRef]
64. Karayiannis, N.C.; Foteinopoulou, K.; Laso, M. The structure of random packings of freely jointed chains of tangent hard spheres. *J. Chem. Phys.* **2009**, *130*, 164908. [CrossRef] [PubMed]
65. Karayiannis, N.C.; Foteinopoulou, K.; Laso, M. Contact network in nearly jammed disordered packings of hard-sphere chains. *Phys. Rev. E* **2009**, *80*, 011307. [CrossRef] [PubMed]

66. Laso, M.; Karayiannis, N.C.; Foteinopoulou, K.; Mansfield, M.L.; Kroger, M. Random packing of model polymers: Local structure, topological hindrance and universal scaling. *Soft Matter* **2009**, *5*, 1762–1770. [CrossRef]
67. Foteinopoulou, K.; Karayiannis, N.C.; Laso, M.; Kroger, M.; Mansfield, M.L. Universal Scaling, Entanglements, and Knots of Model Chain Molecules. *Phys. Rev. Lett.* **2008**, *101*, 265702. [CrossRef]
68. Karayiannis, N.C.; Kroger, M. Combined Molecular Algorithms for the Generation, Equilibration and Topological Analysis of Entangled Polymers: Methodology and Performance. *Int. J. Mol. Sci.* **2009**, *10*, 5054–5089. [CrossRef]
69. Karayiannis, N.C.; Foteinopoulou, K.; Laso, M. The role of bond tangency and bond gap in hard sphere crystallization of chains. *Soft Matter* **2015**, *11*, 1688–1700. [CrossRef]
70. Karayiannis, N.C.; Foteinopoulou, K.; Laso, M. Jamming and crystallization in athermal polymer packings. *Philos. Mag.* **2013**, *93*, 4108–4131. [CrossRef]
71. Karayiannis, N.C.; Foteinopoulou, K.; Abrams, C.F.; Laso, M. Modeling of crystal nucleation and growth in athermal polymers: Self-assembly of layered nano-morphologies. *Soft Matter* **2010**, *6*, 2160–2173. [CrossRef]
72. Karayiannis, N.C.; Foteinopoulou, K.; Laso, M. Entropy-Driven Crystallization in Dense Systems of Athermal Chain Molecules. *Phys. Rev. Lett.* **2009**, *103*, 045703. [CrossRef]
73. Ramos, P.M.; Herranz, M.; Foteinopoulou, K.; Karayiannis, N.C.; Laso, M. Entropy-Driven Heterogeneous Crystallization of Hard-Sphere Chains under Unidimensional Confinement. *Polymers* **2021**, *13*, 1352. [CrossRef] [PubMed]
74. Pant, P.V.K.; Theodorou, D.N. Variable Connectivity Method For The Atomistic Monte-Carlo Simulation Of Polydisperse Polymer Melts. *Macromolecules* **1995**, *28*, 7224–7234. [CrossRef]
75. Ramos, P.M.; Karayiannis, N.C.; Laso, M. Off-lattice simulation algorithms for athermal chain molecules under extreme confinement. *J. Comput. Phys.* **2018**, *375*, 918–934. [CrossRef]
76. Siepmann, J.I.; Frenkel, D. Configurational Bias Monte-Carlo—A New Sampling Scheme for Flexible Chains. *Mol. Phys.* **1992**, *75*, 59–70. [CrossRef]
77. Laso, M.; de Pablo, J.J.; Suter, U.W. Simulation of Phase-Equilibria for Chain Molecules. *J. Chem. Phys.* **1992**, *97*, 2817–2819. [CrossRef]
78. Mavrantzas, V.G.; Boone, T.D.; Zervopoulou, E.; Theodorou, D.N. End-bridging Monte Carlo: A fast algorithm for atomistic simulation of condensed phases of long polymer chains. *Macromolecules* **1999**, *32*, 5072–5096. [CrossRef]
79. Karayiannis, N.C.; Mavrantzas, V.G.; Theodorou, D.N. A novel Monte Carlo scheme for the rapid equilibration of atomistic model polymer systems of precisely defined molecular architecture. *Phys. Rev. Lett.* **2002**, *88*, 105503. [CrossRef]
80. Karayiannis, N.C.; Giannousaki, A.E.; Mavrantzas, V.G.; Theodorou, D.N. Atomistic Monte Carlo simulation of strictly monodisperse long polyethylene melts through a generalized chain bridging algorithm. *J. Chem. Phys.* **2002**, *117*, 5465–5479. [CrossRef]
81. Faken, D.; Jónsson, H. Systematic analysis of local atomic structure combined with 3D computer graphics. *Comput. Mater. Sci.* **1994**, *2*, 279–286. [CrossRef]
82. Martin, A.V.; Kozlov, A.; Berntsen, P.; Roque, F.G.; Flueckiger, L.; Saha, S.; Greaves, T.L.; Conn, C.E.; Hawley, A.M.; Ryan, T.M.; et al. Fluctuation X-ray diffraction reveals three-dimensional nanostructure and disorder in self-assembled lipid phases. *Commun. Mater.* **2020**, *1*, 40. [CrossRef]
83. Steinhardt, P.J.; Nelson, D.R.; Ronchetti, M. Bond-orientational order in liquids and glasses. *Phys. Rev. B* **1983**, *28*, 784–805. [CrossRef]
84. Larsen, P.M.; Schmidt, S.; Schiotz, J. Robust structural identification via polyhedral template matching. *Modell. Simul. Mater. Sci. Eng.* **2016**, *24*. [CrossRef]
85. Tanemura, M.; Hiwatari, Y.; Matsuda, H.; Ogawa, T.; Ogita, N.; Ueda, A. Geometrical analysis of crystallization of the soft-core model. *Prog. Theor. Phys.* **1977**, *58*, 1079–1095. [CrossRef]
86. Anikeenko, A.V.; Medvedev, N.N.; Aste, T. Structural and entropic insights into the nature of the random-close-packing limit. *Phys. Rev. E* **2008**, *77*, 031101. [CrossRef] [PubMed]
87. Ackland, G.J.; Jones, A.P. Applications of local crystal structure measures in experiment and simulation. *Phys. Rev. B* **2006**, *73*, 054104. [CrossRef]
88. Cohen, M.H.; Grest, G.S. Liquid-glass transition, a free-volume approach. *Phys. Rev. B* **1979**, *20*, 1077–1098. [CrossRef]
89. Egami, T.; Maeda, K.; Vitek, V. Structural Defects In Amorphous Solids—A Computer-Simulation Study. *Philos. Mag. A* **1980**, *41*, 883–901. [CrossRef]
90. Ding, J.; Cheng, Y.Q.; Sheng, H.; Asta, M.; Ritchie, R.O.; Ma, E. Universal structural parameter to quantitatively predict metallic glass properties. *Nat. Commun.* **2016**, *7*, 13733. [CrossRef] [PubMed]
91. Malins, A.; Williams, S.R.; Eggers, J.; Royall, C.P. Identification of structure in condensed matter with the topological cluster classification. *J. Chem. Phys.* **2013**, *139*, 234506. [CrossRef] [PubMed]
92. Stukowski, A. Structure identification methods for atomistic simulations of crystalline materials. *Modell. Simul. Mater. Sci. Eng.* **2012**, *20*, 045021. [CrossRef]
93. Paret, J.; Jack, R.L.; Coslovich, D. Assessing the structural heterogeneity of supercooled liquids through community inference. *J. Chem. Phys.* **2020**, *152*, 144502. [CrossRef] [PubMed]
94. Nye, J.F. *Physical Properties of Crystals: Their Representation by Tensors and Matrices*; Oxford Science Publications: Oxford, UK, 2010.
95. Malgrange, C.; Ricolleau, C.; Schlenker, M. *Symmetry and Physical Properties of Crystals*; Springer: Dordrecht, The Netherlands, 2014. [CrossRef]

96. Giacovazzo, C.; Monaco, H.L.; Artioli, G.; Viterbo, D.; Ferraris, G.; Gilli, G.; Zanotti, G.; Gatti, M. *Fundamentals of Crystallography;* Oxford Science: Oxford, UK, 2005.
97. Laso, M.; Jimeno, N. *Representation Surfaces for Physical Properties of Materials: A Visual Approach to Understanding Anisotropic Materials;* Springer: Berlin/Heidelberg, Germany, 2020.
98. Hales, T.C. A proof of the Kepler conjecture. *Ann. Math.* **2005**, *162*, 1065–1185. [CrossRef]
99. Hales, T.C.; Harrison, J.; McLaughlin, S.; Nipkow, T.; Obua, S.; Zumkeller, R. A Revision of the Proof of the Kepler Conjecture. *Discret. Comput. Geom.* **2010**, *44*, 1–34. [CrossRef]
100. Bernal, J.D.; Finney, J.L. Random close-packed hard-sphere model. 2. Geometry of random packing of hard spheres. *Discuss. Faraday Soc.* **1967**, *43*, 62–69. [CrossRef]
101. Bernal, J.D. Geometry of The Structure of Monatomic Liquids. *Nature* **1960**, *185*, 68–70. [CrossRef]
102. Torquato, S.; Truskett, T.M.; Debenedetti, P.G. Is random close packing of spheres well defined? *Phys. Rev. Lett.* **2000**, *84*, 2064–2067. [CrossRef]
103. Karayiannis, N.C.; Laso, M. Dense and nearly jammed random packings of freely jointed chains of tangent hard spheres. *Phys. Rev. Lett.* **2008**, *100*, 050602. [CrossRef]
104. Hoy, R.S. Jamming of Semiflexible Polymers. *Phys. Rev. Lett.* **2017**, *118*, 068002. [CrossRef]
105. Shakirov, T.; Paul, W. Crystallization in melts of short, semiflexible hard polymer chains: An interplay of entropies and dimensions. *Phys. Rev. E* **2018**, *97*, 042501. [CrossRef] [PubMed]
106. O'Malley, B.; Snook, I. Crystal nucleation in the hard sphere system. *Phys. Rev. Lett.* **2003**, *90*, 085702. [CrossRef] [PubMed]
107. Karayiannis, N.C.; Malshe, R.; de Pablo, J.J.; Laso, M. Fivefold symmetry as an inhibitor to hard-sphere crystallization. *Phys. Rev. E* **2011**, *83*, 061505. [CrossRef]
108. Karayiannis, N.C.; Malshe, R.; Kroger, M.; de Pablo, J.J.; Laso, M. Evolution of fivefold local symmetry during crystal nucleation and growth in dense hard-sphere packings. *Soft Matter* **2012**, *8*, 844–858. [CrossRef]
109. Karayiannis, N.C.; Foteinopoulou, K.; Laso, M. Spontaneous Crystallization in Athermal Polymer Packings. *Int. J. Mol. Sci.* **2013**, *14*, 332–358. [CrossRef] [PubMed]
110. Chen, Y.S.; Yao, Z.W.; Tang, S.X.; Tong, H.; Yanagishima, T.; Tanaka, H.; Tan, P. Morphology selection kinetics of crystallization in a sphere. *Nat. Phys.* **2021**, *17*, 121–127. [CrossRef]
111. Arai, S.; Tanaka, H. Surface-assisted single-crystal formation of charged colloids. *Nat. Phys.* **2017**, *13*, 503–509. [CrossRef]
112. Torquato, S.; Stillinger, F.H. Jammed hard-particle packings: From Kepler to Bernal and beyond. *Rev. Mod. Phys.* **2010**, *82*, 2633–2672. [CrossRef]
113. Wilken, S.; Guerra, R.E.; Levine, D.; Chaikin, P.M. Random Close Packing as a Dynamical Phase Transition. *Phys. Rev. Lett.* **2021**, *127*, 038002. [CrossRef]
114. Rissone, P.; Corwin, E.I.; Parisi, G. Long-Range Anomalous Decay of the Correlation in Jammed Packings. *Phys. Rev. Lett.* **2021**, *127*, 038001. [CrossRef]
115. Alder, B.J.; Wainwright, T.E. Phase Transition For A Hard Sphere System. *J. Chem. Phys.* **1957**, *27*, 1208–1209. [CrossRef]
116. Ni, R.; Dijkstra, M. Effect of bond length fluctuations on crystal nucleation of hard bead chains. *Soft Matter* **2013**, *9*, 365–369. [CrossRef]
117. Johnson, J.K.; Panagiotopoulos, A.Z.; Gubbins, K.E. Reactive canonical monte-carlo—A new simulation technique for reacting or associating fluids. *Mol. Phys.* **1994**, *81*, 717–733. [CrossRef]
118. Laso, M.; Karayiannis, N.C.; Muller, M. Min-map bias Monte Carlo for chain molecules: Biased Monte Carlo sampling based on bijective minimum-to-minimum mapping. *J. Chem. Phys.* **2006**, *125*, 164108. [CrossRef] [PubMed]
119. Uhlherr, A.; Doxastakis, M.; Mavrantzas, V.G.; Theodorou, D.N.; Leak, S.J.; Adam, N.E.; Nyberg, P.E. Atomic structure of a high polymer melt. *Europhys. Lett.* **2002**, *57*, 506–511. [CrossRef]
120. Uhlherr, A.; Leak, S.J.; Adam, N.E.; Nyberg, P.E.; Doxastakis, M.; Mavrantzas, V.G.; Theodorou, D.N. Large scale atomistic polymer simulations using Monte Carlo methods for parallel vector processors. *Comput. Phys. Commun.* **2002**, *144*, 1–22. [CrossRef]
121. Kroger, M. Shortest multiple disconnected path for the analysis of entanglements in two- and three-dimensional polymeric systems. *Comput. Phys. Commun.* **2005**, *168*, 209–232. [CrossRef]
122. Caraglio, M.; Enzo, C.M.; Orlandini, E. Physical Links: Defining and detecting inter-chain entanglement. *Sci. Rep.* **2017**, *7*, 1–10. [CrossRef]
123. Tzoumanekas, C.; Theodorou, D.N. Topological analysis of linear polymer melts: A statistical approach. *Macromolecules* **2006**, *39*, 4592–4604. [CrossRef]
124. Everaers, R.; Sukumaran, S.K.; Grest, G.S.; Svaneborg, C.; Sivasubramanian, A.; Kremer, K. Rheology and microscopic topology of entangled polymeric liquids. *Science* **2004**, *303*, 823–826. [CrossRef] [PubMed]
125. Shanbhag, S.; Larson, R.G. Identification of Topological Constraints in Entangled Polymer Melts Using the Bond-Fluctuation Model. *Macromolecules* **2006**, *39*, 2413–2417. [CrossRef]

International Journal of
Molecular Sciences

MDPI

Article

Computational-Driven Epitope Verification and Affinity Maturation of TLR4-Targeting Antibodies

Bilal Ahmad [1], Maria Batool [1,2], Moon-Suk Kim [1] and Sangdun Choi [1,2,*]

[1] Department of Molecular Science and Technology, Ajou University, Suwon 16499, Korea; bilalpharma77@gmail.com (B.A.); mariabatool.28@gmail.com (M.B.); moonskim@ajou.ac.kr (M.-S.K.)
[2] S&K Therapeutics, Woncheon Hall 135, Ajou University, Suwon 16499, Korea
* Correspondence: sangdunchoi@ajou.ac.kr; Tel.: +82-31-219-2600; Fax: +82-31-219-1615

Abstract: Toll-like receptor (TLR) signaling plays a critical role in the induction and progression of autoimmune diseases such as rheumatoid arthritis, systemic lupus erythematous, experimental autoimmune encephalitis, type 1 diabetes mellitus and neurodegenerative diseases. Deciphering antigen recognition by antibodies provides insights and defines the mechanism of action into the progression of immune responses. Multiple strategies, including phage display and hybridoma technologies, have been used to enhance the affinity of antibodies for their respective epitopes. Here, we investigate the TLR4 antibody-binding epitope by computational-driven approach. We demonstrate that three important residues, i.e., Y328, N329, and K349 of TLR4 antibody binding epitope identified upon in silico mutagenesis, affect not only the interaction and binding affinity of antibody but also influence the structural integrity of TLR4. Furthermore, we predict a novel epitope at the TLR4-MD2 interface which can be targeted and explored for therapeutic antibodies and small molecules. This technique provides an in-depth insight into antibody–antigen interactions at the resolution and will be beneficial for the development of new monoclonal antibodies. Computational techniques, if coupled with experimental methods, will shorten the duration of rational design and development of antibody therapeutics.

Keywords: antibody; epitope; molecular dynamics; mutation; toll-like receptor

Citation: Ahmad, B.; Batool, M.; Kim, M.-S.; Choi, S. Computational-Driven Epitope Verification and Affinity Maturation of TLR4-Targeting Antibodies. *Int. J. Mol. Sci.* **2021**, *22*, 5989. https://doi.org/10.3390/ijms22115989

Academic Editor: Małgorzata Borówko

Received: 26 April 2021
Accepted: 29 May 2021
Published: 1 June 2021

1. Introduction

Toll-like receptors (TLRs), as featured pattern recognition receptors, are proven to operate as germline-encoded proteins recognizing conserved pathogen-associated molecular patterns [1]. Dysregulation of cellular activities by microbes or their products through TLR signaling affects both innate and adaptive immune responses [2]. Excessive activation of TLRs disrupts immune homeostasis and is triggered by persistent induction of proinflammatory cytokines and chemokines, thereby subsequently leading to the initiation of various inflammatory and autoimmune disorders such as systemic lupus erythematosus, sepsis, psoriasis, atherosclerosis and asthma [3]. TLR4, the first mammalian Toll protein characterized in humans [4], recognizes damage-associated molecular patterns in the debris released by injured tissues and necrotic cells as well as pathogen-associated molecular patterns, and is thus associated with the development of several acute and chronic disorders such as sepsis [5]. TLR4 in association with interleukin (IL) 29 plays a crucial role in synovial inflammation. IL-29 upregulates synovial-fibroblast TLR4, which enhances synovium inflammation in rheumatoid arthritis (RA). Elevated expression may be due to elevated numbers of macrophages that penetrate the synovium and promote RA [6,7]. Numerous studies have shown that TLR4 stimulates the expression of many proinflammatory cytokines that play a crucial part in myocardial inflammation, including myocarditis, myocardial infarction, ischemia-reperfusion, and heart failure [8]. Ample evidence confirms the participation of TLR4 in neuroinflammation, where activation of this receptor stimulates microglial expression and activation of NF-κB, as well as the induction of inflammatory cytokines IL-1β, IL-6 and tumor necrosis factor (TNF) α [9,10]. The involvement of TLR4 in the pathogenesis and aggravation of these diseases highlights its importance as a potential drug target. In the TLR family, TLR4 remains a priority drug target and has been extensively studied for its therapeutic potential in several inflammatory disorders; many of these therapeutics are undergoing clinical trials [11].

Antibody-based medicines are currently the most widely used form of biotherapeutics, and their market expanded rapidly in recent years with increasing numbers of such modalities receiving FDA approval. Compared to small-molecule drugs, monoclonal antibodies (mAbs) are considered more selective and highly effective and have become mainstream therapies for autoimmune and other hard-to-cure diseases. Moreover, they have revolutionized the treatment of cancers, where inflammation is regarded as a crucial factor [12]. Nonetheless, small-molecule drugs that lack specificity are likely to have off-target effects. These drugs can interact unexpectedly and undesirably with tissues, cells and cellular components. The application of small-molecule drugs has a greater number of adverse effects due to their lower specificity as compared to mAbs. On the contrary, the clinical approval rate is higher for mAbs, and their development is easier and considerably faster than sorganic small-molecule compounds.

The development of high-affinity mAbs for therapeutic purposes is still a holy grail of molecular engineering [13]. Researchers are using numerous empirical procedures, including site-directed mutagenesis, complementarity-determining region (CDR)-grafting, and phage display techniques to enhance the target-binding affinity of mAbs [14]. Computational approaches are coming onto the scene and facilitating the development and affinity enhancement of mAbs. Prior sequence and structural information on both antigens and mAbs, and their epitope and paratope insights, are of great value for enhancing the affinity and stability of mAbs. These computational methods involve structural bioinformatic techniques such as homology modeling, protein–protein docking, protein interface analysis and molecular dynamics (MD) for rational design of mAbs. They provide detailed insights into binding and unbinding mechanisms as well as structure kinetics of protein-protein, protein-ligand and antigen-antibody that can be used to guide ligand, protein, peptide and mAb design [15]. The modelling of mAb-antigen interaction, and applying attraction or repulsion filters for masking the nonparatope residues as in Cluspro program, substantially improves the antigen-antibody docking complexes [16]. Thus, false positive of an epitope can be removed. The computational techniques alone can be used for affinity maturation of selected mAbs. Techniques such as AbDesign, RosettaAntibodyDesign, OptCDR and OptMAVEn are categorized as ab initio protocols for the design of novel paratopes [17,18]. Machine learning and deep learning methods can help to design CDRs for human IgG antibodies with target affinity that is superior to that of candidates derived from phage display panning experiments [19,20]. Thus far, a few effective TLR4-targeting mAbs have been investigated in preclinical and clinical studies [21,22]. NI-0101, previously known as Hu 15C1, has been developed in BALB/c mice and proved to efficiently block the signaling of lipopolysaccharide-triggered TLR4 both in vitro and in vivo [22,23]. NI-0101 is a humanized immunoglobulin (Ig) G1κ mAb, which is engineered to interfere with the dimerization of TLR4 and to abrogate its downstream signaling. NI-0101 has been proved effective in lipopolysaccharide-treated healthy volunteers [24]; nonetheless, recent clinical findings suggest that to cure RA, inhibition of TLR4 alone is not sufficient [25]. A molecular understanding of antigen–antibody interaction and identification of their epitope–paratope hotspots are indispensable for affinity enhancement of known mAbs. Taking advantage of the epitope knowledge provided by Greg Elson's group through alanine mutagenesis [22], Loyau et al. utilized structural, computational, and phage display techniques to enhance TLR4-binding affinity of Hu 15C1 and suggested a C2E3 derivative with better TLR4-binding affinity [26]. Overall, these studies predicted the TLR4-mAb binding interface through site-directed mutational analyses. Nevertheless, the conformational changes that occur in the TLR4 structure owing to these mutations have been overlooked. These mutation-driven conformational changes obstruct TLR4-mAb interaction. In this study, we used computational mutagenesis, molecular docking, MD simulations and molecular mechanics to delineate the dynamic binding interface of Hu 15C1 and its derivative C2E3 toward TLR4. The crucial data generated during this study regarding TLR4 epitopes will be helpful for designing efficient mAbs to block TLR4 signaling and to curb the associated immune complications.

2. Results

2.1. Computational TLR4 Epitope Mutagenesis and Network Analysis

Hu 15C1 is a potent anti-TLR4 mAb, which, owing to its efficacy, has reached clinical trials and is being evaluated in RA patients [25]. The epitope of mAb Hu 15C1 has been mapped by alanine mutagenesis and has been found to be located in the ectodomain of TLR4 near the dimerization interface [27]. This epitope has also been investigated in another study [26], in which a new derivative antibody, C2E3, was designed. By exploring the TLR4 structure, researchers hypothesize that the epitope-constituting residues suggested by these studies are crucial for preserving the structural integrity of the TLR4 backbone (Figure 1A). With the aim of gaining insights into TLR4 structure and mAb binding at the atomic level of specific residues, computational alanine scanning, and an MD simulation were performed here. Six hot spot residues, i.e., Y328, N329, K349, K351, E369 and D371, were individually mutated to alanine and vetted for changes in the interactions between TLR4 and a mAb.

Moreover, to comprehend the multiple interaction types involved in the residue network in wild-type TLR4 (TLR4^{wt}) and its muteins, network analysis was performed. Substantial changes in the hydrogen bond network and in van der Waals interactions were observed at the mutation sites, and these changes led to a distinct conformational change in the epitope (Figure S1). Topological parameters of the residue interaction network (RIN) are given in Figure S1H. In the RIN of TLR4^{wt} and its muteins, there was a considerable variation in the node degree distribution. Thus, mutating these residues could distort the structure of TLR4 and make the precise Hu 15C1 epitope ambiguous.

2.2. MD of TLR4^{wt} and Muteins

Behaviors of TLR4 and its muteins were evaluated through conventional MD. Root mean square deviations (RMSDs) of Cα atoms of muteins and TLR4^{wt} were calculated as a function of time. Comparative RMSD plots suggested that TLR4^{wt} remained stable, while all the muteins underwent rigorous fluctuations (Figure 1B). The muteins manifested RMSDs in the following order: TLR4^{Y328A} > TLR4^{N329A} > TLR4^{D371A} > TLR4^{K349A} > TLR4^{E369A} > TLR4^{K351A} > TLR4^{wt}. Additionally, to probe the effect of these mutations on TLR4 structure, root mean square fluctuation (RMSF) analysis was carried out, which confirmed that muteins TLR4^{Y328A}, TLR4^{N329A}, and TLR4^{K349A} underwent more fluctuations as compared to the other muteins and TLR4^{wt} structures (Figure 1D). Detailed analyses then suggested (discussed below) that these mutations affect the overall structure of TLR4, particularly the proposed epitope region. Owing to these mutation-driven substantial deviations in structural integrity, we concluded that these residues are more important for preserving TLR4 folding. Furthermore, the compactness of TLR4^{wt} and its muteins was determined by measuring the radius of gyration (R_g). Differences in R_g between TLR4^{wt} and muteins are presented in Figure 1C. TLR4^{Y328A}, TLR4^{N329A} and TLR4^{D371A} showed significant increases in R_g, indicating a decrease in the compactness of the system. In the second run of MD simulation the RMSD and R_g (Figure S2) showed almost a similar trend except TLR4^{K349A} which underwent a bit higher RMSD fluctuation during 125–150 ns time-point (Figure S2D). To investigate the robustness of these results, an amber AMBER99SB-ILDN force field was implemented in all cases. All the trajectories showed a similar trend but with a lower RMSD than in the CHARMM36 force-field except TLR4^{Y328A} which showed fluctuation from the 65–85 ns time-point (Figure S2B). In compactness, a similar trend was followed with lower R_g value as compared to the CHARMM36 force-field

(Figure S2E,F). The two force-fields generally produce very similar overall RMSD and Rg. It has been found that CHARMM36 tends to produce more structure variation shifted towards disordered conformations, while as AMBER99SB-ILDN over stabilizes the native fold [28,29].

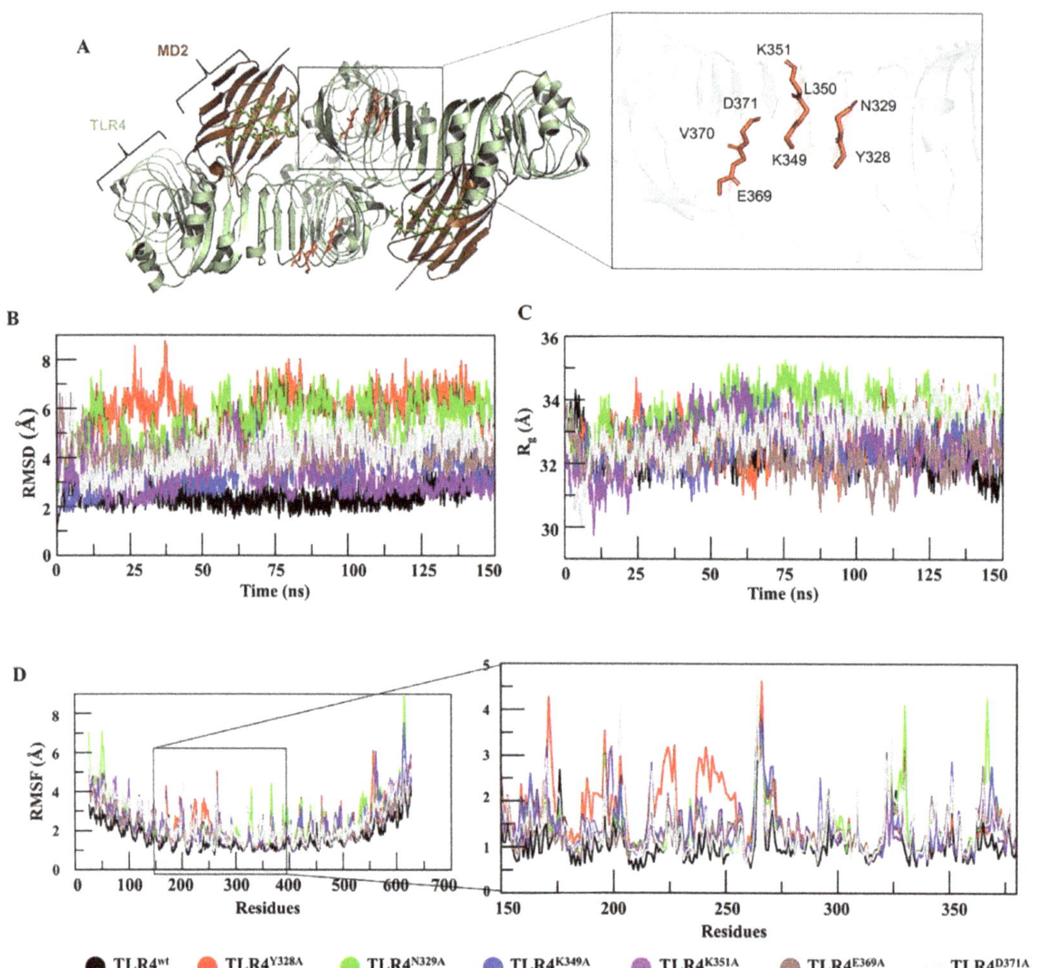

Figure 1. Characterization of an epitope using MD. (**A**) Epitope location in the TLR4–MD2 dimer interface and the residues of the epitope are labelled. (**B**) The mutational effect of these residues on TLR4 stability was analyzed by RMSD, where $TLR4^{Y328A}$ (red), $TLR4^{N329A}$ (green), and $TLR4^{D371A}$ (grey) deviated substantially from $TLR4^{wt}$ (black) while RMSD of $TLR4^{K349A}$ (blue) remained constant. (**C**) R_g determining the compactness of a protein was analyzed for $TLR4^{wt}$ and its muteins. For $TLR4^{Y328A}$ (red) and $TLR4^{N329A}$ (green), the compactness was lost because their R_g was higher than that of $TLR4^{wt}$ (black). (**D**) RMSF of $TLR4^{wt}$ and its muteins was measured. $TLR4^{N328A}$, $TLR4^{Y329A}$, and $TLR4^{K349A}$ showed a fluctuation more vividly in the epitopic region as illustrated in the zoomed view.

2.3. Mutation-Induced Conformational Changes in TLR4 Structure

To assess harmonic motions, NMA of $TLR4^{wt}$ and muteins was performed. $TLR4^{wt}$ showed an outward-motion tendency at both N and C termini; by contrast, $TLR4^{Y328A}$ featured an inward movement at both termini (Figure 2A,B). The Y328A mutation seemingly changes the rigidity of the convex surface of TLR4 and allows the N and C termini to bend inward and come closer. A conformational change of such magnitude could affect the dimerization interface of TLR4 monomers in the $(TLR4–MD2)^2$ complex (where MD2 is myeloid differentiation factor 2). The terminal regions of $TLR4^{N329A}$ showed a vertical motion, inducing structural torsion, which resulted in a reduction in the number of internal hydrogen bonds and deformation of the solenoid structure of TLR4 (Figure 2C). $TLR4^{K349A}$ underwent motions similar to those of $TLR4^{wt}$ (Figure 2D). This finding was in line with the RMSF and RMSD plots, where $TLR4^{K349A}$ did not show significant deviation and featured $TLR4^{wt}$-like behavior. Other muteins, $TLR4^{K351A}$, $TLR4^{E369A}$ and $TLR4^{D371A}$, showed a peculiar motion; however, the magnitude of these movements was not significant (Figure S3).

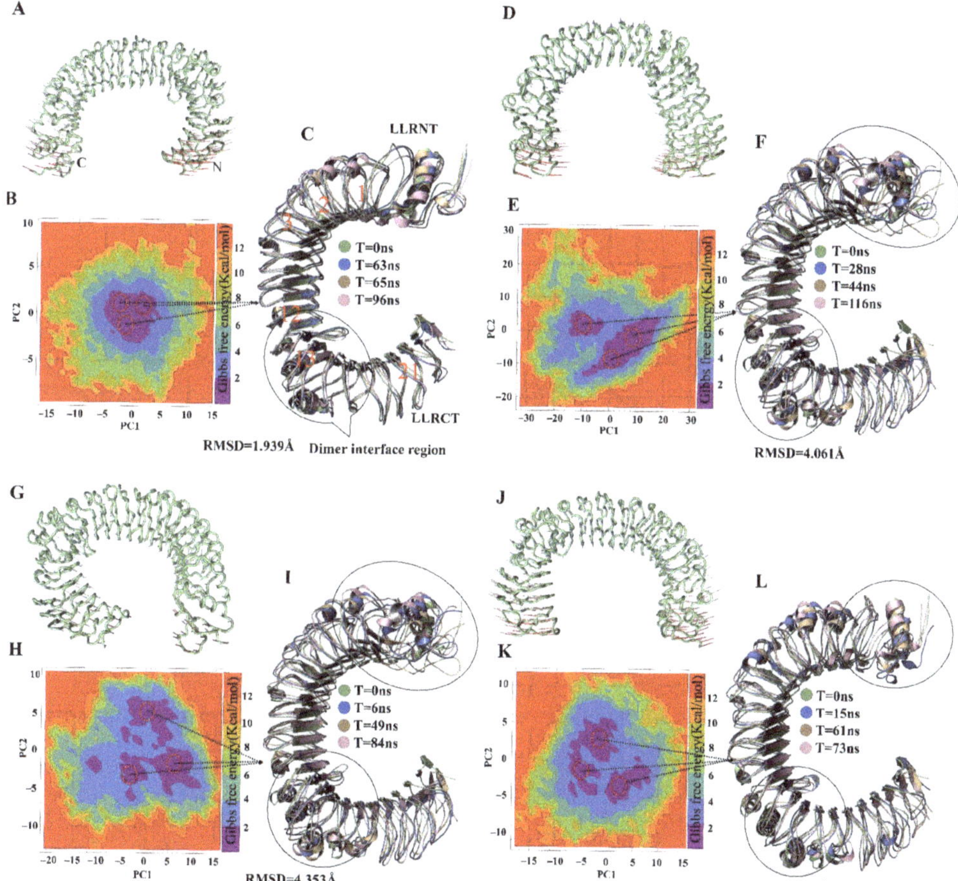

Figure 2. Motion and FEL of $TLR4^{wt}$ and its muteins. The motion of N and C termini in a porcupine plot of (**A**) $TLR4^{wt}$ is outward and in (**D**) it is inward with large magnitude, and for (**G**) $TLR4^{Y329A}$ and (**J**) $TLR4^{K349A}$, it is irregular. The FEL was computed using PC1 and PC2 as reaction coordinates for (**B**) $TLR4^{wt}$, where all conformations remained confined to one minimum. (**E**) $TLR4^{Y328A}$ conformations split into two minima with higher energy barrier. (**H**) $TLR4^{Y329A}$ and (**K**) $TLR4^{K349A}$ split into three minima. A representative lowest-energy structure from local minima indicated by red circles was superimposed (**C**,**F**,**I**,**L**) with the respective first or 0 ns frame structure in $TLR4^{wt}$ and mutants shown along with the RMSD values. The black circle exemplifies the variation in structure at the epitopic region or dimer interface and termini.

2.4. Visualization and Identification of Principal Motion of TLR4^{wt} and Muteins

The dominant motions in TLR4^{wt} and muteins were determined through principal component analysis (PCA) in which the first five eigenvectors captured most of MD motions. Most of the dynamic structural information for each system was captured by 5–6 eigenvectors with considerable fluctuations, while fluctuations of the remaining eigenvectors had low amplitude. The first three eigenvectors accounted for 63.9%, 67.4%, 75.2%, 63.6%, 75.0%, 65.2% and 74.7% of principal motions for TLR4^{wt}, TLR4^{Y328A}, TLR4^{N329A}, TLR4^{K349A}, TLR4^{K351A}, TLR4^{E369A}, and TLR4^{D371A}, respectively (Figure S4). The first three eigenvectors of TLR4^{wt} and muteins were projected to determine interconformer relations (Figure S4). In Figure 2 and Figure S2, continuous color representation from red to blue indicates the change in conformation. The distribution of conformers in TLR4^{wt} remained mostly in two subspaces, i.e., red and blue. In the case of TLR4^{Y328A}, the conformers were scattered into a subspace owing to periodic jumps between different conformations. Unlike TLR4^{wt}, TLR4^{N329A} conformers dispersed and covered a large subspace, as was the case for TLR4^{Y328A}. In the RMSD graph, TLR4^{Y328A} and TLR4^{N329A} showed instability because greater fluctuations were observed in PCA. The graph revealed continuous periodic jumps from one state to another, representing the instability of the alanine muteins of TLR4^{Y328} and TLR4^{N329}. The transition between different conformations in these two muteins makes the epitope less exposed due to conformational changes. The other muteins showed different behavior in the PCA plots (Figure S4).

2.5. Energetics of Conformation Transitions

PCA was carried out to extract the predominant motions of the TLR4^{wt} and mutein structures; these data can provide a better picture of conformational changes and structural evolution along the trajectory of the MD simulation. The first two principal components were employed to calculate Gibbs free energy of each system via plotting in the free-energy landscape (FEL). A plateau in an FEL plot indicates the state of a protein corresponding to its energy and structural transition to the lowest-energy state along a simulation path. An overall view suggested that TLR4^{wt} remains in a single large global minima basin in the dark blue area (Figure 2B). Structure coordinates from the energy minima were randomly sampled to check conformational changes. Three samples were superimposed over the first frame. There was no substantial change in these structures, and their average RMSD was recorded and found to be 1.939 Å, which meant that TLR4^{wt} retains its conformation during the simulation.

In contrast, the energy plateau for TLR4^{Y328A}, TLR4^{N329A} and TLR4^{K349A} was found to split into two, four and three energy minima, respectively, separated by high energy barriers (Figure 2E,H,K). Besides, the representative structures randomly sampled from these energy minima showed a significant change in their N and C termini. Moreover, conformational changes in leucine-rich repeat (LRR) regions were observed. LRR13 is in the proximity to the dimeric interface of TLR4 and manifested a significant difference in the muteins. The α-helix near the dimer interface also underwent substantial conformational changes. Nonetheless, TLR4^{K351A}, TLR4^{E369A}, and TLR4^{D371A} structures extracted at different time points from corresponding deep blue regions of the FEL did not show any difference in structural topology and RMSD (Figure S3).

2.6. The Influence of Mutations on the TLR4-mAb Interaction and Binding Affinity

We noticed that the mutations caused various structural and conformational changes in TLR4, thereby increasing internal energy, and decreasing stability. Moreover, we noted that mutations TLR4^{Y328A}, TLR4^{N329}, and TLR4^{K349A}, which mask the epitopes, influenced the structural integrity of TLR4. Taking into consideration the effect of these three mutations on TLR4, the mutated structures were docked with Hu 15C1 Fab and were simulated for 150 ns to gain insights into the detailed structural dynamics and comparative changes in binding affinity. Unlike the TLR4^{wt}–Hu 15C1 complex, mutant complexes yielded a continuous rise in the RMSD plot (Figure 3A). The RMSD rise in these mutant complexes showed a similar trend when the AMBER99SB-ILDN force-field was implemented (Figure S5). This observation suggested that these site-specific point mutations led to structural changes in TLR4 that subsequently affect the binding and stability of the complex. Given that these mutations break the hydrogen bond network in the surroundings and partially mask epitope hotspots, we propose that these residues have an indirect impact on the binding to Hu 15C1.

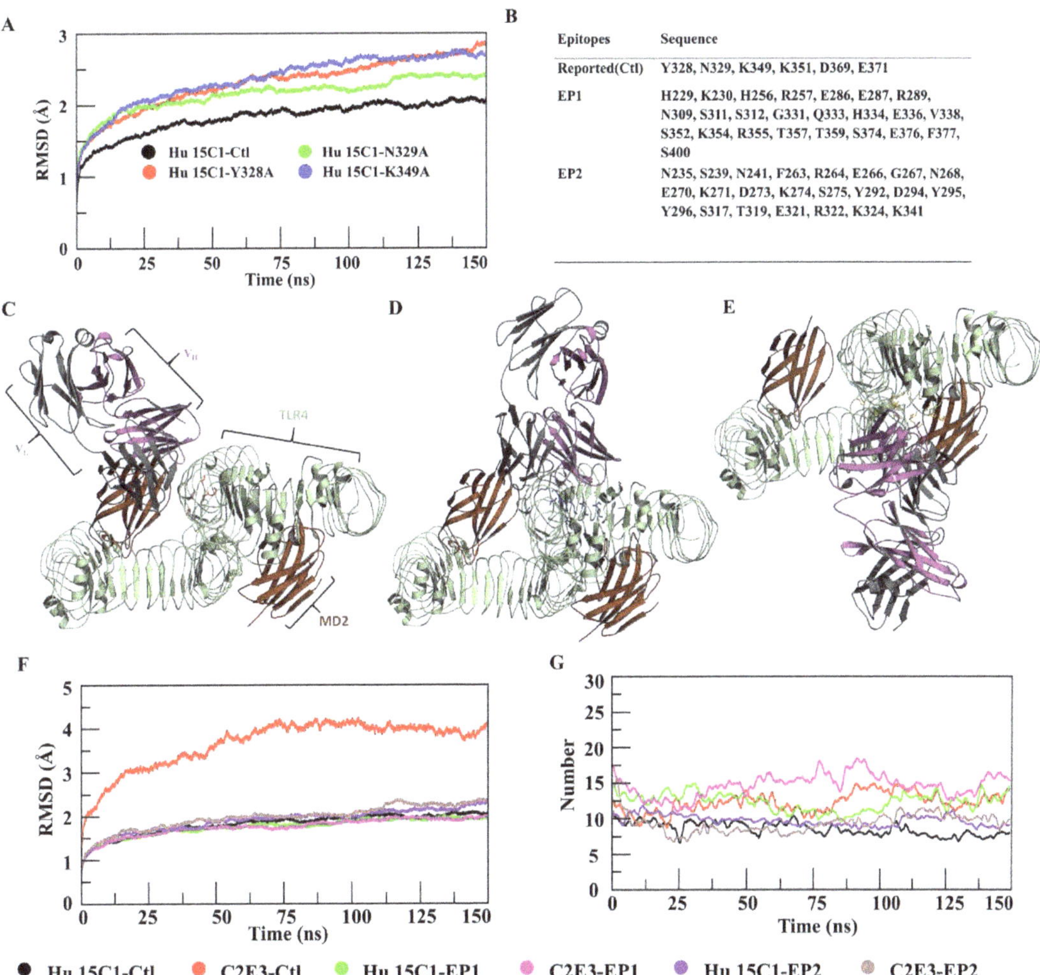

B

Epitopes	Sequence
Reported(Ctl)	Y328, N329, K349, K351, D369, E371
EP1	H229, K230, H256, R257, E286, E287, R289, N309, S311, S312, G331, Q333, H334, E336, V338, S352, K354, R355, T357, T359, S374, E376, F377, S400
EP2	N235, S239, N241, F263, R264, E266, G267, N268, E270, K271, D273, K274, S275, Y292, D294, Y295, Y296, S317, T319, E321, R322, K324, K341

Figure 3. Epitope prediction and docking mode and stability analysis. (**A**) RMSDs of four complexes in which mutated complexes showed an increasing trend. (**B**) Residues for the construction of three conformational epitopes. Modes of docking of Hu 15C1 with TLR4 at sites (**C**) Ctl, (**D**) EP1, and (**E**) EP2 with superimposition of the human TLR4–MD2 dimer. RMSDs of six complexes are illustrated in (**F**), where C2E3–Ctl (red) showed a maximum deviation, and all remaining complexes were stable. The number of hydrogen bonds during the simulation was determined in (**G**), where C2E3–EP1 (magenta) featured the highest, C2E3–Ctl (red) and Hu 15C1–EP1 (green) moderate, and Hu 15C1–Ctl (black), Hu 15C1–EP2 (blue), and C2E3–EP2 (brown) the lowest number of hydrogen bonds formed during the simulation.

To investigate the influence of these mutations on the binding affinity of TLR4 for Hu 15C1, binding free energies of TLR4–mAb complexes were calculated. As expected, we observed a decrease in the binding energy of mutein complexes as compared to the

TLR4^{wt}–Hu 15C1 complex (Table 1). Total binding free energy in each mutant system diminished by ~7–9%, which confirmed the decrease in the affinity of the mAb. The decrease in binding energy of the mutein complexes occurred because the interaction was reduced when residues, i.e., Y328, N329, and K349, were mutated into alanine in mutein complexes. The mutated residues do not interact with the antibody as in TLR4^{wt}–Hu 15C1, which entails that the energy contributed by the mutated residue is lost. Thus, the decrease in the total binding energy of the mutein complexes was observed, suggesting that the original amino acids in this position is energetically vital for antigen binding.

Table 1. Binding free energy of Hu 15C1 and C2E3 toward TLR4.

Complex-Type	Vdw Energy	Electrostatic Energy	Polar Solvation	SASA	Binding Energy
Hu 15C1-Ctl	−379.741 +/− 18.392	−3222.642 +/− 100.808	776.278 +/− 54.509	−42.161 +/− 3.383	−2868.266 +/− 98.367
C2E3-Ctl	−458.220 +/− 22.391	−3339.736 +/− 98.973	1059.671 +/− 65.271	−58.917 +/− 4.214	−2797.202 +/− 97.581
Hu 15C1-Y328A	−341.401 +/− 20.569	−3232.006 +/− 172.050	1139.37 +/− 100.126	−56.347 +/− 3.599	−2490.384 +/− 169.412
Hu 15C1-N329A	−330.692 +/− 20.298	−3360.506 +/− 84.052	1377.683 +/− 124.152	−63.289 +/− 3.401	−2376.804 +/− 131.607
Hu 15C1-K349A	−336.232 +/− 20.298	−3260.506 +/− 84.052	1377.673 +/− 124.152	−55.389 +/− 3.401	−2274.454 +/− 131.607
Hu 15C1-EP1	−514.827 +/− 27.258	−4178.843 +/− 126.407	1398.977 +/− 85.365	−63.568 +/− 3.580	−3358.261 +/− 96.088
C2E3-EP1	−619.706 +/− 28.893	−6359.128 +/− 190.555	1996.806 +/− 109.523	−86.952 +/− 3.889	−5068.979 +/− 115.771
Hu 15C1-EP2	−385.218 +/− 19.617	−2301.460 +/− 71.852	1004.632 +/− 113.944	−48.778 +/− 3.287	−1730.824 +/− 122.321
C2E3-EP2	−344.892 +/− 19.082	−3348.181 +/− 107.983	885.182 +/− 71.196	−41.990 +/− 3.310	−2849.881 +/− 61.910

2.7. Epitope Prediction and Interaction Analysis of the Epitope–Paratope Interface for Rational Design of Antibodies

The hotspot residues, as mentioned above, affect the conformation of TLR4, thereby reducing mAb-binding affinity. Therefore, we further investigated TLR4 epitopes and remapped them using conformational CDR information on Hu 15C1 and its derivative antibody, C2E3. Conformational epitopes on TLR4 were predicted based on an antibody–antigen score. Among the identified epitopes, EP1 and EP2 were selected and compared with a control (referred as Ctl) site. The residues of these epitopes are listed in Figure 3B and the Ramachandran plot of Ctl, EP1, and EP2 is shown in Figure S6. Ctl residues overlap with a few residues of EP1. Ctl is located near the TLR4 dimeric interface, whereas EP2 is located at the TLR4-MD2 interface. Hu 15C1 and C2E3 Fabs were docked to EP1, EP2, and Ctl epitope patches. In each resultant complex, the docked pose of a representative member from the most populated cluster was selected, as depicted in Figure 3C–E and Figure S5A–F. Superposition of the docking complexes revealed overlapping of one of mAb's chain at sites Ctl and EP1 (Figure S7G,H).

To assess the stability of the TLR4–mAb complexes, MD simulations were performed and then analyzed. Hu 15C1–Ctl, Hu 15C1–EP1 and C2E3–EP1 had a similar pattern of the RMSD of backbone atoms, indicating that these complexes are highly stable. Nevertheless, C2E3–EP2 yielded a deviation in the last 75 ns, whereas C2E3–Ctl manifested a continuous increase in RMSD throughout the trajectory (Figure 3F). The marked RMSD increase could be explained by the finding that C2E3 induces a surge in the energy of epitopic or CDR residues upon TLR4 binding (data not shown). A similar tendency was observed in the RMSF plots of these complexes (Figure S7I,J). Residues 200–410 of TLR4 and 150–200 of the antibody in TLR4–C2E3 underwent more fluctuation compared to similar residues in the other complexes. The robustness of these finds was investigated under the effect of the AMBER99SB-ILDN force-field. The RMSD graph (Figure S8) of all the complexes showed a similar trend except C2E3–Ctl, although with a continuous increase in RMSD but much lower than when simulated with CHARMM36 (Figure S8B). Again, the lower RMSD in AMBER99SB-ILDN could be due to the overstabilization of native conformation. Nevertheless, C2E3–EP1 showed a similar trend in RMSD but slightly higher than in the CHARMM36 force field (Figure S8D).

To investigate the antigen–antibody interaction, we performed a dynamic analysis of the interaction of Hu 15C1 and C2E3 with TLR4; this approach provides the basis for rational design of mAbs. Primarily, the number of hydrogen bonds between each mAb and

TLR4 was determined with a cutoff of <5 Å (Figure 3G). In Hu 15C1–Ctl, Hu 15C1–EP1, and C2E3–EP1, the total number of hydrogen bonds was 12–15 and persisted during the simulation. By contrast, in Hu 15C1–EP2 and C2E3–EP2, the hydrogen bond number during the simulation was within the range of 7–12.

To illustrate the distribution of interacting residues of CDRs in the antibodies toward TLR4, the radial distribution function and minimum distance were computed. In a comparison between the interactions of the two mAbs with the TLR4, the CDRs of the C2E3 antibody showed stronger affinity because the distribution of CDR residues interacting with epitopes was higher in comparison with Hu 15C1 (Figure 4). The molecular association of pairwise residue contacts within 5 Å toward TLR4 at three epitopic sites was greater for C2E3, indicating its stronger binding in comparison with Hu 15C1. The peak distribution of Hu 15C1 was higher at Ctl with a peak amplitude above 20 for the Ser368–Gly91 molecular association. The height of distribution peaks for Gly394–Lys310, Glu396–Asp310, and Thr373–Pro312 reached 20 and formed at shorter distances (Figure 4A). In comparison with Hu 15C1, the molecular association peaks formed for C2E3 toward TLR4 at Ctl had a lower amplitude. The pronounced peak of Lys376–Ser101 was above 15 and formed within the 5 Å distance. On the other hand, the rest of the peaks were smaller and formed at above 5 Å; therefore, the pairwise antibody–antigen residue contacts were weak (Figure 4B). The minimum distance for most of the interactions of the Hu 15C1–Ctl complex persisted during the simulation (Figure S9A and Table S1).

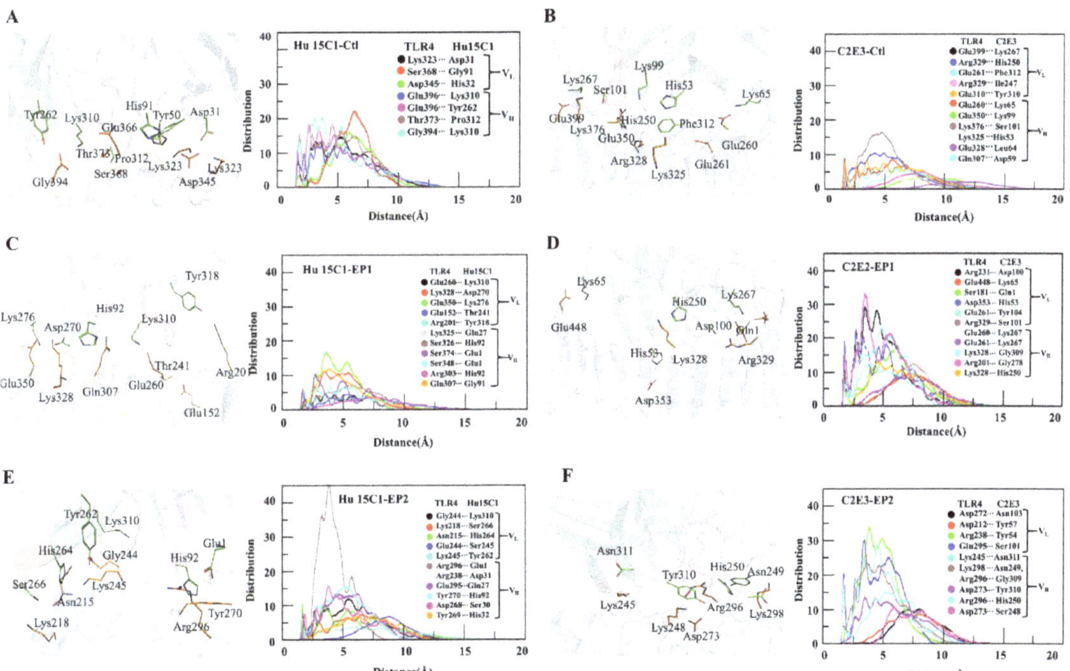

Figure 4. Distribution of minimum distances between the residues of Hu 15C1 and C2E3 interacting with TLR4 at sites (**A**,**B**) Ctl, (**C**,**D**) EP1, and (**E**,**F**) EP2.

The minimum distance during the simulation in the C2E3–Ctl complex remained above 4 Å for some interacting residues (Figure S9B and Table S1). Both mAbs Hu 15C1 and C2E3 at the EP1 site were found to be buried inside, as evidenced by the docking pose (Figure 3C–E) where the maximum surface interacts with TLR4 at this site; therefore, a maximal interaction between the interacting residues was recorded. The highest peaks in the C2E3–EP1 complex formed at the shortest distance of <5 Å. The most pronounced peaks were rendered by Arg231–Asp100 and Arg201–Gly278 with an amplitude of 30, whereas the peaks formed by Asp353–His53, Glu260–Lys267, and Lys328–Gly278 were >20 in height (Figure 4D). In the Hu 15C1–EP1 complex, the molecular association peaks of Glu350–Lys278, Gln307–Gly91, and Lys328–Asp270 were above 10 and formed at a distance of <5 Å, while the amplitude of the rest of the contacts remained below 10 and formed at a distance of >5 Å (Figure 4C). The interacting residues in both complexes mostly remained intact except for Glu152–Thr241, Arg201–Tyr318, and Glu307–Gly91 in Hu 15C1–EP1 and Arg201–Ser278 in C2E3–EP1 during the simulation (Figure S9C,D and Table S2). The Hu 15C1 distribution of interactions remained below 10 except for the pronounced Arg296–Glu1 peak, which showed an amplitude up to 40 at the EP1 site. C2E3 featured a higher distribution of residue interactions at the EP2 site because Arg238–Tyr54, Glu295–Ser101 and Lys245–Asn311 peak amplitudes were approximately 30, and the peaks formed within the 5 Å distance (Figure 4E,F). In Hu 15 C1–EP2 and C2E3–EP2, the minimum distance between the interacting residues was registered above 5 Å for a few interactions (Figure S9E,F and Table S3). The radial distribution functions for Hu 15C1 and C2E3 at the EP1 site were more vigorous as the pronounced peaks formed at <5 Å distance, and the minimum distance of most of the interacting residues of Hu 15C1 and C2E3 at the EP2 site was >4 Å. In contrast to the EP1 site, the gap remained <4 Å during the simulation (Figure 4 and Figure S6). The interface interaction analysis revealed that at the EP1 site, most of the mAb–TLR4 interactions are energetically favorable, as suggested by the peak distribution at this epitope.

The interactions associated with the formation of paratope–epitope interface residues resulted in a free-energy change for the binding reaction. To compare the binding affinity for TLR4 between Hu 15C1 and C2E3 at epitopic sites Ctl, EP1 and EP2, total binding energies were computed by the molecular mechanics Poisson–Boltzmann surface area (MMPBSA) method. The total binding energies of complexes Hu 15C1–TLR4 and C2E3–TLR4 were −3358.261 and −5068.979 kJ/mol, respectively, at the EP1 site (Table 1), thus ensuring their stability in comparison to the Ctl site with a total binding energy of −2868.266 kJ/mol toward Hu 15C1 and −2797.202 kJ/mol toward C2E3. In Table 1, the binding free energies imply that the affinity of Hu 15C1 and C2E3 for the EP2 site is less than that for the EP1 site but greater than that for the control site. The binding free energy of both mAbs toward TLR4 at EP1 is much higher than that at EP2. The more negative the binding energy value, the stronger is the binding affinity of the antibody for its specific target antigen [30].

3. Discussion

Knowledge about a protein–protein interaction interface at the atomic level is essential for the development or design of substances intended to treat diseases. Similarly, the understanding of epitope–paratope interactions in an antibody–antigen complex at the atomic level is crucial for rational development of effective therapeutics.

An important step for the characterization of a functional antibody is accurate delineation of an epitope [31]. Insights about an epitope from a therapeutic perspective can be applied to rational design of mAbs [32]. On the other hand, a detailed epitope analysis is a major restraint because individual amino acid residues constituting an epitope are difficult to pinpoint. Given that experimental methods are time and resource consuming, and success is not guaranteed, computational techniques represent a rapid alternative for validation and characterization of an epitope and affinity maturation of mAbs [17].

In the present study, the epitope reported by the Elson group [22] was characterized and validated by in silico alanine mutagenesis. Epitopic residues in TLR4 were mutated to alanine to investigate the impact of each residue on TLR4 conformation and binding to an antibody. Among the six mutated residues, three, i.e., Y328, N329, and K349, were found to cause a conformational change in the TLR4 structure. RIN analysis suggested that these mutations induce a shift in residue topology owing to a loss of contact with the nearest neighbor or the formation of a new network in the protein structure (Figure S1).

Because the epitope of TLR4 is conformational, the loss of the contact with the surrounding residues owing to a mutation will make the epitope ambiguous. The compactness of these muteins, as denoted by R_g, showed a significant increase (Figure 2C). The higher R_g values suggest that the mutations decreased structural stability. The MD simulation analysis of these mutations revealed the protein's unstable behavior because RMSDs of the muteins uncovered variable behavior during the simulation, and residues from 175 to 400 (Hu 15C1–binding epitope) [26] of the mutated structures oscillated with a higher amplitude, as seen in the RMSF plot (Figure 1C). Gibbs FEL analysis indicated the energetic importance of these mutated residues for the stability of the TLR4 structure. The FEL plot contains splitting of the energy landscape with higher energy barriers, and similar trends were observed for all three mutations. The different conformations assumed by TLR4 owing to the mutations led to conformational changes in the epitopes and resulted in abrogation of the antibody binding, as illustrated by the PCA (Figure S4). The experimental data on the site-directed mutagenesis showed that one of these mutations, i.e., the K349 mutation, alters Hu 15C1 binding [26]. This result may be due to the conformational change in TLR4, which makes the Hu 15C1–binding epitope ambiguous. The impact of the mutation on TLR4 conformation, flexibility, and stability, as evidenced by NMA, showed that the protein folding energy may be increased and stability decreased. The motion of the termini of the muteins, causing twisting and deformity in the protein structure, leads to an increase in internal energy (Figure 2). Steric hindrance or torsion depending upon the direction of the motion induces an increase in the tension on intramolecular bonding, because of which internal hydrogen bonds between the residues break, which results in instability of the TLR4 solenoid structure. From thermodynamic and biophysical points of view, a conformational change in the TLR4 structure is due to the mutation and can make the conformational epitope ambiguous for Hu 15C1 binding. These observed impacts including steric clashes, hydrogen bond network disruption or abrogation of antibody binding, can lead to gain or loss in function of the receptor [33,34]. Moreover, Tsukamoto et al., assessed the functional role of TLR4 activation by a dual-luciferase NF-κB reporter assay, and demonstrated that the overexpression of $TLR4^{wt}$ induced NF-κB activation was reduced upon transfection of the mutated plasmid in which LLR13 residues (TLR4 residing epitope) were mutated [27]. From these studies, we can assume that the mutation-driven structure change makes the antibody unable to recognize the epitope, and these residues play a functional role in TLR4 activation as they are essential for dimerization. We believe that the evidence is in support of our findings that mutations are responsible for structural change, which impacts the function of the protein [34,35].

To ascertain whether the mutations affected the binding affinity of Hu 15C1 and overall stability of the Hu 15C1–TLR4 complex, Hu 15C1 was docked with muteins $TLR4^{Y328A}$, $TLR4^{N329A}$ and $TLR4^{K349A}$, and with $TLR4^{wt}$. The backbone RMSD of the simulated mutein mAb-docked complexes uncovered an increasing deviation trend and consequently revealed the unstable nature of these mutein complexes (Figure 3A). Moreover, the binding-energy data showed a significant decline of binding affinity as compared to Ctl (Table 1). The hampering of mAb binding can be explained by the conformational changes in the TLR4 structure because of these mutations. The findings were next confirmed by PCA and FEL data, which suggested that the decrease in binding affinity may be due to the epitope-masking phenomenon. Several studies have shown that minimal structural changes in an antigen are sufficient to prevent antibody binding [36,37]. The effect of these mutations on binding affinity can influence the function of mAbs. Hu 15C1 did not manifest cross-reactivity in other species because of the ambiguity of the epitope owing to several mutations in TLR4 in these species [26]. Computationally assisted mutagenesis of an antigen or antibody, and a dynamic analysis of these muteins, are beneficial during antibody design for mutated epitopes and for maturation of mAbs. Recently, Takefumi et al. investigated a single-amino-acid substitution to improve antigen–antibody interaction by alanine scanning of interfacial residues. Furthermore, using MD, they found that the enthalpic improvement can be attributed to the stabilization of antigen–antibody interfaces [38]. The dynamics at atomic-level resolution of antigen–antibody interactions provide accurate insights for antibody development. To further investigate the allosteric binding of mAbs Hu 15C1 and C2E3 to TLR4, computationally assisted re-epitoping was performed. TLR4 is an important protein for the immune response. It may possess multiple binding sites because immune-response proteins have multiple and overlapping binding sites [39]. The Ctl epitope is located in the LLR13 region near the TLR4 dimeric interface [27]. One of the predicted epitopes, EP1, overlaps with Ctl. In our analysis of the paratope–epitope interaction interface for therapeutic and specificity purposes, Hu 15C1 and C2E3 were docked to two predicted epitopes. RMSD data on MD simulation trajectories of these complexes revealed that Hu 15C1 and C2E3 are highly stable and buried at EP1 (Figure 3C–F). In contrast to Ctl, this site provides a maximal surface for the interaction with mAbs Hu 15C1 and C2E3. The interface interaction analysis of Hu 15C1 and C2E3 at sites Ctl, EP1 and EP2 indicated that the molecular association of TLR4-mAb residues ensured the highest-peak formation at a shorter distance at the EP1 site, which means that these contacts are energetically preferred for both mAbs. Moreover, ionic and hydrogen bonds persisted during the simulation at the EP1 site, suggesting that Hu 15C1 and C2E3 are strongly bonded at the EP1 site in contrast to the other epitopic sites (Figure 4). The bonding of contacting residues is energetically favored because the distance between the interacting residues remains <4 Å during the simulation (Figure S9). The C2E3–Ctl interface interaction data showed an extended distance between the interacting residues during the simulation and energetically unfavorable interactions because the distribution peaks were short and formed at a distance of >5 Å, which could be the reason for the increased RMSD values. The binding-energy data support this finding. We observed the highest binding energy of Hu 15C1 at Ctl in contrast to the other epitopic sites; similarly, for C2E3, binding energy was the lowest. The salt bridge formation at the EP1 site for Hu 15C1 and C2E3 was greater, which provided extra stability to the complex. Salt bridge formation with a favorable geometrical orientation gives stability to the overall structure [40]. The binding energy data (Table 1) uncovered higher binding energy of C2E3 in comparison with Hu 15C1. It has been experimentally proved that C2E3 has stronger affinity than Hu 15C1 [41]. In short, the epitope analyses provided a comprehensive view of the TLR4 conformational changes that occur due to the mutations and indicated that they may yield epitope ambiguity. These outcomes will make the development of new mAbs easier and faster and will elucidate epitope structure for designing highly specific mAbs. For designing CDRs and their rationalization, the analyses of antibody–antigen interface interaction are important. Investigating the dynamic effect at the atomic level in the paratope–epitope interaction at three different TLR4 epitopic sites, i.e., Ctl, EP1, and EP2, for Hu 15C1 and C2E3, provided an in-depth understanding of these interactions. We believe that these findings will be helpful for rational design of high-affinity mAbs and for affinity maturation of already known mAbs Hu 15C1 and C2E3. Moreover, this study gives insights into a novel epitope at the TLR4-MD2 interface, indicating that heterodimerization of TLR4 and MD2 may be disrupted and that TLR4 signaling can be inhibited. In 2011, a group reported a MD2 mimicking peptide (MD2-I), which binds to the same site and disrupts the TLR4/MD2 interactions. MD2-I specificity has been evaluated in macrophages and animal models [42]. In addition to this, a few studies have also shown the importance of the druggability of this site [43–45].

mAbs have become an essential therapeutic tool, but their rationalization is a challenging phase and a crucial step in the development of highly specific mAbs. Comparative epitope analysis and antigen–antibody interface analysis via MD simulation offers an atomic-level understanding of the dynamic behavior of the antibody–antigen interactions, which is essential for antibody development. Furthermore, in comparison with other methods, the MMPBSA technique has an edge for the evaluation of affinity maturation and for clarification of the changes in binding affinity. However, TLR4 is a multidomain protein, but its full-length crystal structure is not available; therefore, we used the crystal structure of an isolated external domain in our study. The isolated domains can be less stable than the full-length multidomain protein but are fully functional [46]. In this study, we used computational methods for epitope verification and affinity maturation of TLR4-targeting antibodies. However, due to the limited facilities, we could not validate our findings experimentally, which can be performed in future studies. Therefore, for efficient mAb design, it is important to couple MD simulations with machine learning and deep learning methods. The structure from conformational space of the MD simulations can be utilized to find new plausible mAb conformations complementing pre-existing ones. We expect that the coupling of these computational methods with experimental methods will not only reduce the cost and time but also offer an efficient approach to the development of rationalized mAbs.

In this study, we demonstrated a computational-driven approach for the epitope verification and affinity maturation of TLR4 antibodies. By using network analysis and implicit MD simulation with FEL, we were able to determine the change in hydrogen bond network and van der Waals interactions in Y328, N329 and K349 among six mutated epitopic residues affecting structural integrity and the energy landscape of the TLR4 and, consequently, the antibody affinity. Further, we predicted the novel epitope located in the MD2 binding site region, which could be explored for new therapeutic antibodies. The uniqueness of our approach is that we successfully employed a wide array of computational-driven techniques to determine the dynamics of mAb-TLR4 interactions at the atomic level, and comparative epitope analysis by computing the binding affinity of the TLR4 antibodies. We believe that this computational-driven approach will accelerate the rational design of therapeutic antibodies. However, experimental validation of the present computational methods remains to be observed until more mAb-antigen are studied and experimentally verified. Moreover, coupling the present pipeline with deep learning and machine learning techniques can further enhance the feasibility of the method. Overall, this work shows that network analysis, MD simulation and MMPBSA techniques can be used as an exploratory methodology for the study of antibody rationalization.

4. Materials and Methods

4.1. In Silico Structure Reconstruction and Mutation of TLR4

The 3D coordinates of TLR4 (Protein Data Bank [PDB] ID: 3FXI) and TLR4-specific mAbs Hu 15C1 (PDB ID: 4R7D) and C2E3 (PDB ID: 4R7N) were retrieved from the RCSB PDB database. A monomeric TLR4 structure was first extracted from the TLR4–MD2 complex and optimized in Molecular Operating Environment (MOE) 2019.01. After the removal of water related to crystallization, hydrogen atoms and partial charges were added, incomplete and broken side chains were remodeled and hybridization events were adjusted at default structure preparation parameters in MOE. TLR4 muteins were created by the computational alanine-scanning method. The residues with low energy rotamers were selected and replaced with the amino acid alanine by means of a protein builder tool in the MOE suite. A similar protocol was used to generate TLR4–mAb complexes for the muteins followed by energy minimization to optimize the mutein structures.

4.2. Construction of the Interaction Network

Network description is widely used to analyze network topology and dynamics of complex systems. The RIN was constructed from a graph-based model of protein structure and topological residue connectivity, where residues are nodes and detected interactions are represented as edges. The RIN was constructed in RINalyzer with the help of the structure viz module implemented in Cytoscape 3.7.2 as per the protocol of Guilaume et al. [47]. The cutoff distance was set to <5 Å for residue interaction. The topological network parameters were calculated as an undirected network using a network analyzer [48].

4.3. Epitope Prediction and CDR Annotations

Conformational epitopes were predicted with the Epipred tool of the SAbPred web server [49,50]. Based on combined conformational matching of the antibody–antigen structures via geometric fitting and knowledge-based asymmetrical antibody–antigen scoring, epitopes of the antigen are listed rank-wise. The score of an epitope is given by:

$$\textit{Epitope Score} = \sum d(n)\mathbf{Pr}(T_{ab}, T_{ag}) \quad (1)$$

where T_{ab} and T_{ag} are the amino acid types of the antibody and antigen residues, respectively, that belong to node n. The stereochemistry and geometry of selected epitopes was verified by Ramachandran plot via MOE's inbuilt utility.

CDRs of antibodies retrieved from the PDB database are annotated according to the numbering scheme of Chothia and Lesk [51,52]. Instead of whole Fab regions, CDRs of the mAbs were considered ligand sites in the docking and simulation.

4.4. Molecular Docking and MD Simulation

Molecular docking of Hu 15C1 and C2E3 Fabs to TLR4 was performed on the Haddock 2.2 web server on the basis of the information derived from experimental and bioinformatics data [53]. The two mAbs were docked with the two newly predicted epitopes (discussed above) and a previously reported epitope (Ctl). In all three cases, the docked solutions generated by the Haddock server were clustered by their RMSDs. The RMSD measures the average distance of backbone atoms of the superimposed docked solutions, signifying the proximity between solutions. The most populated cluster with a lower RMSD was selected, and solutions with the lowest RMSD within the selected cluster were chosen for the MD analysis. The docking results obtained via Haddock were next validated by means of an antigen–antibody docking package implemented in the MOE suite.

MD simulations capture the dynamic behavior of proteins and other biomolecules in full atomic detail and provide an accurate insight into antigen-antibody interactions at very fine temporal resolution. The MD simulations were performed in GROMACS 2019.3 [54] on Intel E5-2680 and Intel E3-1275 with an Nvidia GeForce GTX 1060 graphics processing unit. TLR4, its muteins, and mAb–TLR4 complexes were solvated using the TIP3P water model in a cubic box; the dimensions of the box boundaries were extended by 10 Å from protein atoms. Na^+ and Cl^- counter ions were added to neutralize the charge of the simulation system and energy minimization was performed using the CHARMM36 (Chemistry at Harvard macromolecular mechanics) and AMBER99SB-ILDN force-fields and a steep-descent algorithm [55]. V-rescale and Berendsen coupling schemes were employed for temperature and pressure equilibration procedures, respectively. After the temperature and pressure equilibration, MD simulations were carried out for 150 ns for each system in the CHARMM36 and AMBER99SB-ILDN force-fields.

4.5. Binding-Free-Energy Calculations

The MMPBSA approach was used to analyze free-energy interactions between the mAbs and TLR4. The enthalpy of the system was computed via molecular mechanics of MMPBSA, whereas the effect of the polar and nonpolar part of the solvent effect on free energy was determined via the Poisson–Boltzmann equation and calculation of the surface area. The basic equation is

$$\Delta G_{bind} = \Delta E_{MM} + \Delta\Delta G_{sol} - T\Delta S \quad (2)$$

where ΔG_{bind} is binding free energy, ΔE_{MM} represents the intramolecular energy difference in a vacuum, $\Delta\Delta G_{sol}$ is the solvation energy difference, T denotes absolute temperature and ΔS is the entropy change. In GROMACS, the built-in g_mmpbsa tool and APBS were called for the MMPBSA calculations, and the last 50 ns of an MD simulation trajectory of each complex with 10-frame intervals was extracted for the energy calculations. For the g_mmpbsa run, the dielectric constant of the aqueous solvent was set to 80, and the interior dielectric constant was set to 4; the surface tension constant g was set to 0.022 kJ/mol.

4.6. Principal Component Analysis and Free-Energy Landscape

PCA reduces a multidimensional dataset with many variables to a few 'principal components' that still preserve most of the differences between the data. In MD, PCA helps to elucidate the essential dynamics of the system in a low-dimensional FEL. PCA ultimately provides a view of the atoms (in the MD simulations) that move anisotropically to maximize the variance. To assess conformational flexibility of TLR4 and its muteins, PCA was performed in GROMACS, and the results were analyzed by means of the bio3D R package [56]. Before the PCA, translational and rotational motions were eliminated from the MD trajectory of all systems toward the average geometric center of the protein by least-square fit superimposition onto a reference structure [57]. To generate a covariance matrix, configurational space is reconstructed using a simple linear transformation in Cartesian coordinate space. Covariance matrix diagonalization generates an eigenvector set, where its vector components describe the direction of the motion, and the corresponding eigenvalue represents the magnitude of its energetic contribution to the motion. The FEL characterizes the state of a protein by exploring its conformations via an MD sampling technique and finds where Gibbs free energy for a sample is minimal. The gmx_sham tool implemented in GROMACS was used to predict the FEL through plotting of principal components PC1 and PC2 in the academic version of the Mathematica software (version 11.3; Wolfram Research, Inc., Champaign, IL, USA).

The DynaMut [58] machine-learning–based web server was employed to predict the impact of the mutations on TLR4 structure stability and flexibility. This tool predicts the harmonic motion of the wild-type and mutant TLR4 systems. By means of normal mode analysis (NMA), the nontrivial mode of a molecular motion of TLR4 and its mutants was studied and its vector representation was visualized in Pymol 2.3.

Supplementary Materials: Supplementary materials can be found at https://www.mdpi.com/article/10.3390/ijms22115989/s1.

Author Contributions: B.A. and S.C.: Conceptualization, methodology, data curation, and writing-original draft preparation. B.A. and M.B.: Investigation and formal analysis. M.-S.K. and S.C.: Supervision. All the authors: writing, reviewing and editing. All authors have read and agreed to the published version of the manuscript.

Funding: This work was supported by the National Research Foundation of Korea (NRF-2020R1F1A1 071517, 2019M3D1A1078940, and 2019R1A6A1A11051471).

Data Availability Statement: The structure models and simulation trajectories are available upon request (sangdunchoi@ajou.ac.kr). All remaining data are contained within the manuscript.

Acknowledgments: We gratefully acknowledge Asma Achek for the great help she provided in manuscript editing, and her valuable suggestions that improved the quality of the paper.

Conflicts of Interest: The authors declare that the research was conducted in the absence of any commercial or financial relationships that could be construed as a potential conflict of interest.

References

1. Takeuchi, O.; Akira, S. Pattern recognition receptors and inflammation. *Cell* **2010**, *140*, 805–820. [CrossRef]
2. O'Neill, L.A.; Bryant, C.E.; Doyle, S.L. Therapeutic targeting of Toll-like receptors for infectious and inflammatory diseases and cancer. *Pharmacol. Rev.* **2009**, *61*, 177–197. [CrossRef]
3. Shah, M.; Kim, G.Y.; Achek, A.; Cho, E.Y.; Baek, W.Y.; Choi, Y.S.; Lee, W.H.; Kim, D.J.; Lee, S.H.; Kim, W.; et al. The alphaC helix of TIRAP holds therapeutic potential in TLR-mediated autoimmune diseases. *Biomaterials* **2020**, *245*, 119974. [CrossRef]
4. Medzhitov, R.; Preston-Hurlburt, P.; Janeway, C.A., Jr. A human homologue of the Drosophila Toll protein signals activation of adaptive immunity. *Nature* **1997**, *388*, 394–397. [CrossRef] [PubMed]
5. Singer, M.; Deutschman, C.S.; Seymour, C.W.; Shankar-Hari, M.; Annane, D.; Bauer, M.; Bellomo, R.; Bernard, G.R.; Chiche, J.D.; Coopersmith, C.M.; et al. The Third International Consensus Definitions for Sepsis and Septic Shock (Sepsis-3). *JAMA* **2016**, *315*, 801–810. [CrossRef] [PubMed]
6. Xu, D.; Yan, S.; Wang, H.; Gu, B.; Sun, K.; Yang, X.; Sun, B.; Wang, X. IL-29 Enhances LPS/TLR4-Mediated Inflammation in Rheumatoid Arthritis. *Cell. Physiol. Biochem.* **2015**, *37*, 27–34. [CrossRef] [PubMed]

7. Achek, A.; Shah, M.; Seo, J.Y.; Kwon, H.K.; Gui, X.; Shin, H.J.; Cho, E.Y.; Lee, B.S.; Kim, D.J.; Lee, S.H.; et al. Linear and Rationally Designed Stapled Peptides Abrogate TLR4 Pathway and Relieve Inflammatory Symptoms in Rheumatoid Arthritis Rat Model. *J. Med. Chem.* **2019**, *62*, 6495–6511. [CrossRef]
8. Yang, Y.; Lv, J.; Jiang, S.; Ma, Z.; Wang, D.; Hu, W.; Deng, C.; Fan, C.; Di, S.; Sun, Y.; et al. The emerging role of Toll-like receptor 4 in myocardial inflammation. *Cell Death Dis.* **2016**, *7*, e2234. [CrossRef]
9. Yao, L.; Kan, E.M.; Lu, J.; Hao, A.; Dheen, S.T.; Kaur, C.; Ling, E.A. Toll-like receptor 4 mediates microglial activation and production of inflammatory mediators in neonatal rat brain following hypoxia: Role of TLR4 in hypoxic microglia. *J. Neuroinflammation* **2013**, *10*, 23. [CrossRef]
10. Teng, W.; Wang, L.; Xue, W.; Guan, C. Activation of TLR4-mediated NFkappaB signaling in hemorrhagic brain in rats. *Mediat. Inflamm.* **2009**, *2009*, 473276. [CrossRef]
11. Anwar, M.A.; Shah, M.; Kim, J.; Choi, S. Recent clinical trends in Toll-like receptor targeting therapeutics. *Med. Res. Rev.* **2019**, *39*, 1053–1090. [CrossRef] [PubMed]
12. Haabeth, O.A.; Bogen, B.; Corthay, A. A model for cancer-suppressive inflammation. *Oncoimmunology* **2012**, *1*, 1146–1155. [CrossRef]
13. Beck, A.; Wurch, T.; Bailly, C.; Corvaia, N. Strategies and challenges for the next generation of therapeutic antibodies. *Nat. Rev. Immunol.* **2010**, *10*, 345–352. [CrossRef] [PubMed]
14. Lu, R.M.; Hwang, Y.C.; Liu, I.J.; Lee, C.C.; Tsai, H.Z.; Li, H.J.; Wu, H.C. Development of therapeutic antibodies for the treatment of diseases. *J. Biomed. Sci.* **2020**, *27*, 1. [CrossRef] [PubMed]
15. Bruce, N.J.; Ganotra, G.K.; Kokh, D.B.; Sadiq, S.K.; Wade, R.C. New approaches for computing ligand-receptor binding kinetics. *Curr. Opin. Struct. Biol.* **2018**, *49*, 1–10. [CrossRef] [PubMed]
16. Kozakov, D.; Hall, D.R.; Xia, B.; Porter, K.A.; Padhorny, D.; Yueh, C.; Beglov, D.; Vajda, S. The ClusPro web server for protein-protein docking. *Nat. Protoc.* **2017**, *12*, 255–278. [CrossRef] [PubMed]
17. Norman, R.A.; Ambrosetti, F.; Bonvin, A.; Colwell, L.J.; Kelm, S.; Kumar, S.; Krawczyk, K. Computational approaches to therapeutic antibody design: Established methods and emerging trends. *Brief. Bioinform.* **2019**. [CrossRef]
18. Yamashita, T. Toward rational antibody design: Recent advancements in molecular dynamics simulations. *Int. Immunol.* **2018**, *30*, 133–140. [CrossRef]
19. Liu, G.; Zeng, H.; Mueller, J.; Carter, B.; Wang, Z.; Schilz, J.; Horny, G.; Birnbaum, M.E.; Ewert, S.; Gifford, D.K. Antibody complementarity determining region design using high-capacity machine learning. *Bioinformatics* **2020**, *36*, 2126–2133. [CrossRef] [PubMed]
20. Friedensohn, S.; Neumeier, D.; Khan, T.A.; Csepregi, L.; Parola, C.; de Vries, A.R.G.; Erlach, L.; Mason, D.M.; Reddy, S.T. Convergent selection in antibody repertoires is revealed by deep learning. *bioRxiv* **2020**. [CrossRef]
21. Hatterer, E.; Shang, L.; Simonet, P.; Herren, S.; Daubeuf, B.; Teixeira, S.; Reilly, J.; Elson, G.; Nelson, R.; Gabay, C.; et al. A specific anti-citrullinated protein antibody profile identifies a group of rheumatoid arthritis patients with a toll-like receptor 4-mediated disease. *Arthritis Res. Ther.* **2016**, *18*, 224. [CrossRef] [PubMed]
22. Dunn-Siegrist, I.; Leger, O.; Daubeuf, B.; Poitevin, Y.; Depis, F.; Herren, S.; Kosco-Vilbois, M.; Dean, Y.; Pugin, J.; Elson, G. Pivotal involvement of Fcgamma receptor IIA in the neutralization of lipopolysaccharide signaling via a potent novel anti-TLR4 monoclonal antibody 15C1. *J. Biol. Chem.* **2007**, *282*, 34817–34827. [CrossRef] [PubMed]
23. Shang, L.; Daubeuf, B.; Triantafilou, M.; Olden, R.; Depis, F.; Raby, A.C.; Herren, S.; Dos Santos, A.; Malinge, P.; Dunn-Siegrist, I.; et al. Selective antibody intervention of Toll-like receptor 4 activation through Fc gamma receptor tethering. *J. Biol. Chem.* **2014**, *289*, 15309–15318. [CrossRef]
24. Monnet, E.; Lapeyre, G.; Poelgeest, E.V.; Jacqmin, P.; Graaf, K.; Reijers, J.; Moerland, M.; Burggraaf, J.; Min, C. Evidence of NI-0101 pharmacological activity, an anti-TLR4 antibody, in a randomized phase I dose escalation study in healthy volunteers receiving LPS. *Clin. Pharmacol. Ther.* **2017**, *101*, 200–208. [CrossRef] [PubMed]
25. Monnet, E.; Choy, E.H.; McInnes, I.; Kobakhidze, T.; de Graaf, K.; Jacqmin, P.; Lapeyre, G.; de Min, C. Efficacy and safety of NI-0101, an anti-toll-like receptor 4 monoclonal antibody, in patients with rheumatoid arthritis after inadequate response to methotrexate: A phase II study. *Ann. Rheum. Dis.* **2020**, *79*, 316–323. [CrossRef] [PubMed]
26. Loyau, J.; Didelot, G.; Malinge, P.; Ravn, U.; Magistrelli, G.; Depoisier, J.F.; Pontini, G.; Poitevin, Y.; Kosco-Vilbois, M.; Fischer, N.; et al. Robust Antibody-Antigen Complexes Prediction Generated by Combining Sequence Analyses, Mutagenesis, In Vitro Evolution, X-ray Crystallography and In Silico Docking. *J. Mol. Biol.* **2015**, *427*, 2647–2662. [CrossRef] [PubMed]
27. Tsukamoto, H.; Yamagata, Y.; Ukai, I.; Takeuchi, S.; Okubo, M.; Kobayashi, Y.; Kozakai, S.; Kubota, K.; Numasaki, M.; Kanemitsu, Y.; et al. An inhibitory epitope of human Toll-like receptor 4 resides on leucine-rich repeat 13 and is recognized by a monoclonal antibody. *FEBS Lett.* **2017**, *591*, 2406–2416. [CrossRef]
28. Kamenik, A.S.; Handle, P.H.; Hofer, F.; Kahler, U.; Kraml, J.; Liedl, K.R. Polarizable and non-polarizable force fields: Protein folding, unfolding, and misfolding. *J. Chem. Phys.* **2020**, *153*, 185102. [CrossRef]
29. Jana, K.; Kepp, K.P. Force-Field Benchmarking by Alternatives: A Systematic Study of Ten Small α- and β-Proteins. *bioRxiv* **2020**, 974477. [CrossRef]
30. Du, X.; Li, Y.; Xia, Y.L.; Ai, S.M.; Liang, J.; Sang, P.; Ji, X.L.; Liu, S.Q. Insights into Protein-Ligand Interactions: Mechanisms, Models, and Methods. *Int. J. Mol. Sci.* **2016**, *17*, 144. [CrossRef]

31. Kringelum, J.V.; Lundegaard, C.; Lund, O.; Nielsen, M. Reliable B cell epitope predictions: Impacts of method development and improved benchmarking. *PLoS Comput. Biol.* **2012**, *8*, e1002829. [CrossRef] [PubMed]
32. Kazi, A.; Chuah, C.; Majeed, A.B.A.; Leow, C.H.; Lim, B.H.; Leow, C.Y. Current progress of immunoinformatics approach harnessed for cellular- and antibody-dependent vaccine design. *Pathog. Glob. Health* **2018**, *112*, 123–131. [CrossRef] [PubMed]
33. Fowler, D.M.; Araya, C.L.; Fleishman, S.J.; Kellogg, E.H.; Stephany, J.J.; Baker, D.; Fields, S. High-resolution mapping of protein sequence-function relationships. *Nat. Methods* **2010**, *7*, 741–746. [CrossRef] [PubMed]
34. Studer, R.A.; Dessailly, B.H.; Orengo, C.A. Residue mutations and their impact on protein structure and function: Detecting beneficial and pathogenic changes. *Biochem. J.* **2013**, *449*, 581–594. [CrossRef]
35. Schaefer, C.; Rost, B. Predict impact of single amino acid change upon protein structure. *BMC Genomics* **2012**, *13* (Suppl. 4), S4. [CrossRef]
36. Sela-Culang, I.; Kunik, V.; Ofran, Y. The structural basis of antibody-antigen recognition. *Front. Immunol.* **2013**, *4*, 302. [CrossRef]
37. Chiu, M.L.; Goulet, D.R.; Teplyakov, A.; Gilliland, G.L. Antibody Structure and Function: The Basis for Engineering Therapeutics. *Antibodies* **2019**, *8*, 55. [CrossRef] [PubMed]
38. Yamashita, T.; Mizohata, E.; Nagatoishi, S.; Watanabe, T.; Nakakido, M.; Iwanari, H.; Mochizuki, Y.; Nakayama, T.; Kado, Y.; Yokota, Y.; et al. Affinity Improvement of a Cancer-Targeted Antibody through Alanine-Induced Adjustment of Antigen-Antibody Interface. *Structure* **2019**, *27*, 519–527.e515. [CrossRef]
39. Zhao, L.; Wong, L.; Lu, L.; Hoi, S.C.; Li, J. B-cell epitope prediction through a graph model. *BMC Bioinform.* **2012**, *13* (Suppl. 17), S20. [CrossRef]
40. Kumar, S.; Nussinov, R. Close-range electrostatic interactions in proteins. *Chembiochem* **2002**, *3*, 604–617. [CrossRef]
41. Loyau, J.; Malinge, P.; Daubeuf, B.; Shang, L.; Elson, G.; Kosco-Vilbois, M.; Fischer, N.; Rousseau, F. Maximizing the potency of an anti-TLR4 monoclonal antibody by exploiting proximity to Fcgamma receptors. *MAbs* **2014**, *6*, 1621–1630. [CrossRef] [PubMed]
42. Liu, L.; Ghosh, N.; Slivka, P.F.; Fiorini, Z.; Hutchinson, M.R.; Watkins, L.R.; Yin, H. An MD2 hot-spot-mimicking peptide that suppresses TLR4-mediated inflammatory response in vitro and in vivo. *Chembiochem* **2011**, *12*, 1827–1831. [CrossRef]
43. Nishitani, C.; Mitsuzawa, H.; Sano, H.; Shimizu, T.; Matsushima, N.; Kuroki, Y. Toll-like receptor 4 region Glu24-Lys47 is a site for MD-2 binding: Importance of CYS29 and CYS40. *J. Biol. Chem.* **2006**, *281*, 38322–38329. [CrossRef]
44. Nishitani, C.; Mitsuzawa, H.; Hyakushima, N.; Sano, H.; Matsushima, N.; Kuroki, Y. The Toll-like receptor 4 region Glu24-Pro34 is critical for interaction with MD-2. *Biochem. Biophys. Res. Commun.* **2005**, *328*, 586–590. [CrossRef] [PubMed]
45. Re, F.; Strominger, J.L. Separate functional domains of human MD-2 mediate Toll-like receptor 4-binding and lipopolysaccharide responsiveness. *J. Immunol.* **2003**, *171*, 5272–5276. [CrossRef]
46. Bhaskara, R.M.; Srinivasan, N. Stability of domain structures in multi-domain proteins. *Sci. Rep.* **2011**, *1*, 40. [CrossRef] [PubMed]
47. Brysbaert, G.; Mauri, T.; de Ruyck, J.; Lensink, M.F. Identification of Key Residues in Proteins Through Centrality Analysis and Flexibility Prediction with RINspector. *Curr. Protoc. Bioinform.* **2019**, *65*, e66. [CrossRef]
48. Assenov, Y.; Ramirez, F.; Schelhorn, S.E.; Lengauer, T.; Albrecht, M. Computing topological parameters of biological networks. *Bioinformatics* **2008**, *24*, 282–284. [CrossRef]
49. Dunbar, J.; Krawczyk, K.; Leem, J.; Marks, C.; Nowak, J.; Regep, C.; Georges, G.; Kelm, S.; Popovic, B.; Deane, C.M. SAbPred: A structure-based antibody prediction server. *Nucleic Acids Res.* **2016**, *44*, W474–W478. [CrossRef]
50. Krawczyk, K.; Liu, X.; Baker, T.; Shi, J.; Deane, C.M. Improving B-cell epitope prediction and its application to global antibody-antigen docking. *Bioinformatics* **2014**, *30*, 2288–2294. [CrossRef]
51. Chothia, C.; Lesk, A.M. Canonical structures for the hypervariable regions of immunoglobulins. *J. Mol. Biol.* **1987**, *196*, 901–917. [CrossRef]
52. Murzin, A.G.; Brenner, S.E.; Hubbard, T.; Chothia, C. SCOP: A structural classification of proteins database for the investigation of sequences and structures. *J. Mol. Biol.* **1995**, *247*, 536–540. [CrossRef]
53. Ritleng, P.; Loubiere, R.; Marcelet, B. [Suppurative pseudo-lithiasic canaliculitis]. *Ophtalmologie* **1989**, *3*, 1–3. [PubMed]
54. Abraham, M.J.; Murtola, T.; Schulz, R.; Páll, S.; Smith, J.C.; Hess, B.; Lindahl, E. GROMACS: High performance molecular simulations through multi-level parallelism from laptops to supercomputers. *SoftwareX* **2015**, *1*, 19–25. [CrossRef]
55. Huang, J.; Rauscher, S.; Nawrocki, G.; Ran, T.; Feig, M.; de Groot, B.L.; Grubmuller, H.; MacKerell, A.D., Jr. CHARMM36m: An improved force field for folded and intrinsically disordered proteins. *Nat. Methods* **2017**, *14*, 71–73. [CrossRef]
56. Skjaerven, L.; Yao, X.Q.; Scarabelli, G.; Grant, B.J. Integrating protein structural dynamics and evolutionary analysis with Bio3D. *BMC Bioinform.* **2014**, *15*, 399. [CrossRef]
57. Amadei, A.; Linssen, A.B.; de Groot, B.L.; van Aalten, D.M.; Berendsen, H.J. An efficient method for sampling the essential subspace of proteins. *J. Biomol. Struct. Dyn.* **1996**, *13*, 615–625. [CrossRef] [PubMed]
58. Rodrigues, C.H.; Pires, D.E.; Ascher, D.B. DynaMut: Predicting the impact of mutations on protein conformation, flexibility and stability. *Nucleic Acids Res.* **2018**, *46*, W350–W355. [CrossRef] [PubMed]

International Journal of *Molecular Sciences*

MDPI

Article

Exploring the Binding Mechanism of PF-07321332 SARS-CoV-2 Protease Inhibitor through Molecular Dynamics and Binding Free Energy Simulations

Bilal Ahmad [1], Maria Batool [1,2], Qurat ul Ain [1], Moon Suk Kim [1] and Sangdun Choi [1,2,*]

[1] Department of Molecular Science and Technology, Ajou University, Suwon 16499, Korea; bilalpharma77@gmail.com (B.A.); mariabatool.28@gmail.com (M.B.); ainne.w@gmail.com (Q.u.A.); moonskim@ajou.ac.kr (M.S.K.)

[2] S&K Therapeutics, Campus Plaza 418, Ajou University, Suwon 16502, Korea

* Correspondence: sangdunchoi@ajou.ac.kr; Tel.: +82-31-219-2600; Fax: +82-31-219-1615

Abstract: The novel coronavirus disease, caused by severe acute respiratory coronavirus 2 (SARS-CoV-2), rapidly spreading around the world, poses a major threat to the global public health. Herein, we demonstrated the binding mechanism of PF-07321332, α-ketoamide, lopinavir, and ritonavir to the coronavirus 3-chymotrypsin-like-protease ($3CL^{pro}$) by means of docking and molecular dynamic (MD) simulations. The analysis of MD trajectories of $3CL^{pro}$ with PF-07321332, α-ketoamide, lopinavir, and ritonavir revealed that $3CL^{pro}$–PF-07321332 and $3CL^{pro}$–α-ketoamide complexes remained stable compared with $3CL^{pro}$–ritonavir and $3CL^{pro}$–lopinavir. Investigating the dynamic behavior of ligand–protein interaction, ligands PF-07321332 and α-ketoamide showed stronger bonding via making interactions with catalytic dyad residues His41–Cys145 of $3CL^{pro}$. Lopinavir and ritonavir were unable to disrupt the catalytic dyad, as illustrated by increased bond length during the MD simulation. To decipher the ligand binding mode and affinity, ligand interactions with SARS-CoV-2 proteases and binding energy were calculated. The binding energy of the bespoke antiviral PF-07321332 clinical candidate was two times higher than that of α-ketoamide and three times than that of lopinavir and ritonavir. Our study elucidated in detail the binding mechanism of the potent PF-07321332 to $3CL^{pro}$ along with the low potency of lopinavir and ritonavir due to weak binding affinity demonstrated by the binding energy data. This study will be helpful for the development and optimization of more specific compounds to combat coronavirus disease.

Keywords: COVID-19; SARS-CoV-2; PF-07321332; α-ketoamide; 3CL protease; main protease

Citation: Ahmad, B.; Batool, M.; Ain, Q.u.; Kim, M.S.; Choi, S. Exploring the Binding Mechanism of PF-07321332 SARS-CoV-2 Protease Inhibitor through Molecular Dynamics and Binding Free Energy Simulations. *Int. J. Mol. Sci.* **2021**, *22*, 9124. https://doi.org/10.3390/ijms22179124

Academic Editor: Małgorzata Borówko

Received: 31 July 2021
Accepted: 21 August 2021
Published: 24 August 2021

1. Introduction

The recent outbreak of coronavirus disease 2019 (COVID-19) caused by the novel severe acute respiratory syndrome coronavirus 2 (SARS-CoV-2) affected more than 192 million people and killed 4.1 million across the globe (as of 23 July 2021) [1,2]. The genome sequence of SARS-CoV-2 is closely related to SARS-CoV and β-coronavirus, sharing 79.5% and 96.2% sequence identity, respectively [3–5]. Coronaviruses belong to the Coronaviridae family and are found in birds and mammals [6]. The genomic organizations of SARS-CoV and MERS-CoV are similar [7] and both viruses comprise two polypeptides, pp1a and pp1ab. These polypeptides are processed into nonstructural proteins that play a fundamental part in the replication of these viruses, and the whole process is mediated by two kinds of main proteases: 3-chymotrypsin like protease ($3CL^{pro}$) and papain-like proteases [8,9].

$3CL^{pro}$, also called the main protease, consists of 306 amino acid residues, and is known to cleave at 11 sites in the polyproteins. The cleaving phenomenon leads to the formation of a helicase, single-stranded-RNA–binding protein, RNA-dependent RNA polymerase, 2′-*O*-ribose methyltransferase, endoribonuclease, and exoribonuclease [10,11]. $3CL^{pro}$ forms a homodimer, and each subunit consists of three domains. The substrate-binding site of $3CL^{pro}$ contains four subsites, S1′, S1, S2, and S4, and is highly conserved among all coronaviruses [3]. The active site of $3CL^{pro}$ is located in a cleft between domains I and II, while domain III helps with the formation of the dimer and is connected to domain II via a long loop (residues 184–199) [3,8]. A $3CL^{pro}$ antagonist will be highly specific to SARS-CoV-2 and will have minimal side effects because $3CL^{pro}$ shares no homology with human proteases [3,12]. Owing to the essential role of $3CL^{pro}$ in the transcription and replication of the viral genome and strong conservation of its binding-pocket residues, it is considered an ideal drug target in SARS-CoV-2 and other coronaviruses.

To date, more than 26% of world population has received at least one dose of COVID-19 vaccine [13]. The COVID-19 pandemic requires not only prevention via vaccine but also drug treatment. Antiretroviral drugs have been tested in past human coronavirus infections and against SARS-CoV-2, but a recent clinical trial of lopinavir and ritonavir failed to show any clinical benefit in COVID-19 disease. Protease inhibitors are used to treat HIV/AIDS and hepatitis C, but PF-07321332 (Figure 1A) is the first oral protease inhibitor to target the SARS-CoV-2 virus. In addition to PF-07321332, α-ketoamide (Figure 1B) has been reported to show in vitro inhibition of $3CL^{pro}$ [14]. The bespoke clinical candidate PF-07321332 (NCT04756531) is a potent protease inhibitor with potent antiviral activity against SARS-CoV-2.

The development of new drugs or vaccines is time-consuming; therefore, a drug-repurposing strategy was approved for the treatment of COVID-19 at this critical time [15]. The use of FDA-approved anti–HIV-I drugs lopinavir and ritonavir (Figure 1C,D) in combination against MERS-CoV and SARS-CoV has been reported earlier [16–19]. This combination of lopinavir and ritonavir is currently in a phase II trial along with interferon β-1b against MERS-CoV [16]. In contrast, for the treatment of MERS, this regimen is in phase IV of clinical trials in combination with Arbidol® hydrochloride and oseltamivir (NCT04255017).

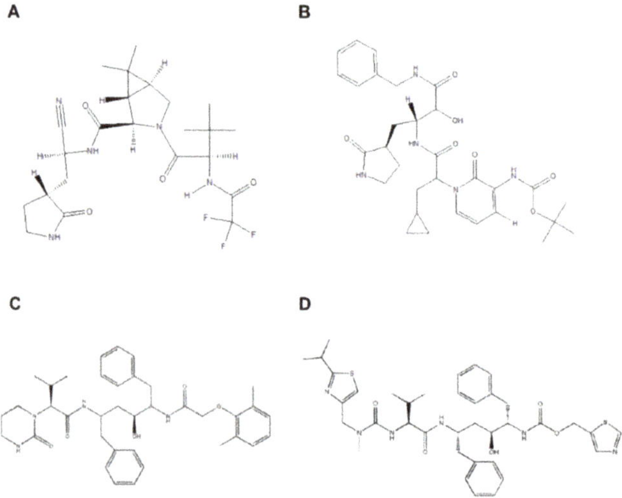

Figure 1. 2D structure representation of (**A**) PF-07321332, (**B**) α-ketoamide, (**C**) lopinavir, and (**D**) ritonavir.

PF-07321332 is a $3CL^{pro}$ inhibitor with potent in vitro antiviral activity against SARS-CoV-2 and other coronaviruses (NCT04756531). Protease inhibitors interfere with the cutting step of the protease enzyme reaction. These drugs interact with the protease and block its polypeptide cutting ability. Lopinavir is used as a retroviral-protease inhibitor for the treatment of HIV-I infection and is usually co-administered with ritonavir, which increases the half-life of lopinavir and inactivates cytochrome P450 3A4 [20,21]. PF-07321332 is administered in combination with low doses of ritonavir as a booster to increase the bloodstream levels of PF-07321332. In the present study, the binding mechanisms of PF-07321332, α-ketoamide, lopinavir, and ritonavir were explored through molecular dynamics (MD) simulations and binding energy calculations. This study offers a deeper understanding of the mechanism of binding of these drugs to $3CL^{pro}$ and will be helpful to identify potent protease inhibitor candidates for SARS-CoV-2 and other coronaviruses. In future, our data can be used to identify novel drug candidates using supervised machine learning methods [22].

2. Results

2.1. Comparative Analysis of PF-07321332, α-ketoamide, Lopinavir, and Ritonavir Binding to $3CL^{pro}$

The crystal structure of $3CL^{pro}$ in complex with PF-07321332, lopinavir, and ritonavir is not yet available. Therefore, the docked complex of PF-07321332, lopinavir, and ritonavir with $3CL^{pro}$ was generated using the Molecular operating environment (MOE) software. Among the top 10 docking solutions, the best ligand conformation was selected based on the docking score and binding pose. In addition, the structure of α-ketoamide co-crystalized with $3CL^{pro}$ (PDB ID: 6Y2F) was retrieved from the RCSB PDB database.

To elucidate the mechanism of binding of PF-07321332 (PF), α-ketoamide (keto), lopinavir (lop), and ritonavir (rit) with $3CL^{pro}$, all complexes were subjected to 100 ns simulations in Gromacs. To assess the stability of all four systems, root mean square deviation (RMSD) values were calculated for each ligand and protein individually and in complexed form, as presented in Figure 2. RMSD of apo-$3CL^{pro}$ at 45 ns indicated a deviation up to 3 Å, after which the system was stabilized and oscillated in the mean range of 2.5 Å (Figure 2A). The protein in $3CL^{pro}$–PF (Figure 2B) remained stable throughout simulation with an average RMSD of 2.7 Å, while in $3CL^{pro}$–keto, the protein fluctuated in a 30–50 ns interval. The protein in $3CL^{pro}$–rit showed RMSD rising above 3 Å after 60 ns with higher fluctuation at 90 ns. The RMSD protein graph of $3CL^{pro}$–lop and $3CL^{pro}$–rit deviated from the apo-form. The $3CL^{pro}$–rit showed sinusoidal behavior until 60 ns and higher peak in the 60–80 ns time interval. Nevertheless, the RMSD of the protein in $3CL^{pro}$–lop increased gradually from 60 ns onwards (Figure 2B). It also manifested fewer fluctuations initially than apo-$3CL^{pro}$ did until 40 ns, and then both systems yielded somewhat similar patterns up to 75 ns, and suddenly, the protein in $3CL^{pro}$–rit showed a remarkable deviation between 75 ns and 100 ns (Figure 2B). While assessing the RMSD of all four complexes (Figure 2C), we noticed that the $3CL^{pro}$–PF complex remained stable during the simulation. The RMSD of $3CL^{pro}$–keto complex fluctuated between 20–50 ns and then gradually stabilized. $3CL^{pro}$–rit and $3CL^{pro}$–lop complexes showed similar trend as their proteins. The RMSD graph of the ligands (Figure 2D) indicated that lopinavir and ritonavir featured greater deviations in comparison with other two ligands, whereas PF manifested the least deviations. The ligand in the $3CL^{pro}$–keto complex showed high jumps around 70 ns but quickly gained stability with a RMSD value of 1.5 Å. RMSD of the ligand in $3CL^{pro}$–lop showed increasing trend from 45 ns onwards. The ligand in $3CL^{pro}$–rit manifested a sudden increase in RMSD after 20 ns and fluctuated between 3 and 4.5 Å but attained stability later. We concluded that inhibitors PF and α-ketoamide imparted stability to the protein in their complexes.

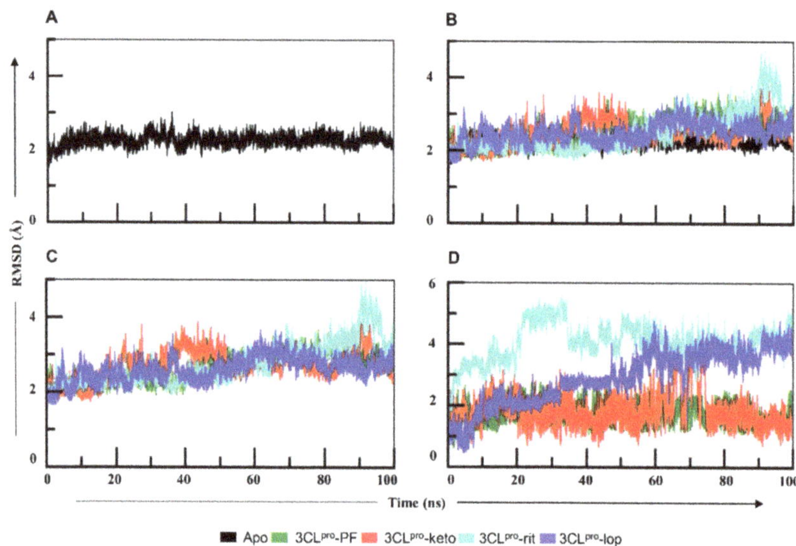

Figure 2. (**A**) The root mean square deviation (RMSD) graph of the apo form of 3CL^{pro} during the 100 ns simulation. The graph suggests that protein was stable during the whole simulation, with an average RMSD of 2.2 Å. (**B**) The RMSD graphs of all four proteins in complexes in comparison with the apo form. The protein in 3CL^{pro}–PF and 3CL^{pro}–keto remained stable whereas in 3CL^{pro}–lop and 3CL^{pro}–rit, it showed more fluctuations than in the other two 3CL^{pro} systems. (**C**) The RMSD graph of 3CL^{pro} complexes. The 3CL^{pro}–rit and 3CL^{pro}–lop complexes showed similar trend as the proteins. (**D**) RMSD graphs of the ligands in complexes with 3CL^{pro}. The ligands PF-07321332 and α-ketoamide seemed stable, whereas both lopinavir and ritonavir in 3CL^{pro} complexes featured noticeable fluctuations. PF—PF-07321332; keto—α-ketoamide; rit—ritonavir; lop—lopinavir.

To analyze the dynamic behavior of protein residues, root mean square fluctuation (RMSF) values were assessed. As compared with the apo form, RMSF (Figure 3A) of the protein in 3CL^{pro}–rit indicated that amino acid residues 45–75 in the spanning region fluctuated by more than 3 Å. By contrast, in 3CL^{pro}–lop, the residues of 3CL^{pro} between positions 45 and 75 fluctuated in the range similar to that of its apo form, i.e., 2.5 Å (Figure 3A). The rest of the protein residues of both complexes oscillated equally relative to the apoprotein. Overall, the residues of the 3CL^{pro} in complex with the ligands PF and α-ketoamide did not undergo any fluctuation, while in 3CL^{pro}–rit complex, the protein featured the greatest fluctuations in comparison with the apo form.

To assess the compactness of protein structure, we calculated the radius of gyration (Rg). This parameter of the protein in 3CL^{pro}–PF (Figure 3B) remained in the steady state and close to that of the apoprotein during the simulations, which means that the folding of the protein in 3CL^{pro}–PF complex remained undisturbed. Figure 3B illustrates that the apoprotein followed a similar trend and oscillated at ~22 Å. The Rg of the protein in 3CL^{pro}–keto fluctuated between 20–50 ns; thereafter, it retained steady state for the rest of the simulation. Rg of 3CL^{pro}–rit remained steady up to 50 ns, then it gradually increased. 3CL^{pro}–lop fluctuated with the highest peak of 23 Å near time point 39 ns, indicating that protein folding was disturbed. Lopinavir and ritonavir might have led to the unfolding in the 3CL^{pro} and thus exhibited increasing trend in Rg of the protein.

Next, the number of hydrogen bonds was calculated to evaluate the protein–ligand interactions during the simulations. The PF formed three hydrogen bonds (Figure 3C), and this remained consistent throughout the time trajectory. The α-ketoamide formed two bonds initially but their number dropped to one after 50 ns. The lopinavir initially formed three hydrogen bonds with 3CL^{pro}, but their number gradually decreased, and the hydrogen bonds disappeared around 48 ns. Lopinavir formed two bonds in the beginning and ritonavir formed one, but these interactions were inconsistent throughout the MD simulations (Figure 3C). Therefore, the hydrogen bonds formed by lopinavir and ritonavir with 3CL^{pro} are inconsistent and weak as compared with the interactions of 3CL^{pro} with PF-07321332 and α-ketoamide.

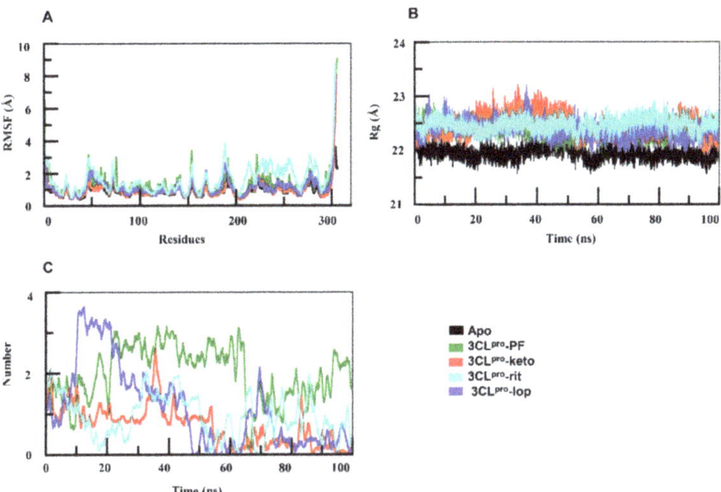

Figure 3. (**A**) Root mean square fluctuation graphs of apo-3CL^{pro} and of 3CL^{pro} complexes. In comparison with the apo-form, regions spanning amino acid residues 45–75 and 150–200 in both docked complexes showed deviations. (**B**) Radius of gyration of the 3CL^{pro} in complexes was higher than that of its apo-form. (**C**) Number of hydrogen bonds in the four complexes. PF—PF-07321332; keto—α-ketoamide; rit—ritonavir; lop—lopinavir.

2.2. Interaction Analysis of PF-07321332, α-ketoamide, Lopinavir, and Ritonavir with 3CL^{pro}

To evaluate the ligand–protein interactions, we computed the minimum distance between interacting atoms of the ligand and protein residues. Our analysis revealed the dynamics of the interaction of PF-07321332, α-ketoamide, lopinavir, and ritonavir with 3CL^{pro} (Figure 4). In 3CL^{pro}–PF, three hydrogen bonds were formed (Table 1). The distance of Cys145–O2, Glu166–O3, and Gln189–H20 bonds remained within 4 Å (Figure 4A). The catalytic dyad residue Cys145 made strong bonds with atoms O2 and H9 as they remained intact during the simulation. This showed that the PF-07321332 has high affinity towards 3CL^{pro}. The 4-methoxylindole group interacted through a hydrogen bond with the backbone of Glu166, a key residue located in the middle of the binding site. The study of Glu166 mutation, carried out in SARS-CoV 3CL protease (96% identity to SARS-CoV-2 3CL^{pro}), has corroborated the role of this key residue in the substrate-induced dimerization [23]. The trifluoroacetamide moiety of the PF-07321332 compound involved in the Gln189–H20 bond accommodated the cyclopropyl-proline moiety to occupy the central portion of the binding site. Subsequently, the pyrrolidone moiety rearrangement led to bond formation of reactive nitrile group with Cys145. According to the analysis of the entire trajectory of 3CL^{pro}–PF complex, PF-07321332 contacted Leu141 and Gln142. Leu141, Gln142, and Cys145 were part of the oxyanion loop (residues 138–145), which is lining the glutamine binding pocket and is presumably involved in the stabilization of the tetrahedral acyl transition state [24]. The ligand in 3CL^{pro}–keto made three hydrogen bond interactions (Glu166–O48, Cys145–O41, and Met165–O48) and one arene interaction, His41–H32 (Figure 4B). The cyclopropyl moiety facilitated ligand binding in the shallow substrate-binding pocket at the surface of the 3CL^{pro}. The α-keto group of the inhibitor interacted with Cys145. The study reported that thiohemiketal formation occurred in a reversible reaction by the nucleophilic attack of the catalytic Cys145 on the ligand [14]. In our study, α-ketoamide occupied the substrate binding space and its carbonyl oxygen formed hydrogen bond with the main-chain oxygen of Glu166. All these interactions kept fluctuating as the distance between the interacting entities continuously changed due to the bulky nature of the ligand. Similarly, in the $3CL^{pro}$–rit complex, three hydrogen bonds and an arene-type interaction were formed (Figure 4D). The length of His41–H6 and Gly143–O3 bonds remained within 7 Å during the initial 50 ns and continuously increased onwards. In the arene-type interaction, the distance between Pro168 and the thiazolyl group fluctuated within 5–10 Å during the simulation. The hydrogen bond Thr190–S2 appeared during the initial 10 ns, and the bond length was within 5 Å, but it diminished after 10 ns (Figure 4D). Two hydrogen bonds formed in $3CL^{pro}$–lop. Initially, the bond length of His41–O38 and Glu166–O20 (Table 1) was less than 5 Å, but after 50 ns, an increase in the bond length was detected because of ligand fluctuations, as depicted in the RMSD graph (Figures 2D and 4C).

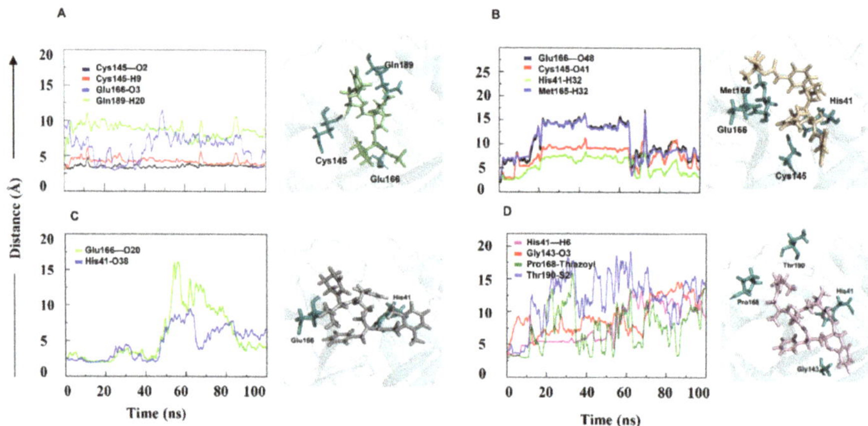

Figure 4. The minimum distance graph of protein–ligand interactions. The interaction distance between the protein and ligand in 3CLpro–PF (**A**) and $3CL^{pro}$–keto (**B**) remained constant. Interaction in $3CL^{pro}$–lop (**C**) initially remained constant but showed major fluctuations afterwards. In $3CL^{pro}$–rit (**D**), the lengths of all four bonds continuously increased. PF—PF-07321332; keto—α-ketoamide; rit—ritonavir; lop—lopinavir.

Table 1. Residue and ligand atom interactions with the bond type and energy in $3CL^{pro}$ complexes.

Complex	Residue	Ligand Atom	Bond Type	Energy
$3CL^{pro}$–PF	Cys145	O2	Hydrogen	−5.04
	Cys145	H9	Hydrogen	−5.04
	Glu166	O3	Hydrogen	−4.41
	Gln189	H20	Hydrogen	−3.01
$3CL^{pro}$–keto	Glu166	O48	Hydrogen	−0.70
	Cys145	O41	Hydrogen	−0.80
	Met165	O48	Hydrogen	−0.80
	His41	H32	Arene	−0.90
$3CL^{pro}$–lop	His41	O38	Hydrogen	−4.1
	Glu166	O20	Hydrogen	−1.6
$3CL^{pro}$–rit	His41	H6	Arene	−3.0
	Gly143	O3	Hydrogen	−2.2
	Pro168	Thiozyl group	Arene	−0.8
	Thr190	S2	Hydrogen	−0.8

2.3. Binding Free Energy Calculation

Protein–ligand binding affinity is essential in understanding molecular recognition. The bespoke antiviral clinical candidate PF-07321332 (NCT04756531) is a potent $3CL^{pro}$ inhibitor with antiviral activity against SARS-CoV-2. The relative binding free energy of PF-07321332, α-ketoamide, lopinavir, and ritonavir with $3CL^{pro}$ was computed. To calculate the binding affinity of these ligands with $3CL^{pro}$, we extracted 1000 frames from the simulation trajectories. The g_mmpbsa (Table 2) computation revealed that the binding affinity of PF-07321332 for $3CL^{pro}$ was higher in comparison with α-ketoamide, lopinavir, and ritonavir. Relative to α-ketoamide, PF-07321332 showed more considerable binding energy toward $3CL^{pro}$, almost by 28 kJ/mol, and three times more than lopinavir and ritonavir. These data showed that PF-07321332 had a stronger binding affinity for $3CL^{pro}$.

Table 2. Binding energy of $3CL^{pro}$ and HIV-I protease complexes as computed via the g_mmpbsa method.

Complex	VdW Energy	Electrostatic	Polar Solvation	SASA	Total Binding Energy (kJ/mol)
$3CL^{pro}$–PF	−135.294 +/− 16.258	−39.293 +/− 25.780	96.026+/− 21.556	−23.441 +/− 2.526	−102.002 +/− 21.336
$3CL^{pro}$–keto	−76.478+/− 25.052	−40.348 +/− 30.471	50.648 +/− 49.365	−7.887 +/− 3.036	−74.065 +/− 22.579
$3CL^{pro}$–lop	−110.529 +/− 28.28	−65.740 +/−26.09	168.011 +/− 33.31	−25.239 +/−2.40	−33.497 +/− 26.52
$3CL^{pro}$–rit	−75.994 +/− 60.285	−55.115 +/− 23.09	91.803 +/− 18.798	−4.219 +/− 8.798	−43.525 +/− 29.171

3. Discussion

Combating the COVID-19 pandemic requires both prevention, including vaccine-based prevention, and targeted treatment for those who get infected. Given the way SARS-CoV-2 is evolving into new genetic variants and the worldwide impact of COVID-19, it is likely that access to treatment alternatives is crucial, now and beyond the pandemic. The frequent mutations aid the virus to infect different hosts and increase the virulence and transmissibility of SARS-CoV-2. The RNA viruses, including SARS-CoV-2, develop one mutation per replication cycle, which means RNA viruses may adapt to new environment, but these viruses are also restricted in their ability to change their genomes because they must keep their mutation rate low in order to survive [25]. Several mutations have been reported in the viral genome so far; 106 major amino acid substitutions were found to be responsible for increased virulence and transmissibility of the virus [26]. Most of these missense mutations occurred in structural proteins of SARS-CoV-2, such as spike (S) [27] and nucleocapsid (N) proteins. Only one point mutation, G15S, has been predicted using in silico studies in $3CL^{pro}$, but this residue is far from the active site and has a destabilizing effect on the protein structure [26]. Therefore, a virus comprising such mutations with a destabilizing effect would not last for long and would have reduced virulence. Only stabilizing mutations, such as in the S protein, increase the virulence and transmissibility of the virus [28]. A recent study has reported that the significant key mutations are located at the inverted repeat regions and CpG islands. These findings showed the role of mutations in the instability of the viral genome [29].

In this study, we have explored the binding of PF-07321332, α-ketoamide, lopinavir, and ritonavir to SARS-CoV-2 $3CL^{pro}$ through MD simulation and MMPBSA calculation. PF-07321332 (NCT04756531) and α-ketoamide molecules that specifically bind to and inhibit SARS-CoV-2 $3CL^{pro}$ could be promising alternatives to fight the pandemic [14]. The comparative binding mode analysis of PF-07321332, α-ketoamide, lopinavir, and ritonavir might provide a glimpse into designing rational drugs through the modification of the inhibitors based on the residues in the active site of the enzyme.

To determine the mechanism of binding of PF-07321332, α-ketoamide, lopinavir, and ritonavir to $3CL^{pro}$, molecular docking of these molecules was done. The best docking solutions for PF-07321332, lopinavir, and ritonavir were selected based on the docking score and the best binding pose of the ligand. To understand the binding mechanism of PF-07321332 and α-ketoamide along with the two antiretroviral drugs, lopinavir and ritonavir, to $3CL^{pro}$, MD simulations were performed for 100 ns. Backbone RMSDs of the $3CL^{pro}$–PF and $3CL^{pro}$–keto complexes were stable, whereas the $3CL^{pro}$–lop complex featured a deviation at 60–80 ns and the $3CL^{pro}$–rit complex a deviation between 75 ns and 100 ns (Figure 2B). RMSD deviation of the $3CL^{pro}$ protein was also analyzed in the ligand-bound form. The RMSD of $3CL^{pro}$ (Figure 2C) was found to be stabler in association with either PF-07321332 or α-ketoamide than with lopinavir or ritonavir. In contrast to the other systems, $3CL^{pro}$ in complex with ritonavir featured marked deviation between 88 and 95 ns of the simulation. Both ligands PF-07321332 and α-ketoamide are stabler with $3CL^{pro}$ in comparison with lopinavir and ritonavir because later two underwent marked deviation during the simulation (Figure 2D). The recent clinical trials have also showed these antiretrovirals are not efficacious as they did not significantly accelerate clinical improvements in serious COVID-19 patients [30]. Our binding energy data also showed low affinity of these drugs towards $3CL^{pro}$. To assess the impact of ligand binding on protein residues, RMSF of $3CL^{pro}$ were analyzed in all four complexes. With either lopinavir or ritonavir, amino acid residues 45–65 and 145–200 showed greater fluctuations in comparison with the apo-form (Figure 3A). These fluctuating regions mainly consist of binding-site residues that are interconnected by a hydrogen bond network and are involved in catalytic dyad formation between His41 and Cys145 [31–33]. However, the RMSF of the protein in $3CL^{pro}$–PF showed similar trend as its apo-form. Compactness, which determines stability of a protein, was evaluated for $3CL^{pro}$ using Rg. The compactness of $3CL^{pro}$ in complex with PF-07321332 and α-ketoamide remained in steady state the same as the one of its apo-form (Figure 3B). Rg of $3CL^{pro}$ in its complex with ritonavir or lopinavir was higher than that of apo-$3CL^{pro}$. Nonetheless, $3CL^{pro}$ compactness after PF-07321332 and α-ketoamide binding remained constant throughout the simulation.

One of the most interesting examples of MD application is the drug discovery area, where this method drives experiments [34,35]. To analyze the interactions of the four molecules with $3CL^{pro}$, the minimum distance between the atoms of interacting residues of each protein and ligand (PF-07321332, α-ketoamide, lopinavir, and ritonavir) were calculated as a function of simulation time to analyze the dynamic behavior of the interacting groups. In the interaction analysis, PF-07321332 and α-ketoamide showed stronger interactions with $3CL^{pro}$, and these interactions remained intact during the simulation because the minimum distance between the interacting groups stayed almost unchanged. The interactions, i.e., Cys145–O2 and Cys145–H9 in $3CL^{pro}$–PF (Figure 4A), remained intact during the simulation. This showed that PF-07321332 was strongly bonded to $3CL^{pro}$ and capable to disrupt the His41–Cys145 catalytic dyad, which, together with the N-terminus residues 1 to 7, is thought to have a vital role in proteolytic activity [36]. The Glu166 anchor hooks the ligand tightly to the central region of the binding site, which facilitate the formation of further interactions with other residues. Gly143 of SARS-CoV-2 $3CL^{pro}$ was reported to be the most favorable residue to form hydrogen bonds with ligands, followed by Glu166, Cys145, and His163 [37]. The amides of Gly143, Cys145, and Ser144 form the cysteine protease's canonical "oxyanion hole". The interaction of PF-07321332 and α-ketoamide with the catalytic site residues may cause the distortion of the oxyanion hole in the reaction mechanism, and it may lead to the inhibition of $3CL^{pro}$ in SARS-CoV-2. Compared with PF-07321332 and α-ketoamide, both antiretroviral drugs lopinavir and ritonavir manifested weaker interactions with $3CL^{pro}$, judging by binding energy values (Table 2). PF-07321335 interacts with Cys145, thereby disrupting His41–Cys145 catalytic dyad. The α-ketoamide has the same mechanism as reported earlier carries; a nucleophilic attack of the catalytic Cys145 on the α-keto group of the inhibitor, disrupting the His41–Cys145 catalytic dyad [14]. Our binding energy data (Table 2) revealed that PF-07321335 has higher affinity towards $3CL^{pro}$. This is in accordance with PF-07321335 showing a potent inhibition of $3CL^{pro}$ (NCT04756531). Lopinavir and ritonavir showed weaker interactions because only two residues, His41 and Glu166, were found to be involved in hydrogen bonding. The minimum distance of 5 Å between these interacting entities persisted up to 45 ns; after that, it fluctuated. In contrast with PF-07321335 and α-ketoamide, lopinavir and ritonavir engaged in interactions with $3CL^{pro}$, but these interactions varied substantially. Some chemical substitutions in lopinavir and ritonavir (such as the addition of an α-keto group) can act as electrophiles that may prevent the His41–Cys145 dyad formation by a nucleophilic attack of Cys145. Although in our simulations, lopinavir and ritonavir engaged in interactions with His41 and Cys145 along with other catalytic site residues, these contacts fluctuated considerably, as evident from our interaction analysis. Coronavirus protease $3CL^{pro}$ lacks a C2-symmetric pocket, through which lopinavir and ritonavir bind to HIV-I protease. The absence of a C2-symmetric pocket in coronavirus $3CL^{pro}$ could be the reason why these drugs do not bind to $3CL^{pro}$, which makes them ineffective in COVID-19 patients. The binding energy data (Table 2) showed that lopinavir and ritonavir have weak affinity to $3CL^{pro}$.

Conclusively, $3CL^{pro}$ is an important drug target, which cleaves the viral polyprotein at 11 different cleavage sites and is required for viral maturation. The druggability of $3CL^{pro}$ has already been demonstrated in various studies. The mutations in the dominant variants of SARS-CoV-2, such as B.1.1.7 and B.1.617, with higher transmissibility rate were studied and it has been demonstrated that most of the mutations have been found in the receptor binding domains or structural proteins, but no mutation has been reported in $3CL^{pro}$ so far [38]. Therefore, $3CL^{pro}$ is an attractive target for anti-COVID-19 drug discovery. In recent studies, many new compounds have been reported as potential inhibitors of $3CL^{pro}$ [14,39–42]. We included four of the potential inhibitors in our study, compared their binding energy and discussed their interactions with the protein to identify the potent inhibitor. We found that the compound PF-07321332, currently in clinical trials, was the best of these four compounds. In addition to these molecules, sterenin M was also reported in a recent computational study [39] to bind to the same active binding site as PF-07321332 and interact with the residues of catalytic dyad, but its total binding energy profile showed that the binding of PF-07321332 with protein was much stabler than the one of sterenin M [39]. In similar studies, the identified compounds bind to the same binding site, but the stability of complex is lower than the one of PF-07321332 [40,41]. Hence, considering the current pandemic situation, it is crucial to find some potent drug candidate with good binding affinity. The MD simulation and MMPBSA are widely used for assessing protein–ligand binding affinity, but there are a few limitations of the MD simulations, such as many biochemical processes including receptor conformational shifts relevant to drug binding occur on time scales that are much longer than those amenable to simulations. In MD simulations, each atom is assigned a fixed partial charge before simulation. However, the electron clouds in surrounding atoms shift continuously according to their environments. A dynamic and responsive representation of atomic partial charges would be more accurate.

4. Conclusions

A detailed analysis of interactions of proteins and ligands represents an optimized method for the rational design of new compounds. Binding energy helps to identify a lead compound. MD, free energy perturbation, meta-dynamics, and other methods are consistently used for studying drug–target binding. Assessment of these methods may help to achieve the most definite target-optimized affinity with improved drug efficacy [43]. Integrating MD simulation with binding-free-energy calculation by means of g_mmpbsa to determine interaction free energy between a ligand and protein is an efficient method for distinguishing between active and inactive molecules [44]. The comparative analysis of PF-07321332, α-ketoamide, lopinavir, and ritonavir via MD simulation provided a detailed insight into the interactions of these compounds with $3CL^{pro}$. We believe that our findings revealed the binding mechanism of PF-07321332 and α-ketoamide and further explained the inability of lopinavir and ritonavir to cause clinical improvement in severe COVID-19 patients by due to their low affinity towards $3CL^{pro}$. We believe the dynamic interaction and binding energy of PF-07321332 and α-ketoamide will be helpful in designing potent inhibitors for $3CL^{pro}$.

5. Materials and Methods

5.1. System Preparation for Molecular Docking and MD Simulations

The crystal structure of SARS-CoV-2 3CL^{pro} (Protein Data Bank [PDB] ID: 6M03) and structures of 3CL^{pro} in complex with 3CL^{pro} (PDB ID 6Y2K) [45,46] were downloaded from the PDB [47]. The 2-D structure of PF-07321335 was drawn in Chemdraw and the three-dimensional (3D) coordinates of lopinavir (PubChem CID: 92727) and ritonavir (PubChem CID: 392622) were retrieved from the PubChem database [48] for molecular docking. All hydrogens and missing atoms were added to the proteins, while crystal water was removed. The MOE software from Chemical Computing Group Canada (Montreal, QC, Canada) [49] was used for energy minimization of 3CL^{pro} and both drugs to remove any steric clashes and to improve the bond lengths and bond angles. The default parameters were used for energy minimization, while the gradient was set to 0.01 rms kcal/mol.

5.2. Molecular Docking of 3CL^{pro} to Lopinavir and Ritonavir

PF-07321335, α-ketoamide, lopinavir, and ritonavir were docked into the binding pocket of 3CL^{pro} using the MOE software [49]. The key residues involved in ligand binding reported in the literature were selected for docking: His41, Asn142, Cys145, His164, Met165, Glu166, Gln189, and Thr190. The triangle matcher placement method was selected with the London dG scoring function, and the induced-fit method was applied for refinement of the binding poses of ligands. A total of 100 conformations were generated for each ligand, and the best pose with the lowest binding energy was selected. Complexes of 3CL^{pro} with PF-07321335, α-ketoamide, lopinavir, and ritonavir were analyzed and saved for MD simulations.

5.3. Building the Ligand Topology

Prior to MD simulations, the ligand topology was generated on the CHARMM general force field (CGenFF) server (University of Maryland, Baltimore, MD, USA) [50]. Ligands, along with their 3D coordinates, were extracted from their respective complexes and saved in mol2 file format, and all hydrogens were added. The mol2 file was uploaded to the CGenFF server, which generated a CHARMM-compatible stream file comprising ligand information, such as atom types, charges, and bond parameters. A Python script was executed to convert the stream files into Gromacs-compatible [51,52] files. These files were used for the MD simulations.

5.4. MD Simulations of Complexes

The dataset for MD simulations contained apo-3CL^{pro} and four 3CL^{pro} complexes. All five MD simulations were carried out using the Gromacs (University of Groningen, Groningen, Netherlands) software for 100 ns each, and the CHARMM36 forcefield [53] was applied to the systems. Periodic boundary conditions were used, and the protein was placed inside a 10 Å cubic box. For solvation of the system, the TIP3P water model [54] was used, and an appropriate amount of counterions was added to neutralize the system. The solvated electroneutral system was subjected to energy minimization by the steepest-descent algorithm, followed by temperature and pressure equilibration steps. The production run was carried out for 100 ns in each system, and data analysis was performed by means of Gromacs built-in tools, MOE, and PyMOL (Schrödinger, LLC. DeLano Scientific, San Francisco, CA, USA) [55,56].

5.5. Binding-Free-Energy Calculation

The molecular mechanics Poisson–Boltzmann surface area (MMPBSA) method was used to compute the binding free energy of all complexes. A thousand frames were extracted from each 100 ns simulation trajectory by means of the *gmx trjconv* utility of

Gromacs. The extracted frames were used for free-binding-energy calculation with the help of the g_mmpbsa tool [44]. The total binding free energy ($\Delta G_{binding}$) was calculated as

$$\Delta G_{binding} = \Delta G_{complex} - (\Delta G_{protein} + \Delta G_{ligand})$$

where $G_{complex}$ is the average free energy of the complex, and $G_{protein}$ and G_{ligand} are the average energy values of the protein and ligand.

Author Contributions: Conceptualization, B.A. and M.B.; methodology, B.A., Q.u.A., and M.B.; formal analysis, B.A., Q.u.A., and M.B.; data curation, B.A., Q.u.A., and M.B.; writing—original draft preparation, B.A., Q.u.A., and M.B.; writing—review and editing, B.A. and M.B.; supervision, S.C.; project administration, M.S.K. and S.C.; funding acquisition, M.S.K. and S.C. All authors have read and agreed to the published version of the manuscript.

Funding: This work was supported by the National Research Foundation of Korea (NRF- 2020R1F1A 1071517, 2019M3D1A1078940, and 2019R1A6A1A11051471).

Institutional Review Board Statement: Not applicable.

Informed Consent Statement: Not applicable.

Data Availability Statement: The structure models and simulation trajectories are available upon request (sangdunchoi@ajou.ac.kr). All remaining data are contained within the manuscript.

Conflicts of Interest: The authors declare no conflict of interest.

References

1. Velavan, T.P.; Meyer, C.G. The COVID-19 epidemic. *Trop. Med. Int. Health* **2020**, *25*, 278–280. [CrossRef]
2. Johns Hopkins Coronavirus Resource Center. COVID-19 Map. Available online: https://coronavirus.jhu.edu/map.html (accessed on 10 August 2020).
3. Dai, W.; Zhang, B.; Su, H.; Li, J.; Zhao, Y.; Xie, X.; Jin, Z.; Liu, F.; Li, C.; Li, Y.; et al. Structure-based design of antiviral drug candidates targeting the SARS-CoV-2 main protease. *Science* **2020**, *368*, 1331–1335. [CrossRef]
4. Zhou, P.; Yang, X.L.; Wang, X.G.; Hu, B.; Zhang, L.; Zhang, W.; Si, H.R.; Zhu, Y.; Li, B.; Huang, C.L.; et al. A pneumonia outbreak associated with a new coronavirus of probable bat origin. *Nature* **2020**, *579*, 270–273. [CrossRef] [PubMed]
5. Lu, R.; Zhao, X.; Li, J.; Niu, P.; Yang, B.; Wu, H.; Wang, W.; Song, H.; Huang, B.; Zhu, N.; et al. Genomic characterisation and epidemiology of 2019 novel coronavirus: Implications for virus origins and receptor binding. *Lancet* **2020**, *395*, 565–574. [CrossRef]
6. Durai, P.; Batool, M.; Shah, M.; Choi, S. Middle East respiratory syndrome coronavirus: Transmission, virology and therapeutic targeting to aid in outbreak control. *Exp. Mol. Med.* **2015**, *47*, e181. [CrossRef] [PubMed]
7. Batool, M.; Shah, M.; Patra, M.C.; Yesudhas, D.; Choi, S. Structural insights into the Middle East respiratory syndrome coronavirus 4a protein and its dsRNA binding mechanism. *Sci. Rep.* **2017**, *7*, 11362. [CrossRef] [PubMed]
8. Anand, K.; Ziebuhr, J.; Wadhwani, P.; Mesters, J.R.; Hilgenfeld, R. Coronavirus main proteinase (3CLpro) structure: Basis for design of anti-SARS drugs. *Science* **2003**, *300*, 1763–1767. [CrossRef]
9. Van Boheemen, S.; de Graaf, M.; Lauber, C.; Bestebroer, T.M.; Raj, V.S.; Zaki, A.M.; Osterhaus, A.D.; Haagmans, B.L.; Gorbalenya, A.E.; Snijder, E.J.; et al. Genomic characterization of a newly discovered coronavirus associated with acute respiratory distress syndrome in humans. *mBio* **2012**, *3*, e00473-12. [CrossRef] [PubMed]
10. Muramatsu, T.; Takemoto, C.; Kim, Y.T.; Wang, H.; Nishii, W.; Terada, T.; Shirouzu, M.; Yokoyama, S. SARS-CoV 3CL protease cleaves its C-terminal autoprocessing site by novel subsite cooperativity. *Proc. Natl. Acad. Sci. USA* **2016**, *113*, 12997–13002. [CrossRef]
11. Thiel, V.; Ivanov, K.A.; Putics, A.; Hertzig, T.; Schelle, B.; Bayer, S.; Weissbrich, B.; Snijder, E.J.; Rabenau, H.; Doerr, H.W.; et al. Mechanisms and enzymes involved in SARS coronavirus genome expression. *J. Gen. Virol.* **2003**, *84*, 2305–2315. [CrossRef]
12. Ren, Z.; Yan, L.; Zhang, N.; Guo, Y.; Yang, C.; Lou, Z.; Rao, Z. The newly emerged SARS-like coronavirus HCoV-EMC also has an "Achilles' heel": Current effective inhibitor targeting a 3C-like protease. *Protein Cell* **2013**, *4*, 248–250. [CrossRef] [PubMed]
13. Ritchie, H.; Mathieu, E.; Rodés-Guirao, L.; Appel, C.; Giattino, C.; Ortiz-Ospina, E.; Hasell, J.; Macdonald, B.; Beltekian, D.; Roser, M. Coronavirus pandemic (COVID-19). 2020. Available online: https://ourworldindata.org/covid-vaccinations (accessed on 31 July 2021).
14. Zhang, L.; Lin, D.; Sun, X.; Curth, U.; Drosten, C.; Sauerhering, L.; Becker, S.; Rox, K.; Hilgenfeld, R. Crystal structure of SARS-CoV-2 main protease provides a basis for design of improved alpha-ketoamide inhibitors. *Science* **2020**, *368*, 409–412. [CrossRef]
15. Kalil, A.C. Treating COVID-19-Off-Label Drug Use, Compassionate Use, and Randomized Clinical Trials During Pandemics. *JAMA* **2020**, *323*, 1897–1898. [CrossRef]

16. Arabi, Y.M.; Alothman, A.; Balkhy, H.H.; Al-Dawood, A.; Al Johani, S.; Al Harbi, S.; Kojan, S.; Al Jeraisy, M.; Deeb, A.M.; Assiri, A.M.; et al. Treatment of Middle East Respiratory Syndrome with a combination of lopinavir-ritonavir and interferon-beta1b (MIRACLE trial): Study protocol for a randomized controlled trial. *Trials* **2018**, *19*, 81. [CrossRef] [PubMed]
17. Chu, C.M.; Cheng, V.C.; Hung, I.F.; Wong, M.M.; Chan, K.H.; Chan, K.S.; Kao, R.Y.; Poon, L.L.; Wong, C.L.; Guan, Y.; et al. Role of lopinavir/ritonavir in the treatment of SARS: Initial virological and clinical findings. *Thorax* **2004**, *59*, 252–256. [CrossRef]
18. Chen, F.; Chan, K.H.; Jiang, Y.; Kao, R.Y.; Lu, H.T.; Fan, K.W.; Cheng, V.C.; Tsui, W.H.; Hung, I.F.; Lee, T.S.; et al. In vitro susceptibility of 10 clinical isolates of SARS coronavirus to selected antiviral compounds. *J. Clin. Virol.* **2004**, *31*, 69–75. [CrossRef]
19. Wu, C.Y.; Jan, J.T.; Ma, S.H.; Kuo, C.J.; Juan, H.F.; Cheng, Y.S.; Hsu, H.H.; Huang, H.C.; Wu, D.; Brik, A.; et al. Small molecules targeting severe acute respiratory syndrome human coronavirus. *Proc. Natl. Acad. Sci. USA* **2004**, *101*, 10012–10017. [CrossRef]
20. Sevrioukova, I.F.; Poulos, T.L. Structure and mechanism of the complex between cytochrome P4503A4 and ritonavir. *Proc. Natl. Acad. Sci. USA* **2010**, *107*, 18422–18427. [CrossRef]
21. Chandwani, A.; Shuter, J. Lopinavir/ritonavir in the treatment of HIV-1 infection: A review. *Ther. Clin. Risk Manag.* **2008**, *4*, 1023–1033. [CrossRef] [PubMed]
22. Batool, M.; Ahmad, B.; Choi, S. A Structure-Based Drug Discovery Paradigm. *Int. J. Mol. Sci.* **2019**, *20*, 2783. [CrossRef]
23. Cheng, S.C.; Chang, G.G.; Chou, C.Y. Mutation of Glu-166 blocks the substrate-induced dimerization of SARS coronavirus main protease. *Biophys J.* **2010**, *98*, 1327–1336. [CrossRef]
24. Fornasier, E.; Macchia, M.L.; Giachin, G.; Sosic, A.; Pavan, M.; Sturlese, M.; Salata, C.; Moro, S.; Gatto, B.; Bellanda, M.; et al. A novel conformational state for SARS-CoV-2 main protease. *bioxRiv* **2021**. [CrossRef]
25. Bartas, M.; Brazda, V.; Bohalova, N.; Cantara, A.; Volna, A.; Stachurova, T.; Malachova, K.; Jagelska, E.B.; Porubiakova, O.; Cerven, J.; et al. In-Depth Bioinformatic Analyses of Nidovirales Including Human SARS-CoV-2, SARS-CoV, MERS-CoV Viruses Suggest Important Roles of Non-canonical Nucleic Acid Structures in Their Lifecycles. *Front. Microbiol.* **2020**, *11*, 1583. [CrossRef] [PubMed]
26. Jacob, J.J.; Vasudevan, K.; Pragasam, A.K.; Gunasekaran, K.; Veeraraghavan, B.; Mutreja, A. Evolutionary Tracking of SARS-CoV-2 Genetic Variants Highlights an Intricate Balance of Stabilizing and Destabilizing Mutations. *mBio* **2021**, *12*, e0118821. [CrossRef]
27. West, A.P., Jr.; Wertheim, J.O.; Wang, J.C.; Vasylyeva, T.I.; Havens, J.L.; Chowdhury, M.A.; Gonzalez, E.; Fang, C.E.; Di Lonardo, S.S.; Hughes, S.; et al. Detection and characterization of the SARS-CoV-2 lineage B.1.526 in New York. *Nat. Commun.* **2021**, *12*, 4886. [CrossRef]
28. Kim, J.S.; Jang, J.H.; Kim, J.M.; Chung, Y.S.; Yoo, C.K.; Han, M.G. Genome-Wide Identification and Characterization of Point Mutations in the SARS-CoV-2 Genome. *Osong Public Health Res. Perspect.* **2020**, *11*, 101–111. [CrossRef] [PubMed]
29. Goswami, P.; Bartas, M.; Lexa, M.; Bohalova, N.; Volna, A.; Cerven, J.; Cervenova, V.; Pecinka, P.; Spunda, V.; Fojta, M.; et al. SARS-CoV-2 hot-spot mutations are significantly enriched within inverted repeats and CpG island loci. *Brief. Bioinform.* **2021**, *22*, 1338–1345. [CrossRef]
30. Cao, B.; Wang, Y.; Wen, D.; Liu, W.; Wang, J.; Fan, G.; Ruan, L.; Song, B.; Cai, Y.; Wei, M.; et al. A Trial of Lopinavir-Ritonavir in Adults Hospitalized with Severe Covid-19. *N. Engl. J. Med.* **2020**, *382*, 1787–1799. [CrossRef]
31. Ul Qamar, M.T.; Alqahtani, S.M.; Alamri, M.A.; Chen, L.L. Structural basis of SARS-CoV-2 3CL(pro) and anti-COVID-19 drug discovery from medicinal plants. *J. Pharm. Anal.* **2020**, *10*, 313–319. [CrossRef]
32. Paasche, A.; Zipper, A.; Schäfer, S.; Ziebuhr, J.; Schirmeister, T.; Engels, B. Evidence for Substrate Binding-Induced Zwitterion Formation in the Catalytic Cys-His Dyad of the SARS-CoV Main Protease. *Biochemistry* **2014**, *53*, 5930–5946. [CrossRef] [PubMed]
33. Xue, X.; Yu, H.; Yang, H.; Xue, F.; Wu, Z.; Shen, W.; Li, J.; Zhou, Z.; Ding, Y.; Zhao, Q.; et al. Structures of two coronavirus main proteases: Implications for substrate binding and antiviral drug design. *J. Virol.* **2008**, *82*, 2515–2527. [CrossRef]
34. Liu, X.; Shi, D.; Zhou, S.; Liu, H.; Liu, H.; Yao, X. Molecular dynamics simulations and novel drug discovery. *Expert Opin. Drug Discov.* **2018**, *13*, 23–37. [CrossRef]
35. Borhani, D.W.; Shaw, D.E. The future of molecular dynamics simulations in drug discovery. *J. Comput. Aided Mol. Des.* **2012**, *26*, 15–26. [CrossRef]
36. Mengist, H.M.; Dilnessa, T.; Jin, T. Structural Basis of Potential Inhibitors Targeting SARS-CoV-2 Main Protease. *Front. Chem.* **2021**, *9*, 622898. [CrossRef] [PubMed]
37. Nguyen, D.D.; Gao, K.; Chen, J.; Wang, R.; Wei, G.W. Unveiling the molecular mechanism of SARS-CoV-2 main protease inhibition from 137 crystal structures using algebraic topology and deep learning. *Chem. Sci.* **2020**, *11*, 12036–12046. [CrossRef]
38. Jukic, M.; Skrlj, B.; Tomsic, G.; Plesko, S.; Podlipnik, C.; Bren, U. Prioritisation of Compounds for 3CL(pro) Inhibitor Development on SARS-CoV-2 Variants. *Molecules* **2021**, *26*, 3003. [CrossRef]
39. Prajapati, J.; Patel, R.; Goswami, D.; Saraf, M.; Rawal, R.M. Sterenin M as a potential inhibitor of SARS-CoV-2 main protease identified from MeFSAT database using molecular docking, molecular dynamics simulation and binding free energy calculation. *Comput. Biol. Med.* **2021**, *135*, 104568. [CrossRef] [PubMed]
40. Uniyal, A.; Mahapatra, M.K.; Tiwari, V.; Sandhir, R.; Kumar, R. Targeting SARS-CoV-2 main protease: Structure based virtual screening, in silico ADMET studies and molecular dynamics simulation for identification of potential inhibitors. *J. Biomol. Struct. Dyn.* **2020**, *4*, 32. [CrossRef]
41. Ram, T.S.; Munikumar, M.; Raju, V.N.; Devaraj, P.; Boiroju, N.K.; Hemalatha, R.; Prasad, P.V.V.; Gundeti, M.; Sisodia, B.S.; Pawar, S.; et al. In silico evaluation of the compounds of the ayurvedic drug, AYUSH-64, for the action against the SARS-CoV-2 main protease. *J. Ayurveda Integr. Med.* **2021**. [CrossRef]

42. Sabbah, D.A.; Hajjo, R.; Bardaweel, S.K.; Zhong, H.A. An Updated Review on SARS-CoV-2 Main Proteinase (M(Pro)): Protein Structure and Small-Molecule Inhibitors. *Curr. Top. Med. Chem.* **2021**, *21*, 442–460. [CrossRef] [PubMed]
43. De Vivo, M.; Masetti, M.; Bottegoni, G.; Cavalli, A. Role of Molecular Dynamics and Related Methods in Drug Discovery. *J. Med. Chem.* **2016**, *59*, 4035–4061. [CrossRef] [PubMed]
44. Kumari, R.; Kumar, R.; Lynn, A. g_mmpbsa—A GROMACS Tool for High-Throughput MM-PBSA Calculations. *J. Chem. Inf. Model.* **2014**, *54*, 1951–1962. [CrossRef]
45. Wong-Sam, A.; Wang, Y.F.; Zhang, Y.; Ghosh, A.K.; Harrison, R.W.; Weber, I.T. Drug Resistance Mutation L76V Alters Nonpolar Interactions at the Flap-Core Interface of HIV-1 Protease. *ACS Omega* **2018**, *3*, 12132–12140. [CrossRef]
46. Olajuyigbe, F.; Demitri, N.; Geremia, S. Investigation of 2-Fold Disorder of Inhibitors and Relative Potency by Crystallizations of HIV-1 Protease in Ritonavir and Saquinavir Mixtures. *Cryst. Growth Des.* **2011**, *11*, 4378–4385. [CrossRef]
47. Berman, H.M.; Westbrook, J.; Feng, Z.; Gilliland, G.; Bhat, T.N.; Weissig, H.; Shindyalov, I.N.; Bourne, P.E. The Protein Data Bank. *Nucleic Acids Res.* **2000**, *28*, 235–242. [CrossRef]
48. Kim, S.; Chen, J.; Cheng, T.; Gindulyte, A.; He, J.; He, S.; Li, Q.; Shoemaker, B.A.; Thiessen, P.A.; Yu, B.; et al. PubChem 2019 update: Improved access to chemical data. *Nucleic Acids Res.* **2019**, *47*, D1102–D1109. [CrossRef] [PubMed]
49. *Molecular Operating Environment (MOE) 2018.0*; Chemical Computing Group ULC: Montreal, QC, Canada, 2018.
50. Yu, W.; He, X.; Vanommeslaeghe, K.; MacKerell, A.D., Jr. Extension of the CHARMM General Force Field to sulfonyl-containing compounds and its utility in biomolecular simulations. *J. Comput. Chem.* **2012**, *33*, 2451–2468. [CrossRef] [PubMed]
51. Pronk, S.; Pall, S.; Schulz, R.; Larsson, P.; Bjelkmar, P.; Apostolov, R.; Shirts, M.R.; Smith, J.C.; Kasson, P.M.; van der Spoel, D.; et al. GROMACS 4.5: A high-throughput and highly parallel open source molecular simulation toolkit. *Bioinformatics* **2013**, *29*, 845–854. [CrossRef]
52. Van Der Spoel, D.; Lindahl, E.; Hess, B.; Groenhof, G.; Mark, A.E.; Berendsen, H.J. GROMACS: Fast, flexible, and free. *J. Comput. Chem.* **2005**, *26*, 1701–1718. [CrossRef]
53. Huang, J.; MacKerell, A.D., Jr. CHARMM36 all-atom additive protein force field: Validation based on comparison to NMR data. *J. Comput. Chem.* **2013**, *34*, 2135–2145. [CrossRef] [PubMed]
54. Boonstra, S.; Onck, P.R.; Giessen, E. CHARMM TIP3P Water Model Suppresses Peptide Folding by Solvating the Unfolded State. *J. Phys. Chem. B* **2016**, *120*, 3692–3698. [CrossRef] [PubMed]
55. Mooers, B.H.M. Shortcuts for faster image creation in PyMOL. *Protein Sci.* **2020**, *29*, 268–276. [CrossRef] [PubMed]
56. *The PyMOL Molecular Graphics System*; Version 2.0; Schrödinger, LLC.: New York, NY, USA; DeLano Scientific: South San Francisco, CA, USA, 2020.

International Journal of
Molecular Sciences

MDPI

Article

A Systematic Approach: Molecular Dynamics Study and Parametrisation of Gemini Type Cationic Surfactants

Mateusz Rzycki [1,2,*], Aleksandra Kaczorowska [2], Sebastian Kraszewski [2] and Dominik Drabik [2,3]

1 Department of Experimental Physics, Faculty of Fundamental Problems of Technology, Wroclaw University of Science and Technology, 50-370 Wroclaw, Poland
2 Department of Biomedical Engineering, Faculty of Fundamental Problems of Technology, Wroclaw University of Science and Technology, 50-370 Wroclaw, Poland; aleksandra.kaczorowska@pwr.edu.pl (A.K.); sebastian.kraszewski@pwr.edu.pl (S.K.); dominik.drabik@uwr.edu.pl (D.D.)
3 Laboratory of Cytobiochemistry, Faculty of Biotechnology, University of Wroclaw, 50-383 Wroclaw, Poland
* Correspondence: mateusz.rzycki@pwr.edu.pl

Abstract: The spreading of antibiotic-resistant bacteria strains is one of the most serious problem in medicine to struggle nowadays. This triggered the development of alternative antimicrobial agents in recent years. One of such group is Gemini surfactants which are massively synthesised in various structural configurations to obtain the most effective antibacterial properties. Unfortunately, the comparison of antimicrobial effectiveness among different types of Gemini agents is unfeasible since various protocols for the determination of Minimum Inhibitory Concentration are used. In this work, we proposed alternative, computational, approach for such comparison. We designed a comprehensive database of 250 Gemini surfactants. Description of structure parameters, for instance spacer type and length, are included in the database. We parametrised modelled molecules to obtain force fields for the entire Gemini database. This was used to conduct in silico studies using the molecular dynamics to investigate the incorporation of these agents into model E. coli inner membrane system. We evaluated the effect of Gemini surfactants on structural, stress and mechanical parameters of the membrane after the agent incorporation. This enabled us to select four most likely membrane properties that could correspond to Gemini's antimicrobial effect. Based on our results we selected several types of Gemini spacers which could demonstrate a particularly strong effect on the bacterial membranes.

Keywords: gemini; molecular dynamics; force field; parametrisation; antimicrobial; membranes

Citation: Rzycki, M.; Kaczorowska, A.; Kraszewski, S.; Drabik, D. A Systematic Approach: Molecular Dynamics Study and Parametrisation of Gemini Type Cationic Surfactants. *Int. J. Mol. Sci.* **2021**, *22*, 10939. https://doi.org/10.3390/ijms222010939

Academic Editor: Małgorzata Borówko

Received: 27 August 2021
Accepted: 8 October 2021
Published: 10 October 2021

1. Introduction

Antimicrobial resistance against available antibiotics has been acknowledged as one of the most serious problems in medicine nowadays. This resulted in a surge of new research works related to the synthesis of novel compounds that could serve as a potential modern-generation groups of antimicrobial particles. One of these groups are Gemini surfactants (initially referred to as bis-surfactants), which are heavily reported for their antimicrobial effect [1]. In recent years, Gemini surfactants have been heavily addressed in the world of science. Over the past five years, more than 130 articles dealing with the subject of Gemini surfactants have been published, among which researchers determined the methods of synthesis of new compounds, their physicochemical properties and even their potential use or application.

Gemini surfactants have unique structural properties. They consist of two amphiphilic groups connected by a spacer at the head level, which can be both hydrophilic and hydrophobic [2,3]. They have at least two hydrophobic chains and two ionic or polar groups. There is a great variety in their structure e.g., short and long methylene groups can be used as a linker, stiff (stilbene), polar (polyether) and nonpolar (aliphatic) groups can be used as a linker [2,3]. The ionic group can be positive (ammonium) or negative (phosphorus, sulfur, carboxylase), while the polar non-ionic groups can be polyether or sugar. Most Gemini surfactants have a symmetrical structure with two identical polar groups and two identical chains (but there are also Gemini that are asymmetrical or with three polar groups or chains) [4]. A universal scheme of Gemini is presented in Figure 1.

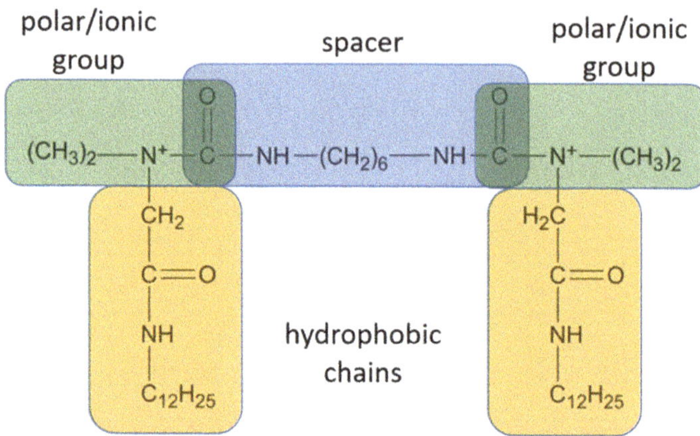

Figure 1. General scheme of Gemini structure classification presented on QAS-type spacer Gemini molecule.

A major part of synthesised Gemini surfactants has performed exquisite antibacterial properties against both Gram-positive and Gram-negative bacteria [5–7]. The most common antibacterial particles of this group are based on quaternary ammonium salts (QAS). Such salts prevent the development of bacteria and fungi; therefore, they are used on a large scale for cleaning, maintenance and disinfection. There were several attempts to evaluate the effectiveness of antibacterial activity of Gemini surfactants. However, these are usually limited to the compound structure—it is directly associated with the type and length of spacer in the molecule and/or the length of hydrophobic chains [8–10]. Numerous scientists proved that number of carbon atoms is correlated with the antimicrobial activity [11,12]. It has been established that a greater number of carbons in the molecule's structure increases its antibacterial activity, and the presence of 12 carbon atoms cause the greatest antibacterial response. It was proposed that the shorter chains might not interact with the hydrophobic region of the bilayer as smoothly and immediately as the longer ones [13]. However, very long tails might curve and twist disqualifying the interactions with negatively charged membrane surfaces by covering cationic head groups. Although it is believed that the major element in the surfactant antimicrobial properties is connected to the hydrophobic chain. It was confirmed that the head group type and structure are also essential factors of biological activity as in the case of QAS molecules [14]. Moreover, Moran et al. revealed that the structure of the hydrophilic core also plays an important role in antimicrobial effects [15,16].

Nevertheless, as mentioned earlier, all works focus only on the structural differences of Gemini surfactants. Furthermore, the conclusions are usually limited to one subgroup of Gemini compounds, hence when analysed more globally, are often mutually contradictory. The reported antimicrobial activity is based on minimum inhibitory concentration (MIC), which strongly depends on the protocol used [17]. The studies reporting the interactions and the effect of Gemini on membranes—with particular emphasis on their properties and potential rupture—are scarce in the literature. There are only available few studies on commercially available Gemini surfactants such as octenidine (OCT) [18–20]. This is quite surprising as membrane destruction was emphasised as one of the potential targets for antimicrobial effect [21,22]. To this end, in our work we have focused on systematic theoretical studies of Gemini agents. Specifically, we have reviewed available literature and recreated the structure of the synthesised Gemini particle groups. This was followed by the classification of molecules into subgroups and the parametrisation of the compounds to create force fields for molecular dynamics studies. As a result, we obtained 250 valid force fields of Gemini class surfactants. Finally, we have selected valid representatives of the subgroups and investigated their interaction with lipid membranes. The selection of representatives was partially determined by the conclusions of structural studies. We ended with 25 selected particles used in molecular dynamics studies. The model membrane system was based on the inner membrane of *E. coli*. In this work we report the theoretical effect of Gemini class surfactants on properties and behaviour of the membrane. Additionally, the incorporation and behaviour of the molecules were also assessed. Based on our systematic characterisation of membrane system, we selected four parameters that were strongly affected by Gemini agents' incorporation. Those were: area compressibility, bending rigidity, lateral diffusion coefficient and surface tension. This selection of impactful parameters allowed us to make a preliminary selection of Gemini molecules groups that could show strong antimicrobial effect from those analysed. This work could provide a means for more detailed studies of Gemini class surfactants and their interaction with lipid membrane models. Such systematic computational analysis provides in silico method to select, from the group of molecules, the ones that are most likely candidates for antimicrobial compounds. It can result in decreasing the amount of expensive synthesis work, which can restrain this type of studies. In a further perspective, it could help in initial scanning of the molecules and facilitate comparison between different MIC studies to determine valid candidates for next-generation antimicrobial substances.

2. Results and Discussion

2.1. Parametrisation

Variety of different Gemini type molecules are synthesised and characterised every year in various literature reports. However, usually their antimicrobial effectiveness is described by a single MIC experiment using various protocols and bacteria families, after which they are left forgotten. Perhaps the new antimicrobial agents, more effective than currently available, have been already synthesised. Due to the shortcomings of the MIC experiments and the inability of systematic comparison it could be impossible to use them. Furthermore, these molecules have a specific biophysical effect on membranes, although are rarely used in molecular dynamic studies due to the missing of an appropriate parametrisation. To this end, we have collected the structures of synthesised Gemini molecules from a significant number of recent literature reports [1,3,7,10,23–49]. Using SCIGRESS software, these structures were designed and preliminarily optimised in the water solvent. It was followed by their equilibration and determination of the Hessian matrix was carried out using Gaussian software. Finally, data from both geometry and a Hessian matrix were used for parametrisation of modelled particles and the force fields creation. This approach was successfully used beforehand to create force fields for various particles [18,50,51]. In Supporting Materials (SM) we have delivered the detailed base of modelled 250 particles (see Microsoft Excel datasheet) with optimised force fields (see included zip file). Force fields are ready-to-use in NAMD software however, a detailed description on how to prepare them for GROMACS users was also included. Molecules were divided into groups based on the origin of the spacer. Each molecule is characterised by the molecular scheme, segment name, spacer formula, length of the spacer, length and formula of chain components and the presence of organic salt. Additionally, based on the modelled molecule structure, partition coefficient (logP) and critical micelle concentration (CMC) were determined. Several molecules were presented as a preview in Table 1 while the total selection is included in Table S1. The theoretical value of logP could be useful for molecule selection as it can indicate whether the molecule incorporates into the membrane in the first place. On the other hand, CMC value may suggest the aggregation behaviour of investigated agents. However, it should be noted that the algorithm is based on phenomenological values hence CMC should be only considered as an approximation.

Table 1. Representation of selected modelled molecules from detailed base included in Supporting Materials.

Group	Scheme	ID	Seg Name	Linker	Linker Length (n)	Chain Com-pound (R1 or R2)	Number of Carbons in R1/R2 (m)	Number of Carbons from N+	Chemical Formula	Organic Salt	log10 (CMC)	Ref.
Alkyl Bisp	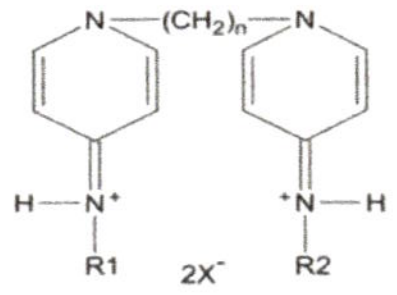	Alk_6_12	A6G	$(CH_2)_6$	6	$C_{12}H_{25}$	12	12	$C_{40}H_{72}N_4$	Br	−4.09	[1]
Aryl Bisp	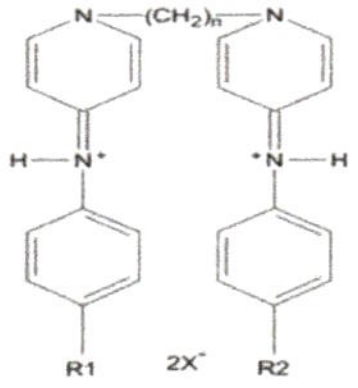	Ary_8_00b	AC2	$(CH_2)_8$	8	F	0	6	$C_{30}H_{34}F_2N_4$	Cl	−3.49	[1]
fQAS	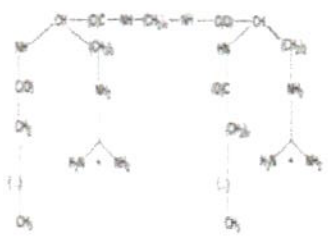	fQS12_12	F0A	$C_{12}H_{16}F_8$	12	$C_{12}H_{25}$	12	12	$C_{40}H_{78}F_8N_2$	Br-	−6.65	[45]
Ring	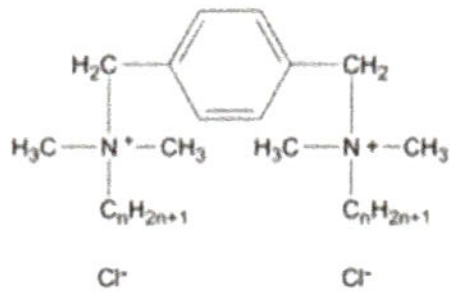	Rin_01_12b	RI1	C_6H_4	1	$C_{12}H_{25}$	12	12	$C_{36}H_{70}N_2$	Cl	−3.63	[46]

Table 1. *Cont.*

Group	Scheme	ID	Seg Name	Linker	Linker Length (n)	Chain Compound (R1 or R2)	Number of Carbons in R1/R2 (m)	Number of Carbons from N+	Chemical Formula	Organic Salt	log10 (CMC)	Ref.
Ester	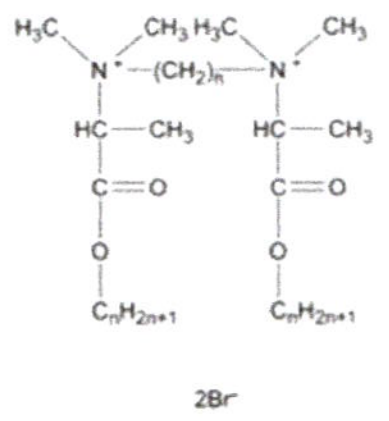	Est_2_12	E2D	C_2H_4	2	$C_{12}H_{25}$	12	15	$C_{36}H_{74}N_2O_4$	Br	−4.21	[35]
Ionic	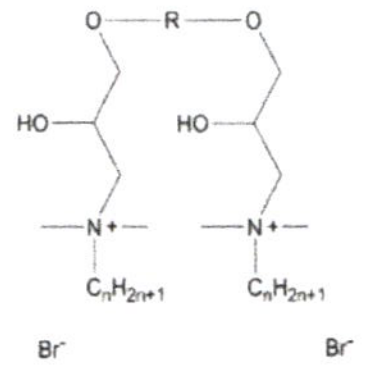	Ion_4_12	I4C	$CH_2CH_2OCH_2C4H_2$		$C_{12}H_{25}$	12	12	$C_{38}H_{82}N_2O_5$	Br	−4.8	[10]
QAS	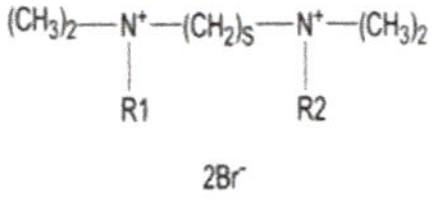	QAS_4_12	Q4A	C_4H_6	4	$C_{12}H_{25}$	12	12	$C_{32}H_{70}N_2$	Br/Cl	−2.61	[27, 36]
tQAS	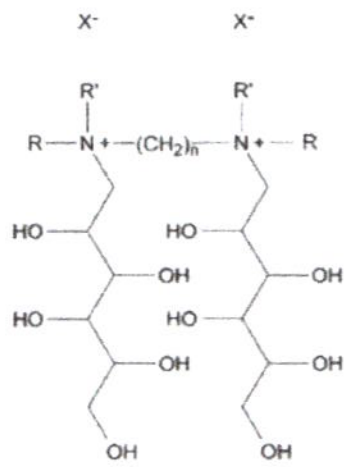	tQS_4_1002	T4F	$(CH_2)_4$	4	R1: $C_{10}H_{21}$ R2: C_2H_5	10 & 2	18	$C_{40}H_{86}N_2O_{10}$	I	−5,82	[49]

2.2. Membrane Characterisation

As previously stated, Gemini molecules are well-known for their antimicrobial activity. It was considered by Epand et al. [21] as well as shown in the OCT studies, that this effect is related to membrane disruption [18]. The effect of those Gemini molecules on the membrane properties was assessed to determine the properties that most likely correspond to the antimicrobial effect. A number of molecules were selected to investigate the effect of Gemini particles on membrane's behaviour. Specifically, at least one molecule from each group was selected. Since in several works [52–56] it was reported that strongest antimicrobial effect was observed for Gemini agents with chain length equal to 12 carbon atoms, such condition was adapted during molecule selection from the group. All of the investigated molecules, except diGalactose (dGl), were incorporated into the membrane during the simulation time. The dGl molecule fluctuated over the bilayer surface, maintaining a 30 ± 4 Å distance from phosphorous atoms in lipid heads. The explanation of a lack of incorporation for dGl most likely lies in the negative logP of the molecule. A detailed location of system components such as lipid fragments or Gemini molecules has been presented in the partial density chart in Figure S1–S5 in SM. Selected screenshots of the systems with anchored molecules are presented in Figure 2. The membrane composition was selected in such a way to most accurately reflect the inner membrane of *E. coli* [21,57].

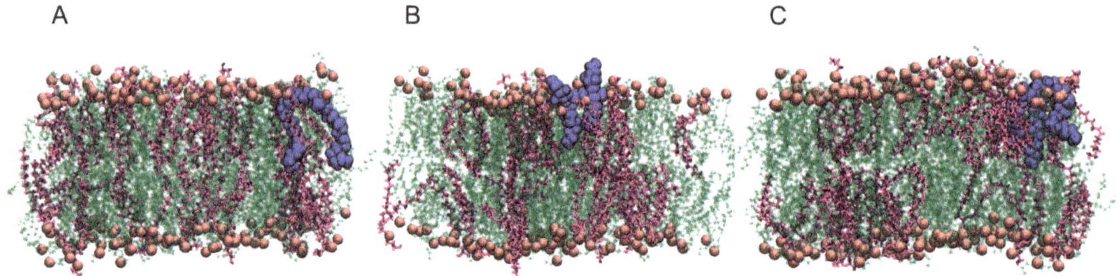

Figure 2. Selected screenshots of the systems with incorporated molecules, (**A**) Ole, (**B**) tQS, (**C**) hQS.

Membrane with incorporated Gemini molecules was thoroughly analysed to determine its properties such as area per lipid, membrane thickness, interdigitation, penetration depth, lateral diffusion, bending and tilt rigidities, area compressibility and surface tension. From such set of parameters four most likely candidates were chosen that could correspond to the antimicrobial effect. Those are membrane compressibility, bending rigidity, lateral diffusion and surface tension. The detailed characterisation of Gemini effect on the membrane, including all of the investigated parameters, is presented in SM Table S2. In this work, we additionally simulated the OCT molecule and we used it as a positive control. We assume that the effect on the membrane of these commercially available molecules could serve as a guidepost regarding desirable changes in selected properties. Since other molecules might indicate a much different mode of action therefore in our selection, we took into account the possible different mechanisms. Such a mechanism could induce a different magnitude of parameter change. As a result, we were also considering, in our selections, the extremum parameter changes, not only guided by the tendencies given by the effect of OCT. Concerning our analysis, we selected the four most changeable bilayer parameters reflecting membrane-agent interaction and potential antimicrobial activity. We deliver a total set of parameters in SM Table S2. The rest of the determined parameters were not selected due to insignificant differences between the analysed systems. Membrane thickness was, in general, determined to be between 39 and 41 Å. The difference of 2 Å between extreme particles with uncertainty equal to 1 Å was enough to exclude this parameter as an influential one. Similarly, tilt bending ranged from 9.9 up to 10.9 fold K_bT with an uncertainty of 0.3 fold K_bT. For APL, when the leaflet in which the Gemini agent incorporated is analysed, the values range between 58 and 62 Å^2. Only two cases are extreme, which are 67.0 ± 3.3 Å^2 for aryl bispiridine (Ary) and 64.8 ± 2.7 Å^2 for glucose (Glu). Interestingly, both of those spacers were highlighted as possible antimicrobial based on analysis of other significant parameters.

The area compressibility is one of the most robust parameters in our dataset, hence we predict that it may be an adequate property reflecting antimicrobial effect on the membrane. This mechanical parameter quantifies the energetic cost associated with the membrane's area stretching and/or compressing. For high values of area compressibility, the membrane is resistant to external pressure. For low values the membrane loses its resistivity. Both cases can result in inability of proper cell function. The determined values of area compressibility of bilayers with incorporated Gemini molecules are presented in Figure 3. The area compressibility of positive control—membrane with incorporated OCT—is almost seven times higher than in the case of the model membrane. Interestingly, membrane area compressibility with incorporated Adamantane (Adm) is higher than in the case of positive control. Four other molecules from the ester (Est), higher quaternary ammonium salt (hQAS), oligomeric QAS (o-QAS) and pyridine (Pyr) group also induced significant growth in the area compressibility. Furthermore, two Gemini molecules had a decreasing effect on the area compressibility of the membrane. Specifically, those from Saccharide (Sch) and Alkyl Bispyridinamine (Alk) group.

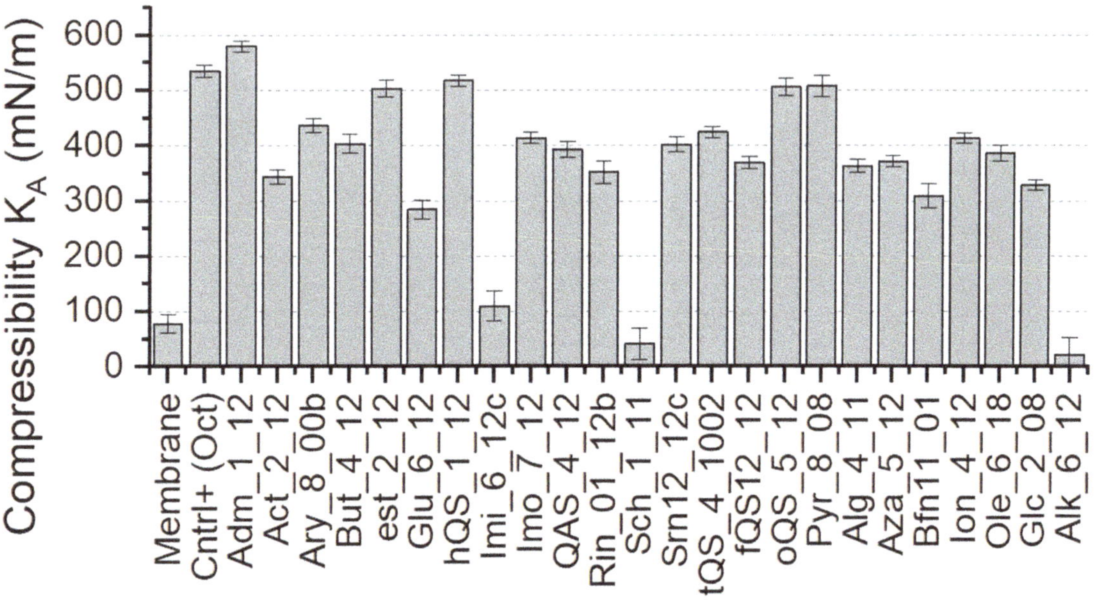

Figure 3. Determined values of compressibility (K_A) for membrane system with incorporated Gemini molecules.

In our previous work we highlighted that mechanical parameter such as bending rigidity play an important role in OCT mode of action [18]. To this end we selected this parameter as a likely and promising candidate that correspond to antimicrobial activity. Briefly, bending rigidity quantifies the energetic cost associated with the membrane bending. The determined values of bending rigidity of membranes with incorporated Gemini agents are presented in Figure 4. First, the difference between pure membrane and the positive control is not statistically significant. To our knowledge, the OCT effect on bending rigidity is closely related to the aggregation properties rather than the effect of a single molecule action [18]. Nevertheless, several Gemini molecules significantly affected the bending behaviour of membranes. Considering OCT as a positive control we found the activity of several agents such as ionic (Ion), alginine (Alg), Pyr, butene (But) and glucose (Glu) as not statistically significant, hence similar to OCT. If, however, the strongest difference between pure membrane is considered, aryl bispiridine (Ary), imino (Imo), oQAS, gluconamid (Glc) were the most active candidates.

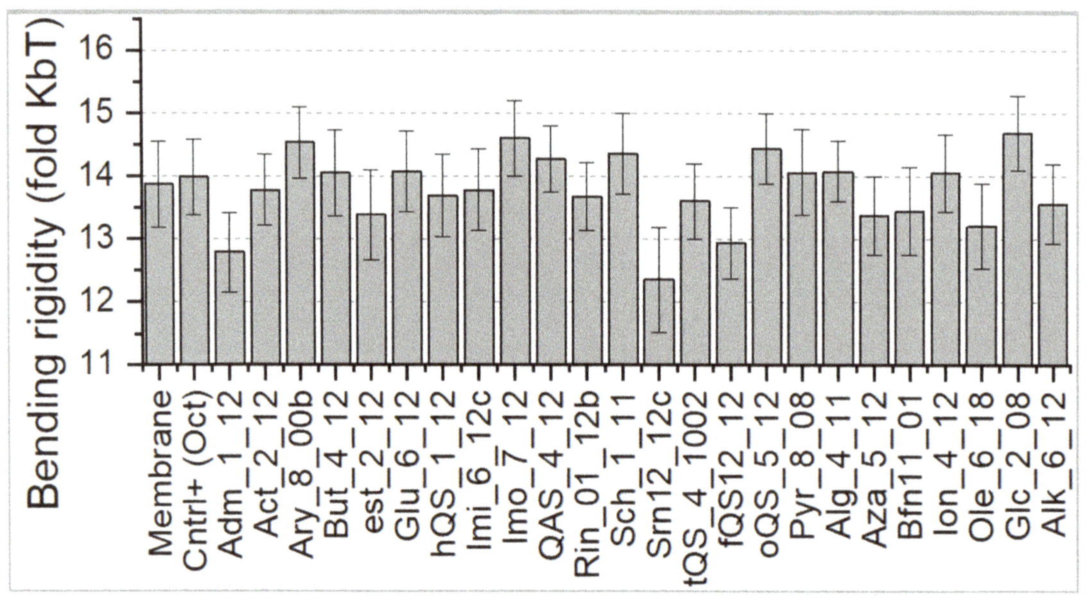

Figure 4. Calculated values of bending rigidity for membrane systems with incorporated Gemini molecules.

Lateral diffusion is a property that defines the mobility of lipid molecules in the membrane plane. Changes in lateral diffusion of the lipids could result in changes of movement of proteins, which can affect the activity of transporters and channel proteins [58]. Such an effect on proteins could significantly impede the functioning of microbe cells and correspond to antimicrobial effect of Gemini molecules. Although this dependency is not strictly related to the Gemini's destructive effect on the membrane, it also should be considered as a factor. The values of lateral diffusion of lipid molecules are presented in Figure 5. Our positive control indicates that decreasing effect on lateral diffusion should be desirable however, the antimicrobial effect of OCT is based on membrane disruption and cannot be considered as a factor in this case. Results obtained for model membrane are in strong agreement with the GUVs *E. coli* mimicking studies, i.e., experimental diffusion coefficient equals D = 6.09 $\mu m^2/s$ [59]. Interestingly, different antimicrobial agent-thymol induced growth in lipids mobility, supporting the agent translocation [59]. To this end we selected three lowest and three highest values of lateral diffusion for selected agents that influence this membrane property. Gemini molecules that strongly increased the lateral diffusion of lipids on membranes were But, imidazolium (Imi) and Pyr. On the other hand, Gemini molecules that strongly decreased the lateral diffusion were o-QAS, Imo and Glu.

Finally, the surface tension was determined for membranes with incorporated Gemini agents. Briefly, surface tension is defined as a cohesive force that keeps the cell membrane intact. Hence, its fluctuations may be very informative and extremely important for the determination of the antimicrobial mode of action based on the membrane disruption. Values of the membrane surface tension influenced by Gemini detergents are presented in Figure 6. The surface tension of membrane treated with OCT was twice as high than in the model membrane's case. Interestingly, a significant number of investigated molecules had much stronger effect on membrane surface tension compared to the positive control. Three groups with the highest surface tension fluctuations were Ary, Aza and Alk. Additionally, the activity of several molecules such as hQS, Imo and Pyr led to decreased membrane tension. This should also be considered as change that could result in antimicrobial effect. Pure membrane exhibited natively certain surface tension hence, any strong deviation from this value could result in disruption of biological processes on the membrane.

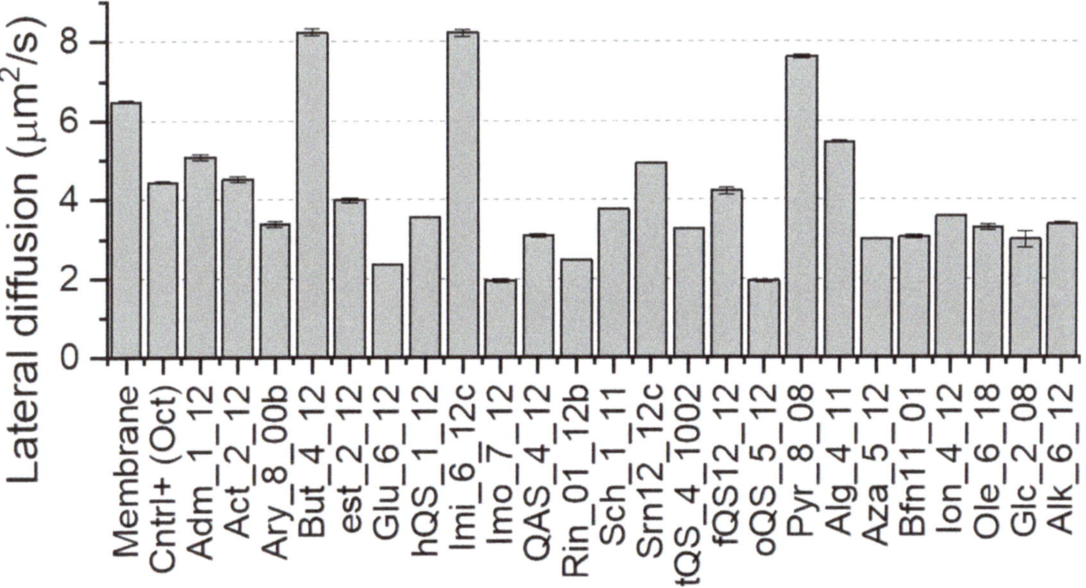

Figure 5. Determined values of lateral diffusion (2D) for membrane systems with incorporated Gemini molecules.

Taking into account our results oQAS, Pyr, Imo groups from the Gemini family may exhibit strong antimicrobial effects. These molecules act as prominent candidates since three from four selected membrane parameters were significantly affected. Wang et al. in their experimental work [60] reported that presence of oxygen atom in oQAS spacer chain introduces higher flexibility and reduction of coulombic repulsion allows long side alkyl chain to tighter aggregation. This stays in line with our results since the oQAS molecules deeply penetrate the bilayer affecting membrane diffusion and mechanical properties. Moreover, Wettig et al. [61] highlighted unique transfection properties of Imo compared to other synthesised molecules since additional flexibility from extra methylene unit between nitrogen centres and readily protonated imino group is present. In our opinion, given molecule properties may influence the membrane resulting in limited tension and diffusion. Interestingly, both oQAS and Imo agents have similarities in structures (latter has additional nitrogen and methylene units in the spacer region) and induce comparable change in the membrane's properties. Similarly, Quagliotto et al. [31] in the experimental work reported that increased Pyr concentration reduced the surface tension. This is in accordance with our theoretical approach where we observed significant limitations in membrane surface tension. Other vital candidates, that were selected based on two from four parameters, are Ary, Glu, hQAS and Alk. In the experimental work, Bailey et al. [1] concluded that Ary and Alk agents showed antimicrobial effectiveness, according to MIC. However, the latter showed weaker activity when compared to Ary. The authors emphasised that in the case of alkyl series the most effective agents are those with 22 up to 30 carbon atoms in the molecule. These could influence the character of membrane–molecule interactions and thus result in the fluctuation of membrane parameters. Our results also highlighted hQAS, which may be associated with molecule rigid spacer and three-charged headgroup indicating affinity to negative membranes [24]. Finally, Kumar et al. [38] reported that Glu shows excellent surface-active properties and low cytotoxicity, which stay in agreement with our findings based on membrane parameters variation. This selection was presented in Table 2. Despite suggested membrane thinning in reported experimental works we did not observe significant occurrence nor changes in acyl chain interdigitation in our studies (see Table S2) [62,63]. Moreover, in a significant part of analysed molecules, we observed their preferential localisation in the carbonyl-glycerol region. This was influenced by neither how long the alkyl chain nor the spacer were (see Figure S1–S5). Nevertheless, experimental comparison studies using uniformed protocol are required to confirm whether selected parameters directly correspond to the discussed antimicrobial effect.

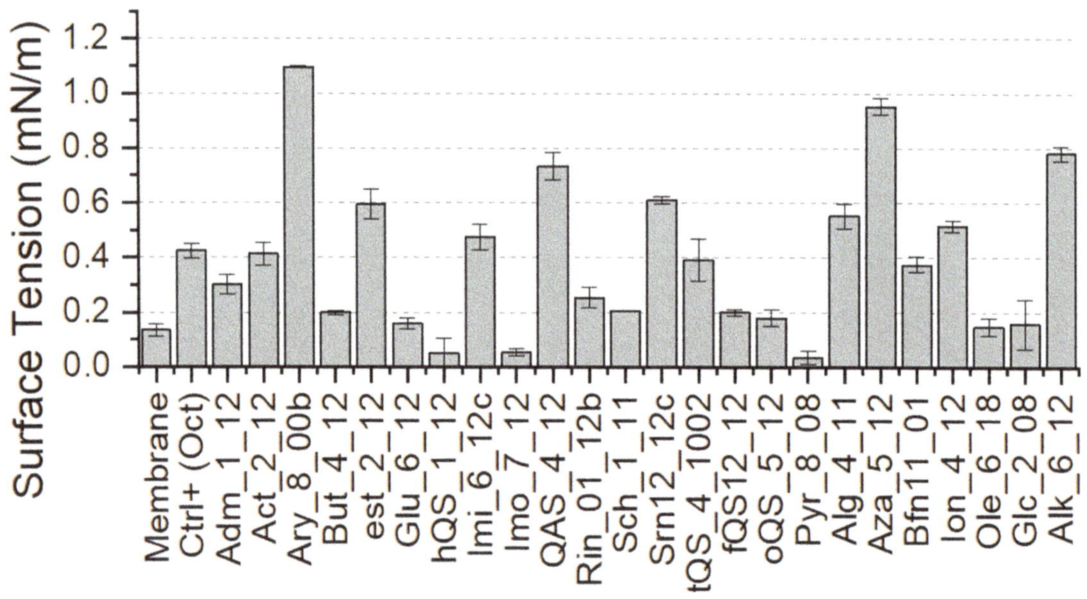

Figure 6. Surface tension, which was determined for membrane systems after the incorporation of investigated Gemini molecules.

Table 2. Selected potential antimicrobial candidates from each parameter group. It is suggested to compare those in experimental studies. Frequently appearing molecules were bolded to emphasise their repetition between different parameters consideration.

Compressibility K_A [mN/m]	Bending Rigidity [fold K_BT]	Lateral Diffusion [$\mu m^2/s$]	Surface Tension [mN/m]
o-QAS	**o-QAS**	**o-QAS**	Aza
Pyr	Ion	**Pyr**	**Pyr**
Adm	**Imo**	**Imo**	**Imo**
Est	**But**	**But**	
Sch	**Ary**	Imi	**Ary**
	Glu	**Glu**	
hQAS	Alg		**hQAS**
Alk	Glc		**Alk**

3. Materials and Methods

3.1. Molecule Parametrisation

Quantum level calculations were performed using the Gaussian 2016 software package [64]. The equilibrium geometry of investigated Gemini molecules was calculated using density functional theory (DFT) (B3LYP)/6-31++G (d) level of theory; first with Loose Self Consistent Field (SCF) procedure, then with Tight. The solvent effect was taken into consideration using the integral equation formalism of the polarisable continuum model IEFPCM. Temperature was set to 300 K. Supplementary analysis based on the construction of the Hessian matrix (the matrix of second derivatives of the energy with respect to geometry) was also performed for further use in the force field parameterisation. The specific geometric and electronic data, such as bond lengths, angles, dihedrals and charge distribution were extracted from a Hessian matrix. The charge distribution was determined from the RESP charge calculations as being the most adapted to reproduce the molecular behaviour with the subsequently used CHARMM force field. For logP determination, the octanol/water partitioning coefficient was calculated using SCIGRESS software (SCIGRESS, Molecular modeling software, FQS Poland, ver. FJ-3.3.3). For CMC determination, the algorithm proposed by Mozrzymas was used [65]. It is based on phenomenological values and second-order connectivity index, that was determined using SCIGRESS software. Molecule schemes were prepared using MoleculeSketch (v. 2.2.3).

3.2. Molecular Dynamics Simulations

The all-atom models of the membranes were generated using CHARMM-GUI membrane builder [66]. The bacterial membrane model consisted of 80% PYPE, 15% PYPG, 5% PVCL2 [21,57]. The lipid bilayer was solvated with TIP3P water molecules (100 water molecules per lipid) and 240 mM NaCl were added based on literature data [67].

MD simulations were performed using the GROMACS (version 2020.4) package with the CHARMM36 force field [68,69]. Membrane systems were first minimised with the steepest descent algorithm for energy minimisation. Further calculations were carried out in the NPT ensemble (constant Number of particles, Pressure and Temperature) with Berendsen thermostat and barostat using semi-isotropic coupling at T = 303.15 K with time constant $\tau = 1$ ps and p = 1 bar with $\tau = 5$ ps. The primary part of the NPT calculations was performed using the leap-frog integrator with a 1 fs timestep. Afterwards, for the further NPT ensemble at T = 303.15 K, $\tau = 1$ ps and p = 1 bar, $\tau = 5$ ps, a Nose-Hoover thermostat [70] and Parrinello-Rahman barostat [71] were used. The second part of long-run production was carried out for 500 ns using the leap-frog integrator. Chemical bonds between hydrogen and heavy atoms were constrained to their equilibrium values with the LINCS algorithm, while long-range electrostatic forces were evaluated using the particle mesh Ewald (PME) method [72] with the integration timestep of 2 fs. Based on simulated pure membranes, the behaviour of Gemini surfactants was investigated. Molecules were placed on average 2.5 nm above the membrane leaflet and the same MD procedure was employed. For visualisation purpose, Visual Molecular Dynamics (VMD) was used [73].

3.3. Membrane System Characteristics

Membrane Thickness and Area per Lipid. Both area per lipid and membrane thickness were determined using self-made MATLAB scripts (Matlab R2019a). Briefly, for each leaflet Z-position on all phosphorus atoms were averaged, and distance between average Z-positions between each of leaflets was calculated for each frame. The final membrane thickness value is an average over analysed trajectory. Similarly, for each frame position of phosphorus atoms (or Gemini atom on the Z-level corresponding to phosphorus atoms) each leaflet was subjected to Voronoi tessellation. The average area for all lipid molecules was calculated for each leaflet and frame and was averaged over the analysed trajectory.

Bending rigidity and Tilt rigidity. Both bending rigidity and tilt rigidities were determined using self-made MATLAB scripts that were based on the works of Doktorova et al. [74]. Briefly, a probability distribution for both tilt and splay are determined for all lipids over all analysed time steps. Tilt is defined as an angle between the lipid director (vector between lipid head–midpoint between C2 and P atoms–and lipid tail–midpoint between 16th carbon atoms) and bilayer normal. Lipid splay Sr is defined as divergence of an angle formed by the directors of neighbouring lipids providing that they are weakly correlated.

Compressibility. Compressibility was determined using self-made MATLAB scripts based on the work of Doktorova et al. [75]. Briefly, a real-space analysis of local thickness fluctuations is sampled from the simulations for carbon atoms. This is followed by determination of reference surface and calculation of potential mean force from fluctuations from whole analysed trajectory to determine compressibility for given leaflet.

Lateral Diffusion Coefficient. The diffusion coefficient from the 2D mean square displacement (MSD) equation was calculated Diffusion Coefficient Tool [76] from the slope of the MSD curve through Einstein's relation. This relation is presented in Equation (1), where M (t) is the MSD at a range of lag time tau and E represents the dimensionality (XY). For the computation accuracy, only phosphorous atoms (in the range of 20 Å from surfactant) of all lipids were considered.

$$D(\tau) = \frac{M(\tau)}{2E\tau} \tag{1}$$

Interdigitation. For all provided systems the fluctuation of lipid interdigitation was determined using MEMBPLUGIN available in VMD software [77]. It is given in length units and reflects the interdigitation between opposite leaflets in the system—unless no interdigitation occurs it is equal to zero.

Penetration Depth. The depth of surfactant penetration was measured with respect to the membrane centre. From the last 50 ns of the trajectory the positions of the deepest placed carbon atoms on each alkyl chain were taken and evaluated with respect to the distance between phosphorous atoms divided by two, which represent the membrane centre.

Surface tension. The surface tension of membranes with anchored Gemini surfactants was computed using gmx energy function build-in GROMACS software using pressure tensor (Pxx, Pyy, Pzz values according to the Irving-Kirkwood method [78–80] and Equation (2), where L is the length of the simulation box in z dimension and represents an ensemble average given from gmx energy.

$$\gamma = \frac{L}{2}\langle P_{zz} - \frac{P_{xx} + P_{yy}}{2}\rangle \tag{2}$$

Significance test. Significance tests were performed using OriginLab OriginPro 9.0 software. Specifically, one-way ANOVA was performed and was supplemented with post-hoc Tukey test to determine significance between individual populations.

4. Conclusions

In this work, we optimised and parametrised 250 Gemini molecules. We described each of those molecules with theoretical values of logP and log (CMC) as well as provided a detailed description of those molecules in the attached spreadsheet. Additionally, we included those parametrised force fields in SM for future simulation studies. This may be remarkably helpful in further antimicrobial action studies, as a significant number of Gemini cationic molecules with various spacers were modelled and parametrised. Such systematic summarisation may be extensively used not only for theoretical studies but also for experimental ones with the aim to deliver comprehensive knowledge and molecular mechanism of surfactant effectiveness. Furthermore, we selected 25 molecules from various groups and simulated their behaviour in systems with membrane mimicking the inner membrane of *E. coli*. This detailed characterisation of parameters allowed us to extract four types of parameters—area compressibility, bending rigidity, lateral diffusion coefficient and membrane surface tension—that could correspond to the antimicrobial effect of those molecules. Based on our preliminary screening we concluded that the type of Gemini molecules that could exhibit strong antimicrobial effects are oQAS, Pyr, Imo. Additionally, other possible candidates are Ary, Glu, hQAS and Alk. In this work we proposed and deliver a uniform theoretical approach to compare Gemini surfactant effectiveness. Nevertheless, this systematic approach should be confirmed experimentally to provide solid biological relevance.

Supplementary Materials: The following are available online at https://www.mdpi.com/article/10.3390/ijms222010939/s1: (1) Gemini molecules database (.xls). (2) Complete force-fields for Gemini molecules from the database (.zip). (3) Supporting materials (.docx) including system density profiles of investigated agents, table with membrane system characterisation and instruction for NAMD FF to GROMACS FF conversion.

Author Contributions: Conceptualisation, M.R. and D.D.; methodology, M.R., A.K. and D.D.; validation, M.R. and D.D.; formal analysis, M.R. and D.D.; investigation, M.R., A.K. and D.D.; resources, S.K.; data curation, M.R., A.K. and D.D.; writing—original draft preparation, M.R. and D.D.; writing—review and editing, M.R., A.K. and D.D.; visualisation, M.R. and D.D.; supervision, D.D.; project administration, S.K.; funding acquisition, S.K. All authors have read and agreed to the published version of the manuscript.

Funding: This work was possible thanks to the financial support from the National Science Centre (Poland) grant No. 2015/19/B/NZ7/02380. Additionally, D.D. acknowledge support from National Science Center (grant number 2018/30/E/NZ1/00099).

Data Availability Statement: Most of the data are available in the manuscript supplementary information, including force fields. The simulation data presented in this study are available on request from the corresponding author due to GBs files sizes.

Acknowledgments: We would like to thank both Kamila Szostak-Paluch and Beata Hanus-Lorenz from Wroclaw University of Science and Technology for valuable discussion regarding structures of cationic Gemini molecules.

Conflicts of Interest: The authors declare no conflict of interest.

References

1. Bailey, D.M.; DeGrazia, C.G.; Hoff, S.J.; Schulenberg, P.L.; O'Connor, J.R.; Paris, D.A.; Slee, A.M. Bispyridinamines: A new class of topical antimicrobial agents as inhibitors of dental plaque. *J. Med. Chem.* **2002**, *27*, 1457–1464. [CrossRef]
2. Brycki, B.E.; Kowalczyk, I.H.; Szulc, A.; Kaczerewska, O.; Pakiet, M. Multifunctional gemini surfactants: Structure, synthesis, properties and applications. *Appl. Charact. Surfactants, Reza Najjar* **2017**, 97–155. [CrossRef]
3. Hait, S.; Moulik, S. Gemini surfactants: A distinct class of self-assembling molecules. *Curr. Sci.* **2002**, *82*, 1101–1111.
4. Zana, R. Dimeric and oligomeric surfactants. Behavior at interfaces and in aqueous solution: A review. *Adv. Colloid Interface Sci.* **2002**, *97*, 205–253. [CrossRef]
5. Koziróg, A.; Brycki, B.; Pielech-Przybylska, K. Impact of cationic and neutral gemini surfactants on conidia and hyphal forms of Aspergillus brasiliensis. *Int. J. Mol. Sci.* **2018**, *19*, 873. [CrossRef]
6. Kuperkar, K.; Modi, J.; Patel, K. Surface-active properties and antimicrobial study of conventional cationic and synthesized symmetrical gemini surfactants. *J. Surfactants Deterg.* **2012**, *15*, 107–115. [CrossRef]
7. Menger, F.M.; Mbadugha, B.N.A. Gemini surfactants with a Disaccharide Spacer. *J. Am. Chem. Soc.* **2001**, *123*, 875–885. [CrossRef]
8. Zhong, X.; Guo, J.; Fu, S.; Zhu, D.; Peng, J. Synthesis, surface property and antimicrobial activity of cationic gemini surfactants containing adamantane and amide groups. *J. Surfactants Deterg.* **2014**, *17*, 943–950. [CrossRef]
9. Pernak, J.; Branicka, M. The properties of 1-alkoxymethyl-3-hydroxypyridinium and 1-alkoxymethyl-3-dimethylaminopyridinium chlorides. *J. Surfactants Deterg.* **2003**, *6*, 119–123. [CrossRef]
10. Li, H.; Yu, C.; Chen, R.; Li, J.; Li, J. Novel ionic liquid-type Gemini surfactants: Synthesis, surface property and antimicrobial activity. *Colloids Surf. A Physicochem. Eng. Asp.* **2012**, *395*, 116–124. [CrossRef]
11. Wang, L.; Qin, H.; Ding, L.; Huo, S.; Deng, Q.; Zhao, B.; Meng, L.; Yan, T. Preparation of a novel class of cationic gemini imidazolium surfactants containing amide groups as the spacer: Their surface properties and antimicrobial activity. *J. Surfactants Deterg.* **2014**, *17*, 1099–1106. [CrossRef]
12. Tawfik, S.M.; Abd-Elaal, A.A.; Shaban, S.M.; Roshdy, A.A. Surface, thermodynamic and biological activities of some synthesized Gemini quaternary ammonium salts based on polyethylene glycol. *J. Ind. Eng. Chem.* **2015**, *30*, 112–119. [CrossRef]
13. Bao, Y.; Guo, J.; Ma, J.; Li, M.; Li, X. Physicochemical and antimicrobial activities of cationic gemini surfactants with polyether siloxane linked group. *J. Mol. Liq.* **2017**, *242*, 8–15. [CrossRef]

14. Gilbert, P.; Moore, L.E. Cationic antiseptics: Diversity of action under a common epithet. *J. Appl. Microbiol.* **2005**, *99*, 703–715. [CrossRef] [PubMed]
15. Koch, A.L. Bacterial wall as target for attack past, present, and future research. *Clin. Microbiol. Rev.* **2003**, *16*, 673–687. [CrossRef] [PubMed]
16. Morán, C.; Clapés, P.; Comelles, F.; García, T.; Pérez, L.; Vinardell, P.; Mitjans, M.; Infante, M.R. Chemical structure/property relationship in single-chain arginine surfactants. *Langmuir* **2001**, *17*, 5071–5075. [CrossRef]
17. Schuurmans, J.; Nuri Hayali, A.; Koenders, B.; ter Kuile, B. Variations in MIC value caused by differences in experimental protocol. *J. Microbiol. Methods* **2009**, *79*, 44–47. [CrossRef] [PubMed]
18. Rzycki, M.; Drabik, D.; Szostak-Paluch, K.; Hanus-Lorenz, B.; Kraszewski, S. Unraveling the mechanism of octenidine and chlorhexidine on membranes: Does electrostatics matter? *Biophys. J.* **2021**, *120*, 3392–3408. [CrossRef]
19. Malanovic, N.; Ön, A.; Pabst, G.; Zellner, A.; Lohner, K. Octenidine: Novel insights into the detailed killing mechanism of Gram-negative bacteria at a cellular and molecular level. *Int. J. Antimicrob. Agents* **2020**, *56*. [CrossRef]
20. Hübner, N.-O.; Siebert, J.; Kramer, A. Octenidine dihydrochloride, a modern antiseptic for skin, mucous membranes and wounds. *Skin Pharmacol. Physiol.* **2010**, *23*, 244–258. [CrossRef]
21. Epand, R.M.; Epand, R.F. Bacterial membrane lipids in the action of antimicrobial agents. *J. Pept. Sci.* **2011**, *17*, 298–305. [CrossRef]
22. Epand, R.M.; Epand, R.F. Lipid domains in bacterial membranes and the action of antimicrobial agents. *Biochim. Biophys. Acta* **2009**, *1788*, 289–294. [CrossRef] [PubMed]
23. Silva, S.G.; Oliveira, I.S.; do Vale, M.L.C.; Marques, E.F. Serine-based gemini surfactants with different spacer linkages: From self-assembly to DNA compaction. *Soft Matter* **2014**, *10*, 9352–9361. [CrossRef] [PubMed]
24. Han, Y.; Wang, Y. Aggregation behavior of gemini surfactants and their interaction with macromolecules in aqueous solution. *Phys. Chem. Chem. Phys.* **2011**, *13*, 1939–1956. [CrossRef] [PubMed]
25. Ding, Z.; Hao, A. Synthesis and surface properties of novel cationic gemini surfactants. *J. Dispers. Sci. Technol.* **2010**, *31*, 338–342. [CrossRef]
26. Di Meglio, C.; Rananavare, S.B.; Sönke Svenson, A.; Thompson, D.H. Bolaamphiphilic phosphocholines: Structure and phase behavior in aqueous media. *Langmuir* **1999**, *16*, 128–133. [CrossRef]
27. Datta, S.; Biswas, J.; Bhattacharya, S. How does spacer length of imidazolium gemini surfactants control the fabrication of 2D-Langmuir films of silver-nanoparticles at the air–water interface? *J. Colloid Interface Sci.* **2014**, *430*, 85–92. [CrossRef] [PubMed]
28. Brycki, B.; Szulc, A. Gemini alkyldeoxy-D-glucitolammonium salts as modern surfactants and microbiocides: Synthesis, antimicrobial and surface activity, biodegradation. *PLoS ONE* **2014**, *9*, e84936. [CrossRef]
29. Sheikh, M.S.; Khanam, A.J.; Matto, R.U.H.; Kabir-Ud-Din. Comparative study of the micellar and antimicrobial activity of gemini-conventional surfactants in pure and mixed micelles. *J. Surfactants Deterg.* **2013**, *16*, 503–508. [CrossRef]
30. Obłąk, E.; Piecuch, A.; Guz-Regner, K.; Dworniczek, E. Antibacterial activity of gemini quaternary ammonium salts. *FEMS Microbiol. Lett.* **2014**, *350*, 190–198. [CrossRef]
31. Quagliotto, P.; Barolo, C.; Barbero, N.; Barni, E.; Compari, C.; Fisicaro, E.; Viscardi, G. Synthesis and characterization of highly fluorinated gemini pyridinium surfactants. *European, J. Org. Chem.* **2009**, *2009*, 3167–3177. [CrossRef]
32. Marín-Menéndez, A.; Montis, C.; Díaz-Calvo, T.; Carta, D.; Hatzixanthis, K.; Morris, C.J.; McArthur, M.; Berti, D. Antimicrobial nanoplexes meet model bacterial membranes: The key role of cardiolipin. *Sci. Rep.* **2017**, *7*, 1–13. [CrossRef]
33. Brycki, B.; Kowalczyk, I.; Kozirog, A. Synthesis, molecular structure, spectral properties and antifungal activity of polymethylene-α,ω-bis(n,n- dimethyl-n-dodecyloammonium bromides). *Molecules* **2011**, *16*, 319–335. [CrossRef] [PubMed]
34. Gospodarczyk, W.; Kozak, M. Interaction of two imidazolium gemini surfactants with two model proteins BSA and HEWL. *Colloid Polym. Sci.* **2015**, *293*, 2855–2866. [CrossRef]
35. Pietralik, Z.; Kołodziejska, Ż.; Weiss, M.; Kozak, M. Gemini surfactants based on bis-imidazolium alkoxy derivatives as effective agents for delivery of nucleic acids: A structural and spectroscopic study. *PLoS ONE* **2015**, *10*, e0144373. [CrossRef]
36. Pietralik, Z.; Krzysztoń, R.; Kida, W.; Andrzejewska, W.; Kozak, M. Structure and conformational dynamics of DMPC/dicationic surfactant and DMPC/dicationic surfactant/DNA systems. *Int. J. Mol. Sci.* **2013**, *14*, 7642–7659. [CrossRef] [PubMed]
37. Kamboj, R.; Singh, S.; Bhadani, A.; Kataria, H.; Kaur, G. Gemini imidazolium surfactants: Synthesis and their biophysiochemical study. *Langmuir* **2012**, *28*, 11969–11978. [CrossRef] [PubMed]
38. Kumar, V.; Chatterjee, A.; Kumar, N.; Ganguly, A.; Chakraborty, I.; Banerjee, M. d-Glucose derived novel gemini surfactants: Synthesis and study of their surface properties, interaction with DNA, and cytotoxicity. *Carbohydr. Res.* **2014**, *397*, 37–45. [CrossRef] [PubMed]
39. Sakai, K.; Umezawa, S.; Tamura, M.; Takamatsu, Y.; Tsuchiya, K.; Torigoe, K.; Ohkubo, T.; Yoshimura, T.; Esumi, K.; Sakai, H.; et al. Adsorption and micellization behavior of novel gluconamide-type gemini surfactants. *J. Colloid Interface Sci.* **2008**, *318*, 440–448. [CrossRef]
40. Massi, L.; Guittard, F.; Levy, R.; Duccini, Y.; Géribaldi, S. Preparation and antimicrobial behaviour of gemini fluorosurfactants. *Eur. J. Med. Chem.* **2003**, *38*, 519–523. [CrossRef]
41. Łuczyński, J.; Frąckowiak, R.; Włoch, A.; Kleszczyńska, H.; Witek, S. Gemini ester quat surfactants and their biological activity. *Cell. Mol. Biol. Lett.* **2013**, *18*, 89–101. [CrossRef]

42. Tatsumi, T.; Zhang, W.; Nakatsuji, Y.; Miyake, K.; Matsushima, K.; Tanaka, M.; Furuta, T.; Ikeda, I. Preparation, surface-active properties, and antimicrobial activities of bis(alkylammonium) dichlorides having a butenylene or a butynylene spacer. *J. Surfactants Deterg.* **2001**, *4*, 271–277. [CrossRef]
43. Tatsumi, T.; Imai, Y.; Kawaguchi, K.; Miyano, N.; Ikeda, I. Antimicrobial activity of cationic gemini surfactant containing an oxycarbonyl group in the lipophilic portion against gram-positive and gram-negative microorganisms. *J. Oleo Sci.* **2014**, *63*, 137–140. [CrossRef]
44. Ronsin, G.; Kirby, A.J.; Rittenhouse, S.; Woodnutt, G.; Camilleri, P. Structure and antimicrobial activity of new bile acid-based gemini surfactants. *J. Chem. Soc. Perkin Trans. 2* **2002**, *7*, 1302–1306. [CrossRef]
45. Piętka-Ottlik, M.; Lewińska, A.; Jaromin, A.; Krasowska, A.; Wilk, K.A. Antifungal organoselenium compound loaded nanoemulsions stabilized by bifunctional cationic surfactants. *Colloids Surf. A Physicochem. Eng. Asp.* **2016**, *510*, 53–62. [CrossRef]
46. Pérez, L.; Torres, J.L.; Manresa, A.; Solans, C.; Infante, M.R. Synthesis, aggregation, and biological properties of a new class of gemini cationic amphiphilic compounds from arginine, bis (Args). *Langmuir* **1996**, *12*, 5296–5301. [CrossRef]
47. Zhong, X.; Guo, J.; Feng, L.; Xu, X.; Zhu, D. Cationic Gemini surfactants based on adamantane: Synthesis, surface activity and aggregation properties. *Colloids Surf. A Physicochem. Eng. Asp.* **2014**, *441*, 572–580. [CrossRef]
48. Machuca, L.M.; Reno, U.; Plem, S.C.; Gagneten, A.M.; Murguía, M.C.; Machuca, L.M.; Reno, U.; Plem, S.C.; Gagneten, A.M.; Murguía, M.C. N-acetylated gemini surfactants: Synthesis, surface-active properties, antifungal activity, and ecotoxicity bioassays. *Adv. Chem. Eng. Sci.* **2015**, *5*, 215–224. [CrossRef]
49. Silva, S.G.; Alves, C.; Cardoso, A.M.S.; Jurado, A.S.; de Lima, M.C.P.; Vale, M.L.C.; Marques, E.F. Synthesis of gemini surfactants and evaluation of their interfacial and cytotoxic properties: Exploring the multifunctionality of serine as headgroup. *Eur. J. Org. Chem.* **2013**, *2013*, 1758–1769. [CrossRef]
50. Barucha-Kraszewska, J.; Kraszewski, S.; Ramseyer, C. Will C-Laurdan dethrone Laurdan in fluorescent solvent relaxation techniques for lipid membrane studies? *Langmuir* **2013**, *29*, 1174–1182. [CrossRef] [PubMed]
51. Drabik, D.; Chodaczek, G.; Kraszewski, S. Effect of amyloid-β monomers on lipid membrane mechanical parameters–potential implications for mechanically driven neurodegeneration in Alzheimer's disease. *Int. J. Mol. Sci.* **2021**, *22*, 18. [CrossRef]
52. Shaban, S.M.; Aiad, I.; Moustafa, H.Y.; Hamed, A. Amidoamine Gemini surfactants based dimethylamino propyl amine: Preparation, characterization and evaluation as biocide. *J. Mol. Liq.* **2015**, *212*, 907–914. [CrossRef]
53. Pérez, L.; Garcia, M.T.; Ribosa, I.; Vinardell, M.P.; Manresa, A.; Infante, M.R. Biological properties of arginine-based gemini cationic surfactants. *Environ. Toxicol. Chem.* **2002**, *21*, 1279–1285. [CrossRef] [PubMed]
54. Pringle, M.J.; Brown, K.B.; Miller, K.W. Can the Lipid Theories of Anesthesia Account for the Cutoff in Anesthetic Potency in Homologous Series of Alcohols? *Mol. Pharmacol.* **1981**, *19*, 49–55. [PubMed]
55. Chernomordik, L.; Kozlov, M.M.; Zimmerberg, J. Lipids in biological membrane fusion. *J. Membr. Biol.* **1995**, *146*, 1–14. [CrossRef] [PubMed]
56. Piecuch, A.; Obłąk, E.; Guz-Regner, K. Antibacterial activity of alanine-derived gemini quaternary ammonium compounds. *J. Surfactants Deterg.* **2016**, *19*, 275–282. [CrossRef] [PubMed]
57. Rzycki, M.; Kraszewski, S.; Drabik, D. Towards mimetic membrane system. In *International Conference on Computational Science;* Springer: Cham, Switzerland, 2021; pp. 551–563. [CrossRef]
58. Ramadurai, S.; Duurkens, R.; Krasnikov, V.V.; Poolman, B. Lateral diffusion of membrane proteins: Consequences of hydrophobic mismatch and lipid composition. *Biophys. J.* **2010**, *99*, 1482–1489. [CrossRef]
59. Sharma, P.; Parthasarathi, S.; Patil, N.; Waskar, M.; Raut, J.S.; Puranik, M.; Ayappa, K.G.; Basu, J.K. Assessing barriers for antimicrobial penetration in complex asymmetric bacterial membranes: A case study with thymol. *Langmuir* **2020**, *36*, 8800–8814. [CrossRef]
60. Wang, X.; Wang, J.; Wang, Y.; Yan, H.; Li, P.; Thomas, R. Effect of the nature of the spacer on the aggregation properties of gemini surfactants in an aqueous solution. *Langmuir* **2003**, *20*, 53–56. [CrossRef] [PubMed]
61. Wettig, S.D.; Wang, C.; Verrall, R.E.; Foldvari, M. Thermodynamic and aggregation properties of aza- and imino-substituted gemini surfactants designed for gene delivery. *Phys. Chem. Chem. Phys.* **2007**, *9*, 871–877. [CrossRef] [PubMed]
62. Neubauer, D.; Jaśkiewicz, M.; Bauer, M.; Olejniczak-Kęder, A.; Sikorska, E.; Sikora, K.; Kamysz, W. Biological and physico-chemical characteristics of arginine-rich peptide gemini surfactants with lysine and cystine spacers. *Int. J. Mol. Sci.* **2021**, *22*, 3299. [CrossRef]
63. Ruiz, A.; Pinazo, A.; Pérez, L.; Manresa, A.; Marqués, A.M. Green catanionic gemini surfactant–lichenysin mixture: Improved surface, antimicrobial, and physiological properties. *ACS Appl. Mater. Interfaces* **2017**, *9*, 22121–22131. [CrossRef] [PubMed]
64. Frisch, M.J.; Trucks, G.W.; Schlegel, H.B.; Scuseria, G.E.; Robb, M.A.; Cheeseman, J.R.; Scalmani, G.; Barone, V.; Petersson, G.A.; Nakatsuji, H. *Gaussian 16, Revision 2016*; Gaussian Inc.: Wallingford, UK, 2016. Available online: https://gaussian.com/citation/ (accessed on 23 July 2019).
65. Mozrzymas, A. Modelling of the critical micelle concentration of cationic gemini surfactants using molecular connectivity indices. *J. Solution Chem.* **2013**, *42*, 2187–2199. [CrossRef] [PubMed]
66. Wu, E.L.; Cheng, X.; Jo, S.; Rui, H.; Song, K.C.; Dávila-Contreras, E.M.; Qi, Y.; Lee, J.; Monje-Galvan, V.; Venable, R.M.; et al. CHARMM-GUI membrane builder toward realistic biological membrane simulations. *J. Comput. Chem.* **2014**, *35*, 1997–2004. [CrossRef] [PubMed]

67. Szatmári, D.; Sárkány, P.; Kocsis, B.; Nagy, T.; Miseta, A.; Barkó, S.; Longauer, B.; Robinson, R.C.; Nyitrai, M. Intracellular ion concentrations and cation-dependent remodelling of bacterial MreB assemblies. *Sci. Rep.* **2020**, *10*, 1–13. [CrossRef]
68. Klauda, J.B.; Venable, R.M.; Freites, J.A.; O'Connor, J.W.; Tobias, D.J.; Mondragon-Ramirez, C.; Vorobyov, I.; MacKerell, A.D.; Pastor, R.W. Update of the CHARMM all-atom additive force field for lipids: Validation on six lipid types. *J. Phys. Chem. B* **2010**, *114*, 7830–7843. [CrossRef] [PubMed]
69. Abraham, M.J.; Murtola, T.; Schulz, R.; Páll, S.; Smith, J.C.; Hess, B.; Lindah, E. Gromacs: High performance molecular simulations through multi-level parallelism from laptops to supercomputers. *SoftwareX* **2015**, *1–2*, 19–25. [CrossRef]
70. Evans, D.J.; Holian, B.L. The Nose–Hoover thermostat. *J. Chem. Phys.* **1998**, *83*, 4069. [CrossRef]
71. Parrinello, M.; Rahman, A. Polymorphic transitions in single crystals: A new molecular dynamics method. *J. Appl. Phys.* **1998**, *52*, 7182. [CrossRef]
72. Darden, T.; York, D.; Pedersen, L. Particle mesh Ewald: An N·log(N) method for Ewald sums in large systems. *J. Chem. Phys.* **1998**, *98*, 10089. [CrossRef]
73. Humphrey, W.; Dalke, A.; Schulten, K. VMD: Visual molecular dynamics. *J. Mol. Graph.* **1996**, *14*, 33–38. [CrossRef]
74. Doktorova, M.; Harries, D.; Khelashvili, G. Determination of bending rigidity and tilt modulus of lipid membranes from real-space fluctuation analysis of molecular dynamics simulations. *Phys. Chem. Chem. Phys.* **2017**, *19*, 16806–16818. [CrossRef] [PubMed]
75. Doktorova, M.; LeVine, M.V.; Khelashvili, G.; Weinstein, H. A new computational method for membrane compressibility: Bilayer mechanical thickness revisited. *Biophys. J.* **2019**, *116*, 487–502. [CrossRef] [PubMed]
76. Giorgino, T. Computing diffusion coefficients in macromolecular simulations: The diffusion coefficient tool for VMD. *J. Open Source Softw.* **2019**, *4*, 1698. [CrossRef]
77. Guixà-González, R.; Rodriguez-Espigares, I.; Ramírez-Anguita, J.M.; Carrió-Gaspar, P.; Martinez-Seara, H.; Giorgino, T.; Selent, J. MEMBPLUGIN: Studying membrane complexity in VMD. *Bioinformatics* **2014**, *30*, 1478–1480. [CrossRef]
78. Irving, J.H.; Kirkwood, J.G. The statistical mechanical theory of transport processes. IV. The equations of hydrodynamics. *J. Chem. Phys.* **2004**, *18*, 817. [CrossRef]
79. Bacle, A.; Gautier, R.; Jackson, C.L.; Fuchs, P.F.J.; Vanni, S. Interdigitation between triglycerides and lipids modulates surface properties of lipid droplets. *Biophys. J.* **2017**, *112*, 1417–1430. [CrossRef]
80. Bourasseau, E.; Homman, A.-A.; Durand, O.; Ghoufi, A.; Malfreyt, P. Calculation of the surface tension of liquid copper from atomistic Monte Carlo simulations. *Eur. Phys. J. B* **2013**, *86*, 251. [CrossRef]

International Journal of
Molecular Sciences

Article

Osmosis-Driven Water Transport through a Nanochannel: A Molecular Dynamics Simulation Study

Changsun Eun

Department of Chemistry, Hankuk University of Foreign Studies, Yongin 17035, Korea; ceun@hufs.ac.kr

Received: 11 August 2020; Accepted: 26 October 2020; Published: 28 October 2020

Abstract: In this work, we study a chemical method to transfer water molecules from a nanoscale compartment to another initially empty compartment through a nanochannel. Without any external force, water molecules do not spontaneously move to the empty compartment because of the energy barrier for breaking water hydrogen bonds in the transport process and the attraction between water molecules and the compartment walls. To overcome the energy barrier, we put osmolytes into the empty compartment, and to remove the attraction, we weaken the compartment-water interaction. This allows water molecules to spontaneously move to the empty compartment. We find that the initiation and time-transient behavior of water transport depend on the properties of the osmolytes specified by their number and the strength of their interaction with water. Interestingly, when osmolytes strongly interact with water molecules, transport immediately starts and continues until all water molecules are transferred to the initially empty compartment. However, when the osmolyte interaction strength is intermediate, transport initiates stochastically, depending on the number of osmolytes. Surprisingly, because of strong water-water interactions, osmosis-driven water transport through a nanochannel is similar to pulling a string at a constant speed. Our study helps us understand what minimal conditions are needed for complete transfer of water molecules to another compartment through a nanochannel, which may be of general concern in many fields involving molecular transfer.

Keywords: molecular dynamics simulation; osmosis; water transport; nanochannel; carbon nanotube; graphene; osmolyte; compartment

1. Introduction

Water molecules are polar molecules that strongly interact with each other, and they form a hydrogen network, whereas nonpolar molecules do not have such a strong interaction. This difference in the intermolecular interaction strength can be manifested in their physical states and transport behavior. For example, under ambient conditions, water molecules are in the liquid state but their corresponding nonpolar molecules, which are artificial molecules made from water molecules by removing electric charges, show gaseous behavior [1]. Thus, in this case, when we transfer molecules from one compartment to another compartment, water molecules do not spontaneously move to the other compartment without any external forces, while nonpolar molecules can move via diffusion. In particular, for the latter case, we intensively studied the nonequilibrium transport behavior of nonpolar molecules driven by diffusion and osmotic pressure, and the equilibrium states, using molecular dynamics (MD) simulations [1]. However, such detailed research on nonequilibrium water transport has not been performed. Therefore, here, to study the case of water, we use the same system that we employed in our previous work for nonpolar molecules [1] but use water molecules instead of nonpolar molecules.

Not only is water transport interesting in terms of the need to better understand the fundamental nature of molecular transport for strongly interacting molecules, but also, it is of concern in natural and engineering systems involving water transport. One example of a natural system is that in a cell, water transport occurs to regulate osmolality through aquaporins [2–4]. Another example in water engineering is water transport in reverse osmosis to obtain pure water, where water passes through a membrane with carbon nanotubes under high pressure [5]. In these examples, water moves from one location to another. Therefore, understanding the mechanisms by which water moves is essential in controlling water flow in real cases as well as advancing our knowledge.

Why do water molecules move? To properly address this question, specifically, let us assume we have two nanoscale compartments connected by a nanochannel, in which one compartment is filled with water and the other is empty. How can one transfer all the water molecules from the filled compartment to the empty compartment? In principle, mass transfer, including water transport, occurs when the chemical potential exhibits a spatial difference [6]; thus, the key idea is to create a desirable chemical potential difference. In particular, for the transfer to the empty compartment, one needs to lower the chemical potential in the empty compartment or raise the chemical potential in the filled compartment. For this, one can use a chemical method to create osmotic pressure, such as in aquaporins [4] and carbon nanotube membranes [7], by inserting osmolytes into the empty compartment [1], or use a mechanical method to create a pressure difference, such as in reverse osmosis, by applying external pressure [5,8–10]. In this work, we focus on the chemical method and discuss what specific conditions are needed for our system to induce complete water transfer to the empty compartment. We also address the kinetic characteristics during water transport, along with the kinetically stable states.

This paper is organized as follows. In Section 2, we introduce our model systems to study the water transport and explain the details of MD simulations. In Section 3, we investigate when water transfer occurs by examining the number of osmolytes and their interaction strength with water molecules. We also discuss the kinetically stable and equilibrium states appearing in the transfer due to the osmolytes and the associated kinetic properties such as the waiting time and transition time for the transport, order of the transition rate law, and transport rate. In Section 4, we summarize our findings and the implications of our work.

2. Computational Models and Methods

To study the transport of water molecules from one place to another, we consider a simple system that has two compartments connected by a carbon nanotube (CNT), which was used in our previous work for the study of nonpolar molecule transport [1] (see Figure 1). In contrast to the previous work, in this work, we include 884 water molecules instead of nonpolar molecules. To construct the compartments and nanochannel, we use graphene plates whose x–y dimensions are 3.03×3.22 nm and an armchair (6,6) CNT with a length of 4 nm. To connect the spaces in the compartments and the CNT, we make small holes in the graphene plates next to the CNT. The gap created by the graphene plate and the CNT in Figure 1a is too small for a water molecule to pass through. Figure 1a shows the associated lengths along the z axis. To model the graphene plates and the CNT, we use the AMBER force field with $\varepsilon = 0.3598$ kJ/mol and $\sigma = 0.3400$ nm for the 6–12 Lennard-Jones (LJ) parameters [11,12]. To model water molecules, we employ TIP3P water [13]. The electrostatic interactions are calculated using the particle-mesh Ewald (PME) method [14]. For LJ interactions, if not specified, the Lorentz-Berthelot rule [15] is used for the LJ parameters, i.e., the sigma (σ) and epsilon (ε) parameters. In this study, the cutoff distance for the electrostatic and LJ interactions is 1.4 nm.

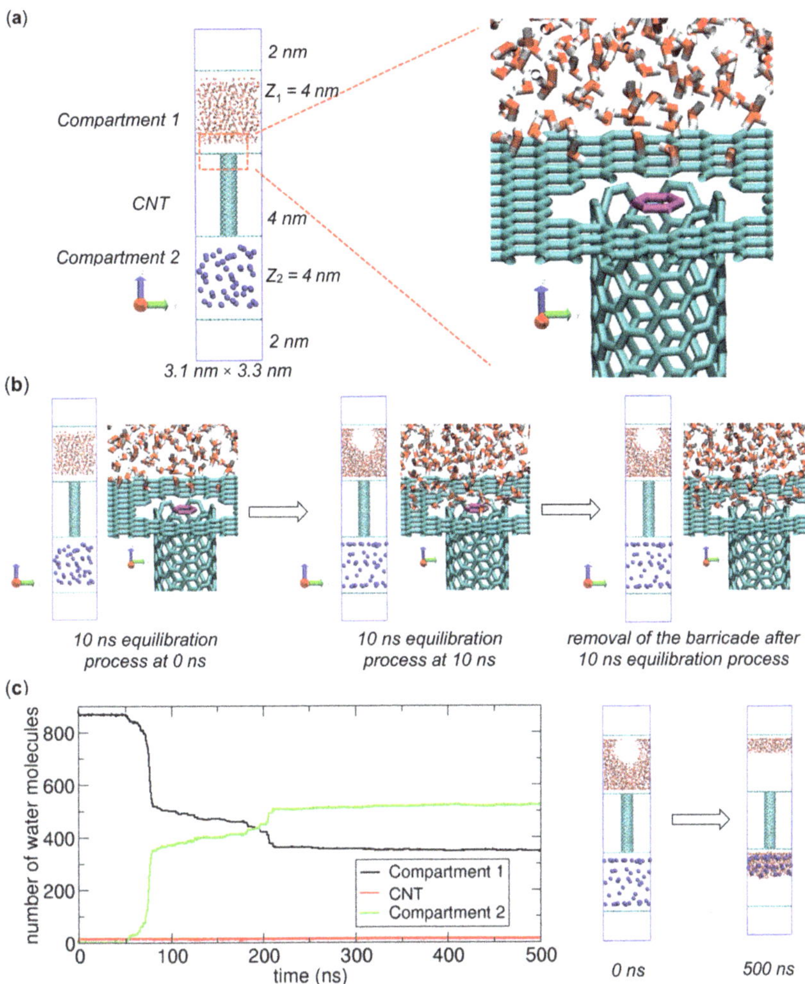

Figure 1. Simulation of water molecule transport from a compartment filled with water molecules (Compartment 1) to another spacious compartment (Compartment 2) through a carbon nanotube (CNT) in the presence of 50 osmolytes with IS_{WO} = 10. Here, osmolytes are represented by blue spheres. (**a**) Initial configuration of the system, in which all water molecules are in Compartment 1 (left), and a zoomed-in view around the boundary between Compartment 1 and the CNT (right), where a molecular barricade, i.e., a six-carbon ring (violet), is located at a fixed position on the boundary. From the zoomed-in view, a gap appears to exist between the CNT and the graphene with a hole such that water might pass through it, but in reality, the gap is too small for water to pass through it. Note that the blue lines indicate the periodic boundary of the system. (**b**) Configurations at 0 ns (left) and 10 ns (center) from the 10 ns equilibration process in the presence of the barricade (violet), and the configuration (right) in which the six-carbon ring has been removed after the equilibration process. The last configuration is used for the simulation of water transport. (**c**) Numbers of water molecules in Compartment 1 (black), the CNT (red), and Compartment 2 (green) during a transport process (left) along with MD simulation snapshots at 0 and 500 ns (right).

To simulate water transport, we first prepare an initial state of the system in which the water molecules are only in Compartment 1 and not in the CNT or Compartment 2. Then, using the

GROMACS package [16], we perform two-step MD simulations with periodic boundary conditions in all three spatial directions and a time step of 2.0 fs for equilibration (see Figure 1b) and transport (see Figure 1c) processes. In the simulations, we fix the positions of the compartments and CNT using the freeze group option of GROMACS [15], while water molecules are free to move. Before the water molecules are allowed to move to the CNT and then to Compartment 2, for the equilibration process, we run an NVT MD simulation for 10 ns at 300 K with a separator, i.e., a molecular barricade, composed of a 6-carbon ring at the boundary between Compartment 1 and the CNT (see the violet hexagons in Figure 1a,b). With the barricade, all water molecules remain in Compartment 1. More details of the system can be found in our previous work [1]. Here, to maintain the temperature, we employ the modified Berendsen algorithm named the V-rescale thermostat [17] with a coupling constant of 0.1 ps, which is implemented in GROMACS. During the simulation, the molecules are equilibrated in Compartment 1. After 10 ns, for the transport process, we remove the barricade and run an NVT MD simulation at 300 K to study the water transport between the two compartments.

To induce water transfer, we use a chemical force created by osmolytes. The osmolyte considered here is a nanoparticle that cannot pass through the CNT and remains in the compartment it was originally placed in. In the presence of osmolytes that interact with water molecules in Compartment 2, the chemical potential in Compartment 2 is lower than that in Compartment 1. Therefore, theoretically, if the chemical potential in Compartment 2 is sufficiently low, water transport to Compartment 2 occurs. To model osmolytes, in this work, we use the same osmolytes that we used in our previous work [1]. The reference osmolyte is a larger-size AMBER carbon atom that has the same LJ epsilon (ε) value as an AMBER carbon atom but a larger LJ sigma (σ) value (0.7 nm) than the AMBER carbon atom (0.34 nm). In our previous work, we demonstrated that the value of 0.7 nm is too large for the reference osmolyte to pass through the CNT while water molecules can, which implies that the CNT can be regarded as a semipermeable membrane in osmosis.

Since the strength of the interaction between an osmolyte and a water molecule can significantly affect the chemical potential of water in Compartment 2, we prepare various osmolytes by modifying the ε value of the reference osmolyte for the LJ interaction between the oxygen atom of water (TIP3P oxygen atom) and the osmolyte atom. Note that the ε value of the reference osmolyte is 0.47837 kJ/mol. Specifically, we modify the strength by multiplying this reference value by a multiplication factor. We call this multiplication factor the relative water-osmolyte interaction strength (IS_{WO}). Note that notation IS_{WO} corresponds to notation IS_{MO} in our previous work [1]. In this study, we consider IS_{WO} values of 0.1, 1 (reference), 5, 10, 20 and 50. Figure 1 shows one example of systems containing 50 osmolytes with $IS_{WO} = 10$.

To analyze the transfer of water molecules from the water-filled compartment to the water-empty compartment, we calculate the numbers of water molecules residing in Compartment 1 (N_1), the CNT (N_{CNT}) and Compartment 2 (N_2). One example is shown in Figure 1c. As water transport occurs, N_1 decreases while N_2 increases. N_1 and N_2 show opposite time-transient behaviors, which indicates that the plots of N_1 and N_{CNT} are sufficient to completely determine N_2. Note that $N_1 + N_{CNT} + N_2 = 884$ and $N_{CNT} << N_1, N_2$ since only a small number of water molecules can occupy the CNT. In this paper, all figures, including Figure 1, are prepared using the VMD [18] (http://www.ks.uiuc.edu/Research/vmd/; University of Illinois at Urbana-Champaign, Urbana, IL, USA) and xmgrace (http://plasma-gate.weizmann.ac.il/Grace/) programs.

3. Results and Discussion

3.1. Necessary Conditions for Water Transport

3.1.1. General Thermodynamic Considerations

In general, for water transport from Compartment 1 to Compartment 2 to occur, thermodynamic conditions must be created, where the chemical potential of water in Compartment 2 (μ_2) is lower than that in Compartment 1 (μ_1), i.e., $\mu_1 > \mu_2$. This chemical difference induces mass transfer [6]. However, since the two compartments are not directly connected but are connected through a nanochannel, i.e., a CNT, the CNT plays a role. Thus, we require more thermodynamic conditions related to the CNT. In other words, for the transport of water molecules to Compartment 2, water molecules must easily enter the CNT, which implies that the chemical potential of water in the CNT (μ_{CNT}) should be less than μ_1, i.e., $\mu_1 > \mu_{CNT}$. Similarly, for water molecules to easily escape from the CNT and enter Compartment 2, the condition $\mu_{CNT} > \mu_2$ must be satisfied. Finally, these conditions can be summarized into one inequality: $\mu_1 > \mu_{CNT} > \mu_2$. Therefore, the key to the transfer of water from Compartment 1 to Compartment 2 is that this inequality holds until all water molecules have moved to Compartment 2. However, since the chemical potential difference decreases as water transport continues, the initial difference should be sufficiently large; when more water molecules move to Compartment 2, the force of osmolytes dragging water molecules from Compartment 1 to Compartment 2 is expected to be reduced because of the increase in number of water molecules surrounding osmolytes in Compartment 2 (screening effect). In the next sections, starting from a reference model in our previous work [1], we individually discuss the necessary conditions for water transport.

3.1.2. Transport of Water Molecules in the Absence of Osmolytes

As discussed in Section 3.1.1, our model system is based on the model system in our previous work [1] used for the transport of weakly interacting molecules (or charge-removed water molecules). In the previous work, we showed that even in the absence of osmolytes, transport from Compartment 1 to Compartment 2 occurs due to the entropic force ($\mu_1 > \mu_2$), and at equilibrium ($\mu_1 = \mu_2$), the number of molecules in each compartment is proportional to the size of the compartment. For example, when Compartment 1 and Compartment 2 have the same size, at equilibrium, the number of molecules in Compartment 1 is the same as that in Compartment 2. To determine if we can observe the same phenomena with water molecules, we use the model system used in our previous work, but we replace charge-removed water molecules with water molecules.

The preparation and simulation of systems in the absence of osmolytes are basically the same as those in the presence of osmolytes depicted in Figure 1. Figure 2a shows schematics of the equilibration procedure before a production run. After the 10 ns equilibration, we prepare the initial configuration for the production run and run a 500 ns NVT MD simulation. To determine if transport occurs, we calculate the numbers of water molecules in Compartment 1, the CNT, and Compartment 2 and plot them as functions of time in Figure 2b. The results indicate that transport does not occur, as the number of molecules in Compartment 2 is essentially zero (see the inset of Figure 2b). Apparently, the strong interactions between water molecules prohibit water molecules in Compartment 1 from moving to Compartment 2 because in the absence of such interactions, almost half of the molecules spontaneously move to Compartment 2, as we observed in our previous work [1]. In other words, to initiate water transport, some water molecules in Compartment 1 should dissociate from a group of molecules in Compartment 1 and then transfer to Compartment 2, but since the molecules tend to associate with other water molecules, they tend to remain in Compartment 1, which means that $\mu_1 < \mu_2$. Therefore, in contrast to the entropy-driven transport of charge-removed water, for water transport, an external force must be used to satisfy $\mu_1 > \mu_2$. In this work, as an external force, we use an osmotic force created by osmolytes in Compartment 2, which will be discussed in Section 3.1.3.

Notably, the inset of the plot in Figure 2b shows that water molecules occupy the CNT, which occurs because $\mu_1 > \mu_{CNT}$ [11]. Another interesting feature from the snapshots in Figure 2b is that a cavity is formed in Compartment 1 and stably maintained, due to the attractive interaction between the wall of the compartment and water molecules. Previously, the related hydrophilicity of graphene plates was discussed [19]. One may think that this cavity is compatible with the water transport through the CNT, but it can lead to early termination of the transport process before a majority of water molecules are transferred to Compartment 2, which will be explained in Section 3.1.4.

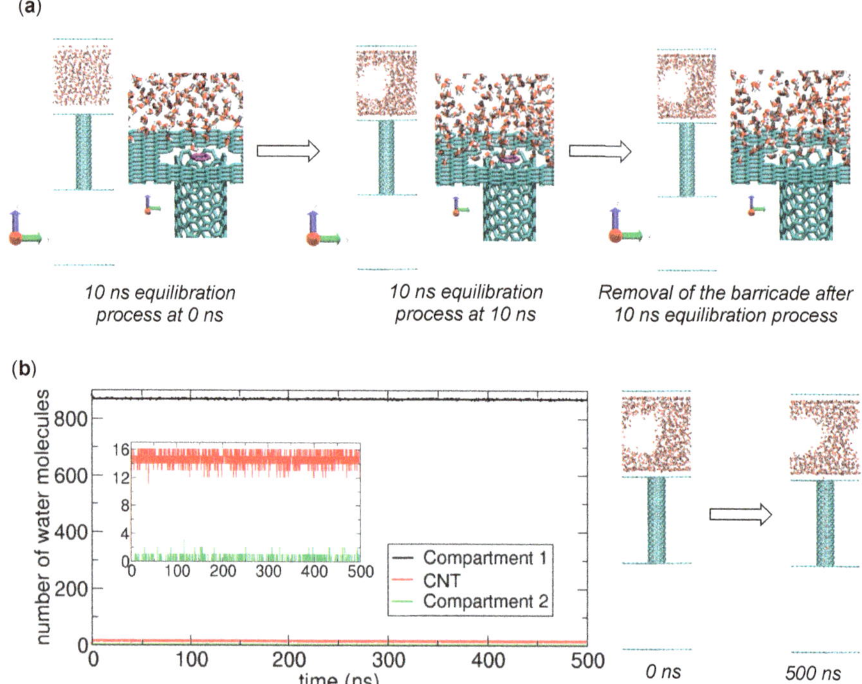

Figure 2. Simulation of water molecule transport from a filled compartment (Compartment 1) to another empty compartment (Compartment 2) through a carbon nanotube (CNT) in the absence of osmolytes. (**a**) Configurations at 0 ns (left) and 10 ns (center) from a 10 ns equilibration process in the presence of a molecular barricade (violet) at the boundary between Compartment 1 and the CNT, and the configuration (right) in which the barricade has been removed, after the equilibration process. The last configuration is used for the simulation of water molecule transport in the next process. (**b**) Results of water transport simulation. Changes in the numbers of water molecules in Compartment 1 (black), the CNT (red), and Compartment 2 (green) during a 500 ns NVT simulation (left) along with snapshots at 0 and 500 ns (right). The inset shows a zoomed-in view of the changes in the CNT and Compartment 2.

3.1.3. Transport of Water Molecules in the Presence of Osmolytes

As discussed in Section 3.1.2, without osmolytes, the chemical potential of water in Compartment 2 is higher than that in Compartment 1: $\mu_1 < \mu_2$. Therefore, to create a transport-inducible environment satisfying $\mu_1 > \mu_2$, we can place osmolytes in Compartment 2. A sufficient number of osmolytes and a sufficient strength of the interaction with water molecules will induce water transport. To obtain a general idea of the effect of osmolytes, we examine some representative cases with various numbers of osmolytes (N_O) and various strengths of the interaction (IS_{WO}). The simulation results for these cases are presented in Figure 3.

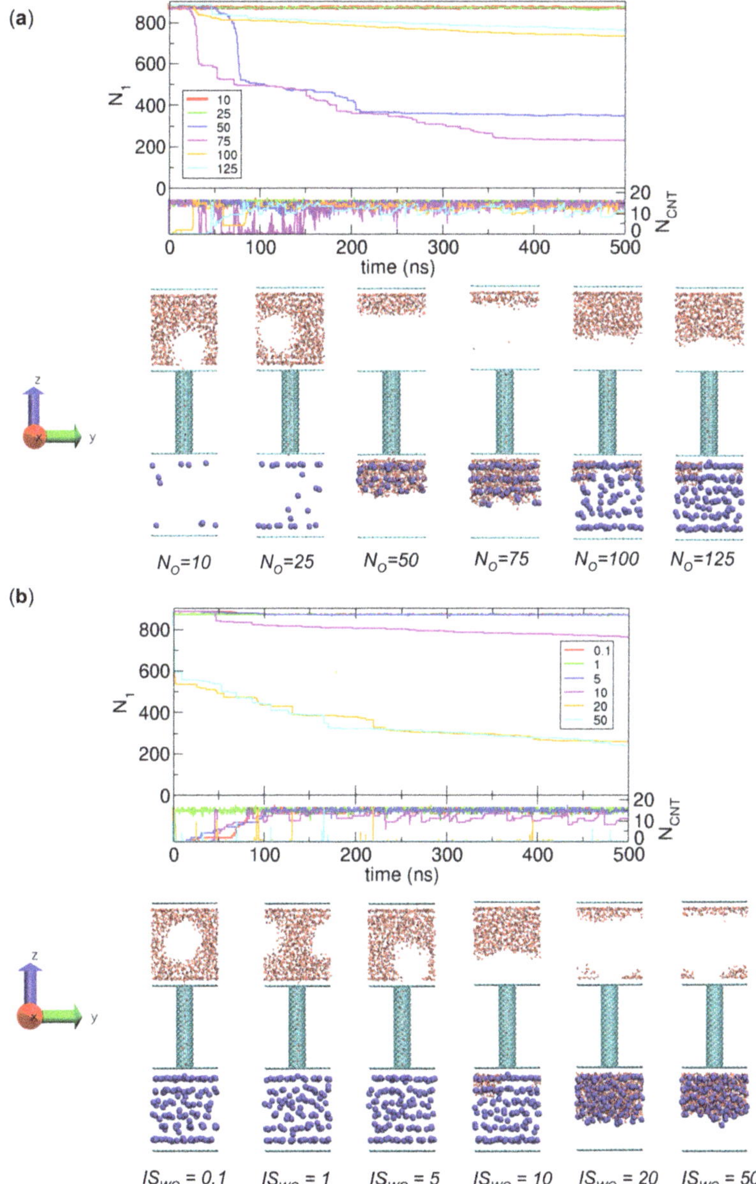

Figure 3. Influence of osmolytes on water transport. (**a**) Numbers of water molecules in Compartment 1 (N_1) and the CNT (N_{CNT}) as functions of time for the cases with IS_{WO} = 10 and N_O = 10 (red), 25 (green), 50 (blue), 75 (violet), 100 (orange), and 125 (cyan), and their final configurations at 500 ns. (**b**) Numbers of water molecules in Compartment 1 (N_1) and the CNT (N_{CNT}) as functions of time for the cases with N_O = 125 and IS_{WO} = 0.1 (red), 1 (green), 5 (blue), 10 (violet), 20 (orange), and 50 (cyan).

In Figure 3a, we fix the value of IS_{WO} as 10 and consider several values of N_O: N_O = 10, 25, 50, 75, 100, and 125. These specific values of N_O are selected because they were used for the transport study of charge-removed water molecules in our previous work [1], and it is necessary to use the same values in this work for the comparison to understand the effect of the interaction strength between transported molecules, as will be discussed in Section 3.2.5. From the simulations, we observe water transport for the cases of N_O = 50, 75, 100, and 125. One may expect that as N_O increases, more water molecules would be transferred to Compartment 2, but surprisingly, we obtain a different result in that the numbers of water molecules in Compartment 2 for the cases of N_O = 50 and 75 are larger than those for the cases of N_O = 100 and 125. This is related to the cavity initially formed in Compartment 1 due to the compartment-water interaction. During transport, the cavity readily breaks a group of water molecules into two disconnected groups: one group is next to the lower graphene plate and CNT, and the other group is next to the upper graphene plate. The water molecules in the former group transfer to the CNT but the water molecules in the latter group remain in Compartment 1. Therefore, the cases of N_O = 100 and 125 have fewer transferred water molecules because the numbers of residual water molecules in the latter group are greater than those in the cases of N_O = 50 and 75.

In contrast to Figure 3a, in Figure 3b, we fix the value of N_O as 125 and consider various values of IS_{WO}, IS_{WO} = 0.1, 1, 5, 10, 20, and 50. These specific values of IS_{WO} were used in our previous work with charge-removed water molecules [1]. In this case, when IS_{WO} = 10, 20, and 50, we observe water transport to Compartment 2. However, because of the breakage due to the cavity during transport, we also observe residual water molecules next to the upper graphene plate of Compartment 1, as shown in Figure 3a. Additionally, we note that because of the strong interaction with IS_{WO} = 20 and 50, all water molecules in the CNT are transferred to Compartment 2, and thus, the CNT is empty; presumably, in this case, $\mu_{CNT} >> \mu_2$.

3.1.4. Transport of Water Molecules in Compartments that Weakly Interact with Water

From the discussion in Section 3.1.3, we clearly must remove the cavity in Compartment 1 to transport more water molecules to Compartment 2. Since the cavity is created because of the attraction between the compartment wall (normal graphene plates) and water molecules, one method to remove the cavity is to weaken the compartment-water interaction. To implement this idea, we reduce the LJ interaction between carbon atoms in the graphene plates and water molecules via the Lorentz-Berthelot rule [15], while the interaction between carbon atoms in the CNT and water molecules remains the same as in the original system. Specifically, we reduce the standard value of ε for carbon, i.e., 0.3598 kJ/mol, by 10 times, so ε = 0.03598 kJ/mol. We expect that this reduction guarantees the condition $\mu_1 > \mu_{CNT}$. With this weak compartment-wall interaction, we perform MD simulations and examine extensive cases with N_O=10, 25, 50, 75, 100, and 125 and IS_{WO} = 0.1, 1, 5, 10, 20 and 50 to find the necessary osmolyte conditions for water transport. The simulation results are displayed in Figure 4, which is the main figure of this work.

Before we study the effect of osmolytes for the new systems above, we run a simulation for the system in the absence of osmolytes to understand the effect due solely to the weak interactions between compartments and water molecules. The configurations in Figure 4a show that the weak interactions remove the cavity in Compartment 1, and the water molecules in Compartment 1 are next to the lower graphene plate, not the upper graphene plate. This occurs because in Compartment 1 the water-water and water-CNT interactions are stronger than the reduced water-graphene interactions. Additionally, because of the reduction in the compartment-water interaction, water molecules are less likely to move from the CNT to Compartment 2, i.e., $\mu_{CNT} < \mu_2$. Therefore, as shown in Figure 4a, water molecules remain in Compartment 1 and the CNT; thus, in this case, osmolytes are required to induce water transport to Compartment 2, as shown in Section 3.1.3.

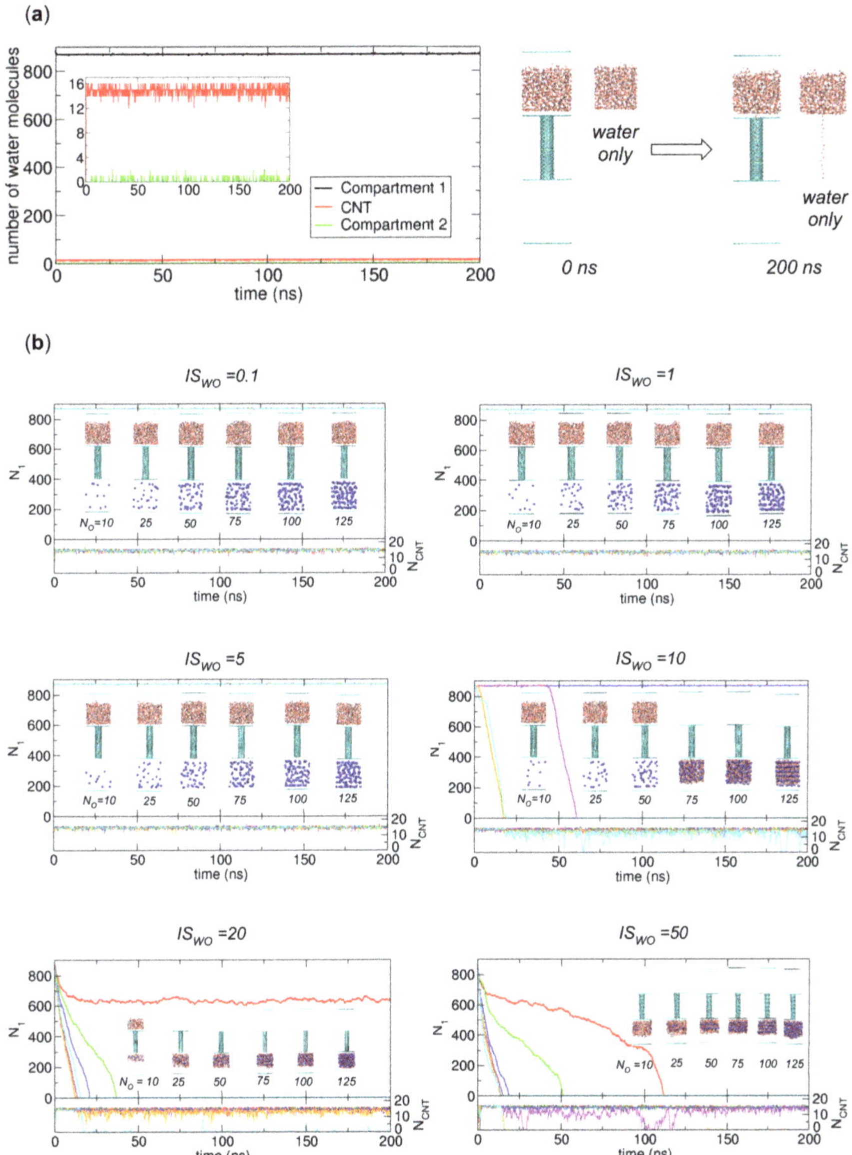

Figure 4. Simulation of water molecule transport in the systems composed of compartment walls that weakly interact with water molecules in the (**a**) absence and (**b**) presence of osmolytes with N_O = 10 (red), 25 (green), 50 (blue), 75 (violet), 100 (orange), and 125 (cyan) and IS_{WO} = 0.1, 1, 5, 10, 20, and 50. For each system, the numbers of water molecules in Compartment 1 (N_1) and the CNT (N_{CNT}) are plotted as functions of time. The insets show the final configurations of systems at 200 ns. Note that for all cases, the cavities in Figures 2 and 3 disappear.

When we place osmolytes in Compartment 2 and weaken the compartment-water interactions, we observe complete or nearly complete water transport in some cases in the sense that all water molecules are removed from Compartment 1 and transferred to CNT or Compartment 2. In these cases, the osmotic effect is significant. From Figure 4b, the cases showing (nearly) complete water transfer are the cases with N_O = 75, 100, and 125 for IS_{WO} = 10, N_O = 25, 50, 75, 100, and 125 for IS_{WO} = 20, and N_O = 10, 25, 50, 75, 100, and 125 for IS_{WO} = 50. One exceptional case is that with N_O = 10 and IS_{WO} = 20 in that only some of the water molecules in Compartment 1 are transferred, and some residual water molecules remain in Compartment 1. Including this exceptional case, we later analyze all the cases in terms of kinetically stable states.

Additionally, we consider the possibility of water transport between the compartments without osmolytes. To address this possibility, it is the best to discuss with a free energy profile of system as a function of N_1 or N_2, but it is generally challenging to obtain an accurate free energy profile, and it requires a lot of computer resources. Therefore, to obtain a general idea of the shape of the free energy profile, we prepare various systems, which represent intermediate states that can be observed in the water transport, by preparing the initial configurations with N_1 = 884, 829, 774, 663, 442, 221, 110, 55 and 0. Then, we perform the MD simulations to observe the time evolution of states. The simulation results are summarized in Figure 5. Figure 5a indicates that the intermediate states go to either the state where water molecules occupy only Compartment 1 and CNT (State A; the cases with N_1 = 884, 829, 774, 663, and 442) or the state where they occupy only Compartment 2 and CNT (State B; the cases with N_1 = 221, 110, 55, and 0). The final configurations in Figure 5b clearly show that there are only these two stable states. From Figure 5a, we also observe that initially when $N_1 > N_2$ or $N_2 > N_1$, the initial state quickly goes to State A or State B, respectively. However, initially when $N_1 = N_2$, N_1 is largely fluctuating before the state eventually goes to State A or State B; in fact, this fluctuation is similar to the one observed when the two groups pull against each other at opposite ends of a rope with similar strength in a tug of war. Based on the results in Figure 5, we expect that free energy profile has the minima near N_1 = 884 (State A) and 0 (State B) and a higher value (possibly the maximum) near N_1 = 442, the slope near the minima is relatively large, and the slope near N_1 = 442 is small, which enables a large fluctuation in N_1. This shape of free energy profile implies that there is a free energy barrier in the transition between stable States A and B. Thus, we need an external force such as osmotic pressure to induce water transport. For charge-removed water molecules [1], the free energy profile is expected to show the opposite behavior with the minimum near N_1 = 442; therefore, it is interesting to see how the free energy profile changes with the strength of interaction among transported molecules, which will be a subject of future study.

3.2. Water Transport Analysis

3.2.1. Potential Energy Change

In Section 3.1.4, we observe water transport from the 200 ns simulations. Now, the following question arises: what is the driving force for the water transport? Since this is an NVT simulation, the appropriate free energy for the system is the Helmholtz free energy $A = E - TS$, where E, T, and S are the internal energy, temperature, and entropy, respectively [6]. Therefore, to systematically understand the Helmholtz free energy change ΔA, we can consider the energetic contribution related to the first term E and the entropic contribution related to the second term TS. However, from Figures 2 and 3, we understand that the energetic contribution is dominant over the entropic contribution because if the entropic change is dominant, water molecules should be transferred to Compartment 2 due to the translational entropy and the mixing entropy with osmolytes. Therefore, we focus on the internal energy change ΔE, which is the sum of the potential energy change and kinetic energy change. Furthermore, since we study the system under a constant temperature, we expect that the change due to the kinetic energy is not significant. Thus, we expect that the potential energy change is responsible for the water transport. We calculate the potential energy change to determine if the change is correlated with the water transport, to explain the driving force for the transport.

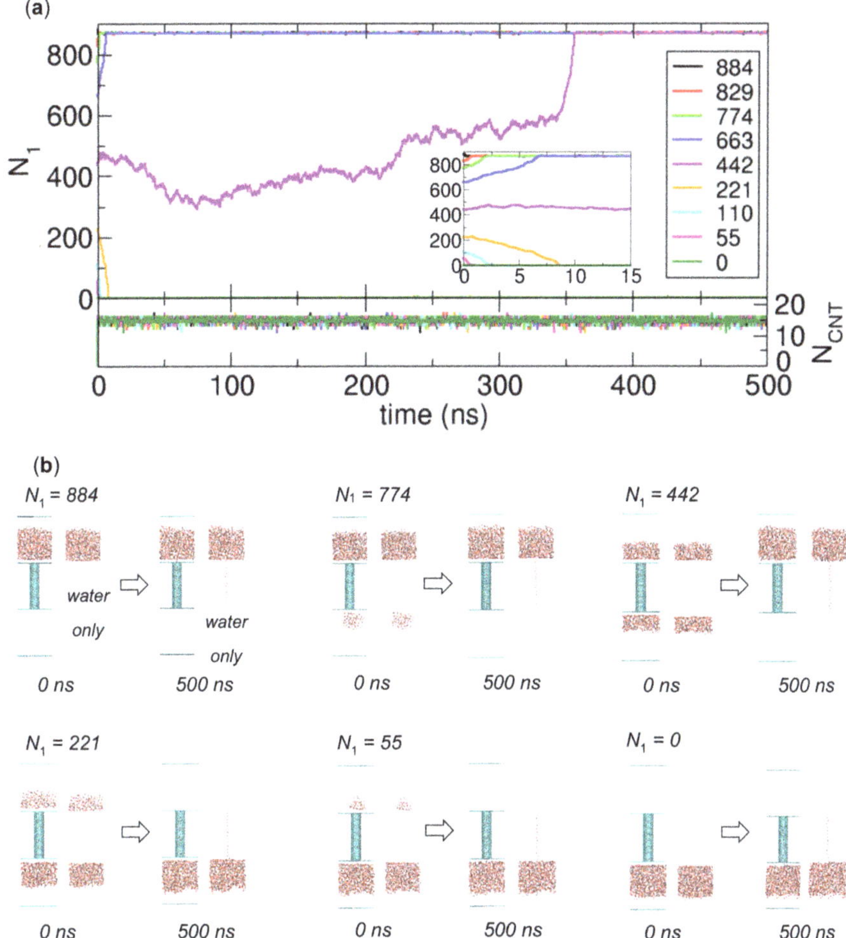

Figure 5. Transport simulation of water molecules in various systems of compartment walls that weakly interact with water molecules in the absence of osmolytes. Various initial configurations with N_1 = 884, 829, 774, 663, 442, 221, 110, 55 and 0 are prepared from the 10 ns equilibration process. (**a**) Numbers of water molecules in Compartment 1 (N_1) and the CNT (N_{CNT}) as functions of time. The inset show the plots for a time interval from 0 ns to 15 ns. (**b**) Initial and final configurations obtained from the 500 ns MD simulations for various initial configurations with N_1 = 884, 774, 442, 221, 55 and 0.

As shown in Figure 6, we calculate the potential energy change for the cases of IS_{WO} = 10, 20, and 50, where water transport is observed. In this calculation, for convenience, for each value of IS_{WO}, we set the average potential energy over the time interval between 100 ns and 200 ns for the N_O = 10 case to zero. The potential energies for the other cases are calculated based on this reference. As we expect, when water transport occurs, the potential energy change is significant and is highly correlated with the change in N_1, which implies that the potential energy change due to the mixing of water molecules and osmolytes is the major driving force for water transport driven by osmosis.

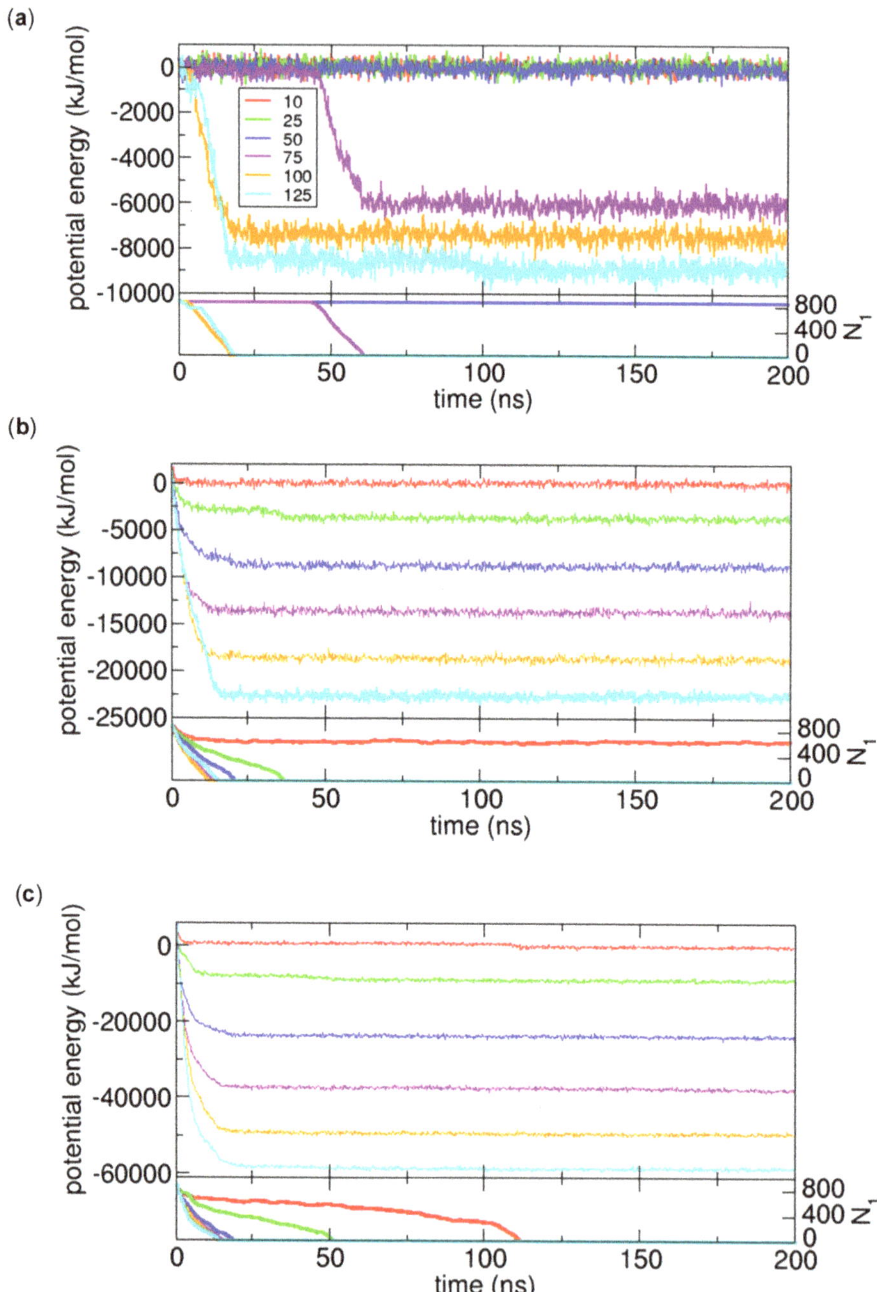

Figure 6. Changes in potential energy and in the number of water molecules in Compartment 1 (N_1) as functions of time for the cases of N_O = 10, 25, 50, 75, 100, and 125 when (**a**) IS_{WO} = 10 (**b**) IS_{WO} = 20, and (**c**) IS_{WO} = 50.

3.2.2. Kinetically Stable States

From Figure 4, we observe kinetically stable states in terms of the numbers of water molecules in Compartment 1, the CNT and Compartment 2. Specifically, we find three main stable states, some of which, we believe, are global equilibrium states. In other words, the initial state in which water molecules are only in Compartment 1 transitions to one of three states: in one state, water occupies only Compartment 1 and the CNT (State I); in another state, water occupies only the CNT and Compartment 2 (State II); and in the other state, water occupies only Compartment 2 (State III). Here, State I can be considered a metastable state in the presence of osmolytes. Specifically, when we compare a separated state in which water molecules are in Compartment 1 and osmolytes are in Compartment 2 (State I) with a mixed state in which water molecules and osmolytes are in Compartment 2 (State II and State III), the mixed state is more energetically favorable than State I because of the strong interaction between water molecules and osmolytes. Therefore, the observation of State I is probably due to the kinetic barrier for the transition to State II or State III. In Figure 4, the only exception to the above three states is the stable state in the case of $N_O = 10$ and $IS_{WO} = 20$, in which water molecules exist in all regions. We will discuss this state later, which we believe is another metastable state.

When the interaction between osmolytes and water molecules is weak, i.e., $IS_{WO} = 0.1$, 1, and 5, the systems reach State I, and transfer to Compartment 2 does not occur. Additionally, when the interaction is intermediate, i.e., $IS_{WO} = 10$, and the number of osmolytes is not sufficient to induce water transport ($N_O = 10$, 25, and 50), the system also reaches State I. However, when $N_O = 75$, 100, and 125, the stable states are State II. Moreover, when the interaction is strong, i.e., $IS_{WO} = 20$, 50, even a small number of osmolytes can induce water transport and the states also become State II. Specifically, the cases with $N_O = 25$, 50, 75, and 100 for $IS_{WO} = 20$ and $N_O = 10$, 25, 50, and 75 for $IS_{WO} = 50$ show State II as a stable state. Here, interestingly, the addition of more osmolytes induces complete transfer of water molecules to Compartment 2, which means that the stable states are State III. Specifically, the cases with $N_O = 125$ for $IS_{WO} = 20$ and $N_O = 100$ and 125 for $IS_{WO} = 50$ show State III as a stable state. The observations for the stable states can be summarized in Table 1.

Table 1. Kinetically stable states observed from the simulations in Figure 4.

IS_{WO} \ NO	10	25	50	75	100	125
0.1	State I	State I	State I	State I	State I	State I
1	State I	State I	State I	State I	State I	State I
5	State I	State I	State I	State I	State I	State I
10	State I	State I	State I	State II	State II	State II
20	metastable	State II	State II	State II	State II	State III
50	State II	State II	State II	State II	State III	State III

Table 1 shows that when the interaction strength IS_{WO} and number of osmolytes N_O increase, the kinetically stable state shifts from State I to State II to State III. For example, from the right-most column ($N_O = 125$) in Table 1, we easily note that State I, State II, and State III appear in order when IS_{WO} increases. Similarly, if we examine the states along the row of $IS_{WO} = 10$ (the states with the same IS_{WO} value of 10) in the table, we find that as N_O increases, State I and State II appear in order. Here, one interesting question is whether we can observe State III if we increase the number of osmolytes. To address this question, we could insert more osmolytes into Compartment 2. However, since the compartment space is limited in terms of the number of osmolytes, we use an alternative method in which we increase the ratio of the number of osmolytes to the number of water molecules (N_O/N_{total}) by reducing the total number of water molecules instead of increasing the number of osmolytes.

To determine if higher ratios of N_O/N_{total} in the cases of $IS_{WO} = 10$ induce State III, we prepare systems with fewer water molecules based on the system of $N_O = 125$, $IS_{WO} = 10$ and $N_{total} = 884$ ($N_O/N_{total} = 0.141$). In other words, from the final state at 200 ns shown in Figure 4, we remove some of the water molecules, so that $N_{total} = 850$, 800, 750, 700, 650, and 600, which implies that

N_O/N_{total} = 0.147, 0.156, 0.167, 0.179, 0192, and 0.208, respectively. From the 200 ns simulations of the systems, we calculate the occupancy ratio of water in Compartment 1 (N_1/N_{total}) and the number of water molecules in the CNT (N_{CNT}). With the cases of N_{total} =884 in Figure 4, we summarize the simulation results as a function of N_O/N_{total} in Figure 7. The results clearly indicate that at high values of N_O/N_{total} (>~0.17), the stable states are State III. In other words, N_1/N_{total} and N_{CNT} are practically zero. This result may suggest that in the original system of 884 water molecules, if we include more osmolytes, i.e., the ratio of N_O/N_{total} increases, we will observe State III as a stable state.

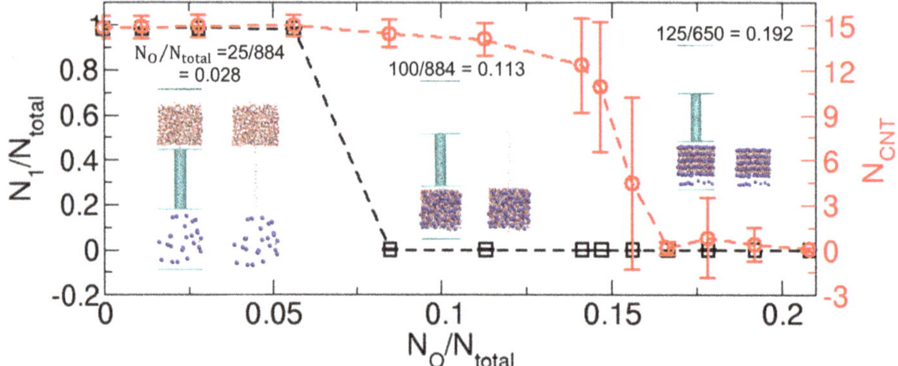

Figure 7. Kinetically stable states observed for a range of N_O/N_{total}, the ratio of the number of osmolytes (N_O) to the total number of water molecules (N_{total}). The stable states are characterized by the occupancy ratio of water molecules in Compartment 1 (N_1/N_{total}) and the number of water molecules in the CNT (N_{CNT}), averaged over the time interval from 100 ns to 200 ns in the 200 ns simulations. The insets illustrate three representative configurations of States I, II, and III, whose values of N_O/N_{total} are 0.028, 0.113, and 0.192, respectively, which are taken at 200 ns from the 200 ns simulations. The error bars indicate the standard deviations.

Another interesting feature from Figure 7 is that the number of water molecules in the CNT largely fluctuates in the transition regime between State II and State III, whose N_O/N_{total} value is approximately 0.15. This fluctuation in N_{CNT} is a typical behavior observed in the wetting-dewetting transition [20].

One remaining issue that we have to address is whether the state observed in the case of N_O = 10 and IS_{WO} = 20 is a global equilibrium state or a metastable state. Here, a metastable state means that the state eventually transitions to another more stable state if we wait. Considering the stable states observed for other values of N_O (see the row of IS_{WO} = 20 in Table 1), one would guess that State II is a more stable (equilibrium) state.

To examine whether State II is a more stable state for the case of N_O = 10 and IS_{WO} = 20, we first prepare five such initial states by removing some osmolytes from the final states (State II) with N_O = 25, 50, 75, 100, and 125 and IS_{WO} = 20 at 200 ns in Figure 4, and run MD simulations to determine whether the states characterized as State II are stable. The steady state in terms of N_1 and N_{CNT} in Figure 8a indicates that State II is kinetically stable. The relevant configurations for the originally observed state and State II are displayed in Figure 8b,c, respectively. Then we compare the potential energies between the original state and State II, as displayed in Figure 8d. From the comparison, we see that the potential energies of State II are lower than the potential energy of the original state observed in Figure 4. In particular, since the potential energy difference (~700 kJ/mol on average) is significant, we conclude that the stable state in Figure 4 is a metastable state, and State II is a global equilibrium state. The physical reason for the stability in the metastable state is probably associated with the state being stuck in a local energy minimum, which implies that for further investigations, we may need to examine the spatial arrangements of osmolytes and water molecules in detail.

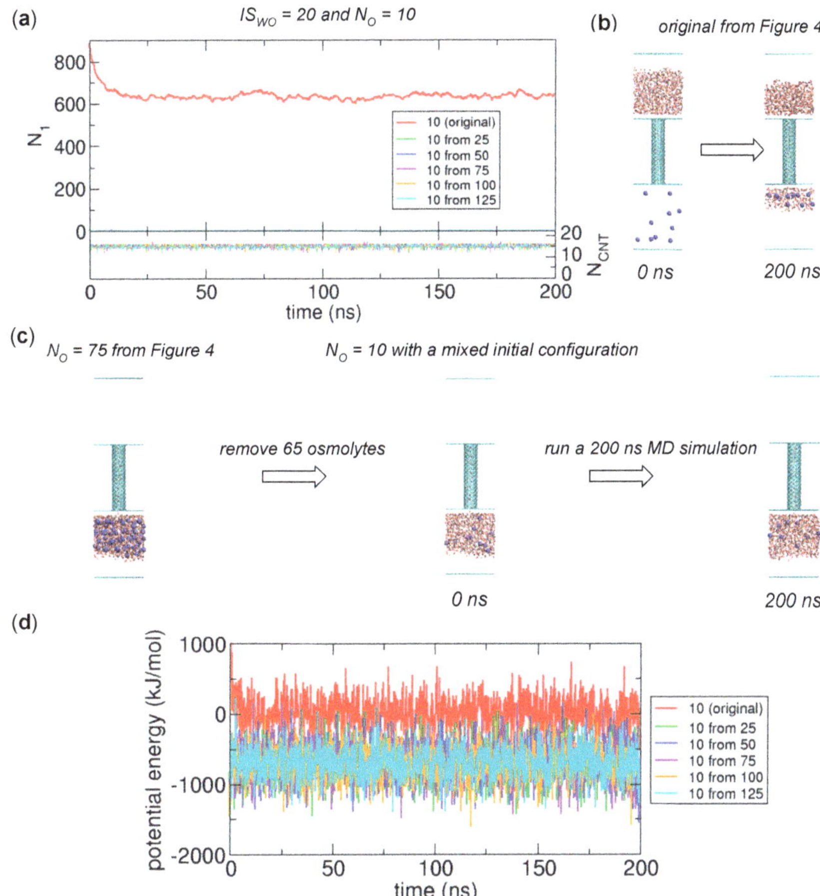

Figure 8. Equilibrium states for the case of $N_O = 10$ and $IS_{WO} = 20$ from the MD simulations with mixed initial configurations. The mixed initial configurations are prepared by removing some osmolytes from the final configurations at 200 ns in Figure 4 for the cases of N_O = 25, 50, 75, 100, and 125 and IS_{WO} = 20. (**a**) Numbers of water molecules in Compartment 1 (N_1) and the CNT (N_{CNT}) as functions of time for the cases with the original initial configuration (red) and mixed configurations (other colors). (**b**) Original initial configuration and final configuration at 200 ns obtained from the simulations in Figure 4. (**c**) Mixed initial configuration prepared from the final configuration with $N_O = 75$ at 200 ns in Figure 4. (**d**) Changes in potential energy as functions of time. In this plot, the average value of the potential energy for the original case over the time interval from 100 to 200 ns is set to zero.

Finally, if we consider the water transport process as a process to reach an equilibrium state, we can regard the water transport as a transition process from the initial state to State II or State III. In particular, we can call the transition to State III complete transfer of water because all water molecules are transferred from Compartment 1 to Compartment 2 due to the osmolytes.

3.2.3. Stochastic Nature of Water Transport Occurrence

Figure 4 shows that water transport occurs at a very early stage of the simulations when IS_{WO} is large (20 or 50), while transport does not occur when IS_{WO} is small (0.1, 1, or 5). However, when

IS_{WO} is intermediate (10), water transport can occur or not depending on the number of osmolytes N_O; in Figure 4, it occurs only when N_O is greater than 75.

For IS_{WO} = 10, to better understand the N_O criterion for the occurrence of water transport, we examine more systems between the two systems with N_O = 75 (transport observed) and 50 (no transport) in Figure 4, which correspond to N_O = 70, 65, 60, 55, and 50. For each system, we perform five independent simulations. The simulation results are shown in Figure 9.

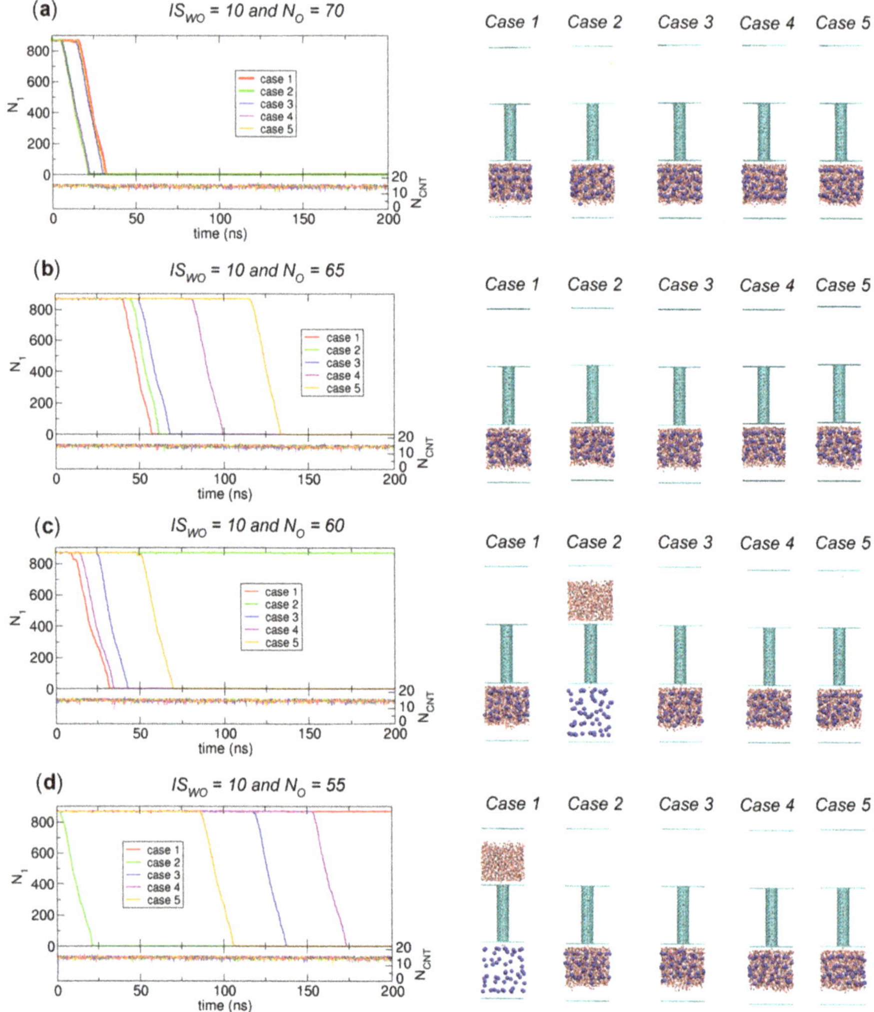

Figure 9. *Cont.*

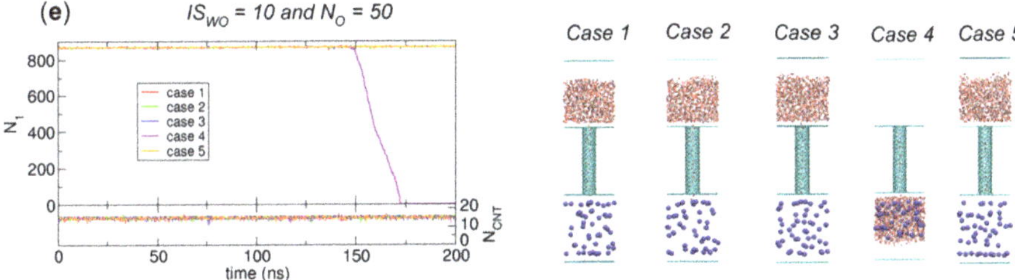

Figure 9. Results of 200 ns simulations for the systems of N_O = (**a**) 70, (**b**) 65, (**c**) 60, (**d**) 55, and (**e**) 50 with IS_{WO} = 10. For each system, we perform five independent simulations, and we plot the numbers of water molecules in Compartment 1 (N_1) and the CNT (N_{CNT}) as functions of time (left) with the final configurations at 200 ns from the five simulations (right).

Interestingly, for the systems of N_O = 65, 60, and 55, the occurrence clearly shows a stochastic nature, which means that simulations with a given N_O can exhibit water transport or not, and moreover, in the cases where water transport occurs, the waiting time for water transport to occur could vary. For example, when N_O = 55, four cases of five display water transport, and their waiting times are 0.2 ns (Case 2), 117.8 ns (Case 3), 153.8 ns (Case 4), and 86.4 ns (Case 5).

For N_O = 70, however, we observe water transport in all five cases, and the waiting times are relatively narrowly distributed (16.2 ns (Case 1), 6.0 ns (Case 2), 14.2 ns (Case 3), 6.2 ns (Case 4), and 16.0 ns (Case 5)). In contrast, for the cases of N_O = 50, in only one case (Case 4) is water transport observed. From this comparison, as N_O decreases, the waiting time increases in that the waiting time is at least 200 ns for the case with no transport. The observation with N_O = 50 gives rise to a question: if we consider a longer simulation time, such as 1000 ns, than the value of 200 ns in Figure 9, can we observe more cases showing water transport? To address this question, we further examine longer simulations for more cases in which one might expect no water transport from the 200 ns simulation results in Figure 4.

We perform 1000 ns simulations for N_O = 52, 50, 45, 40, and 25. For each N_O, we perform five independent simulations. We display the simulation results in Figure 10. The simulation results for N_O = 50 and 45 indicate that up to 200 ns, only two cases of five and only one case exhibit water transport, respectively, but up to 1000 ns, for each N_O, four cases of five show water transport. Therefore, when we increase the simulation time, the probability for water transport increases. Figure 10 also shows that when N_O decreases, the number of water transport occurrences is reduced: five, four, four, three, and zero occurrences for N_O = 52, 50, 45, 40, and 25, respectively. Therefore, from the above discussion, we see the general trend that water transport is more likely to be observed when we include more osmolytes and wait longer; moreover, the waiting time is reduced when more osmolytes are added. Thus, although we do not observe water transport for N_O = 25 in Figure 10, its observation may be possible if we wait much longer.

Finally, related to the discussion of kinetically stable states in Figure 7, since the CNT is occupied by water at all times in Figures 9 and 10, the transition due to the water transport in Figures 9 and 10 is from the initial state to State II. Figure 10 also shows that only the system of N_O = 25, which corresponds to N_O/N_{total} = 0.028 (= 25/884), does not show water transport. Therefore, to make Figure 7 more accurate, we must adjust the N_O/N_{total} criterion value to distinguish between States I and II. However, since determining the exact value of N_O/N_{total} for the boundary requires large-size ensembles and long-time simulations, which is not our major interest, we do not attempt to update Figure 7 based on the results in Figures 9 and 10. Again, the main point of Figure 7 is that mainly three kinetically stable states are observed.

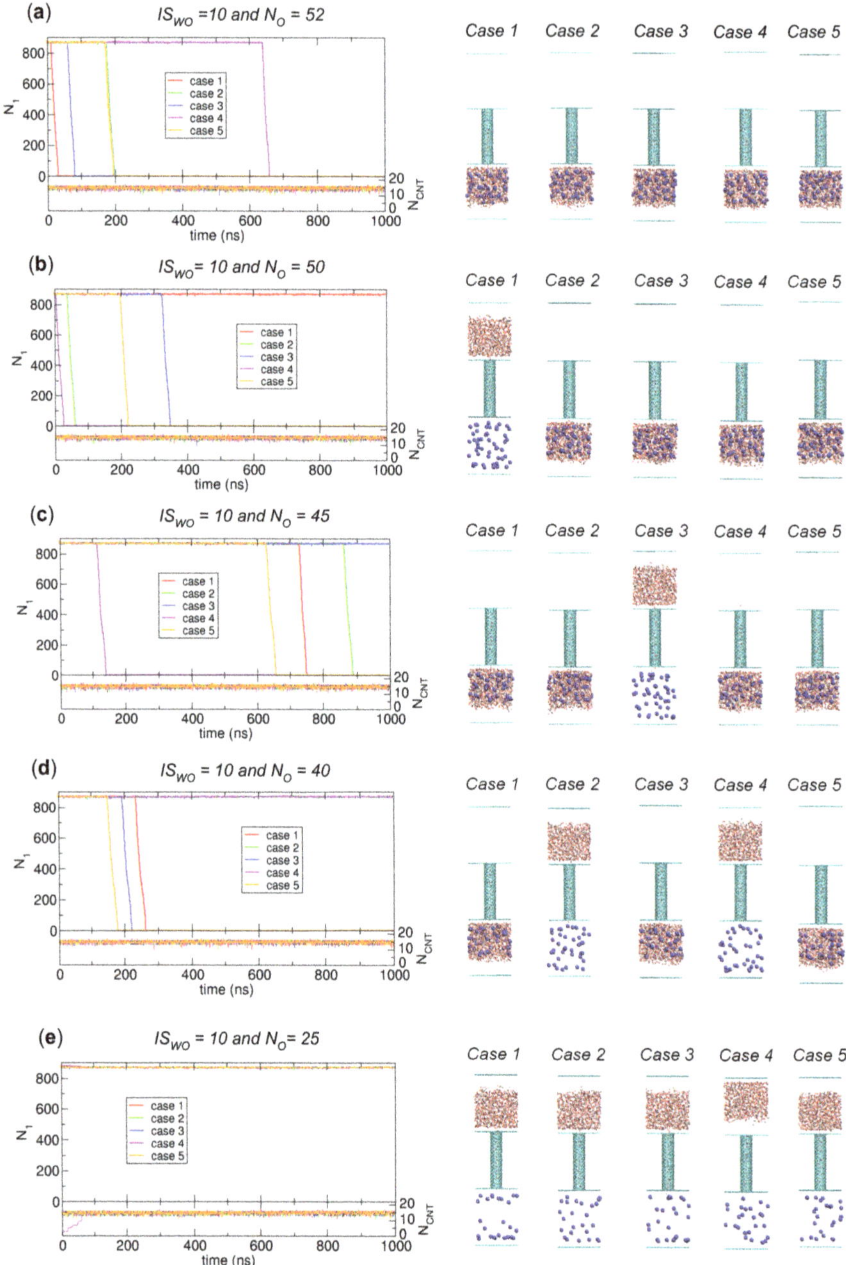

Figure 10. Results of 1000 ns simulations for the systems of N_O = (**a**) 52, (**b**) 50, (**c**) 45, (**d**) 40, and (**e**) 25 with IS_{WO} = 10. For each system, we perform five independent simulations, and we plot the numbers of water molecules in Compartment 1 (N_1) and the CNT (N_{CNT}) as functions of time (left) along with the final configurations at 1000 ns from the five simulations (right).

3.2.4. Characteristics of the Transition due to Water Transport

Here, we further study the details of water transport observed in Figures 4, 9 and 10. Interestingly, during water transport, the number of water molecules in Compartment 1 appears to linearly decrease with time. To quantify this time-transient behavior, we fit the data to a linear curve. One example is shown in Figure 11a for the case of N_O = 75 and IS_{WO} = 10 (also see Figure 4), where the data of N_1 versus time t are best fitted with a linear curve using curve fitting in the xmgrace program. The resulting linear curve is $N_1 = -52.3t + 3179.6$. This fitting is only valid for the time interval corresponding to the water transport. Remarkably, in this case, the correlation coefficient is greater than 0.99. Moreover, we calculate the transition time, which is the time for the transition from the initial state to State II or State III due to water transport. In this case, the transition time is 17.0 ns (= 60.8 ns − 43.8 ns). Additionally, from the linear fitting, we can determine the slope, which gives the transport rate. In this case, the transport rate is 52.3 water molecules/ns.

Since the transition time, linearity of the time-transient behavior, and transport rate can depend on the number of osmolytes N_O, we calculate them as functions of N_O. The results are shown in Figure 11. Interestingly, according to Figure 11b, when the number of osmolytes is sufficiently large (N_O > ~60), the transition time is ~ 20 ns almost regardless of the interaction strength IS_{WO}. Thus, if the osmotic force is sufficiently large beyond a certain value, the transition time reaches a limit, which is ~20 ns in this case. However, more osmolytes can reduce the waiting time for water transport, as we discuss in 3.2.3. When the number of osmolytes is smaller (N_O < ~60), the transition time depends on both N_O and IS_{WO}. In other words, when the osmotic force decreases by reducing N_O or IS_{WO}, the transition time increases, or no transition is observed.

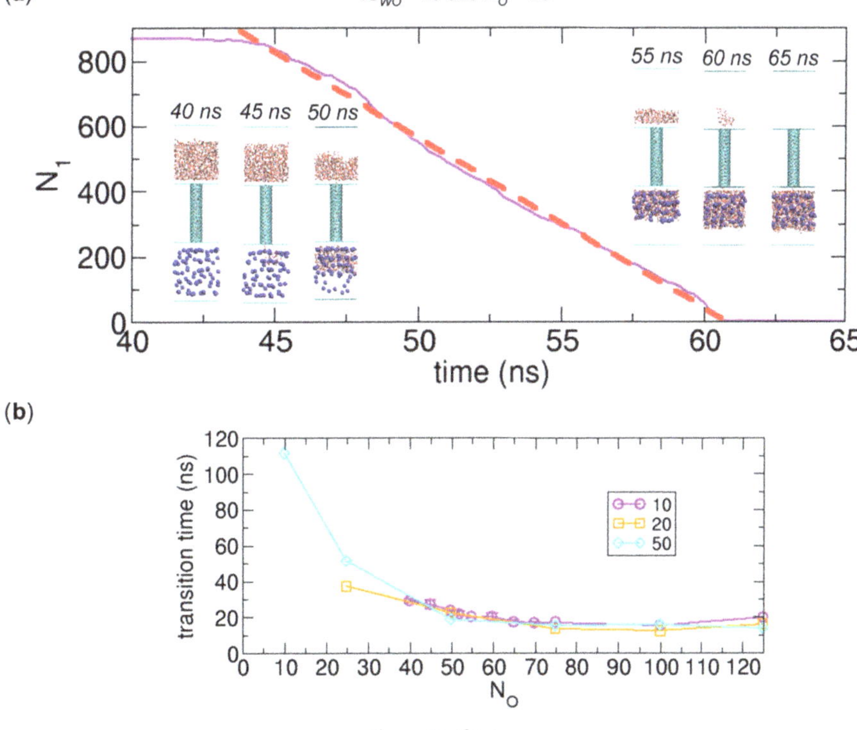

Figure 11. *Cont.*

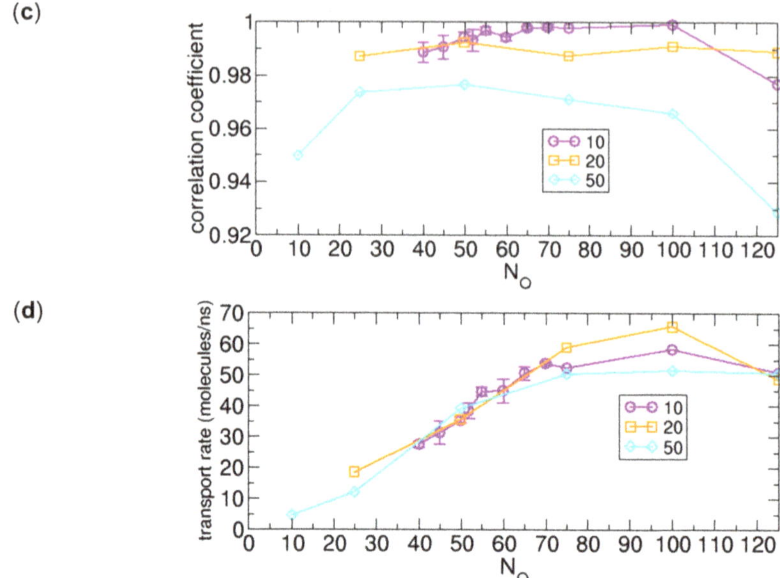

Figure 11. Detailed analysis of water transport observed in Figure 4, Figure 9, and Figure 10. (**a**) Linear curve fitting for the data of N_1 versus time t (ns). The curve fitting is applied only to the water transport region. The resulting linear curve is $N_1 = -52.3t + 3179.6$ and the correlation coefficient is 0.998. The configurations in the insets are provided to show what occurs during water transport. (**b**) Transition time from the initial state to State II or State III due to water transport (or duration of water transport) as functions of the number of osmolytes (N_O) for IS_{WO} = 10, 20, and 50. (**c**) Correlation coefficient calculated from the linear curve fitting as a function of N_O for IS_{WO} = 10, 20, and 50. (**d**) Transport rate obtained from the slopes of the fitted linear curves as a function of N_O for IS_{WO} = 10, 20, and 50. When multiple data sets for water transport are available for one system (e.g., N_O = 50 and IS_{WO} = 10 in Figures 9 and 10), we calculate the average over the multiple sets and the associated standard deviation and plot them in the figures.

In Figure 11c, we quantitatively analyze the linearity of the time-transient behavior in water transport by calculating the correlation coefficient from the linear curve fitting to the data of N_1 versus time t. Surprisingly, when IS_{WO} = 10, the correlation coefficient is very close to 1, which implies a constant flow of water during the transport. However, as IS_{WO} increases or when N_O is large, the transient behavior slightly deviates from linearity. This may reflect the nonlinear nature of interactions between molecules involved in water transport.

As shown in Figure 11d, we calculate the transport rate from the slope obtained from the linear fitting shown in Figure 11a. Noticeably, the three plots for IS_{WO} = 10, 20, and 50 are similar, and the common features are that when $N_O > \sim 60$, the transport rate is almost a constant value (~53 water molecules/ns) with small variations and when $N_O < \sim 60$, it decreases as N_O decreases. Note that the plots in Figure 11b,d are inversely related to each other. The similarity of the transport rates regardless of IS_{WO} for the region of $\sim 40 < N_O < \sim 60$ is an interesting feature, but to better understand its physical origin, further detailed analysis may be required, which is beyond the scope of this work.

3.2.5. Effect of Interactions between Water Molecules on Transport

To understand how the interaction between water molecules affects the transport through a nanochannel, we compare the transport of water molecules with the transport of charge-removed water molecules in terms of the number of molecules in Compartment 1 as a function of time. The results are displayed in Figure 12. Here, water and charge-removed water molecules represent strongly interacting and weakly interacting transported molecules, respectively. Charge-removed water molecules are prepared by removing the electric charges of water molecules or setting the electric charges to zero. The transport of charge-removed water molecules was discussed in detail in our previous work [1].

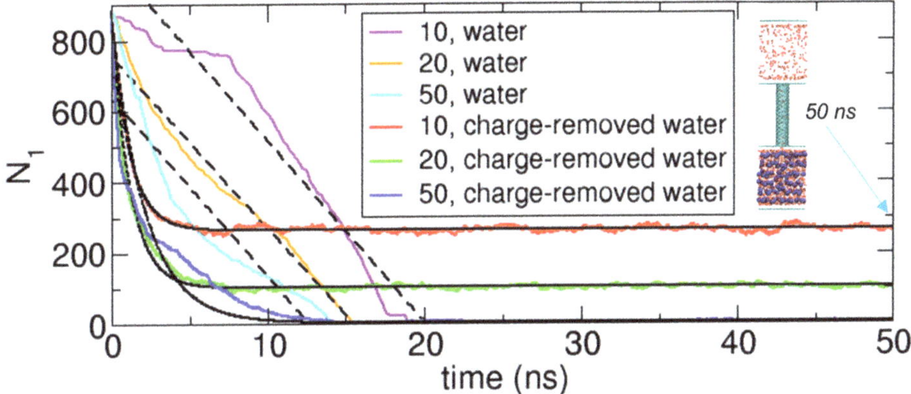

Figure 12. Comparison of the transport of water molecules with the transport of charge-removed water molecules [1] in terms of the numbers of molecules in Compartment 1 (N_1) as functions of time for the cases of N_O =125 and IS_{WO} = 10, 20, and 50. The black lines represent the linear fitting for the transport of water molecules and the exponential fitting for the transport of charge-removed water molecules. The inset figure shows the equilibrium state at 50 ns for the charge-removed water case of N_O = 125 and IS_{WO} = 10.

In Figure 12, first, we can compare the cases of water with the cases of charge-removed water in terms of equilibrium states. At equilibrium, while the number of molecules in Compartment 1 (N_1) for water is zero, (N_1 for charge-removed water is not zero when IS_{WO} = 10 and 20; only when IS_{WO} = 50 does N_1 fluctuate near zero. Without electrostatic interactions, the interactions between charge-removed water molecules are weak, and thus, their physical behavior is similar to that of a gas. Therefore, if the interaction between molecules and osmolytes is not sufficiently strong, then the molecules can fill up the space, and as a result, when IS_{WO} = 10 and 20, some water molecules remain in Compartment 1 (see the inset of Figure 12). When IS_{WO} increases, N_1 is reduced, and especially when IS_{WO} = 50, N_1 is near zero. However, for water, because of the strong interactions, its behavior is liquid-like, and aggregation of water molecules is energetically favorable. Moreover, because of the osmolytes in Compartment 2, being in Compartment 2 is more favorable than being in Compartment 1 for water molecules. Therefore, combining these factors, we understand that water molecules transfer to Compartment 2, and thus, N_1 for water is zero.

The difference in strength of the interaction between transported molecules can also explain the different time-transient behaviors in Figure 12. As we discuss in Section 3.2.4, the time-transient behavior of water is well fitted to a linear function of time, although some deviations from linearity occur when the interaction between water molecules and osmolytes is very strong. This linearity implies a constant transport rate of water molecules, which means that this rate does not depend on the concentration of water in Compartment 1 (zeroth-order rate). Physically, this occurs because water molecules tend to aggregate, so the local concentration of water is almost constant regardless of the available space in Compartment 1 (see the inset of Figure 11a). In other words, near the CNT, water molecules are continuously supplied for transport, compensating for the loss of molecules due to transport. In a sense, water transport through a nanochannel driven by osmosis is similar to pulling a string at a constant speed.

However, as we discussed in our previous work [1], the time-transient behavior of charge-removed water is well fitted to an exponential function of time, not a linear function, because the transport rate is proportional to the concentration (first-order rate). For the transport to occur, charge-removed water molecules in Compartment 1 should approach the CNT, and the probability of going into the CNT is proportional to the number of molecules in Compartment 1. Thus, as more molecules are transferred to Compartment 2, the concentration of molecules in Compartment 1 is reduced, and as a result, the transport rate decreases with time. Particularly, if we assume that the rate is proportional to the concentration of charge-removed water, then we mathematically show that the transient behavior exactly follows an exponential function of time. Therefore, the comparison in Figure 12 indicates that the interaction between transported molecules apparently affects the transport kinetics.

4. Summary and Conclusions

In this work, we have studied the transport of water molecules from a nanometer-sized compartment (Compartment 1) to another same-size empty compartment (Compartment 2) through a CNT using MD simulations. If the transported molecules are gas molecules, which weakly interact with each other, then transport would spontaneously occur such that almost half of the gas molecules would be in one compartment and the other half would be in the other compartment. However, for water molecules, transport to Compartment 2 does not occur without any external force. This is because the chemical potential of water in Compartment 2 is higher than that in Compartment 1 and the CNT.

To induce water transport, we used an osmotic force as an external force by introducing osmolytes in Compartment 2. However, the attractive interactions between Compartment 1 and water molecules can terminate the water transport before all water molecules in Compartment 1 are transported such that some of the water molecules remain next to the walls of Compartment 1. Thus, to transport more water molecules, we weakened the compartment-water interaction by reducing the LJ interaction parameter. With osmolytes in Compartment 2 and weak compartment-water interactions, we were able to observe complete or nearly complete transfer of water molecules to Compartment 2 from extensive MD simulation studies.

To systematically study the effect of osmolytes, we considered various osmotic environments characterized by the number of osmolytes (N_O) and the strength of the interaction between osmolytes and water molecules (IS_{WO}). The extensive case studies indicate that N_O and IS_{WO} appear to have effective threshold values for water transport to occur within a certain time frame in that when N_O and IS_{WO} are sufficiently large, water transport is always observed in the simulations. However, the occurrence of water transport is intrinsically stochastic, and thus, in practice, the chance to observe water transport depends on the length of the simulation time and the number of simulations. For example, for a given system, if one performs more simulations for a longer simulation time, then observation of water transport is more probable.

As a result of water transport, the system transitions from the initial state to a stable state. We found that depending on N_O and IS_{WO}, mainly three kinetically stable states (States I, II, and III) are observed. In the presence of osmolytes in Compartment 2, it is more energetically favorable for water molecules to be in Compartment 2 with osmolytes (State II or State III depending on the occupancy of the CNT) than in Compartment 1 (State I). Therefore, the water transport leading to State II or State III is thermodynamically favorable, but the stability of State I observed in our simulations indicates the existence of a kinetic barrier for the transition to State II or State III, which is the physical origin of the stochastic behavior in the occurrence of water transport.

We also investigate the transition process from the initial state to State II or State III, in which complete or nearly complete transfer of water molecules to Compartment 2 occurs. As expected, as the number of osmolytes increases, the transition time decreases while the transport rate increases. Interestingly, the interaction strength of osmolytes IS_{WO} appears to be crucial in the initiation of the process, but once the transition process is initiated, the number of osmolytes N_O contributes more to determining the transport rate. Additionally, as IS_{WO} and N_O increase, the transport rate seems to saturate.

One of the most interesting findings from this work is that the kinetics of water transport is zeroth-order, while the kinetics of charge-removed water transport is first-order. Here, water and charge-removed water (water without electric charge) represent strongly and weakly interacting molecules, respectively. In other words, the strength of the interaction between the transported molecules can significantly affect the transient behavior of the transport. Physically, the zeroth-order kinetics in water transport means a constant flow of water, which is possible due to the strong attractions between water molecules; one molecule is followed by another, and in this sense, water transport through a nanochannel by osmosis is similar to pulling a string at a constant speed.

In this work, we used a minimal model system that has only the components necessary for producing water transport through a nanochannel by osmosis. Therefore, based on this basic model, we can extend the study to various related topics by modifying the model. For example, in the basic model, we used nonpolar osmolytes, and therefore the interactions between water molecules and osmolytes were simply described by the LJ interactions. However, in reality, osmolytes could be polar or charged molecules. Therefore, in future studies, water transport driven by charged osmolytes could be an interesting subject. Another interesting subject would be the water transport and equilibrium states for the system containing osmolytes in Compartment 1 as well as in Compartment 2.

Funding: This work was supported by the National Research Foundation of Korea (NRF) grant funded by the Korea government (MSIT) (No. 2020R1F1A1070163) and the Hankuk University of Foreign Studies Research Fund of 2020.

Conflicts of Interest: The author declares no conflict of interest.

References

1. Eun, C. Equilibration of molecules between two compartments through a nanochannel in the presence of osmolytes: A molecular dynamics simulation study. *Phys. Chem. Chem. Phys.* **2019**, *21*, 21136–21151. [CrossRef] [PubMed]
2. Agre, P. The aquaporin water channels. *Proc. Am. Thorac. Soc.* **2006**, *3*, 5–13. [CrossRef]
3. Gomes, D.; Agasse, A.; Thiébaud, P.; Delrot, S.; Gerós, H.; Chaumont, F. Aquaporins are multifunctional water and solute transporters highly divergent in living organisms. *Biochim. Biophys. Acta—Biomembr.* **2009**, *1788*, 1213–1228. [CrossRef] [PubMed]
4. Verkman, A.S. Aquaporins. *Curr. Biol.* **2013**, *23*, R52–R55. [CrossRef] [PubMed]
5. Corry, B. Designing carbon nanotube membranes for efficient water desalination. *J. Phys. Chem. B* **2008**, *112*, 1427–1434. [CrossRef] [PubMed]
6. Chandler, D. *Introduction to Modern Statistical Mechanics*; Oxford University Press: Oxford, UK, 1987.
7. Kalra, A.; Garde, S.; Hummer, G. Osmotic water transport through carbon nanotube membranes. *Proc. Natl. Acad. Sci. USA* **2003**, *100*, 10175–10180. [CrossRef] [PubMed]
8. Thomas, J.A.; McGaughey, A.J.H. Reassessing fast water transport through carbon nanotubes. *Nano. Lett.* **2008**, *8*, 2788–2793. [CrossRef] [PubMed]
9. Wang, L.; Dumont, R.S.; Dickson, J.M. Nonequilibrium molecular dynamics simulation of pressure-driven water transport through modified CNT membranes. *J. Chem. Phys.* **2013**, *138*, 124701. [CrossRef] [PubMed]
10. Liu, B.; Wu, R.; Baimova, J.A.; Wu, H.; Law, A.W.K.; Dmitriev, S.V.; Zhou, K. Molecular dynamics study of pressure-driven water transport through graphene bilayers. *Phys. Chem. Chem. Phys.* **2016**, *18*, 1886–1896. [CrossRef]
11. Cornell, W.D.; Cieplak, P.; Gould, I.R.; Merz, K.M.; Ferguson, D.M.; Spellmeyer, D.C.; Fox, T.; Caldwell, J.W.; Kollman, P. A Second Generation Force Field for the Simulation of Proteins, Nucleic Acids, and Organic Molecules. *J. Am. Chem. Soc.* **1995**, *117*, 5179–5197. [CrossRef]
12. Hummer, G.; Rasaiah, J.C.; Noworyta, J.P. Water conduction through the hydrophobic channel of a carbon nanotube. *Nature* **2001**, *414*, 6860. [CrossRef] [PubMed]
13. Jorgensen, W.L.; Chandrasekhar, J.; Madura, J.D.; Impey, R.W.; Klein, M.L. Comparison of simple potential functions for simulating liquid water. *J. Chem. Phys.* **1983**, *72*, 926–935. [CrossRef]

14. Essmann, U.; Perera, L.; Berkowitz, M.L.; Darden, T.; Lee, H.; Pedersen, L.G. A smooth particle mesh Ewald method. *J. Chem. Phys.* **1995**, *103*, 8577. [CrossRef]
15. Abraham, M.J.; van der Spoel, D.; Lindahl, E.; Hess, B.; The G. Development Team. GROMACS User Manual Version 2018. 2018. Available online: www.gromacs.org (accessed on 1 January 2019).
16. Abraham, M.J.; Murtola, T.; Schulz, R.; Páll, S.; Smith, J.C.; Hess, B.; Lindah, E. Gromacs: High performance molecular simulations through multi-level parallelism from laptops to supercomputers. *SoftwareX* **2015**, *1*, 19–25. [CrossRef]
17. Bussi, G.; Donadio, D.; Parrinello, M. Canonical sampling through velocity rescaling. *J. Chem. Phys.* **2007**, *126*, 014101. [CrossRef] [PubMed]
18. Humphrey, W.; Dalke, A.; Schulten, K. VMD: Visual molecular dynamics. *J. Mol. Graph.* **1996**, *14*, 33–38. [CrossRef]
19. Eun, C.; Berkowitz, M.L. Molecular dynamics simulation study of interaction between model rough hydrophobic surfaces. *J. Phys. Chem. A* **2011**, *115*, 6059–6067. [CrossRef] [PubMed]
20. Eun, C.; Berkowitz, M.L. Fluctuations in number of water molecules confined between nanoparticles. *J. Phys. Chem. B* **2010**, *114*, 13410–13414. [CrossRef] [PubMed]

International Journal of
Molecular Sciences

MDPI

Article

Molecular Insight into the Possible Mechanism of Drag Reduction of Surfactant Aqueous Solution in Pipe Flow

Yusei Kobayashi *, Hirotaka Gomyo and Noriyoshi Arai

Department of Mechanical Engineering, Keio University, 3-14-1 Hiyoshi, Kohoku-ku, Yokohama 223-8522, Japan; h.gomyo8@gmail.com (H.G.); arai@mech.keio.ac.jp (N.A.)
* Correspondence: kobayashi@mech.keio.ac.jp

Abstract: The phenomenon of drag reduction (known as the "Toms effect") has many industrial and engineering applications, but a definitive molecular-level theory has not yet been constructed. This is due both to the multiscale nature of complex fluids and to the difficulty of directly observing self-assembled structures in nonequilibrium states. On the basis of a large-scale coarse-grained molecular simulation that we conducted, we propose a possible mechanism of turbulence suppression in surfactant aqueous solution. We demonstrate that maintaining sufficiently large micellar structures and a homogeneous radial distribution of surfactant molecules is necessary to obtain the drag-reduction effect. This is the first molecular-simulation evidence that a micellar structure is responsible for drag reduction in pipe flow, and should help in understanding the mechanisms underlying drag reduction by surfactant molecules under nonequilibrium conditions.

Keywords: drag reduction; surfactant molecules; self-assembly; coarse-grained molecular simulation

Citation: Kobayashi, Y.; Gomyo, H.; Arai, N. Molecular Insight into the Possible Mechanism of Drag Reduction of Surfactant Aqueous Solution in Pipe Flow. *Int. J. Mol. Sci.* **2021**, *22*, 7573. https://doi.org/10.3390/ijms22147573

Academic Editor: Małgorzata Borówko

Received: 7 June 2021
Accepted: 12 July 2021
Published: 15 July 2021

1. Introduction

In the 21st century, soft-matter rheology is recognized as a vitally important field with applications to engineering (e.g., food [1,2], cosmetics [3], medical materials [4]), biology (e.g., strain hardening of fibrin [5] and the motion of motor proteins [6,7]), and the global environment (e.g., mantle flow [8,9] and the origin of life [10,11]). However, the behavior of soft matter is difficult to understand because it encompasses phenomena on multiple spatiotemporal scales, and rheology involves the study of inherently nonequilibrium phenomena. Thus, there are major barriers to understanding either separately, much less in combination; soft-matter rheology remains a challenging subject. The pioneering work of De Gennes [12], and Doi and Edwards [13–15] in the late 1970s sparked interest in explaining the rheological properties of entangled polymer melts by advanced physical modeling. Their "tube model" was able to explain, to a certain extent, the relaxation dynamics of entangled polymers. However, quantitative tube-model predictions for complex polymers, including branched and di-block copolymers and blends, are still not possible because they involve molecular details below tube length. In order to predict and understand the rheological properties of actual soft matter, it is essential to incorporate the properties of molecules.

In recent years, computer simulations have been successfully used to reproduce the behavior of molecules inside complex soft matter and to clarify the source of their rheology [16,17]. For example, theoretical expressions describing the plateau moduli of slip-link and slip-spring models were proposed by Uneyama and Masubuchi [18]; reasonable agreement between their theory and simulations has been confirmed. Numerical simulation [19] showed that shear can promote the crystallization of colloidal star polymers in the vicinity of their glass transition, and that a transition from a bcc to an fcc structure can occur.

One of the major unsolved problems in soft-matter rheology is the origin of drag reduction caused by polymers or surfactants, the so-called Toms effect [20], for which a definitive theory has not yet been constructed because molecular-scale details remain unknown. Nevertheless, the Toms effect has many industrial and engineering applications, including district cooling systems, firefighting, and the pipeline transportation of natural gas, water, and crude oil. Since Toms first discovered the effect using polymer solutions [20], extensive and continuing research on drag reduction by additives has been conducted via numerical simulations [21–24] and experiments [25–28]. Among various drag-reducing agents, surfactants have an advantage over polymers from a practical standpoint because surfactant molecules are able to reform micelle structures even after mechanical degradation (except under extreme shear conditions) [29]. The relation between the viscosity behaviors of surfactant aqueous solutions and the formation of micelles was investigated through coarse-grained molecular-dynamics simulations [30–34], but these studies did not provide evidence regarding the frictional coefficient of pipes.

Several possible mechanisms of turbulent drag reduction have been proposed and are summarized in recent reviews [35–37]. In particular, many previous works suggested that a close relationship exists between the viscoelastic behavior of micellar structures and the Toms effect. Nevertheless, despite extensive research conducted on the topic, no universally acceptable mechanism has yet been identified. This is partly because of the usual multiscale problem in soft-matter systems, but also because the direct experimental observation of self-assembled structures of surfactants under nonequilibrium (e.g., turbulent-flow) conditions is an extremely challenging task. In addition, turbulent flow is intrinsically difficult to understand because of the large number of parameters it involves. Hence, most studies provide only phenomenological explanations under certain conditions; a fundamental understanding of the relation between self-assembled structures and the associated drag reduction is still lacking.

In this study, using large-scale dissipative particle dynamics simulation, we study the relationship between the self-assembly of surfactant molecules and their flow properties under pipe flow. Our goal is to understand the mechanism of turbulence suppression in a surfactant aqueous solution from a molecular viewpoint. The structures and distributions of micelles under turbulent flow are investigated, and the necessary conditions to obtain the drag-reduction effect are determined.

2. Model and Methods

2.1. Dissipative Particle Dynamics (DPD) Method

We employed the dissipative particle dynamics (DPD) [38–40] method to study the turbulent drag-reduction effects of a short-chain surfactant aqueous solution in pipe flow using inhouse code. The DPD method can simulate millisecond time scales and micrometer length scales because only the motion of coarse-grained particles (i.e., groups of atoms or molecules) is simulated. To date, many previous studies [41–44] using the DPD method showed that such a coarse-grained model of a surfactant can reproduce self-assembly behavior (e.g., micellar, hexagonal, and lamellar phases) with increasing surfactant concentration.

The fundamental equation of the DPD method is Newton's equation of motion for a particle subject to three types of forces: conservative, dissipative, and random. Details of the DPD method, including the force formula and its application to generic models, are extensively described elsewhere [38–40,45].

2.2. Simulation Model and Conditions

We used a surfactant molecular model (Figure 1a) that contained one hydrophilic head (h) particle and two hydrophobic tail (t) particles. The nearest-neighbor particles in the surfactant molecule were connected by harmonic springs. Spring force $\boldsymbol{F}_{ij}^{\mathrm{S}}$ between the i-th and j-th particles (located at $\boldsymbol{r}_i$ and $\boldsymbol{r}_j$, respectively) is given by

$$\boldsymbol{F}_{ij}^{\mathrm{S}} = -k_{\mathrm{s}}(|\boldsymbol{r}_{ij}| - r_{\mathrm{s}})\boldsymbol{n}_{ij}, \tag{1}$$

where k_s is the spring constant, r_s is the equilibrium bond distance, $\boldsymbol{r}_{ij} = \boldsymbol{r}_j - \boldsymbol{r}_i$, and $\boldsymbol{n}_{ij} = \boldsymbol{r}_{ij}/|\boldsymbol{r}_{ij}|$. In this study, values $k_s = 100\ k_BT/r_c^2$ and $r_s = 0.86\ r_c$ were adopted, where r_c is the cutoff distance. The length of the surfactant molecule calculated from the bond-length distribution and the radial distribution function was approximately 2.5 DPD dimensionless units. The solvent molecular model (Figure 1b) contained a single water (w) particle.

The interaction parameters between any two DPD particles are shown in Table 1. These interactions between any two particles in the solution can be described by the interaction-energy parameters $a_{ww} = a_{tt} = a_{wh} = 25\,k_BT$, $a_{ht} = a_{wt} = 70\,k_BT$, and $a_{hh} = 40\,k_BT$, where w, h, and t represented the water, head group, and tail group, respectively. Hydrophilic and hydrophobic interactions are related to the solubility parameters. For DPD simulations, the interaction (repulsive) parameters between different particles are tuned to reproduce behavior observed in experiments or atomistic simulations. In addition, the interaction parameters for the conservative force between any two particles are related to the Flory–Huggins χ parameters. The choice of these parameters in this study was inspired by the modeling in a previous study of a short surfactant such as cetyltrimethylammonium bromide (CTAB) containing a sodium salicylate (NaSal) solution [46]. This model, with a moderate repulsive force between hydrophilic head groups ($a_{hh} = 40\ k_BT$), can realize both stable threadlike micelle formation and a diffusion coefficient of the surfactant molecules similar to that observed. As the repulsive parameters increase between hydrophilic head groups, the hydration radius may also be estimated to be larger. The same values of interaction parameters were adopted in many studies [43,47–51]. We also examined that our previous bulk simulation of CTAB containing a NaSal solution [47] produced the same results as those in a previous examination that Yamamoto and Hyodo performed [46]. In addition, the surfactant concentration dependence of the self-assembly behavior observed in our previous simulation [43] was consistent with the results of the previously reported experiment [52]. The size (radius and mass) of a single particle has the same value regardless of type [40,53]. The noise amplitude and friction coefficient were set to be 3.0 and 4.5, respectively. The temperature was set at a constant value, i.e., $1.0\,k_BT$.

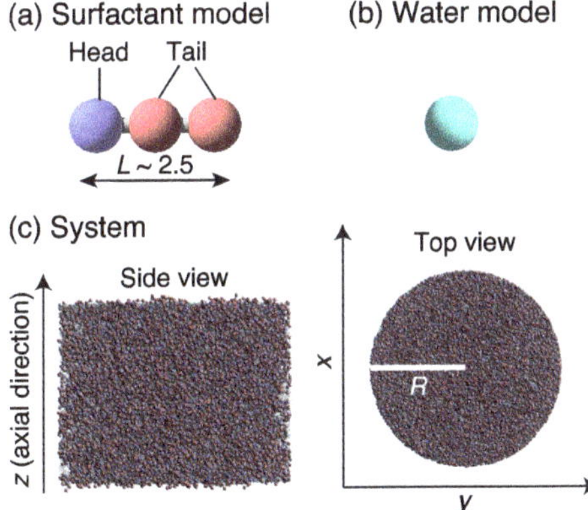

Figure 1. (**a**) Surfactant molecular model composed of one hydrophilic head particle (blue) and two hydrophobic tail particles (red). Length of the surfactant molecule (L), calculated from bond length distribution and the radial distribution function, was approximately 2.5 in the DPD dimensionless unit. (**b**) Water molecular model composed of a single particle (aqua). (**c**) Side and overhead (axial) views of tube system. Inner surface of the cylindrical tube was treated as smooth.

Table 1. Interaction parameters a_{ij} (in $k_B T/r_c$ units) between all pairs.

	h	t	w	Wall
h	40	70	25	25
t	70	25	70	70
w	25	70	25	25
Wall	25	70	25	–

The inner surface of the cylindrical tube was treated as smooth, in agreement with our previous studies [43,48,50,51]. The potential function of the smooth wall was built by summing the DPD force between every solution particle and the wall particles [54]. Integration of this summed force resulted in a force between the DPD particle and the smooth wall (within cutoff distance r_c). The interaction parameters can be seen as a measure of the magnitude of surface energy. The values of the interaction parameter between the hydrophilic wall surface and water, $a_{\mathrm{wall,w}}$, and between the wall and the head group, $a_{\mathrm{wall,h}}$, were both set at $25\,k_B T$. The interaction parameter between wall and tail group, $a_{\mathrm{wall,t}}$, was set at $70\,k_B T$. The radius (R) and length of the tube were 20.0 and 30.0 in dimensionless units, respectively. Density ρ was 5.0; thus, the total number of particles was 188,495. Three surfactant volume fractions (ϕ) were used: 0, 10%, and 30%. The initial configuration for the equilibrium simulations was random (Figure 1c), and a periodic boundary condition was applied in the axial (z) direction of the tube.

For generating pipe flow, the virtual density-gradient method [55] was used. When periodic boundary conditions apply in equilibrium simulations, the original cell is typically attached to copies of itself (image cells) at the boundary to resolve the effects of domain surfaces. In this study, the boundary condition was modified by the elongation and contraction of the image cell, producing a density (or pressure) gradient. As a result, pressure-driven flow was generated. Full details of the procedure are given in [55]. For the range of investigated Reynolds numbers Re, the no-slip boundary condition was satisfied, since we considered the wall surfaces to be hydrophilic in this study. Previous experiments showed that the velocity slip depends on surface hydrophilicity [56], and that the velocities near a hydrophilic microchannel wall agree with those predicted by the no-slip boundary condition [57].

3. Results and Discussion

To obtain initial configurations for the flow simulations, equilibrium simulations of surfactant aqueous solutions were performed at each volume fraction. At rest, spherical and threadlike micelles were observed at $\phi = 0.01$ and $\phi = 0.03$, respectively. These equilibrium morphologies were consistent with those in previous simulation results [50] obtained using a tube model with a 60% smaller radius than the one in this study. Snapshots of these morphologies are shown in Figure S1 of the Supplemental Materials.

Figure 2 shows the frictional coefficient λ of the pipe as a function of Re. The volumetric flow rate, Q, is estimated by applying the cylindrical shell method to a velocity profile [50,51,58] and a generalized Reynolds number is used [59,60], as we focus on the onset point of the transition to turbulence. Here, the power-law parameter, n, is obtained from the relation between the wall shear rate, $\dot{\gamma}_{\mathrm{wall}}$, and the wall shear stress, τ_{wall}, in the steady state (see Figure S2 in the Supplementary Materials). The estimation of flow properties is also described in detail in the Supplementary Materials. For comparison, the theoretical estimate for the drag-reduction rate in laminar flow from the Hagen–Poiseuille law ($\lambda = 64/Re$) is also shown in the figure. For the pure water case ($\phi = 0.00$), the λ values were almost in agreement with the theoretical estimates for $Re \lesssim 250$, but they exceeded the theoretical estimate for a laminar flow with $Re \gtrsim 400$. The main reason for this discrepancy in the transition from laminar to turbulent flow is the compressibility of the DPD fluid. A previous simulation study [61] reported that the onset of the transition to turbulence shifts to a larger Re as compressibility (Mach number Ma) increases.

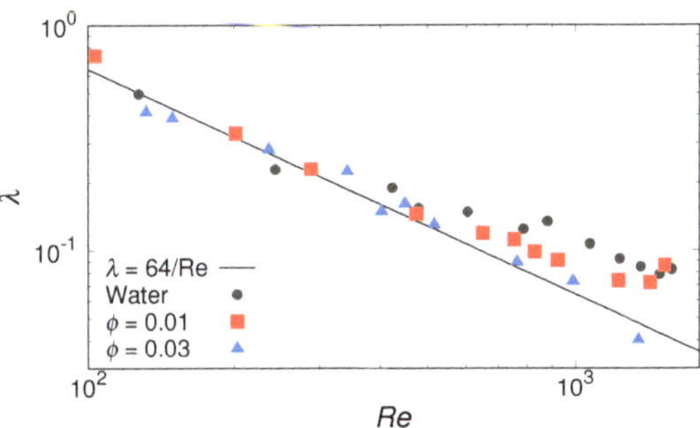

Figure 2. Frictional coefficient of pipe (λ) vs. Reynolds number (Re). Surfactant volume fraction denoted by ϕ. Solid line shows the theoretical estimate for the drag-reduction rate in laminar flow from the Hagen-Poiseuille law. Error bars are smaller than data points.

In this study, we focused on the effect of self-assembled structures of surfactants on the qualitative difference in the onset point of the transition to turbulent flow. When surfactants were added, the transition to turbulence was suppressed for both $\phi = 0.01$ and $\phi = 0.03$. For $\phi = 0.01$, the transition started at a larger Re than that for the pure water case. Further, the frictional coefficients of the pipe at $\phi = 0.01$ were smaller compared to those in the pure water case for $800 \lesssim Re \lesssim 1500$. When increasing the surfactant volume fraction to $\phi = 0.03$, there was near agreement between the theoretical estimate of λ in laminar flow and the simulation results over the entire investigated range of Re; a transition to turbulence was not observed. This ϕ dependence of the frictional coefficient was also confirmed in previous molecular-simulation [62] and experimental [63–65] studies. A saturation concentration of additives ("Virk's asymptote" [66]) may appear; however, only two volume fractions of the surfactant were considered in this study. To confirm that the flow was turbulent and not viscoelastic instability, contour maps of the streamwise velocity averaged over the pipe length at the highest Re for both $\phi = 0.01$ and $\phi = 0.03$ are shown in Figure 3. Blue indicates low-speed streaks, and red indicates high-speed streaks. At $\phi = 0.01$, fast streaks were widely distributed in the radial direction, and distribution behavior was changed as time progressed (Figure 3a,b). At $\phi = 0.03$, contour maps showed typical Poiseuille's flow, and a steady flow was maintained as shown in Figure 3c,d. We also compared the normalized velocity profile for the highest Re at each surfactant volume fraction, as shown in Figure 3e. It was confirmed the influence of turbulent flow in moving the shear gradients to the edge of the pipes, and found that a flattened velocity profile was obtained, similar to "plug flow" at $\phi = 0.01$. Thus, these results correspond with the results of λ vs. Re.

To understand the mechanism of the drag-reduction effect, we next discuss the relation between self-assembly and the transition to turbulent flow. Figure 4 shows representative simulation snapshots of surfactant aqueous solution under pipe flow at $\phi = 0.01$ (panels (a–c)) and $\phi = 0.03$ (panels (d–f)). Here, we consider three flow regimes on the basis of the relation between λ and Re at $\phi = 0.01$. For $Re \lesssim 450$, λ values showed good agreement with the theoretical estimates; this region was defined as the laminar state. For $450 \lesssim Re \lesssim 700$, λ increasingly exceeded the theoretical estimates; this region was defined as the transition state. For $Re \gtrsim 700$, the difference in λ between simulation results and theory was approximately constant; this region was defined as the turbulent state. For comparison, the data for $\phi = 0.03$ were collected at almost the same Re value as for $\phi = 0.01$.

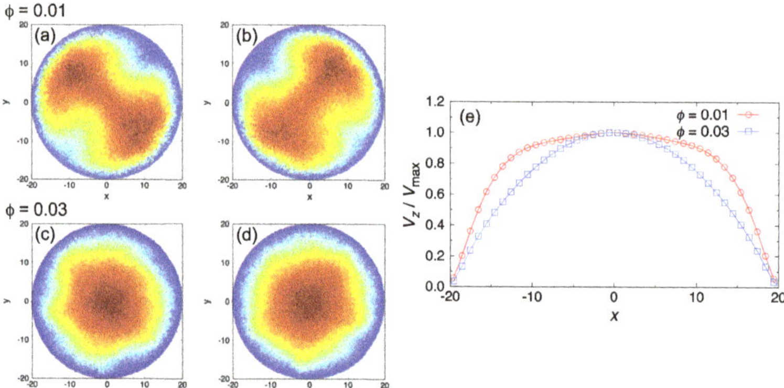

Figure 3. (**a–d**) Snapshots of contour maps of streamwise velocity averaged over the pipe length at the highest *Re*. Blue indicates low-speed streaks, and red indicates high-speed streaks. Two different snapshots (**a**,**b**) for $\phi = 0.01$ at $Re = 1526$, and (**c**,**d**) for $\phi = 0.03$ at $Re = 1344$. (**e**) Comparison of velocity profile for the highest *Re* at each surfactant volume fraction. The vertical axis represents the normalized velocity in the axial (z) direction, V_z/V_{max}, where V_{max} is the maximum velocity of the flow.

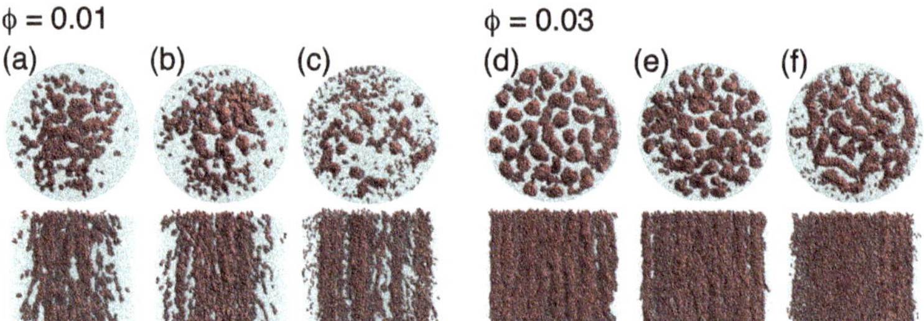

Figure 4. Snapshots of steady-state morphologies of surfactant aqueous solution confined in a hydrophilic tube at volume fractions (**a–c**) $\phi = 0.01$ and (**d–f**) $\phi = 0.03$. (**a**,**d**) Laminar and (**b**,**e**) turbulent regimes correspond to *Re* below 450 or above 700, respectively; region between these values is (**c**,**f**) the transition state. For clarity, hydrophilic head particles are not shown.

At $\phi = 0.01$ in the laminar state, spherical micelles collided with each other and became rodlike, as shown in Figure 4a. This indicates that rodlike micelles maintained laminar flow at higher *Re* when comparing to the pure water case. Previous studies [29,67–69] reported that rodlike micelles are needed for drag reduction; our results support this. For the transition state ($450 \lesssim Re \lesssim 700$), cluster-size probability distribution $P(N_a)$, shown in Figure 5a, showed that flow-enhanced collisions caused the growth of micelles (cluster size $N_a \gtrsim 10^3$) to be formed more than that in the laminar state, but a peak appeared in the distribution in the $1 \lesssim N_a \lesssim 10$ range. As the Reynolds number further increased, the probability of $N_a = 1$ (i.e., of monomers) increased, and the peak of $P(N_a)$ shifted to lower N_a values. Thus, as flow became completely turbulent, micelles became smaller and broke up into monomers. These results suggest that the number and formation of micelles are closely related to the suppression of the turbulent transition. Drag-reduction phenomena depend on the diameter of the tube. Many previous studies [70–74] reported that the tube diameter has an inverse effect on drag-reduction rate. When the tube diameter was increased, larger eddies that cause energy loss were observed. Therefore, in this sense, since the length scale ratio of the micelles to the turbulent eddy size was also a significant factor, it was assumed that the necessary conditions for obtaining the drag-reduction effect that we presented had some impact, even at the same surfactant volume fractions.

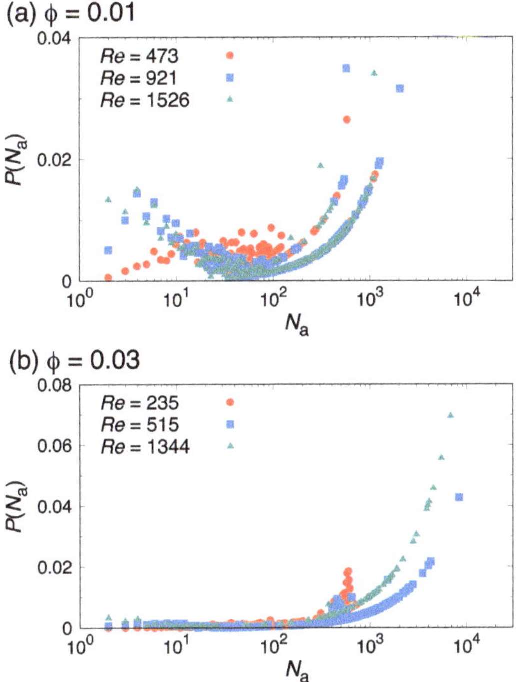

Figure 5. Surfactant cluster size probability distributions $P(N_a)$ at (**a**) $\phi = 0.01$ and (**b**) $\phi = 0.03$ for various Reynolds numbers Re, as indicated.

For a more concentrated system ($\phi = 0.03$) with a relatively low $Re \lesssim 450$ (corresponding to the laminar state when $\phi = 0.01$), threadlike micelles were oriented along the flow (z) direction, as shown in Figure 4d. When the Reynolds number increased to $Re \approx 700$ (corresponding to the transition state when $\phi = 0.01$), the shape of the micelles remained unchanged, as shown in Figure 4e. In contrast to the dilute case ($\phi = 0.01$), only the flow-induced growth of the rodlike micelles appeared; the increase in monomers was not observed (see Figure 5b). For $Re \gtrsim 700$ (corresponding to the turbulent state when $\phi = 0.01$), the rodlike micelles grew further and eventually became sheet-shaped (see top view in Figure 4f). For more quantitative information, we calculated the radius of gyration of a micelle **G** with corresponding eigenvalues $G_1 \geq G_2 \geq G_3$, and then computed the relative shape anisotropy parameter (κ^2), defined as κ^2 is given by

$$\kappa^2 = 1 - 3\frac{G_1G_2 + G_2G_3 + G_3G_1}{(G_1 + G_2 + G_3)^2}. \quad (2)$$

This parameter was bounded between the values of 0 and 1, which corresponded to perfect spherical and linear shapes, respectively. Figure 6 shows the comparison of the relative shape anisotropy parameter (κ^2) distributions for micelles under turbulent flow with different volume volume fractions, ϕ. For $\phi = 0.01$, the distribution at κ^2 = 0.7–0.8 was the largest, and the distribution at κ^2 = 0.9–1.0 that indicated the existence of rodlike micelles was also relatively large. In contrast, for $\phi = 0.03$, κ^2 distribution shifted towards lower values, and a clear decrease in the distribution occurred at κ^2 = 0.9–1.0. Thus, although the size and shape of micelles changed in the range of $Re \gtrsim 700$, the turbulent transition was still effectively suppressed (Figure 2).

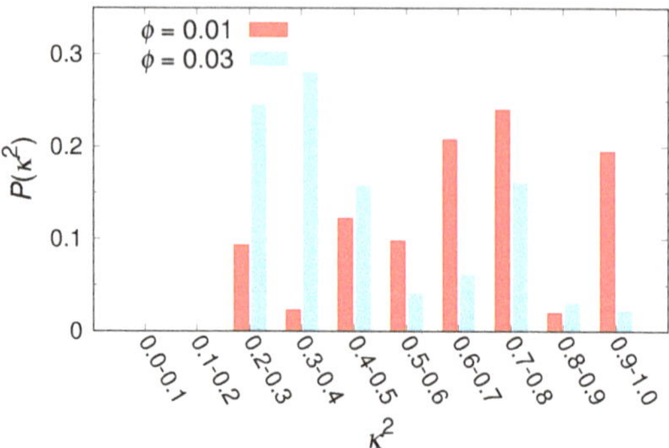

Figure 6. Comparison of relative shape anisotropy parameter (κ^2) distributions for micelles under turbulent flow with different volume fractions, ϕ.

To investigate in detail why a ϕ-dependent delay in the turbulent transition occurred, we computed the density profiles of surfactant molecules in the radial direction in the steady state (Figure 7). For $\phi = 0.01$, we found a distinct difference in density profiles between the turbulent state and the others. For the laminar and transition states, surfactant molecules were distributed within the central region of the tube ($r < 10$), as an adequate number of rodlike micelles still existed. By contrast, the peak of the density profile shifted to $r \approx 15$ for the turbulent state. Thus, for dilute systems, the drag-reduction effect disappeared due to large micelles breaking up into smaller ones and eventually into monomers.

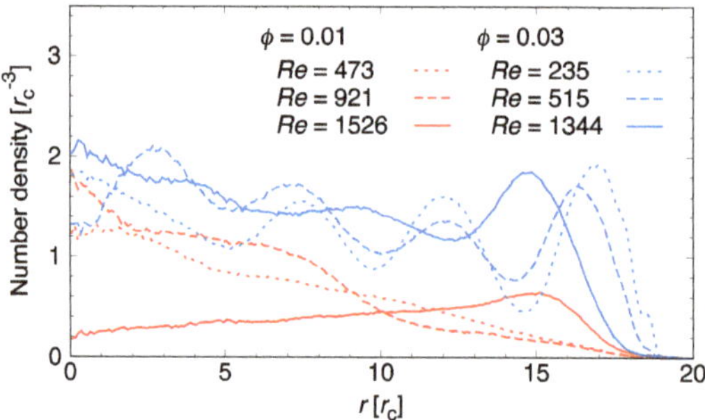

Figure 7. Density profiles of surfactant molecules in radial direction. *Re*: Reynolds number; ϕ: surfactant volume fractions.

For dense systems, there was also a difference in density profiles, particularly for the turbulent state. For $Re \lesssim 700$, several peaks could be seen in the radial direction, reflecting the distribution of orientationally ordered rodlike micelles along the flow direction over the entire radial range. When *Re* was increased over 700 (corresponding to the turbulent state with $\phi = 0.01$), distinct peaks in the range of $r < 15$ disappeared as a result of rodlike micelles changing into sheet-shaped ones.

4. Conclusions

We presented a possible molecular-level mechanism of turbulence suppression in surfactant aqueous solutions based on a large-scale dissipative particle dynamics simulation. Our simulations revealed that the phenomenon of the drag-reduction effect was caused by turbulence suppression, and the number and formation of micelles were closely related to the suppression of the turbulent transition. We established the necessary conditions for obtaining the drag-reduction effect: maintaining (i) a certain minimal size of micellar structures ($N_a \gtrsim 10^3$), even at high *Re*; and (ii) a homogeneous distribution of surfactant molecules in the radial direction of the tube. To the best of our knowledge, our work is the first to show molecular-simulation evidence for the relation between micellar structure and drag reduction in pipe flow. Our findings provide new insights into the mechanism of drag reduction on the molecular level, and may prove valuable for identifying the required synthesis to obtain the drag-reduction effect in a targeted range of Reynolds numbers.

Supplementary Materials: The following are available online at https://www.mdpi.com/article/10.3390/ijms22147573/s1.

Author Contributions: Conceptualization, N.A.; investigation and analysis, Y.K., H.G., and N.A.; resources, Y.K. and N.A.; writing—original-draft preparation, Y.K.; writing—review and editing, Y.K. and N.A.; visualization, Y.K. and H.G.; supervision, N.A.; funding acquisition, Y.K. and N.A. All authors have read and agreed to the published version of the manuscript.

Funding: Y.K. was supported by a research grant from the Keio Leading-Edge Laboratory of Science and Technology, the Keio University Doctorate Student Grant-in-Aid Program, and the Keio Engineering Foundation. N.A. was supported by JSPS KAKENHI grant number 17K14610.

Institutional Review Board Statement: Not applicable.

Informed Consent Statement: Not applicable.

Data Availability Statement: Not applicable.

Conflicts of Interest: The authors declare no conflict of interest.

Abbreviations

The following abbreviations are used in this manuscript:

DPD	Dissipative particle dynamics
CTAB	Cetyltrimethylammonium bromide
NaSal	Sodium salicylate

References

1. Murray, B.S. Interfacial rheology of food emulsifiers and proteins. *Curr. Opin. Colloid Interface Sci.* **2002**, *7*, 426–431. [CrossRef]
2. Tabilo-Munizaga, G.; Barbosa-Cànovas, G.V. Rheology for the food industry. *J. Food Eng.* **2005**, *67*, 147–156. [CrossRef]
3. Gallegos, C.; Franco, J.M. Rheology of food, cosmetics and pharmaceuticals. *Curr. Opin. Colloid Interface Sci.* **1999**, *4*, 288–293. [CrossRef]
4. Townsend, J.M.; Beck, E.C.; Gehrke, S.H.; Berkland, C.J.; Detamore, M.S. Flow behavior prior to crosslinking: The need for precursor rheology for placement of hydrogels in medical applications and for 3D bioprinting. *Prog. Polym. Sci.* **2019**, *91*, 126–140. [CrossRef] [PubMed]
5. Storm, C.; Pastore, J.J.; MacKintosh, F.C.; Lubensky, T.C.; Janmey, P.A. Nonlinear elasticity in biological gels. *Nature* **2005**, *435*, 191–194. [CrossRef] [PubMed]
6. Arai, N.; Yasuoka, K.; Koishi, T.; Ebisuzaki, T.; Zeng, X.C. Understanding molecular motor walking along a microtubule: A themosensitive asymmetric brownian motor driven by bubble formation. *J. Am. Chem. Soc.* **2013**, *135*, 8616–8624. [CrossRef] [PubMed]
7. Shelley, M.J. The dynamics of microtubule/motor-protein assemblies in biology and physics. *Annu. Rev. Fluid Mech.* **2016**, *48*, 487–506. [CrossRef]
8. Karato, S.; Wu, P. Rheology of the upper mantle: A synthesis. *Science* **1993**, *260*, 771–778. [CrossRef]
9. Bürgmann, R.; Dresen, G. Rheology of the lower crust and upper mantle: Evidence from rock mechanics, geodesy, and field observations. *Annu. Rev. Earth Planet. Sci.* **2008**, *36*, 531–567. [CrossRef]
10. Ebisuzaki, T.; Maruyama, S. Nuclear geyser model of the origin of life: Driving force to promote the synthesis of building blocks of life. *Geosci. Front.* **2017**, *8*, 275–298. [CrossRef]
11. Arai, N.; Kobayashi, Y.; Yasuoka, K. A biointerface effect on the self-assembly of ribonucleic acids: A possible mechanism of RNA polymerisation in the self-replication cycle. *Nanoscale* **2020**, *12*, 6691. [CrossRef]
12. de Gennes, P.G. Reptation of a polymer chain in the presence of fixed obstacles. *J. Chem. Phys.* **1971**, *55*, 572. [CrossRef]
13. Doi, M.; Edwards, S.F. Dynamics of concentrated polymer systems. Part 1.—Brownian motion in the equilibrium state. *J. Chem. Soc. Faraday Trans. 2* **1978**, *74*, 1789–1801. [CrossRef]
14. Doi, M.; Edwards, S.F. Dynamics of concentrated polymer systems. Part 2.—Molecular motion under flow. *J. Chem. Soc. Faraday Trans. 2* **1978**, *74*, 1802–1817. [CrossRef]
15. Doi, M.; Edwards, S.F. Dynamics of concentrated polymer systems. Part 3.—The constitutive equation. *J. Chem. Soc. Faraday Trans. 2* **1978**, *74*, 1818–1832. [CrossRef]

16. Zia, R.N. Active and passive microrheology: Theory and simulation. *Annu. Rev. Fluid Mech.* **2018**, *50*, 371–405. [CrossRef]
17. Camerin, F.; Gnan, N.; Ruiz-Franco, J.; Ninarello, A.; Rovigatti, L.; Zaccarelli, E. Microgels at interfaces behave as 2D elastic particles featuring reentrant dynamics. *Phys. Rev. X* **2020**, *10*, 031012. [CrossRef]
18. Uneyama, T.; Masubuchi, Y. Plateau moduli of several single-chain slip-link and slip-spring models. *Macromolecules* **2021**, *54*, 1338–1353. [CrossRef]
19. Ruiz-Franco, J.; Marakis, J.; Gnan, N.; Kohlbrecher, J.; Gauthier, M.; Lettinga, M.P.; Vlassopoulos, D.; Zaccarelli, E. Crystal-to-crystal transition of ultrasoft colloids under shear. *Phys. Rev. Lett.* **2018**, *120*, 078003. [CrossRef] [PubMed]
20. Toms, B.A. Some observation on the flow of linear polymer solutions through straight tubes at large Reynolds numbers. *Proc. 1st Int. Cong. Rheol.* **1948**, *2*, 135–141.
21. Li, J.; Yu, B.; Wang, L.; Li, F.; Hou, L. A mixed subgrid-scale model based on ICSM and TADM for LES of surfactant-induced drag-reduction in turbulent channel flow. *Appl. Therm. Eng.* **2017**, *115*, 1322–1329. [CrossRef]
22. Peaudecerf, F.J.; Landel, J.R.; Goldstein, R.E.; Luzzatto-Fegiz, P. Traces of surfactants can severely limit the drag reduction of superhydrophobic surfaces. *Proc. Natl. Acad. Sci. USA* **2017**, *114*, 7254–7259. [CrossRef] [PubMed]
23. Zhang, W.; Zhang, H.N.; Li, J.; Yu, B.; Li, F. Comparison of turbulent drag reduction mechanisms of viscoelastic fluids based on the Fukagata-Iwamoto-Kasagi identity and the Renard-Deck identity. *Phys. Fluids* **2020**, *32*, 013104.
24. Lopez, J.; Choueiri, G.; Hof, B. Dynamics of viscoelastic pipe flow at low Reynolds numbers in the maximum drag reduction limit. *J. Fluid Mech.* **2019**, *874*, 699–719. [CrossRef]
25. Wakimoto, T.; Araga, K.; Katoh, K. Simultaneous determination of micellar structure and drag reduction in a surfactant solution flow using the fluorescence probe method. *Phys. Fluids* **2018**, *30*, 033103. [CrossRef]
26. Kotenko, M.; Oskarsson, H.; Bojesen, C.; Nielsen, M.P. An experimental study of the drag reducing surfactant for district heating and cooling. *Energy* **2019**, *178*, 72–78. [CrossRef]
27. Liu, D.; Wang, Q.; Wei, J. Experimental study on drag reduction performance of mixed polymer and surfactant solutions. *Chem. Eng. Res. Des.* **2018**, *132*, 460–469. [CrossRef]
28. Tamano, S.; Uchikawa, H.; Ito, J.; Morinishi, Y. Streamwise variations of turbulence statistics up to maximum drag reduction state in turbulent boundary layer flow due to surfactant injection. *Phys. Fluids* **2018**, *30*, 075103. [CrossRef]
29. Zakin, J.L.; Lu, B.; Bewersdorff, H.W. Surfactant drag reduction *Rev. Chem. Eng.* **1998**, *14*, 253–320. [CrossRef]
30. Liu, D.; Liu, F.; Zhou, W.; Chen, F.; Wei, J. Molecular dynamics simulation of self-assembly and viscosity behavior of PAM and CTAC in salt-added solutions. *J. Mol. Liq.* **2018**, *268*, 131–139. [CrossRef]
31. Liu, F.; Zhou, W.; Liu, D.; Chen, F.; Wei, J. Coarse-grained molecular dynamics study on the rheological behaviors of surfactant aqueous solution. *J. Mol. Liq.* **2018**, *265*, 572–577. [CrossRef]
32. Liu, F.; Liu, D.; Zhou, W.; Wang, S.; Chen, F.; Wei, J. Weakening or losing of surfactant drag reduction ability: A coarse-grained molecular dynamics study. *Chem. Eng. Sci.* **2020**, *219*, 115610. [CrossRef]
33. Liu, F.; Liu, D.; Zhou, W.; Chen, F.; Wei, J. Coarse-grained molecular dynamics simulations of the breakage and recombination behaviors of surfactant micelles. *Ind. Eng. Chem. Res.* **2018**, *57*, 9018–9027. [CrossRef]
34. Zhou, J.; Ranjith, P.G. Self-assembly and viscosity changes of binary surfactant solutions: A molecular dynamics study. *J. Colloid Interface Sci.* **2021**, *585*, 250–257. [CrossRef] [PubMed]
35. Asidin, M.A.; Suali, E.; Jusnukin, T.; Lahin, F.A. Review on the applications and developments of drag reducing polymer in turbulent pipe flow. *Chin. J. Chem. Eng.* **2019**, *27*, 1921–1932. [CrossRef]
36. Gu, Y.; Yu, S.; Mou, J.; Wu, D.; Zheng, S. Research progress on the collaborative drag reduction effect of polymers and surfactants. *Materials* **2020**, *13*, 444. [CrossRef] [PubMed]
37. Graham, M.D. Drag reduction and the dynamics of turbulence in simple and complex fluids. *Phys. Fluids* **2014**, *26*, 101301. [CrossRef]
38. Hoogerbrugge, P.; Koelman, J. Simulating microscopic hydrodynamic phenomena with dissipative particle dynamics. *Europhys. Lett.* **1992**, *19*, 155–160. [CrossRef]
39. Español, P.; Warren, P. Statistical mechanics of dissipative particle dynamics. *Europhys. Lett.* **1995**, *30*, 191–196. [CrossRef]
40. Groot, R.D.; Warren, P. Dissipative particle dynamics: Bridging the gap between atomistic and mesoscopic simulation. *J. Chem. Phys.* **1997**, *107*, 4423–4435. [CrossRef]
41. Prinsen, P.; Warren, P.B.; Michels, M.A.J. Mesoscale simulations of surfactant dissolution and mesophase formation. *Phys. Rev. Lett.* **2002**, *89*, 148302. [CrossRef]
42. Mao, R.; Lee, M.T.; Vishnyakov, A.; Neimark, A.V. Modeling aggregation of ionic surfactants using a smeared charge approximation in dissipative particle dynamics simulations. *J. Chem. Theory Comput.* **2015**, *119*, 11673–11683. [CrossRef]
43. Kobayashi, Y.; Arai, N. Self-assembly of surfactant aqueous solution confined in a Janus amphiphilic nanotube. *Mol. Simul.* **2017**, *43*, 1153–1159. [CrossRef]
44. Anderson, R.L.; Bray, D.J.; Regno, A.D.; Seaton, M.A.; Ferrante, A.S.; Warren, P.B. Micelle formation in alkyl sulfate surfactants using dissipative particle dynamics. *J. Am. Chem. Soc.* **2018**, *14*, 2633–2643. [CrossRef] [PubMed]
45. Jury, S.; Bladon, P.; Cates, M.; Krishna, S.; Hagen, M.; Ruddock, J.N.; Warren, P.B. Simulation of amphiphilic mesophases using dissipative particle dynamics. *Phys. Chem. Chem. Phys.* **1999**, *1*, 2051–2056. [CrossRef]
46. Yamamoto, S.; Hyodo, S. Mesoscopic simulation of the crossing dynamics at an entanglement point of surfactant threadlike micelles. *J. Chem. Phys.* **2005**, *122*, 204907. [CrossRef]

47. Arai, N.; Yasuoka, K.; Masubuchi, Y. Spontaneous self-assembly process for threadlike micelles. *J. Chem. Phys.* **2007**, *126*, 244905. [CrossRef]
48. Arai, N.; Yasuoka, K.; Zeng, X.C. Self-assembly of surfactants and polymorphic transition in nanotubes. *J. Am. Chem. Soc.* **2008**, *130*, 7916–7920. [CrossRef]
49. Arai, N.; Yasuoka, K.; Zeng, X.C. Nanochannel with uniform and Janus surfaces: Shear thinning and thickening in surfactant solution. *Langmuir* **2012**, *28*, 2866–2872. [CrossRef]
50. Kobayashi, Y.; Arai, N. Polymodal rheological behaviors induced by self-assembly of surfactants confined in nanotubes. *J. Mol. Liq.* **2019**, *274*, 328–337. [CrossRef]
51. Tsujinoue, H.; Kobayashi, Y.; Arai, N. Effect of the Janus amphiphilic wall on the viscosity behavior of aqueous surfactant solutions. *Langmuir* **2020**, *36*, 10690–10698. [CrossRef]
52. Strey, R.; Schomäcker, R.; Roux, D.; Nallet, F.; Olsson, U. Dilute lamellar and L_3 phases in the binary water-$C_{12}E_5$ system. *J. Chem. Soc. Faraday Trans.* **1990**, *86*, 2253–2261. [CrossRef]
53. Groot, R.D.; Rabone, K.L. Mesoscopic simulation of cell membrane damage, morphology change and rupture by nonionic surfactants. *Biophys. J.* **2001**, *81*, 725–736. [CrossRef]
54. Maddox, M.W.; Gubbins, K.E. A molecular simulation study of freezing/melting phenomena for Lennard-Jones methane in cylindrical nanoscale pores. *J. Chem. Phys.* **1997**, *107*, 9659–9667. [CrossRef]
55. Takenaka, S.; Suga, K.; Kinjo, T.; Hyodo, S. Flow simulations in a sub-micro porous medium by the lattice Boltzmann and the molecular dynamics methods. In Proceedings of the ASME 2009 7th International Conference on Nanochannels, Microchannels, and Minichannels, Pohang, Korea, 22–24 June 2009; pp. 927–936.
56. Zhu, Y.; Granick, S. Rate-dependent slip of Newtonian liquid at smooth surfaces. *Phys. Rev. Lett.* **2001**, *87*, 096105. [CrossRef]
57. Tretheway, D.C.; Meinhart, C.D. Apparent fluid slip at hydrophobic microchannel walls. *Phys. Fluids* **2002**, *14*, L9–L12. [CrossRef]
58. Kobayashi, Y.; Arai, N. Self-assembly and viscosity behavior of Janus nanoparticles in nanotube flow. *Langmuir* **2017**, *33*, 736–743. [CrossRef] [PubMed]
59. Metzner, A.B.; Otto, R.E. Agitation of non-Newtonian fluids. *AIChE J.* **1957**, *3*, 3–10. [CrossRef]
60. Harris, J. A note on the generalized Reynolds number in non-Newtonian flow. *Br. J. Appl. Phys.* **1963**, *14*, 817–818. [CrossRef]
61. van de Meent, J.W.; Morozov, A.; Somfai, E.; Sultan, E.; van Saarloos, W. Coherent structures in dissipative particle dynamics simulations of the transition to turbulence in compressible shear flows. *Phys. Rev. E* **2008**, *78*, 015701(R). [CrossRef]
62. Sultan, E.; van de Meent, J.W.; Somfai, E.; Morozov, A.N.; van Saarloos, W. Polymer rheology simulations at the meso- and macroscopic scale. *Europhys. Lett.* **2010**, *90*, 64002. [CrossRef]
63. Habibpour, M.; Clark, P.E. Drag reduction behavior of hydrolyzed polyacrylamide/xanthan gum mixed polymer solutions. *Pet. Sci.* **2017**, *14*, 412–423. [CrossRef]
64. Nesyn, G.V.; Sunagatullin, R.Z.; Shibaev, V.P.; Malkin, A.Y. Drag reduction in transportation of hydrocarbon liquids: From fundamentals to engineering applications. *J. Pet. Sci. Eng.* **2018**, *161*, 715–725. [CrossRef]
65. Rajappan, A.; McKinley, G.H. Cooperative drag reduction in turbulent flows using polymer additives and superhydrophobic walls. *Phys. Rev. Fluids* **2020**, *5*, 114601. [CrossRef]
66. Virk, P.S. Drag reduction fundamentals. *AIChE J.* **1975**, *21*, 625–656. [CrossRef]
67. Wang, Y.; Yu, B.; Zakin, J.L.; Shi, H. Review on drag reduction and its heat transfer by additives. *Adv. Mech. Eng.* **2011**, *3*, 478749. [CrossRef]
68. Li, F.C.; Yu, B.; Wei, J.J.; Kawaguchi, Y. *Turbulent Drag Reduction by Surfactant Additives*; John Wiley & Sons, Ltd: Hoboken, NJ, USA, 2012; pp. 8–14.
69. Ohlendorf, D.; Interthal, W.; Hoffmann, H. Surfactant systems for drag reduction: Physico-chemical properties and rheological behaviour. *Rheol. Acta* **1986**, *25*, 468–486. [CrossRef]
70. Mohsenipour, A.A.; Pal, R. Drag reduction in turbulent pipeline flow of mixed nonionic polymer and cationic surfactant systems. *Can. J. Chem. Eng.* **2011**, *91*, 190–201. [CrossRef]
71. Karami, H.R.; Mowla, D. Investigation of the effects of various parameters on pressure drop reduction in crude oil pipelines by drag reducing agents. *J. Non-Newton. Fluid Mech.* **2012**, *177–178*, 37–45. [CrossRef]
72. Usui, H.; Itoh, T.; Saeki, T. On pipe diameter effects in surfactant drag-reducing pipe flows. *Rheol. Acta.* **1998**, *37*, 122–128. [CrossRef]
73. Matras, Z.; Kopiczak, B. Intensification of drag reduction effect by simultaneous addition of surfactant and high molecular polymer into the solvent. *Chem. Eng. Res. Des.* **2015**, *96*, 35–42. [CrossRef]
74. Eskin, D. Modeling an effect of pipe diameter on turbulent drag reduction. *Chem. Eng. Sci.* **2017**, *162*, 66–68. [CrossRef]

International Journal of
Molecular Sciences

MDPI

Article

Adsorption on Ligand-Tethered Nanoparticles

Małgorzata Borówko * and Tomasz Staszewski

Department of Theoretical Chemistry, Institute of Chemical Sciences, Faculty of Chemistry, Maria Curie-Skłodowska University, 20-031 Lublin, Poland; staszewski@umcs.pl
* Correspondence: borowko@hektor.umcs.lublin.pl

Abstract: We use coarse-grained molecular dynamics simulations to study adsorption on ligand-tethered particles. Nanoparticles with attached flexible and stiff ligands are considered. We discuss how the excess adsorption isotherm, the thickness of the polymer corona, and its morphology depend on the number of ligands, their length, the size of the core, and the interaction parameters. We investigate the adsorption-induced structural transitions of polymer coatings. The behavior of systems involving curved and flat "brushes" is compared.

Keywords: hairy nanoparticles; adsorption on nanoparticles; nanocarriers; molecular dynamics; computer simulations

check for updates

Citation: Borówko, M.; Staszewski, T. Adsorption on Ligand-Tethered Nanoparticles. *Int. J. Mol. Sci.* **2021**, *22*, 8810. https://doi.org/10.3390/ijms22168810

Academic Editor: Paschalis Alexandridis

Received: 19 July 2021
Accepted: 11 August 2021
Published: 16 August 2021

1. Introduction

For many years, polymer brushes have attracted considerable attention due to their interesting properties and many potential applications [1–11]. The brushes can be synthesized on planar surfaces as well as on nanoparticles. In such polymer coatings, ligands are chemically attached to a substrate. The unique structure of the tethered polymer layers makes them very interesting from a purely practical point of view. Due to the high mobility of the chains, their configurations can change with an environmental stimuli, such as the pH, temperature, and salt concentration [2,3]. These coatings are also thicker and mechanically more stable than physically adsorbed monolayers [1].

Recent advances in polymer chemistry have enabled the synthesis of polymer brushes with well-defined structures and tunable properties [1]. Similarly, developments in nanomaterial chemistry have produced polymer-tethered (hairy) nanoparticles with narrow size distributions and controllable physicochemical characteristics of polymer coronas [2]. These hybrid nanoparticles offer a novel platform for the application of brush systems in diverse fields, such as the production of nanocomposites [4], sensing [5], biotechnology, and biomedicine [6]. The particle-based polymer brush systems can be applied as drug delivery vehicles and as carrier materials for proteins or enzymes [7–9,12–14].

These drug delivery systems are of great interest for the treatment of cancer and other diseases [7–9,12]. Many kinds of drug carriers have been developed [7], including various hairy nanoparticles. Their polymer coatings prevent aggregation and facilitate conjunction with drugs and compounds that can protect the carrier against recognition by the immune system. They can deliver drugs on-site, resulting in higher drug transport efficiency, lower dosages required, and reduced side effects. Hairy nanoparticles can be exploited not only for drug delivery but also for monitoring the treatment's effects or enhancing its efficiency [7,9].

Polymer-tethered magnetic nanoparticles can serve as contrast agents in magnetic resonance imaging [13]. The delivery of drug molecules to the target place consists of three steps, namely the loading of the molecules to a carrier, their transport in the blood, and the triggered release of drugs from a carrier in response to a stimulus encountered on entry into the diseased tissue [7]. Understanding these processes is essential for the rational design of drug delivery systems and other applications of polymer-based carriers.

The great variety of polymeric coatings makes hairy nanoparticles amenable to various loading strategies. The active molecules can form chemical bonds with functional groups in polymer chains or accumulate in the brush via physical adsorption [7]. On the other hand, polymer brushes can be designed to eliminate or significantly reduce the adsorption of biomolecules onto surfaces [10,11]. Hence, it is important to develop insight into the interactions between molecules or colloidal particles and polymer brushes to develop approaches to control the adsorption characteristics.

Regardless of the purely practical motivations for investigating polymer-tethered layers, these systems are extremely interesting from a fundamental point of view. Therefore, polymer brushes on flat substrates were extensively studied using different theoretical methods, including scaling theory, the self-consistent field theory, and the density functional theory [15–19]. Most of the research focused on modeling the morphology of brushes by changing parameters, such as the type of tethers, their lengths, the grafting density, the interactions of chains with the environment, and the temperature. The results were summarized in several reviews [15,16,19].

The morphology of brushes on curved surfaces was less extensively studied. Some works used the similarity between polymer-tethered particles and star polymers [20,21]. A mean-field theory for star polymers was developed by Daoud and Cotton [22]. This theory predicted more stretched chain conformations in the vicinity of the surface due to excluded volume effects and relaxed chain conformations for the peripheral region of the star. Ohno et al. [20,21] extended the Daoud–Cotton model to the polymer-tethered spherical particles of different sizes.

In contrast to flat surfaces, particle curvature implies that the area per chain increases with distance from the surface, and thus the outer ends of the chain have more space. The self-consistent field model and the scaling theory were used to explore the properties of finite chains tethered on curved interfaces [23,24]. Williams and Zhulina [24] described polymer brushes at spherical and cylindrical surfaces immersed in implicit solvents. They analyzed the effects of the surface curvature and the solvent quality of the brush architecture. In turn, Lo Verso et al. [25] used molecular dynamics simulations and density functional theory to study polymers end-grafted to spherical nanoparticles under good solvent conditions.

The behavior of various molecules and particles near brushes was also the subject of theoretical considerations. Most of this research has focused on particles near flat surfaces modified with tethered chains. The adsorption of small molecules [26–30], polymers [31–34], peptides, proteins [35–39] and Janus particles [40,41] on the flat brushes was interpreted in the framework of different approaches. It has been shown how the adsorption depends on the grafting density, the chain length, and interactions between all species [19,26–34]. However, in the case of hairy nanoparticles, research has focused on their interactions with small molecules [42–46] or proteins [47–49].

Grest's group [42–46] used fully atomistic molecular dynamic simulations of spherical particles modified with various ligands solvated in water and organic solvents and on interfaces [46]. Quite recently, Chew et al. [48] studied the hydrophobicity of monolayer-protected gold nanoparticles using atomistic molecular dynamics simulations. They calculated local hydration free energies at the nanoparticle–water interface and found that these energies were correlated with the preferential binding of propane as a representative hydrophobic probe molecule.

It is difficult to theoretically predict the amount of adsorption on hairy particles. The reason lies mainly in the complexity of the problem. Hairy nanoparticles can be treated as "living adsorbents"; their internal morphology changes in a response to interaction with the environment. In addition, the adsorption of fluid molecules induces the reconfiguration of tethered chains [43,45,48,50]. The chains attached to a curved surface are more mobile than those tethered on a flat substrate.

Thus, for hairy particles, these effects become stronger [19–21,23]. Furthermore, nanoparticles are typically dispersed in a solvent, and they can aggregate due to interactions between molecules adsorbed on different hairy particles [50,51]. There are only a few theoretical works devoted to a description of the adsorption on hairy nanoparticles [50,52] or hairy vesicles [53].

Adsorption on hairy nanoparticles was also a subject of experimental studies [52,54–57]. For example, Ballauff et al. [52,54,55] analyzed the impact of several parameters on the adsorption on polyelectrolyte layers. However, Snytska's group [56,57] studied the effects of grafting density on the properties of polymer brushes prepared on flat and on colloidal particle substrates. By changing the grafting density, they tuned the properties of the carrier material, such as swelling, charge, adhesion, and the adsorption of enzyme on grafted brushes. The adsorption of albumin on magnetic submicrospheres with a hairy core-shell structure was also measured [58]. Wang et al. [51] presented the results of research concerning the co-immobilization and separation of proteins using particles modified with polyelectrolytes.

All the above-mentioned studies demonstrated the complexity of the processes underlying the loading of molecules onto nanoparticle-based carriers. Therefore, a thorough description of the adsorption on hairy nanoparticles remains an open challenge. There is still a lack of systematic investigations concerning the correlation between the properties of polymer corona and the efficiency of adsorption on hairy particles and the spatial distribution of adsorbed molecules around the cores.

In this work, we study an idealized coarse-grained model for the adsorption of spherical molecules on polymer-tethered particles by molecular dynamics simulations. Our main purpose is to give a quantitative description of adsorption onto hairy nanoparticles under different conditions and to examine the behavior of spherical brushes in the presence of small particles. We change such parameters as the core diameter, grafting density, length of chains, their flexibility, interactions between all moieties, and the density of adsorbed particles. In this way, we can modulate the thickness of the polymer corona, its internal morphology, and the adsorption of fluid molecules onto a hairy nanoparticle. We want to capture basic factors determining the adsorption properties of hairy particles. Some results are compared with their planar counterparts.

We consider nanoparticles with rather short ligands. The behavior of such nanoparticles is relatively less explored compared with those with very long chains. However, these systems are very interesting from a purely cognitive point of view. Additionally, due to savings in computation time, it is possible to scan a wider range of parameter space. This allows us to capture basic factors determining the adsorption properties of hairy particles.

2. Model and Simulation Methodology

We consider a hairy particle immersed in a fluid consisting of spherical particles P (molecules or small colloids). For computational efficiency, we coarse-grained our system to reduce the number of "atoms" required. A single tethered nanoparticle was modeled as a spherical core with attached f chain molecules. The core diameter equals σ_c, and the particles P have the diameter σ_P. Each chain consists of M tangentially jointed spherical segments of identical diameters σ_s. The chain connectivity is assured by the harmonic segment–segment potentials

$$u_{ss}^{(b)} = k_{ss}(r-\sigma_s)^2, \tag{1}$$

where r is the distance between segments. The first segment of each chain is rigidly fixed to the core at a randomly chosen point on its surface (at the distance $\sigma_{cs} = 0.5(\sigma_c + \sigma_s)$). We assume that the chains are perfectly flexible, and, in each chain, segments are freely connected (model M1).

All the spherical entities interact via the shifted-force Lennard–Jones potential [59]

$$u^{(ij)} = \begin{cases} 4\varepsilon_{ij}\left[(\sigma_{ij}/r)^{12} - (\sigma_{ij}/r)^{6}\right] + \Delta u^{(ij)}(r), & r < r_{cut}^{(ij)}, \\ 0, & \text{otherwise}, \end{cases} \tag{2}$$

where

$$\Delta u^{(ij)}(r) = -(r - r_{cut}^{(ij)})\partial u^{(ij)}(r_{cut}^{(ij)})/\partial r, \tag{3}$$

In the above $r_{cut}^{(ij)}$ denotes the cutoff distance, $\sigma_{ij} = 0.5(\sigma_i + \sigma_j)$ $(i, j = c, s, P)$ and ε_{ij} is the parameter characterizing interaction strengths between spherical species i and j. The indices "c", "s", and "P" correspond to the cores, the chain segments, and the small particles, respectively. All particles are dispersed in an implicit solvent that determines the strength of interactions between them.

We also study hedgehog-like particles with rigid ligands (model M2). Furthermore, we consider also the adsorption of the particles P on a planar surface modified with end-tethered chains using a model analogous to that described above. In this case, however, the solid surface interacts with all particles except the ligand's bonding segments via the hard-wall potential.

We introduce standard units. The diameter of segments is the distance unit, $\sigma_s = \sigma$, and the segment–segment energy parameter, $\varepsilon_{ss} = \varepsilon$, is the energy unit. The mass of a single segment is the mass unity, $m_s = m$. The basic unit of time is $\tau = \sigma\sqrt{(\varepsilon/m)}$. The reduced temperature is $T^* = k_B T/\varepsilon$.

In this study, particles P and chain segments have the same mass, m. The spring constant of the binding potential, $k_{ss} = 1000\varepsilon/\sigma^2$.

In our simulations, core–core, core–chain, and particle–core interactions are purely repulsive, while particle–chain (particle–segment) interactions were attractive. In some simulations, we assume that also segment–segment and particle–particle are attractive. To switch on or switch off attractive interactions, we use the cutoff distance. For attractive interactions, $r_{cut}^{(ij)} = 2.5\sigma_{ij}$, while, for repulsive interactions, $r_{cut}^{(ij)} = \sigma_{ij}$. The grafting density is defined as $\rho_{gr} = f/A$, where A is the area of the substrate.

Molecular dynamics (MD) simulations were performed using the LAMMPS classical MD code [60,61]. The Nose–Hoover thermostat was used to regulate the temperature. The reduced temperature was $T^* = k_B T/\varepsilon = 1$ for all the results reported here. Two series of simulations were carried out, for the hairy nanoparticles and the planar brushes.

In the first case, a cubic simulation box was used with the nanoparticle's core fixed at the center of the simulation cell. The box size varied from $L = 57\sigma$ to $L = 107\sigma$ depending on the system studied. In the case of bigger cores and longer tethers, bigger boxes were used. Standard periodic boundary conditions were introduced in all directions.

The first segment of each tethered ligand was fixed at a distance σ_{cs} away from the center of the nanoparticle. These binding beads are randomly distributed on a spherical surface. All the tethering points were then kept fixed thereafter. The simulation cell always had a size capable of accommodating each hairy particle with stretched ligands, the cloud of adsorbed particles, and a sufficiently large part of the bulk fluid. The sphere inscribed (with the radius $R' = L/2$) in the cube was treated as a virtual adsorption system submerged in the bulk fluid.

A planar brush was simulated in a rectangular box with sizes $L_x = L_y$ and L_z, in directions x, y, z, respectively. The wall located at $z = 0$ was covered by the ligands, while the wall at $z = L_z$ was a bare, hard wall. The ligands were attached to the flat surface in the same way as to the nanoparticle. The distance between these walls was large enough to ensure the existence of the region of a uniform fluid in the middle part of the cell. The system is periodic in the x and y directions. The distance L_z ranged from 40 to 80. In the majority of the runs, the box dimension L_x was 100. In this case, the whole simulation cell is considered as the adsorption system.

The spherical particles, P, were randomly inserted into the system until their desired density, ρ_0 was reached. The initial density of the particles P is defined as $\rho_0 = N/V$, where N is the number of the particles and V is the cell's volume.

Simulations were performed for several dozen sets of parameters characterizing the nanoparticle, for two values of diameter $\sigma_c = 2\sigma$ and 4σ, different lengths of ligands $M = 5$, 10, 15, 20, 30, and numbers of tethered chain $f = 5$, 10, 20, 30. In each case, the density of particles P varied from $\rho_0 = 0.0001$ to $\rho_0 = 0.3$. For selected parameters, simulations were also carried out for the planar surfaces with attached ligands.

To obtain proper equilibrium, we used very long simulations, simulations with different starting configurations for each system, and alternating heating and cooling of the system. We equilibrated the system for at least 10^8 time steps until its total energy reached a constant level, at which it fluctuated around a mean value. The production runs were for at least 10^7 time steps. At the time, data were saved after every 1000 time steps and used for the evaluation of the local densities of chain segments, $\rho_s(z)$, and the fluid particles P, $\rho(r)$, where r was the distance between the core center and a segment or a particle. The results presented here were averaged over at least five statistically independent systems. The process was the same for nanoparticles and planar surfaces.

3. Results and Discussion

3.1. Description of the Studied Systems

We studied the adsorption of small particles on different hairy particles and the flat surfaces modified with tethered chains.

As already mentioned, the simulations were carried for the inert cores. The soft repulsive core–segment and core–particle interactions were assumed. All energy parameters $\varepsilon_{ij} = \varepsilon\ (i,j = c,s,P)$ except for the particle–segment energy parameter, which is varied ($\varepsilon_{Ps} = 1.5\varepsilon$ and 3.0ε). The motivation for choosing these values originates from the fact that similar parameters have already been used for planar brushes [27,28,41] and hairy particles [50]. In the framework of the implicit solvent model, these parameters correspond to the good solvent conditions [19]. Thus, the tethered chains are solvophilic.

We discuss here the results obtained for the nanoparticles with attached flexible (model M1) or rigid ligands (model M2). Most simulations were performed for the systems in which only particle–segment interactions are attractive. However, some simulations were also carried out for the case of attractive segment–segment and particle–particle interactions (models M1a, M2a). Table 1 shows the types of interactions in different models.

Table 1. Types of interactions in different models.

Interactions	Models M1, M2	Models M1a, M2a
segment–core	repulsive	repulsive
particle–core	repulsive	repulsive
segment–segment	repulsive	attractive
particle–particle	repulsive	attractive
particle–segment	attractive	attractive

We begin with a detailed analysis of the behavior of the M1-systems. Then, we present selected results for different models.

Our "basic" model (M1) mimics a typical hairy particle with attached flexible chains. For simplicity, only particle–segment models are assumed to be attractive. First, we study the system behavior for relatively low interactions and set $\varepsilon_{Ps} = 1.5\varepsilon$. In this case, we study the cores of two sizes ($\sigma_c = 4\sigma, 2\sigma$) and the corresponding flat substrate. Different numbers of attached chains (f) and different chain lengths (M) are assumed.

3.2. Excess Adsorption Isotherms

A basic measure of adsorption is the excess adsorption [27,50]. For adsorption on a single spherical hairy particle, the excess adsorption (per particle) is defined as

$$\Gamma = 4\pi \int r^2(\rho(r) - \rho_b)dr, \tag{4}$$

where ρ_b is the reduced density of molecules P in the bulk phase. This value is determined from the analysis of the density profiles. Far away from the center of the core, its impact diminishes, and the density achieves a constant value, ρ_b [27,50].

We normalize the excess adsorption in the following way

$$\Gamma^* = \Gamma / A, \tag{5}$$

where A is the surface area. For nanoparticles, it is the area of the core surface, $A = \pi(\sigma_c^2)$.

In the case of a flat surface, the excess adsorption (per a unit of surface area) is calculated from the well-known equation [27]

$$\Gamma^* = \int (\rho(z) - \rho_b) dz, \tag{6}$$

where z is a distance from the surface. Clearly, Equation (5) becomes a form of Equation (6) for $R_c \to \infty$.

The excess adsorption can be also estimated from the following equation

$$\Gamma = V'(\rho_0 - \rho_b), \tag{7}$$

where V' is the volume of the virtual adsorption system, $V' = V_k = \pi L^3/6$ for the particle, and $V' = V$ for the flat surface. The last equation is used for the experimental measurement of adsorption. In some cases, to analyze the simulation results, we used both of these equations, and the obtained values were consistent.

In Figure 1, we show the excess adsorption isotherms of particles P on different hairy particles as well as on flat brushes. Parts a and b present the results for the core of diameter $\sigma_c = 4\sigma$ and different numbers of attached chains (a) and their lengths (b). Usually, when density ρ_b increases, the excess adsorption rapidly rises, reaches its maximum, and begins to slowly decline. However, for short chains or very small numbers of ligands, an atypical course of excess adsorption isotherms is seen.

After an initial rapid increase, the adsorption drops to a deep minimum and begins to rise again. For higher densities, ρ_b, the excess isotherms have a usual shape. The effect of the "superadsorption" at low densities will be discussed later. An increase in the number of ligands and their length cause an increase in the adsorption. This is a consequence of the assumed model. As the total number of attractive segments increases, the number of potentially available "adsorption sites" also increases.

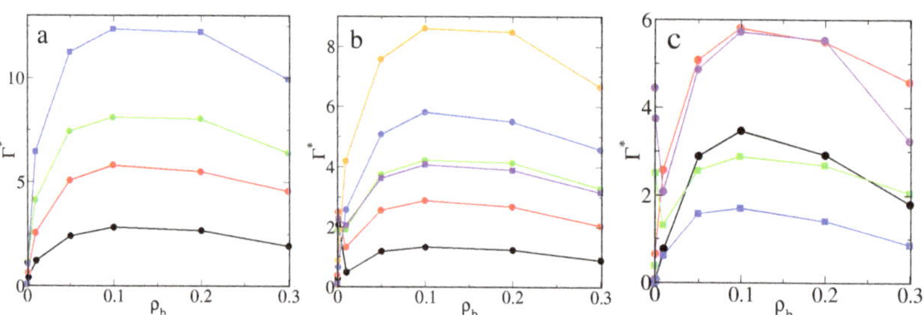

Figure 1. Excess adsorption isotherms for the model M1 and $\varepsilon_{Ps} = 1.5\varepsilon$: (**a**) the hairy particle, $\sigma_c = 4\sigma$, $M = 20$ (circles): $f = 10$ (black), 20 (red), 30 (green), and $M = 30$, $f = 30$ (blue squares), (**b**) the hairy particle, $\sigma_c = 4\sigma$, $f = 20$ (circles): $M = 5$ (black), 10 (red), 15 (green), 20 (blue), 30 (yellow), and $f = 30$, $M = 10$ (violet squares), (**c**) $M = 20$, $\rho_{gr} = 0.398$ (circles): the flat substrate, $f = 20$ (black), the hairy particle, $\sigma_c = 4\sigma$, $f = 20$ (red), the hairy particle, $\sigma_c = 2\sigma$, $f = 5$ (violet), and $f = 20$, $M = 10$ (squares): the flat substrate (blue), the hairy particle, $\sigma_c = 4\sigma$ (green). Symbols correspond to the simulation points. Lines serve as a guide to the eye.

In Figure 1c, the effect of the surface curvature is shown. We compare the excess adsorption isotherms for brushes with the same ligands ($M = 20$) and the same grafting density ($\rho_{gr} = 0.398$). The chains are attached to the bigger ($\sigma_c = 4\sigma$, red line), the smaller core ($\sigma_c = 2\sigma$, violet line), and the flat substrate (black line). The impact of the curvature on the magnitude of adsorption is visible. First, the adsorption (per unit of area) on a flat brush is considerably smaller than on spherical hairy particles. Due to geometrical reasons, a volume of "the adsorbed phase" is larger for a spherical brush, and thus the number of adsorbed particles is greater.

Second, the core size also affects the adsorption. However, this effect is significant only in the case of very low or relatively high bulk densities of adsorbed particles. Note that the local minimum of adsorption at very low density only appears for the smaller core. The number of segments in the polymer layer is much greater for a big core compared to a small one with the same grafting density. Therefore, the excess adsorption on a big particle becomes considerably greater at high densities.

Additionally, in Figure 1c, we present the adsorption isotherms for the same number ($f = 20$) of shorter ligands ($M = 10$) grafted to the flat substrate (black line) and the big core (green line). In the case of flat brushes, an anomaly in excess adsorption does not occur for both longer and shorter chains.

Overall, our simulations show the possibility of controlling adsorption on hairy nanoparticles by changing the core size, the chain length, and the grafting density. For the systems tested here, the adsorption increases as the total number of segments increases. Adsorption on nanoparticles is significantly greater than on flat brushes.

In general, our results are qualitatively consistent with experimental observations [56–58]. However, direct comparison is difficult. There is still a lack of systematic experimental investigations concerning the correlation between the adsorption isotherms and the properties of the hairy nanoparticles. One can find the adsorption isotherms measured for one type of nanoparticles [58] or the values of adsorption estimated at one arbitrary chosen fluid density on different nanoparticles [56].

Marschelke et al. [56] investigated the immobilization of laccase from Trametes versicolor onto hairy particles with a shell built of poly-(2-dimethy laminoethyl methacrylate) (PDMAEMA). This study showed that the polymer loading (adsorption) showed a maximum at different enzyme concentrations depending on the grafting density of the brushes. They also presented the dependence of the polymer loading on the grafting density for different enzyme concentrations.

At a certain low concentration, the polymer loading decreased with the increase of the grafting density. In the case of higher concentrations, the loading increased, achieved a maximum, and decreased. In a wide range of higher concentrations, however, the polymer loading increased with the grafting density as in our simulations.

Adsorption isotherms on hairy nanoparticles are usually estimated only for very dilute solutions and densely grafted substrates [58]. Under such conditions, the adsorption monotonically increases with the increase of the initial concentration of adsorbed particles. The same trend is observed at the beginning of "typical" isotherms obtained from our simulations. However, we also obtained isotherms with the additional maximum at low particle concentration. To the best of our knowledge, this "superadsorption" at the ligand-tethered nanoparticles has not been reported thus far.

3.3. Thickness of Polymer Layer

Adsorption on a brush strongly depends on its structure. On the other hand, the brush morphology varies during adsorption [48,50]. It is well-known that the configuration assumed by the chains depends on the entropic and enthalpic contributions to the free energy. The chains tend to maximize their configurational entropy by adopting a random-walk configuration [19,23]. Simultaneously, they minimize the potential energy by profitable adsorbed particle–segment contacts. The brush structure follows from the competition between these trends.

The average thickness of the polymer layer reflects, to some degree, its structure. In the case of spherical hairy particles, the thickness of the polymer corona is given by [25,43]

$$H = 2(\frac{\int r^3 \rho_s(r)dr}{\int r^2 \rho_s(r)dr} - R_c). \quad (8)$$

For flat substrates, one can obtain [25]

$$H = 2\frac{\int z\rho_s(z)dz}{\int \rho_s(z)dz} \quad (9)$$

We define the relative thickness of the polymer layer as the ratio of the estimated value H to the thickness for completely stretched chains, $H^* = H/M$.

Figure 2 shows the relative average thickness of the bonded phase as a function of the bulk density (ρ_b) for different model systems. In part a, the thicknesses H^* are plotted for rather long chains ($M = 20$) and different numbers of ligands. One can see that the thickness of the polymer corona was strictly correlated with the excess adsorption isotherms. For $f > 10$, all presented curves, similarly to the corresponding excess adsorption isotherms, have the same shape. As the density increased, the H^* fell sharply, then in a narrow region remained almost constant and rose linearly.

At low densities, particles P can deeply penetrate the polymer corona, and their adsorption rapidly increases. The adsorbed particles form bridges between segments and the polymer. An increase in the number of adsorbed particles enhances this effect. However, after reaching a certain threshold density, there is no more room for additional particles and the chains begin to unfold; thus the thickness gradually increases. However, for rather rarely grafted cores with $f = 10$, at extremely low densities, a local maximum is visible (black line). The relative thickness of the polymer corona increased as the number of ligands increased. Due to the effect of the excluded volume, the chains became more stretched.

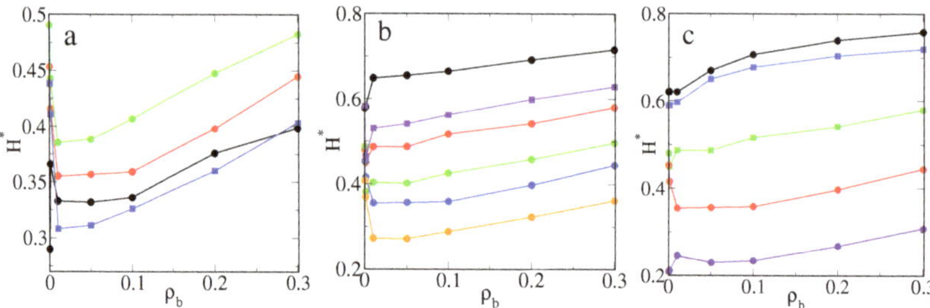

Figure 2. The average thicknesses of the corona for the model M1 and $\varepsilon_{PS} = 1.5\varepsilon$: (**a**) the hairy particle, $\sigma_c = 4\sigma$, $M = 20$ (circles): $f = 10$ (black), 20 (red), 30 (green), and $M = 30$, $f = 30$ (blue squares), (**b**) the hairy particle, $\sigma_c = 4\sigma$, $f = 20$ (circles): $M = 5$ (black), 10 (red), 15 (green), 20 (blue), 30 (yellow), and $f = 30$, $M = 10$ (violet squares), (**c**) $M = 20$ and $\rho_{gr} = 0.398$ (circles): the flat substrate, $f = 20$ (black), the hairy particle, $\sigma_c = 4\sigma$, $f = 20$ (red), the hairy particle, $\sigma_c = 2\sigma$, $f = 5$ (violet), and $f = 20$, $M = 10$ (squares): the flat substrate (blue), the hairy particle, $\sigma_c = 4\sigma$ (green). Symbols correspond to simulation points. Lines serve as a guide to the eye.

Figure 2b displays the impact of the chain length on the functions H^* vs. ρ_b and the fixed number of ligands ($f = 20$). After a sharp minimum at low densities, H^* increased linearly. The same was found for the system with a higher grafting density ($f = 30$) and $M = 10$ (violet squares). However, in the region of densities considered, there was no minimum on the curve plotted for very short chains ($M = 5$). The thickness jumped rapidly and then increased linearly. Comparing these curves, we found that an increase of chain length at the same grafting density led to a considerable decrease of the relative corona thickness. This can be explained by increasing the adsorption of particles and the enhancement of the bridging effects.

In part c, the effect of the curvature on the average brush thickness is presented for systems from Figure 1c. At the same grafting density (full circles) the thickness increased with decreasing surface curvature. In the case of the flat surface, the relationship H^* vs. ρ_b is a slowly increasing function, while, for spherical cores, slight jumps on the curves are visible at very low densities.

Let us briefly discuss the problem of the thickness of polymer-tethered layers. It is well-known that the average height of the flat brush depends mainly on the chain length, the grafting density, and solvent quality [19]. Scaling theories and the self-consistent field theories [19,23,24] predict that the height of the brush increases as the chain length increases; simultaneously, its relative thickness decreases. Thus, our results are qualitatively consistent with theoretical predictions. Similarly, the grafting density increased in accordance with a power-law dependence [19].

Such a relation accurately approximates the data obtained for dense brushes. However, density functional studies [62] showed that the behavior of sparse brushes can be completely different. For very low surface densities of tethered chains, the brush height remained almost constant. In this case, the grafted chains did not practically affect one another, and they assumed unperturbed configurations. For medium-covered substrates, the height fell to a minimum [62]. With a further increase of the grafting density, the brush height increases as in the predicted scaling theories [19]. Although, we also took into account the rather sparse brushes ($f = 5, 10$). In all systems considered here, an increase of the grafting density led to the formation of thicker brushes.

The effect of the adsorption of particles on the brush height was much less studied. We analyzed the impact of the fluid density on the brush height using density functional theory [62,63]. Increasing ρ_b can both increase or decrease the brush height. The course of the functions H vs. ρ_b depends on all interactions in the system [63]. The experimental studies confirmed that the brush height can vary in a complex way during the adsorption.

Marschelke et al. [56] found that the thickness of the swollen PDMAEMA brushes was significantly reduced after the immobilization of enzyme, while the opposite effect was observed for dry brushes. However, Minko et al. [64] reported that the adsorption of enzyme on poly(acrylic acid)-modified (PAA) particles caused the increase of polymer brush thickness in the swollen state. Unfortunately, there is no information about the impact of enzyme concentration on the brush thickness.

A decrease in the brush thickness is commonly explained by the strong attractive interactions between segments [19,23,24]. In our model, this mechanism is completely distinct. Particle–segment interactions play a decisive role to cause the adsorption of particles on "chains". In turn, adsorbed particles influence the chain configurations. The particles form bridges between segments belonging to the same chain and to other chains. Such a bridging effect was observed experimentally [51] The chains wrapped around particles and the polymer corona became more compact. For a sufficiently dense fluid, the adsorbed particles pull segments to the bulk phase so that the height increases. Similar results were obtained for adsorption on nanoparticles with mobile ligands [50].

3.4. Structure of Interfacial Layers Formed on Nanoparticles

To obtain a deeper insight into the structure of the polymer layers, we calculated the one-dimensional density profiles of chain segments, $\rho_s(r)$ and the density profiles of adsorbed particles, $\rho(r)$, around the core.

Figure 3 presents the segment density profiles and the density profiles of adsorbed particles on the big nanoparticle ($\sigma = 4\sigma$) for different numbers of chains and their lengths, (a) $f = 20$, $M = 20$, (b) $f = 20$, $M = 10$, and (c) $f = 10$, $M = 20$, calculated at different initial densities of particles ρ_0.

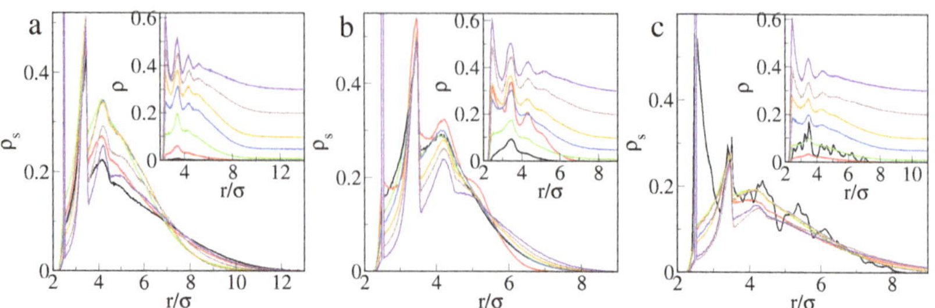

Figure 3. Density profiles of the chain segments and fluid particles (insets) around the core for the model M1 plotted for different initial densities of the fluid $\rho_0 = 0.0001$ (black), 0.001 (red), 0.01 (green), 0.05 (blue), 0.1 (yellow), 0.2 (brown), and 0.3 (violet): (**a**) $f = 20$, $M = 20$, (**b**) $f = 20$, $M = 10$, (**c**) $f = 10$, $M = 20$. Other parameters: $\sigma_c = 4\sigma$, $\varepsilon_{Ps} = 1.5\varepsilon$.

In the segment density profiles for the higher grafting density (Figure 3a,b), one sees three or four peaks. The first peaks correspond to grafted segments that are located directly at $r = 2.5\sigma$. The pronounced structure is also visible further from the core. The second segments in chains show up as rather sharp peaks at $r = 3.5\sigma$. The next peaks at $r = 4.5\sigma$ are much wider. Then, the segment density smoothly decreases. The molecules are attracted by the chains, due to which, they penetrate deeply into the polymer layer. In the fluid density profiles, $\rho(r)$, one sees peaks corresponding to "adsorption on" the subsequent layers of segments. The density of the fluid inside the brush is greater than in the bulk phase.

An initial density of the adsorbed molecules affects the structure of the surface layer. In the case of longer ligands, an increase of the density ρ_0 causes an increase in the density of the particles inside the polymer layer. Such a relation is observed for the "normal" excess adsorption isotherms. For short tethered chains, however, the density profile corresponding to $\rho_0 = 0.001$ (red line) is much greater than that estimated at $\rho_0 = 0.01$ (green line). This reflects the existence of a local maximum at the beginning of the excess adsorption isotherm. At $\rho_0 = 0.001$, three well-pronounced peaks are visible in the density profile of the fluid. In the same part of the brush, the segment density is very high. This suggests the occurrence of a reconfiguration in the polymer corona at this fluid density.

The local maximum at the beginning of excess adsorption isotherm was also found for low grafting density ($f = 10$) and long chains ($M = 20$). Figure 3c depicts the corresponding density profiles. Here, the segment densities in the middle part of the polymer layer were much lower than those for the dense brushes. The segment density was the highest for the lowest density ρ_0. In the center part of the corona, the local fluid density at $\rho_0 = 0.0001$ (black line) was considerably greater than at $\rho_0 = 0.001$ (red line). The one-dimensional segment density profiles, $\rho(r)$, show the densities averaged on a sphere of radius r. They do not show all changes in the shape of the polymer corona.

We monitored the configurations for all studied densities of fluid. The most representative examples for model M1 are presented in Figure 4. In the upper row (a, b, c), the particle with attached long chains ($M = 30$) is presented for different densities of fluid: $\rho_0 = 0.0001$ (a), $\rho_0 = 0.001$ (b), and $\rho_0 = 0.1$ (c). Corresponding configurations for the particles with shorter chains ($M = 10$) are shown in the middle row (c, d, e). In the case of long tethers, at the very low density ($\rho_0 = 0.0001$), the chains are unfolded, and a few particles are arrested inside the polymer layer. The hairy particle has a symmetrical core-shell structure. As more particles are adsorbed, the polymer layer becomes more compact, and, at $\rho_0 = 0.1$, a dense cloud of segments with particles trapped inside is observed.

For the short chains ($M = 10$) at the lowest density ρ_0, the morphology of the polymer layer is similar to the previous results. However, it changes dramatically at $\rho_0 = 0.001$, and a cone-like structure is found. In this case, adsorbed molecules can easily penetrate the polymer layer, and they accumulate near the chains and form bridges between different chains. The cloud of segments becomes highly asymmetrical. A part of the core remains uncovered.

In the bottom row, the snapshots for $f = 10$ and $M = 20$ are shown. In this case, the cone-like structure appears at the lowest density, as the fluid density increases, the chains stretch and transform into a loose asymmetrical structure. Notice that the total number of segments is the same as for the particle drawn in the middle row. This suggests that the cone-like structure can be stable if there are enough adsorbed particles to stick the segments together. As the density ρ_0 increases, the excluded volume effects cause gradual extension of the chains, and this characteristic structure can disappear (see Figure 4f).

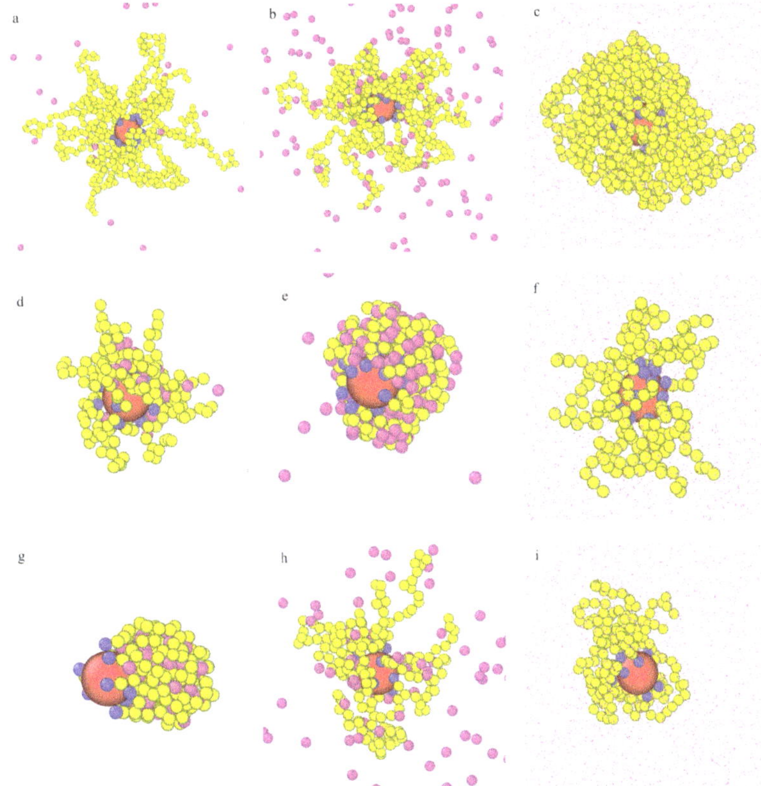

Figure 4. Examples of the equilibrium configurations of hairy particles (model M1, $\varepsilon_{Ps} = 1.5\varepsilon$) immersed in fluids of different densities $\rho_0 = 0.0001$ (**a**,**d**,**g**), 0.001 (**b**,**e**,**h**) , 0.1 (**c**,**f**,**i**) for $f = 20$ and $M = 20$ (**a**–**c**), $f = 20$ and $M = 10$ (**d**–**f**), and $f = 10$ and $M = 20$ (**g**,**h**,**i**). The red sphere represents the core, and blue spheres correspond to bonding segments, yellow spheres and pink spheres represent the remaining segments and fluid particles P, respectively. For clarity, the particles P are represented by points (size ratios are not kept) in parts (**c**,**f**,**i**).

We also found that an increase of curvature of the core favored the formation of cone-like structures (not shown here). We compared the behavior of particles consisting of cores of different sizes with relatively long attached chains ($M = 20$) and the same grafting densities. For the smaller core, the cone-like structure arose and did not for the larger core.

Overall, our results clearly show that adsorption on hairy nanoparticles particles can significantly change the morphology of the polymer layer. The particles were adsorbed mainly inside the brush, and their presence induced changes in the morphology of the polymer layer. In some cases, asymmetric cone-like structures were found. The asymmetry in the polymer coatings was likely to have a significant effect on the aggregation behavior.

Similar effects were reported by Bolintineanu et al. [45] who carried out atomistic molecular dynamics simulations of different alkanethiol-coated gold nanoparticles solvated in water and decane. In some systems, they found significant local bundling of chains on the nanoparticle surface, which resulted in highly asymmetric coatings [45]. Simulations performed by Chew et al. [48] also showed that ligands tended to form "bundles", giving rise to anisotropic structures despite homogeneous surface coatings.

In our previous work [50], we presented the adsorption-induced reconfiguration of the polymer corona built of ligands that could freely move on the core surface. We found that, depending on fluid–chain interactions and the fluid density, isolated hairy particles could be classified as core-shell, octopus-like, and corn-like.

Furthermore, Marschelke et al. [56] proved experimentally that the adsorption of the enzyme affects the polymer swelling and, therefore, leads to the changes in the surface morphology, charge, and adhesion performance of the final polymer–enzyme layer.

3.5. Structure of Interfacial Layers Formed on a Flat Substrate

To investigate a role of curvature of the substrate, we carried out simulation for brushes formed at a flat surface.

Figure 5 illustrates the structure of a bonded layer at the flat surface for the same grafting density as in the case of the hairy particles (see Figure 1c). The segment density profiles have a typical liquid-like structure with the well-pronounced peaks corresponding to subsequent layers of segments. Then, this structure diminishes, and the density of chain segments gradually decreases to zero at the effective brush height. The extent of the layering and the effective brush height increase with increasing the density of adsorbed particles.

In the interior of the polymer layer, the segment density decreases as the fluid density increases. The opposite effect is observed in the outer part of the bonded phase. In the insets, the density profiles of particles are also plotted. The location of peaks in the density profiles of segments and molecules P are the same. As with the hairy particles, molecules are "adsorbed on segments".

The density of adsorbed particles is high near the surface, then it somewhat decreases and, for high bulk densities, increases again at the brush end. The molecules are adsorbed also "on the brush" (ternary adsorption [15,33]). Similar results have been obtained for the flat surfaces modified with tethered short chains using the functional density theory [26–29] and molecular dynamics simulations [41].

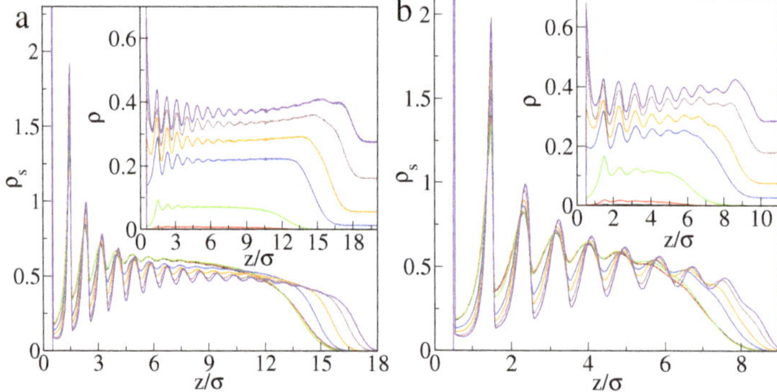

Figure 5. Density profiles of the chain segments and fluid particles (insets) near the flat substrate for the model M1 and different chain lengths: (**a**) $M = 20$, (**b**) $M = 10$ plotted for different initial densities of the fluid $\rho_0 = 0.0001$ (black), 0.001 (red), 0.01 (green), 0.05 (blue), 0.1 (yellow), 0.2 (brown), and 0.3 (violet). Other parameters: $f = 20$, $\varepsilon_{Ps} = 1.5\varepsilon$.

For geometrical reasons, a direct comparison of the density profiles around the core with those calculated near the flat surface is impossible. However, the polymers grafted to the flat surface formed more segment layers, and they were more stretched. This polymer layer was more compact. In contrast to the hairy particles, on flat surfaces, ternary adsorption was found. Particles penetrated less into such a dense polymer layer and accumulated at its outer part and above it.

In Figure 6, we show examples of configurations of flat brushes immersed in a compatible fluid. The grafting density was the same as in the case of the larger core with attached $f = 20$ chains. In part a, the whole simulation box is shown for $M = 10$. The side sections of the systems are also presented for $M = 10$ (b) and $M = 20$ (c). We see that the adsorption on the layer consisting of a longer chain was considerably greater. This was manifested by considerably smaller density in the bulk phase. Ternary adsorption is clearly visible in the snapshots. One can see that there was less available space between chains than in the case of the hairy particles.

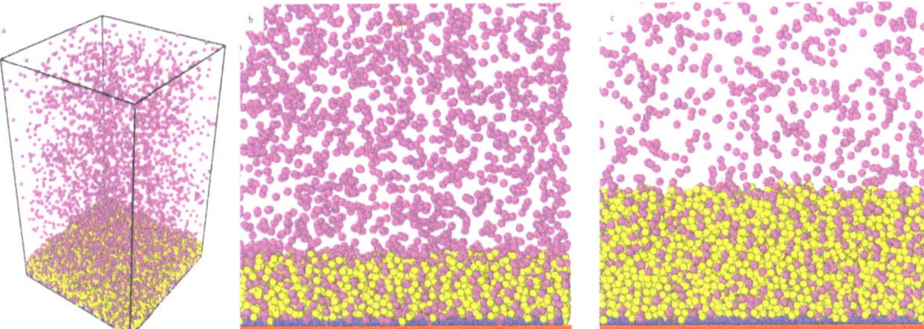

Figure 6. Examples of the equilibrium configurations of the system (M1) involving flat surfaces modified with chains of different lengths: $M = 10$ (**a**,**b**) and $M = 20$. Other parameters: $\rho_{gr} = 0.398$, $\varepsilon_{Ps} = 1.5\varepsilon$. The red line represents the flat substrate. Blue spheres correspond to bonding segments, yellow spheres and pink spheres represent the remaining segments and fluid particles P, respectively. In parts (**b**,**c**), sideways views of simulation boxes are shown.

Our study proves that the curvature of the substrate influenced the adsorption and structural properties of the brushes. Experimental investigations of Marschelke et al. [56,57] showed that there was no direct transferability of the results received from planar to curved substrates. Our simulations confirm their conclusion.

3.6. Other Models for Adsorption on Ligand-Tethered Nanoparticles

As already mentioned, we also performed a few simulations for models M1a, M2, and M2a for $\sigma = 4\sigma$, $M = 10$, and $f = 20$. The results were compared to those obtained for the same geometrical parameters and the basic model M1. We show how the adsorption depended on the degree of flexibility of the ligands, the strength of interactions between the particle and ligand segments, and attractive interactions between the adsorbed particles and between segments.

In Figure 7, excess adsorption isotherms for different model systems are shown. As one can predict, the adsorption was greater for stronger particle–segment interactions. Increasing the energy parameter ε can cause the disappearance of the local maximum at low density in the excess adsorption isotherm (compare the solid blue and green lines).

The attractive particle–particle and segment–segment interactions intensified the adsorption (compare the red and green lines or black and blue lines). Adsorbed particles additionally attract others, and a wider "adsorption layer" is formed. This effect is particularly visible for $\varepsilon_{Ps} = 1.5\varepsilon$ (blue solid line) at high bulk densities. Moreover, it can be seen that, for rigid ligands, adsorption is always greater than for flexible tethered chains.

This effect is minor for $\varepsilon_{Ps} = 1.5\varepsilon$ (black lines) and clearly visible for stronger particle–segment interactions, $\varepsilon_{Ps} = 3.0\varepsilon$, (red lines). The difference in the behavior of particles with rigid and flexible ligands became much more significant for attractive particle–particle and segment–segment interactions (green lines). This brief discussion shows that adsorption was highly dependent on the details of the model used.

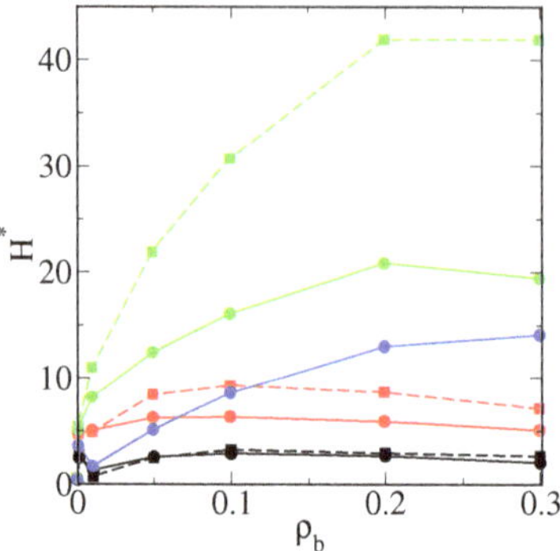

Figure 7. Excess adsorption isotherms on the particle with attached flexible (circles, solid lines) and rigid ligands (squares, dashed lines) for different models of adsorption: models M1 and M2 with $\varepsilon = 1.5\varepsilon$ (black), models M1 and M2 with $\varepsilon = 3.0\varepsilon$ (red), models M1a and M2a with $\varepsilon = 3.0\varepsilon$ (green), and model M1a with $\varepsilon = 1.5\varepsilon$ (blue). Other parameters: $f = 20$, $M = 10$. Symbols correspond to simulation points. Lines serve as a guide to the eye.

It should be stressed that the effect of "superadsorption" at very low densities also occurred for the rigid ligands that did not change their configurations. Thus, it results from a special situation of particles in the confined space near the curved surface between ligands, which act as obstacles. The behavior of particles depends on the resultant of all forces in the systems. In turn, the effective potential is shaped by all pair–pair interactions and changes as the density increases. For a special combination of the system parameters, the effect of "superadsorption" at low densities can be observed. In the case of stronger interactions, this phenomenon does not occur.

Figure 8 presents the density profiles of adsorbed particles and ligand segments at $\rho_0 = 0.1$, where the adsorption is the greatest. To a considerable degree, the structure of the adsorbed fluid replicates the structure of ligands. This was particularly apparent for the hedgehog-like particles (M2). In this case, a series of narrow peaks were observed in the density profiles.

For flexible chains, both the segment density profiles and the fluid density profiles had only a few peaks and decreased continuously in the outer part of the polymer corona. The density of the fluid in the whole surface layer increased for stronger particle–segment interactions and attraction between adsorbed particles. Note that the effect of attraction between particles and between segments was more significant. If these interactions are attractive, the segment density increases not only deep inside the polymer corona but also further from the core.

As mentioned, for flexible tethers, the presence of adsorbed particles influences their configurations. In the framework of the model M1, for stronger particle–segment interactions (compare black and red lines) the segment density was lower near the core, higher for in the middle of the polymer layer, and again slightly higher in the corona periphery. The adsorbed particles push the segments outward from the immediate vicinity of the core. On the other hand, the particles and segments "stick" to each other, allowing the chains to wrap themselves around the particles. The structure of the polymer layer is a result of a complex interplay between these effects.

The attractive interactions between particles and between segments also affected the structure of the polymer layer. For weaker particle–segment interactions ($\varepsilon_{Ps} = 1.5\varepsilon$) attraction between particles caused the segment density to increase inside the polymer corona ($r < 5.5\sigma$) and to decrease outside (black and blue lines). The increase was the strongest in the middle part of the corona. In the case of $\varepsilon_{Ps} = 3.0\varepsilon$, the opposite effect was observed in the interior of the brush (red and green lines). However, for $4.5\sigma < r < 5.5\sigma$, the attractions between particles caused an increase of the segment density. It did not influence the behavior of the outer part of the brush.

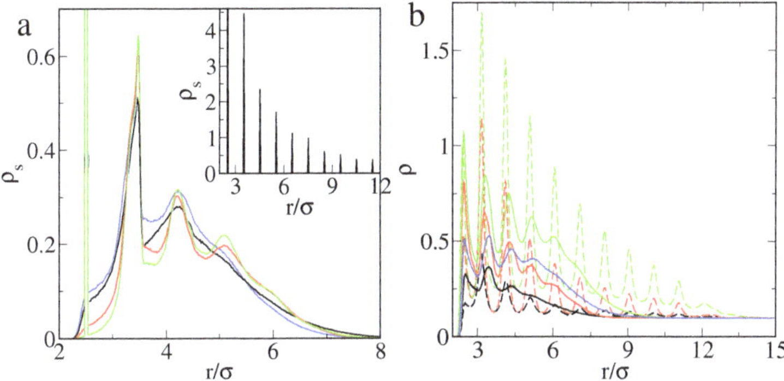

Figure 8. Density profiles of the chain segments (**a**) and fluid particles (**b**) around the core at $\rho_0 = 0.1$ for particles with attached flexible (solid lines) and rigid ligands (dashed lines) plotted for different models and values of ε_{Ps}: models M1 and M2 with 1.5ε (black lines), models M1 and M2 with $\varepsilon_{Ps} = 3.0\varepsilon$ (red lines), models M1a and M2a with $\varepsilon_{Ps} = 3.0\varepsilon$ (green lines), and model M1a with $\varepsilon_{Ps} = 1.5\varepsilon$ (blue line). Other parameters: $\sigma_c = 4\sigma$, $f = 20$, $M = 10$. In the inset of part a, the segment density profile for the hedgehog-like particle is shown.

Figure 9 illustrates the "superadsorption" effect at low densities for hedgehog-like particles. In part a, the fluid density profiles obtained for the model M1 with $\varepsilon_{Ps} = 1.5\varepsilon$ are shown. A change in the sequence of the profiles plotted for increasing densities ρ_0 is clearly visible (see inset). If $\varepsilon_{Ps} = 3.0\varepsilon$, the profiles for $\rho_0 = 0.001$ (red line) and $\rho_0 = 0.01$ (green line) are intertwined. In the latter case, the excess adsorption monotonically increased in the region of low densities.

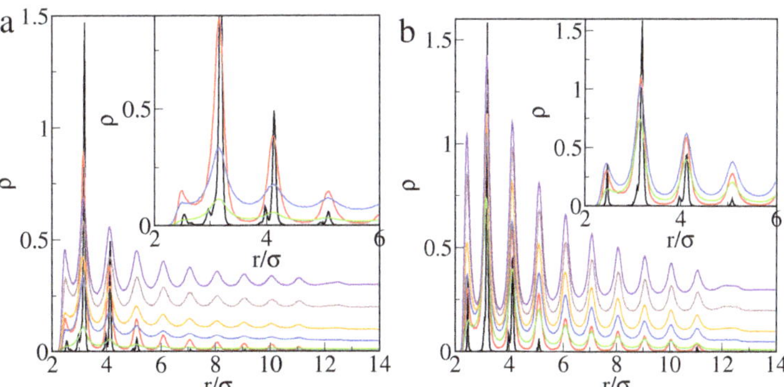

Figure 9. Density profiles of fluid particles around the core plotted for the hedgehog-like particle in the system of the M2-type and $\varepsilon_{Ps} = 1.5\varepsilon$ (**a**), 3.0ε (**b**) at different initial densities of the fluid $\rho_0 = 0.0001$ (black), 0.001 (red), 0.01 (green), 0.05 (blue), 0.1 (yellow), 0.2 (brown), and 0.3 (violet). In the insets, the initial parts of the density profiles are shown on a more accurate scale. Other parameters: $\sigma_c = 4\sigma$, $f = 20$, $M = 10$.

Figure 10 presents selected configurations for the particles with attached flexible ligands. The snapshots for the model M1 and $\varepsilon_{Ps} = 3\varepsilon$ for different densities ρ_0 are shown in the top row. In this case, for $\rho_0 = 0.0001$, the cone-like structure began to be formed. For weaker particle–segment interactions (see Figure 4d–f), the core-shell structure with loosely distributed segments was found at this fluid density. The cone-like structures occurred even for higher densities.

In the bottom row, we present the results for the same particle–segment interactions but for the model M1a. The cone-like structure was observed already at the lowest fluid density. However, at $\rho_0 = 0.1$, the polymer corona became less compact and more symmetrical. Likely, interactions between the adsorbed molecules and molecules in the bulk fluid pull the chains into the surrounding fluid, and the special structure is destroyed. Thus, attractive interactions between fluid molecules considerably affect the structure of hairy particles.

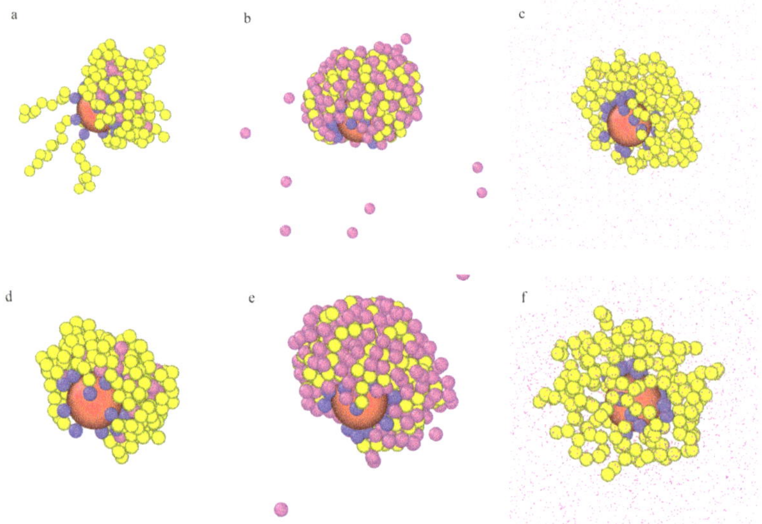

Figure 10. Examples of the equilibrium configurations of hairy particles immersed in fluids of different densities $\rho_0 = 0.0001$ (**a**,**d**), 0.001 (**b**,**e**), 0.1 (**c**,**f**) for model M1 (**a**–**c**), model M1a (**d**–**f**): $f = 20$, $M = 20$ (**a**–**c**), $f = 20$, $M = 10$ (**d**–**f**). Other parameters: $\sigma_c = 4\sigma$, $f = 20$, $M = 10$ and $\varepsilon_{Ps} = 3.0\varepsilon$. The red sphere represents the core, and blue spheres correspond to bonding segments, yellow spheres and pink spheres represent the remaining segments and fluid particles P, respectively. For clarity, particles P are represented by points (size ratios are not kept) in parts (**c**,**f**).

Finally, we discuss the configurations for the particle with stiff ligands (Figure 11). In this case, the core and ligands form a rigid backbone of the particles and no reconfiguration is possible. In part a, a typical configuration at very low density ($\rho_0 = 0.0001$) is shown. Parts a and b illustrate the behavior of the "basic model" M1. When the density ρ_0 increased to $\rho_0 = 0.001$, the particles accumulated near the ligands and between them.

The remaining pictures are for this higher density but for different system parameters. If particle–segment attraction was stronger, the adsorption was markedly greater (Figure 11c). The effect of attractive interactions between particles was even more spectacular as seen in comparing configurations (Figure 11b,d). Particles "condense" in the volume between rigid ligands.

In summary, the behavior of nanoparticles with stiff or flexible ligands is significantly different. The details of the models can influence the results of simulations. Nevertheless, the fundamental features of the systems are quite well imitated by the simplest models M1 or M2.

The study showed that the adsorption of particles on hairy particles is a very complex process that depends on many parameters and relations between them. In general, for the system studied, adsorption rose with increasing the grafting density, the length of chains, and the particle–segment interactions.

The general trends observed in our simulations were consistent with previous experimental results [56–58,64].

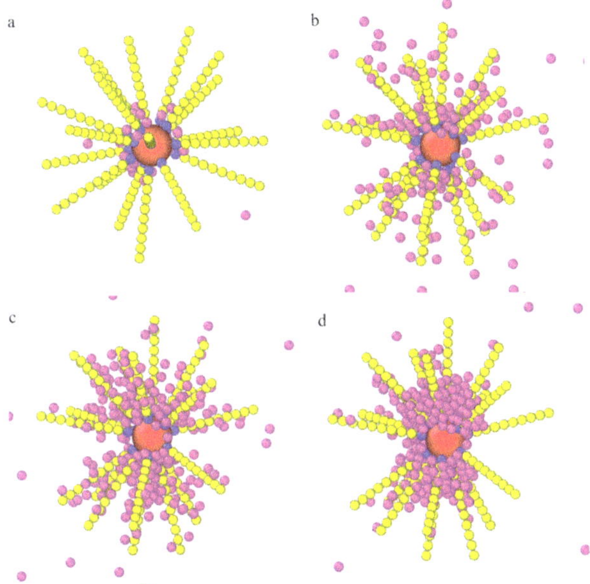

Figure 11. Examples of the equilibrium configurations of the hedgehog-like particles immersed in fluids of densities $\rho_0 = 0.0001$ (**a**) and 0.001 (**b**–**d**) for different models: (**a**,**b**) model M2, $\varepsilon_{Ps} = 1.5\varepsilon$, (**c**) model M2, $\varepsilon_{Ps} = 3.0\varepsilon$, and (**d**) model M2a, $\varepsilon_{Ps} = 3.0\varepsilon$. Other parameters: $\sigma_c = 4\sigma$, $f = 20$, $M = 10$. The red sphere represents the core, blue spheres correspond to bonding segments, and yellow spheres and pink spheres represent the remaining segments and fluid particles P, respectively.

4. Conclusions

This work presents the results of coarse-grained simulations for different polymer-functionalized spherical nanoparticles immersed in an explicit fluid consisting of small particles P. We considered the particles with ligands permanently anchored at randomly chosen points at the core surface. We studied nanoparticles with attached flexible chains and rigid ligands.

First, hairy particles modified with perfectly flexible chains were considered. The simplest possible model in which only particle–segment interactions are attractive while the remaining pair interactions are repulsive was discussed. For this model, we demonstrated that, in a wide region of particle densities, the excess adsorption on a hairy nanoparticle increased for longer chains and higher grafting densities. The adsorption properties of hairy nanoparticles of different sizes and flat surfaces modified with grafted chains were compared. When the grafting density was kept constant, the normalized excess adsorption on the flat brush was considerably smaller.

However, the effect of the core size was negligible in a wide region of densities. This was significant only in the case of very low or relatively high bulk densities of adsorbed particles. For the systems with assumed attractive particle–particle and segment–segment interactions, the excess adsorption was greater than in our "basic" model system, where these interactions were repulsive. Our conclusions regarding the adsorption on hairy particles are in line with previous experimental observations [56–58].

Adsorption on hairy particles changes the morphology of the polymer corona. Our simulations elucidated the mechanism of adsorption-induced nanoparticle shell reconfiguration of the nanoparticle shell. Adsorbed particles form the bridges between segments belonging both to the same chain and different chains [50,51]. The chains form coils with the particles trapped inside and joined together. We computed the average thickness of polymer coatings for different fluid densities and stated that this is strictly correlated with the excess adsorption isotherms.

We analyzed the dependencies of the brush thickness on the density of adsorbed particles and compared them with the theoretical predictions [19,62,63] and experimental results [56]. Depending on the assumed parameters core-shell or cone-like structures of hairy particles were found in the model systems. In the case of the cone-like particles, the segment cloud was highly asymmetrical. The chains with adsorbed particles accumulated on a part of the core surface, while the remainder was uncovered. The polymer layer could be more or less compact. We found continuous adsorption-induced structural transitions in polymer coatings. The cone-like structure transformed into the core-shell structure at higher densities.

The adsorption-induced local bundling of chains on the nanoparticle surface was observed in the simulations carried out for other systems [48,50] as well as in experiments [56]. In the case of short chains, "the superadsorption" was observed at very low particle densities. The excess adsorption isotherm had a low local maximum at a very low density. For higher densities, however, the excess adsorption isotherm had a typical course. Likely, this follows from the superposition of the effect associated with the confinement of particles near the curved surface between ligands (which act as obstacles) and the effect of the interplay between all interactions in the system. This anomaly occurs only for special combinations of parameters and disappears for stronger attractive interactions.

Finally, we discuss adsorption on the hedgehog-like particles with attached rigid ligands. The adsorption on such particles is higher than on the corresponding hairy particles with flexible chains. If particle–particle interactions are attractive, we observe a "condensation" of fluid in "pores" between ligands.

In summary, we demonstrated how the model parameters affected adsorption on hairy nanoparticles. We analyzed the mechanism of adsorption and the structure of the polymer coating with details. The adsorption on hairy particles follows from competition between interactions near the surface and in the bulk phase and the entropy effects associated with the limiting of possible chain configurations near the core. Hairy particles with flexible chains are "living" adsorbents, which makes the research difficult.

Future studies of adsorption on ligand-tethered nanoparticles can proceed on several fronts. The density functional theory of adsorption on flat brushes [26–29,33,34] can be adapted to describe the adsorption on hairy nanoparticles. The theoretical predictions can be compared with the results of our simulations. Moreover, the molecular dynamics simulation of adsorption from different explicit solvents will be performed. We hope that our results will help to rationally design hairy particles that could be carriers of bioactive compounds.

Author Contributions: The manuscript was written through equal contributions from both of the authors. Both authors have given approval to the final version of the manuscript.

Funding: This research received no external funding.

Institutional Review Board Statement: Not applicable.

Conflicts of Interest: The authors declare no conflict of interest.

References

1. Barbey, R.; Lavanant, L.; Paripovic, D.; Schüwer, N.; Sugnaux, C.; Tugulu, S.; Klok, H.A. Polymer Brushes via Surface-Initiated Controlled Radical Polymerization: Synthesis, Characterization, Properties, and Applications. *Chem. Rev.* **2009**, *109*, 5437–5527. [CrossRef]
2. Moffitt, M.G. Self-Assembly of Polymer Brush-Functionalized Inorganic Nanoparticles: From Hairy Balls to Smart Molecular Mimics. *J. Phys. Chem. Lett.* **2013**, *4*, 3654–3666. [CrossRef]
3. Zhao, B.; Zhu, L. Mixed Polymer Brush-Grafted Particles: A New Class of Environmentally Responsive Nanostructured Materials. *Macromolecules* **2009**, *42*, 9369–9383. [CrossRef]
4. Pyun, J.; Matyjaszewski, K. Synthesis of Nanocomposite Organic/Inorganic Hybrid Materials Using Controlled/"Living" Radical Polymerization. *Chem. Mater.* **2001**, *13*, 3436–3448. [CrossRef]
5. Wu, T.; Zou, G.; Hu, J.; Liu, S. Fabrication of Photoswitchable and Thermotunable Multicolor Fluorescent Hybrid Silica Nanoparticles Coated with Dye-Labeled Poly(N-isopropylacrylamide) Brushes. *Chem. Mater.* **2009**, *21*, 3788–3798. [CrossRef]
6. Pengo, P.; Şologan, M.; Pasquato, L.; Guida, F.; Pacor, S.; Tossi, A.; Stellacci, F.; Marson, D.; Boccardo, S.; Pricl, S.; et al. Gold nanoparticles with patterned surface monolayers for nanomedicine: Current perspectives. *Eur. Biophys. J.* **2017**, *46*, 749–771. [CrossRef]
7. Ulbrich, K.; Holá, K.; Šubr, V.; Bakandritsos, A.; Tuček, J.; Zbořil, R. Targeted Drug Delivery with Polymers and Magnetic Nanoparticles: Covalent and Noncovalent Approaches, Release Control, and Clinical Studies. *Chem. Rev.* **2016**, *116*, 5338–5431. [CrossRef]
8. Zhang, L.; Bei, H.P.; Piao, Y.; Wang, Y.; Yang, M.; Zhao, X. Polymer-Brush-Grafted Mesoporous Silica Nanoparticles for Triggered Drug Delivery. *ChemPhysChem* **2018**, *19*, 1956–1964. [CrossRef]
9. Rahman, H.; Hossain, M.R.; Ferdous, T. The recent advancement of low-dimensional nanostructured materials for drug delivery and drug sensing application: A brief review. *J. Mol. Liq.* **2020**, *320*, 114427. [CrossRef]
10. Banerjee, I.; Pangule, R.C.; Kane, R.S. Antifouling Coatings: Recent Developments in the Design of Surfaces That Prevent Fouling by Proteins, Bacteria, and Marine Organisms, *Adv. Mater.* **2011**, *23*, 690–718. [CrossRef]
11. Pavithra, D.; Doble, M. Biofilm formation, bacterial adhesion and host response on polymeric implants—Issues and prevention. *Biomed. Mater.* **2008**, *3*, 034003. [CrossRef] [PubMed]
12. Chivere, V.T.; Kondiah, P.P.D.; Choonara, Y.E.; Pillay, V. Nanotechnology-Based Biopolymeric Oral Delivery Platforms for Advanced Cancer Treatment. *Cancers* **2020**, *12*, 522. [CrossRef]
13. Horák, D.; Babič, M.; Macková, H.; Beneš, M.J. Preparation and properties of magnetic nano- and microsized particles for biological and environmental separations. *J. Sep. Sci.* **2007**, *30*, 1751–1772. [CrossRef]
14. Sheldon, R.A.; van Pelt, S. Enzyme immobilisation in biocatalysis: Why, what and how. *Chem. Soc. Rev.* **2013**, *42*, 6223–6235. [CrossRef]
15. Currie, E.; Norde, W.; Cohen Stuart, M. Tethered polymer chains: Surface chemistry and their impact on colloidal and surface properties. *Adv. Colloid Interface Sci.* **2003**, *100–102*, 205–265. [CrossRef]
16. Descas, R.; Sommer, J.U.; Blumen, A. Grafted Polymer Chains Interacting with Substrates: Computer Simulations and Scaling. *Macromol. Theory Simul.* **2008**, *17*, 429–453. [CrossRef]
17. Ballauff, M.; Borisov, O. Polyelectrolyte brushes. *Curr. Opin. Colloid Interface Sci.* **2006**, *11*, 316–323. [CrossRef]
18. Becker, A.L.; Henzler, K.; Welsch, N.; Ballauff, M.; Borisov, O. Proteins and polyelectrolytes: A charged relationship. *Curr. Opin. Colloid Interface Sci.* **2012**, *17*, 90–96. [CrossRef]
19. Binder, K.; Milchev, A. Polymer brushes on flat and curved surfaces: How computer simulations can help to test theories and to interpret experiments. *J. Polym. Sci. Part B* **2012**, *50*, 1515–1555. [CrossRef]
20. Ohno, K.; Morinaga, T.; Takeno, S.; Tsujii, Y.; Fukuda, T. Suspensions of Silica Particles Grafted with Concentrated Polymer Brush: Effects of Graft Chain Length on Brush Layer Thickness and Colloidal Crystallization. *Macromolecules* **2007**, *40*, 9143–9150. [CrossRef]
21. Ohno, K.; Morinaga, T.; Takeno, S.; Tsujii, Y.; Fukuda, T. Suspensions of Silica Particles Grafted with Concentrated Polymer Brush: A New Family of Colloidal Crystals. *Macromolecules* **2006**, *39*, 1245–1249. [CrossRef]
22. Daoud, M.; Cotton, J.P. Star shaped polymers: A model for the conformation and its concentration dependence. *J. Phys. France* **1982**, *43*, 531–538. [CrossRef]
23. Dan, N.; Tirrell, M. Polymers tethered to curves interfaces: A self-consistent-field analysis. *Macromolecules* **1992**, *25*, 2890–2895. [CrossRef]
24. Wijmans, C.M.; Zhulina, E.B. Polymer brushes at curved surfaces. *Macromolecules* **1993**, *26*, 7214–7224. [CrossRef]
25. Lo Verso, F.; Egorov, S.A.; Milchev, A.; Binder, K. Spherical polymer brushes under good solvent conditions: Molecular dynamics results compared to density functional theory. *J. Chem. Phys.* **2010**, *133*, 184901. [CrossRef] [PubMed]
26. Borówko, M.; Sokołowski, S.; Staszewski, T. A density functional approach to retention in chromatography with chemically bonded phases. *J. Chromatogr. A* **2011**, *1218*, 711–720. [CrossRef]
27. Borówko, M.; Rżysko, W.; Sokołowski, S.; Staszewski, T. Density Functional Approach to Adsorption and Retention of Spherical Molecules on Surfaces Modified with End-Grafted Polymers. *J. Phys. Chem. B* **2009**, *113*, 4763–4770. [CrossRef]
28. Borówko, M.; Sokołowski, S.; Staszewski, T. Adsorption on chemically bonded chain layers with embedded active groups. *Mol. Phys.* **2015**, *113*, 1014–1021. [CrossRef]

29. Borówko, M.; Sokołowski, S.; Staszewski, T. Adsorption from Binary Solutions on the Polymer-Tethered Surfaces. *J. Phys. Chem. B* **2012**, *116*, 3115–3124. [CrossRef]
30. Carignano, M.; Szleifer, I. Prevention of protein adsorption by flexible and rigid chain molecules. *Colloids Surfaces. Biointerfaces* **2000**, *18*, 169–182. [CrossRef]
31. Fang, F.; Szleifer, I. Effect of Molecular Structure on the Adsorption of Protein on Surfaces with Grafted Polymers. *Langmuir* **2002**, *18*, 5497–5510. [CrossRef]
32. Milchev, A.; Egorov, S.A.; Binder, K. Absorption/expulsion of oligomers and linear macromolecules in a polymer brush. *J. Chem. Phys.* **2010**, *132*, 184905. [CrossRef]
33. Borówko, M.; Sokołowski, S.; Staszewski, T. Adsorption of oligomers on the polymer-tethered surfaces. *J. Colloid Interface Sci.* **2011**, *356*, 267–276. [CrossRef] [PubMed]
34. Borówko, M.; Sokołowski, S.; Staszewski, T. Adsorption from Oligomer–Monomer Solutions on the Surfaces Modified with End-Grafted Chains. *J. Phys. Chem. B* **2012**, *116*, 12842–12849. [CrossRef] [PubMed]
35. Leermakers, F.A.M.; Ballauff, M.; Borisov, O.V. On the Mechanism of Uptake of Globular Proteins by Polyelectrolyte Brushes: A Two-Gradient Self-Consistent Field Analysis. *Langmuir* **2007**, *23*, 3937–3946. [CrossRef]
36. Jönsson, M.; Johansson, H.O. Effect of surface grafted polymers on the adsorption of different model proteins. *Colloids Surfaces. Biointerfaces* **2004**, *37*, 71–81. [CrossRef] [PubMed]
37. Halperin, A.; Kröger, M. Ternary Protein Adsorption onto Brushes: Strong versus Weak. *Langmuir* **2009**, *25*, 11621–11634. [CrossRef]
38. Halperin, A. Collapse of Thermoresponsive Brushes and the Tuning of Protein Adsorption. *Macromolecules* **2011**, *44*, 6986–7005. [CrossRef]
39. Tavanti, F.; Pedone, A.; Matteini, P.; Menziani, M.C. Computational Insight into the Interaction of Cytochrome C with Wet and PVP-Coated Ag Surfaces. *J. Phys. Chem. B* **2017**, *121*, 9532–9540. [CrossRef]
40. Borówko, M.; Pöschel, T.; Sokołowski, S.; Staszewski, T. Janus Particles at Walls Modified with Tethered Chains. *J. Phys. Chem. B* **2013**, *117*, 1166–1175. [CrossRef]
41. Staszewski, T.; Borówko, M. Janus dumbbells near surfaces modified with tethered chains. *Adsorption* **2019**, *25*, 459–468. [CrossRef]
42. Lane, J.M.D.; Ismail, A.E.; Chandross, M.; Lorenz, C.D.; Grest, G.S. Forces between functionalized silica nanoparticles in solution. *Phys. Rev. E* **2009**, *79*, 050501. [CrossRef]
43. Peters, B.L.; Lane, J.M.D.; Ismail, A.E.; Grest, G.S. Fully Atomistic Simulations of the Response of Silica Nanoparticle Coatings to Alkane Solvents. *Langmuir* **2012**, *28*, 17443–17449. [CrossRef]
44. Salerno, K.M.; Ismail, A.E.; Lane, J.M.D.; Grest, G.S. Coating thickness and coverage effects on the forces between silica nanoparticles in water. *J. Chem. Phys.* **2014**, *140*, 194904. [CrossRef] [PubMed]
45. Bolintineanu, D.S.; Lane, J.M.D.; Grest, G.S. Effects of Functional Groups and Ionization on the Structure of Alkanethiol-Coated Gold Nanoparticles. *Langmuir* **2014**, *30*, 11075–11085. [CrossRef]
46. Lane, J.M.D.; Grest, G.S. Assembly of responsive-shape coated nanoparticles at water surfaces. *Nanoscale* **2014**, *6*, 5132–5137. [CrossRef]
47. Brancolini, G.; Tozzini, V. Building Minimalist Models for Functionalized Metal Nanoparticles. *Front. Mol. Biosci.* **2019**, *6*, 50. [CrossRef] [PubMed]
48. Chew, A.K.; Dallin, B.C.; Van Lehn, R.C. The Interplay of Ligand Properties and Core Size Dictates the Hydrophobicity of Monolayer-Protected Gold Nanoparticles. *ACS Nano* **2021**, *15*, 4534–4545. [CrossRef]
49. Tavanti, F.; Pedone, A.; Menziani, M.C.; Alexander-Katz, A. Computational Insights into the Binding of Monolayer-Capped Gold Nanoparticles onto Amyloid-β Fibrils. *ACS Chem. Neurosci.* **2020**, *11*, 3153–3160. [CrossRef] [PubMed]
50. Staszewski, T.; Borówko, M. Adsorption-induced co-assembly of hairy and isotropic particles. *Phys. Chem. Chem. Phys.* **2020**, *22*, 8757–8767. [CrossRef]
51. Wang, S.; Chen, K.; Xu, Y.; Yu, X.; Wang, W.; Li, L.; Guo, X. Protein immobilization and separation using anionic/cationic spherical polyelectrolyte brushes based on charge anisotropy. *Soft Matter* **2013**, *9*, 11276–11287. [CrossRef]
52. Henzler, K.; Haupt, B.; Lauterbach, K.; Wittemann, A.; Borisov, O.; Ballauff, M. Adsorption of β-Lactoglobulin on Spherical Polyelectrolyte Brushes: Direct Proof of Counterion Release by Isothermal Titration Calorimetry. *J. Am. Chem. Soc.* **2010**, *132*, 3159–3163. [CrossRef] [PubMed]
53. Aydin, F.; Uppaladadium, G.; Dutt, M. Harnessing steric hindrance to control interfacial adsorption of patchy nanoparticles onto hairy vesicles. *Colloids Surfaces. Biointerfaces* **2016**, *141*, 458–466. [CrossRef] [PubMed]
54. Welsch, N.; Lu, Y.; Dzubiella, J.; Ballauff, M. Adsorption of proteins to functional polymeric nanoparticles. *Polymer* **2013**, *54*, 2835–2849. [CrossRef]
55. Becker, A.L.; Welsch, N.; Schneider, C.; Ballauff, M. Adsorption of RNase A on cationic polyelectrolyte brushes: A study by isothermal titration calorimetry. *Biomacromolecules* **2011**, *12*, 3936–3944. [CrossRef]
56. Marschelke, C.; Raguzin, I.; Matura, A.; Fery, A.; Synytska, A. Controlled and tunable design of polymer interface for immobilization of enzymes: does curvature matter? *Soft Matter* **2017**, *13*, 1074–1084. [CrossRef]
57. Marschelke, C.; Müller, M.; Köpke, D.; Matura, A.; Sallat, M.; Synytska, A. Hairy Particles with Immobilized Enzymes: Impact of Particle Topology on the Catalytic Activity. *ACS Appl. Mater. Interfaces* **2019**, *11*, 1645–1654. [CrossRef]

58. Yan, X.; Kong, J.; Yang, C.; Fu, G. Facile synthesis of hairy core–shell structured magnetic polymer submicrospheres and their adsorption of bovine serum albumin. *J. Colloid Interface Sci.* **2015**, *445*, 9–15. [CrossRef]
59. Toxvaerd, S.; Dyre, J.C. Communication: Shifted forces in molecular dynamics. *J. Chem. Phys.* **2011**, *134*, 081102. [CrossRef]
60. Available online: http://lammps.sandia.gov (accessed on 29 April 2019).
61. Plimpton, S. Fast Parallel Algorithms for Short-Range Molecular Dynamics. *J. Comput. Phys.* **1995**, *117*, 1–19. [CrossRef]
62. Borówko, M.; Patrykiejew, A.; Pizio, O.; Sokołowski, S. Changes in the structure of tethered chain molecules as predicted by density functional approach. *Condens. Matter Phys.* **2011**, *14*, 33604. [CrossRef]
63. Borówko, M.; Sokołowski, S.; Staszewski, T. Adsorption-induced changes of the structure of the tethered chain layers in a simple fluid. *J. Chem. Phys.* **2014**, *140*, 234904. [CrossRef] [PubMed]
64. Kudina, O.; Zakharchenko, A.; Trotsenko, O.; Tokarev, A.; Ionov, L.; Stoychev, G.; Puretskiy, N.; Pryor, S.W.; Voronov, A.; Minko, S. Highly Efficient Phase Boundary Biocatalysis with Enzymogel Nanoparticles. *Angew. Chem. Int. Ed.* **2014**, *53*, 483–487. [CrossRef] [PubMed]

International Journal of
Molecular Sciences

MDPI

Article

Phase Transitions in Two-Dimensional Systems of Janus-like Particles on a Triangular Lattice

Andrzej Patrykiejew

Department of Theoretical Chemistry, Institute of Chemical Sciences, Faculty of Chemistry, MCS University, 20031 Lublin, Poland; andrzej.patrykiejew@poczta.umcs.lublin.pl

Abstract: We studied the phase behavior of two-dimensional systems of Janus-like particles on a triangular lattice using Monte Carlo methods. The model assumes that each particle can take on one of the six orientations with respect to the lattice, and the interactions between neighboring particles were weighted depending on the degree to which their A and B halves overlap. In this work, we assumed that the AA interaction was fixed and attractive, while the AB and BB interactions varied. We demonstrated that the phase behavior of the systems considered strongly depended on the magnitude of the interaction energies between the AB and BB halves. Here, we considered systems with non-repulsive interactions only and determined phase diagrams for several systems. We demonstrated that the phase diagram topology depends on the temperature at which the close-packed systems undergo the orientational order–disorder transition.

Keywords: Janus particles; phase transitions; Monte Carlo simulation

Citation: Patrykiejew, A. Phase Transitions in Two-Dimensional Systems of Janus-like Particles on a Triangular Lattice. *Int. J. Mol. Sci.* **2021**, *22*, 10484. https://doi.org/10.3390/ijms221910484

Academic Editor: Marco Pettini

Received: 6 September 2021
Accepted: 23 September 2021
Published: 28 September 2021

1. Introduction

Janus particles have a surface composed of two chemically different patches, A and B [1,2]. The surface chemical anisotropy, which can be tuned by an appropriate fictionalization, results in orientation-dependent interactions. The chemical composition and the size of patches, influence both self-assembly and the formation of different ordered structures in two- and three-dimensional systems [3–8]. In the region of low and moderate densities, the formation of micelles, vesicles, and worm-like clusters has been observed [9–11]. It was also shown that Janus particles form crystals of different structures and density [11].

The behavior of dense two-dimensional systems of Janus particles has been recently studied by several authors [3,12–14]. Shin and Schweizer [3] used the Kern–Frenkel model [15] and developed a version of self-consistent phonon theory, which predicted the formation of different orientationally ordered hexagonal phases. The structure of these phases was found to be primarily determined by the so-called Janus balance [16], defined by the size of the attractive patch. Shin and Schweizer showed also that such systems may undergo phase transitions between different orientationally ordered phases. Similar orientationally ordered structures were observed by Iwashita and Kimura [4]. On the other hand, experimental study and Monte Carlo simulation of Jiang et al. [5] showed the formation of a glass-like phase, instead of the theoretically predicted zigzag phase [3].

In our recent paper [13], we studied the orientational order–disorder transitions in closely packed two-dimensional systems of Janus particles, using a simple lattice model, which allowed for only six different orientations of each particle. It was demonstrated that the nature of the transition is entirely determined by the sign of the parameter $\epsilon = u_{AA} + u_{BB} - 2u_{AB}$, where u_{AA}, u_{AB} and u_{BB} are the energies of interaction between the nearest neighbor pairs with their AA, AB, and BB halves facing one another. The parameter ϵ determines whether the contacts between the like (AA and BB) or unlike (AB) halves are favored.

When $\epsilon < 0$, the systems were shown to order into the zigzag phase, with the order–disorder transition belonging to the universality class of the three-state Potts model [17].

On the other hand, when $\epsilon > 0$, the ordered phase was found to be different, with the order–disorder transition belonging to the universality class of the four-state Potts model [17]. Here, we should mention that various lattice models (Ising, Potts, etc.) are commonly used to describe the behavior of diverse physical systems [18,19].

We also studied the phase behavior of Janus-like particles [14] using the same lattice model with $u_{AB} = u_{BB} = 0$, i.e., with the interaction potential similar to that proposed by Kern and Frenkel [15]. The model was studied using the Monte Carlo method in the grand canonical ensemble, and two versions of the model were considered. In the first version, the strength of attractive interaction, confined to the A halves of neighboring particles, was assumed to depend on the degree to which they were overlapping. In the second version, it was assumed that the interaction energy between a pair of neighboring particles was the same for any mutual orientations, in which their A patches overlapped to any extent.

It was demonstrated that both versions of the model led to qualitatively different results. In the case of the first version, the self-assembly was found to lead to different stripped structures, depending on the density and the temperature. In particular, we found that, at sufficiently low temperatures, the condensation led from a very dilute lamellar gas phase to the high density ordered zigzag phase. At intermediate temperatures, the system underwent two first-order phase transitions. The first led to the condensation of the gas phase into the partially ordered, phase (Z_2), with kinked stripes that were predominantly ordered along two axes of the lattice.

The second transition occurred between the Z_2 phase and the high density well-ordered zigzag phase (Z). At sufficiently high temperatures, only one continuous transition, between the disordered fluid-like and the ordered zigzag phases, was observed. In the case of the second model, we found only one first-order transition at low temperatures. This transition occurs between a dilute gas-like phase and the ordered phase, which forms a kagome lattice of the density equal to 6/7. A further increase of the density was demonstrated to lead to the reorientation of particles and the formation of dense glass-like structure, similar to that observed by Jiang et al. [5].

Thus, the phase behavior was demonstrated to be sensitive to the magnitude of attractive interaction acting between differently oriented particles.

The primary aim of this work is to discuss the phase behavior of two-dimensional systems of Janus-like particles with the tuned interactions between their different parts. In particular, we were interested in the question of how the phase behavior is affected by the stability of the dense ordered phase and by the strength of attractive interactions between particles.

To this end, we applied the same lattice gas model as used in [13,14,20] and assumed that the interaction between the neighboring Janus particles depends on the degrees to which their different parts overlap. We considered three series of systems in which the A-A interaction was fixed while A-B and B-B interactions were varied. In the first series, the AB and BB interactions were assumed to be the same, $u_{AB} = u_{BB} = u^*$. In the second (third) series, u_{AB} (u_{BB}) was assumed to be equal to zero, while u_{BB} (u_{AB}) was varied.

Only the second series, with $u_{AB} = 0$, appeared to mimic real Janus particles with hydrophilic and hydrophobic parts [2,3,14]. However, the other two series are also of interest, since each of them has demonstrated a little different phase behavior.

2. The Model and Methods

As already mentioned, the model used here is quite similar that considered in [13,14,20]. Thus, the Janus particles placed on a triangular lattice were assumed to be made of two halves A and B, and each particle was assumed to take on one of the six orientations, defined by the angle $\theta(k) = (k-1)(2\pi/6)$ ($k = 1, \ldots, 6$), measured with respect to the x-axis (see Figure 1a). Throughout this work, we assumed that all interactions were short-ranged and limited to the first nearest neighbors.

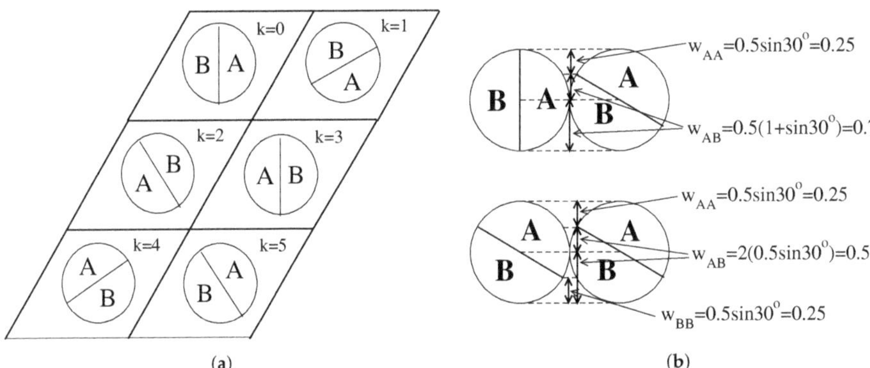

Figure 1. (**a**) The six possible orientations of a particle on a triangular lattice. (**b**) Two representative examples of differently oriented Janus particles, with $(k,l) = (1,6)$ and $(6,6)$ and the separation vector $\boldsymbol{r} = \boldsymbol{a}_1$, demonstrating how the weights determining the interaction energy between neighboring particles were calculated.

The interaction energy between a pair of particles located on adjacent sites i and j, $u(k_i, k_j, \boldsymbol{r}_{ij})$, was assumed to depend on their orientations, k_i and k_j, defined by the angles, $\theta(k_i)$ and $\theta(k_j)$, as well as on the separation vector, $\boldsymbol{r}_{ij}$, of unit length. The triangular lattice is described by three unit vectors: $\boldsymbol{a}_1 = (1,0)$, $\boldsymbol{a}_2 = (0.5, \sqrt{3}/2)$, and $\boldsymbol{a}_3 = (-0.5, \sqrt{3}/2)$, and hence there are six different separation vectors, which are equal to $\pm\boldsymbol{a}_i$ $(i = 1,2,3)$.

To each particle, we assigned the spin vector of unit length, $\boldsymbol{S} = (\cos(\theta), \sin(\theta))$, and hence $u(k_i, k_j, \vec{r}_{ij})$ can be written as $u(\boldsymbol{S}_i, \boldsymbol{S}_j, \boldsymbol{r}_{ij})$. Moreover, we assumed that the energy of interaction between a pair of neighboring particles depends on the degree to which their various halves overlap. This leads to the following expression for $u(\boldsymbol{S}_i, \boldsymbol{S}_j, \boldsymbol{r}_{ij})$:

$$u(\boldsymbol{S}_i, \boldsymbol{S}_j, \boldsymbol{r}_{ij}) = w_{AA}(\boldsymbol{S}_i, \boldsymbol{S}_j, \boldsymbol{r}_{ij})u_{AA} + w_{AB}(\boldsymbol{S}_i, \boldsymbol{S}_j, \boldsymbol{r}_{ij})u_{AB} + w_{BB}(\boldsymbol{S}_i, \boldsymbol{S}_j, \boldsymbol{r}_{ij})u_{BB} \ , \quad (1)$$

where u_{AA}, u_{AB}, and u_{BB} are the interaction energies corresponding to the orientations, in which the AA, AB or BB halves face one another, while $w_{AA}(\boldsymbol{S}_i, \boldsymbol{S}_j, \boldsymbol{r}_{ij})$, $w_{AB}(\boldsymbol{S}_i, \boldsymbol{S}_j, \boldsymbol{r}_{ij})$, and $w_{BB}(\boldsymbol{S}_i, \boldsymbol{S}_j, \boldsymbol{r}_{ij})$ are the weights, determined by the degrees to which the AA, AB, and BB regions overlap for given relative orientations, specified by $\boldsymbol{S}_i$ and $\boldsymbol{S}_j$, and locations, specified by the separation vector $\boldsymbol{r}_{ij}$ (see Figure 1b). There are 12 different values of the pair interaction energy, as summarized in Table 1. In Table 1, we also give the orientations of pairs of neighboring particles corresponding to different values of $u(\boldsymbol{S}_i, \boldsymbol{S}_j, \boldsymbol{r}_{ij})$, for $\boldsymbol{r}_{ij} = \boldsymbol{a}_1$.

Table 1. Possible different values of the interaction energy between a pair of neighboring particles and the pairs of orientations of a given energy for the separation vector $vecr = (1,0)$.

$u_1 =$	u_{AA}	(0,3)
$u_2 =$	u_{AB}	(0,0), (1,5), (2,4), (3,3), (4,2), (5,1)
$u_3 =$	u_{BB}	(3,0)
$u_4 =$	$0.75u_{AA} + 0.25u_{AB}$	(0,2), (0,4), (1,3), (5,3)
$u_5 =$	$0.75u_{AA} + 0.25u_{BB}$	(1,2), (5,4)
$u_6 =$	$0.75u_{BB} + 0.25u_{AB}$	(2,0), (3,1), (3,5), (4,0)
$u_7 =$	$0.75u_{BB} + 0.25u_{AA}$	(2,1), (4,5)
$u_8 =$	$0.75u_{AB} + 0.25u_{AA}$	(0,1), (0,5), (2,3), (4,3)
$u_9 =$	$0.75u_{AB} + 0.25u_{BB}$	(3,2) (5,0), (3,4), (1,0)
$u_{10} =$	$0.5u_{AB} + 0.25(u_{AA} + u_{BB})$	(1,1), (2,2), (4,4), (5,5)
$u_{11} =$	$0.5u_{AA} + 0.5u_{AB}$	(1,4), (5,2)
$u_{12} =$	$0.5u_{BB} + 0.5u_{AB}$	(2,5), (4,1)

To study the phase behavior, we used the Monte Carlo method in the grand canonical ensemble [21]. The Hamiltonian of the model reads

$$H = \frac{1}{2}\sum_{i,j} u(S_i, S_j, \mathbf{r}_{ij}) n_i n_j - N\mu \ , \tag{2}$$

where the sum runs over all pairs of nearest neighbors; $n_i = 1$, when the i-the site is occupied, and 0 otherwise; and N is the total number of particles in the system,

$$N = \sum_{i=1}^{L^2} n_i \tag{3}$$

and μ is the chemical potential. In the system with linear dimension L, the total density is equal to $\rho = N/L^2$, and the densities of differently oriented particles are defined as

$$\rho_k = \frac{1}{L^2}\sum_{i=1}^{L^2} n_i \delta(\theta(k_i) - \theta(k)). \tag{4}$$

Of course,

$$\rho = \frac{1}{L^2}\sum_{l=1}^{6} \rho_l \tag{5}$$

Throughout this work, we assumed that $u_{AA} = -1.0$, with $|u_{AA}|$ taken as the unit of energy, while the values of u_{AB} and u_{BB} were varied. The temperature, the chemical potential, and all other energy-like quantities are expressed in the reduced units.

The simulations were carried out for rhomboid cells of the size $L \times L$, with the standard periodic boundary conditions. Since the systems considered were found to form various ordered structures of different symmetry and density, we considered simulation cells of the sizes suitable to properly accommodate those structures in periodically repeated simulation cells.

The quantities recorded included the averages of the total density, $\langle \rho \rangle$, the densities of differently oriented particles, $\langle \rho_k \rangle$, the potential energy per site, $\langle u \rangle$, the heat capacity

$$C_V = \frac{1}{T^2}[\langle H^2 \rangle - \langle H \rangle^2], \tag{6}$$

and the density susceptibility per site

$$\chi_\rho = \frac{1}{T}[\langle \rho^2 \rangle - \langle \rho \rangle^2] \ . \tag{7}$$

To equilibrate the system, we used 10^6–10^7 Monte Carlo steps and another $5 \cdot 10^6$–10^8 Monte Carlo steps were used to calculate averages. Each Monte Carlo step involved $10 \cdot L^2$ attempts to change the state of the system. In the grand canonical ensemble, the possible changes of the system state involved either the creation of a particle on a randomly chosen site, with also a randomly chosen orientation or the removal of a randomly chosen particle. The simulation at a given temperature usually began at a sufficiently low value of the chemical potential, corresponding to a very low density, and then the chemical potential was gradually increased up to the values at which the nearly entire lattice was filled.

After the recording of such an "ascending" isotherm, we performed the run, starting at a high density, and recorded the "descending" isotherm. This procedure allowed locating the first-order phase transitions. In finite systems, the first-order transitions at low temperatures are usually accompanied by hysteresis loops, due to the presence of metastable states [21,22].

During the equilibration runs, the changes of the recorded quantities were monitored, and the equilibration was assumed to be complete when these quantities ceased to undergo systematic changes and showed only oscillations around average values. In some cases, this was achieved already after 10^5–$5 \cdot 10^6$ Monte Carlo steps; however, usually the equilibration required a considerably larger number of Monte Carlo steps, up to 10^7.

3. Results and Discussion

To begin, we consider the systems with $u_{AB} = u_{BB} = u^*$, assuming that $u^* \in [-1.0, -0.1]$. The isotherms, calculated at different temperatures and for different values of u^*, demonstrated that all these systems exhibit qualitatively the same behavior. Figure 2 presents the isotherms recorded for $u^* = -0.1$, and the systems with $u^* < -0.1$ led to quite similar results and the presence of the first-order transition at sufficiently low temperatures. The transition can be treated as the gas–liquid condensation, and it terminates in the critical point. The critical temperature, $T_c(u^*)$, gradually increases when u^* decreases from -0.1 to -1.0 (see Figure 3). In the particular case of $u^* = -1.0$, the critical temperature takes on the value of about 0.91, as predicted for the isotropic lattice gas model on a triangular lattice [23,24].

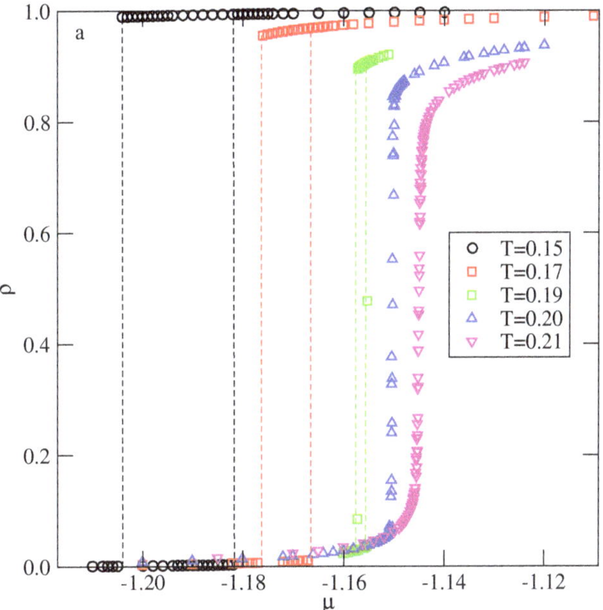

Figure 2. The examples of isotherms recorded for the system with $u^* = -0.1$ at different temperatures (given in the figure).

In [13,20], we show that the close-packed systems, with $\rho = 1.0$ and different values of u^*, undergo the order–disorder transition between the orientationally disordered phase and the ordered zigzag (Z) phase. The transition was demonstrated to be continuous and belonging to the universality class of the three-state Potts model, with the transition temperature decreasing linearly to zero, when u^* decreases toward -1.0. When $u^* = -1.0$, the interactions become isotropic, and hence no orientational order–disorder transition is possible. The locations of the orientational order–disorder transition, $T_o(u^*)$, are also included in Figure 3 and in all systems with $u^* \le -0.1$ the critical temperature is higher than $T_o(u^*)$.

Therefore, at sufficiently low temperatures, the gas should condense directly into the ordered Z phase, while at higher temperatures, but still lower than $T_c(u^*)$, the condensation should lead to the orientationally disordered condensed phase. The transition between the dense orientationally disordered and the ordered zigzag phase is expected to be continuous, just the same as in the close-packed systems. This implies that the line of the orientational order–disorder transition terminates in the critical end point located on the condensed phase branch of the gas-condensed phase coexistence.

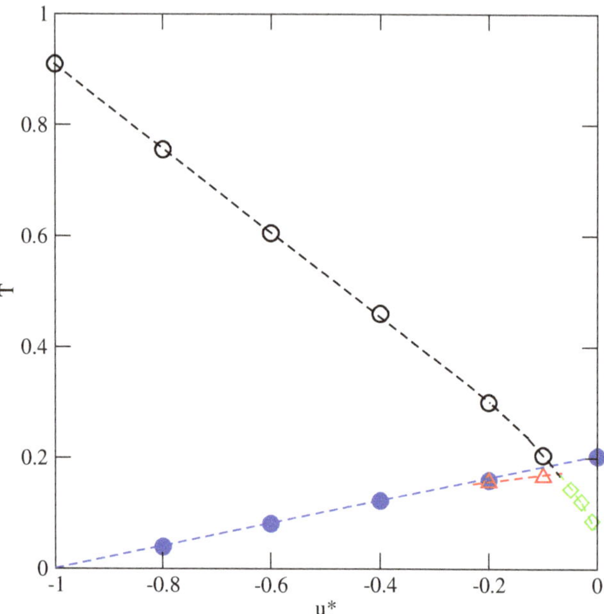

Figure 3. The changes of the critical temperature, $T_c(u^*)$, (circles), the critical end point temperature, $T_{cep}(u^*)$, (triangles) the orientational order−disorder transition temperature in close-packed systems, $T_o(u^*)$ (filled circles), and the tricritical point temperatures, $T_{trc}(u^*)$ (triangles), with u^*.

This scenario was found in the systems characterized by $u^* = -0.1$ and -0.2, in particular, when $u^* = -0.1$. The recorded densities of differently oriented particles, $\langle \rho_k \rangle$, along the isotherms at $T = 0.16$ and 0.17 demonstrated (see Figure 4) that, at $T = 0.16$, the gas condensation led to the orientationally ordered Z phase, in which four orientations were favored (cf. [20]), while, at $T = 0.17$, the gas condensation led to an orientationally disordered liquid.

However, the orientationally disordered liquid phase undergoes the transition to the ordered Z phase at the chemical potential $\mu \approx -1.12$, and the density is equal to about 0.99. From the results obtained at different temperatures, we estimated the phase diagram for this system, which is given in Figure 5. As expected, the line of continuous orientational order–disorder transition meet the gas-condensed phase coexistence at the critical end point, located at $T_{cep}(-0.1) \approx 0.167$, $\mu_{cep}(-0.1) \approx -1.175$, and $\rho_{cep}(-0.1) \approx 0.965$. In the system with $u^* = -0.2$, the critical end point is located at $T_{cep}(-0.2) \approx 0.16$, $\mu_{cep}(-0.2) \approx -1.33$, and at the density $\rho_{cep}(-0.2) \approx 0.997$.

Qualitatively, the same behavior is bound to occur in the systems with lower values of u^*. However, since $T_o(u^*)$ decreases when u^* becomes lower, the critical end point is shifted toward gradually decreasing temperatures, and toward the densities very close to unity. Already in the system with $u^* = -0.2$, the estimated density at the critical end point is very high.

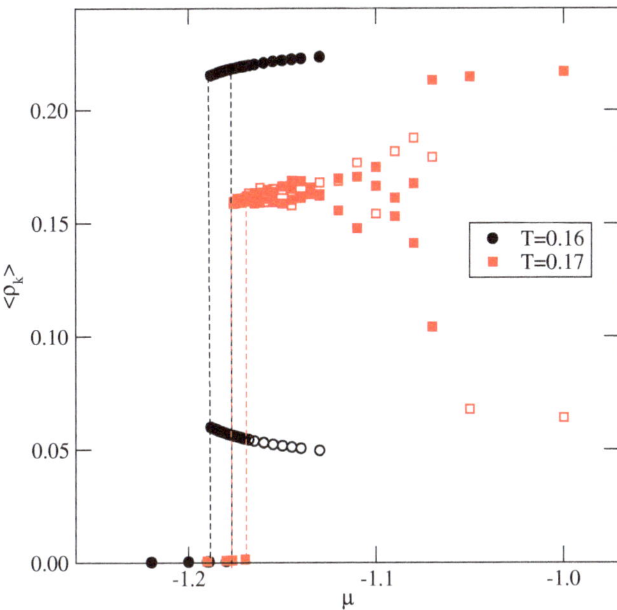

Figure 4. The densities of differently oriented particles along the isotherms obtained for the system with $u^* = -0.1$ at two temperatures, given in the figure. The filled (open) symbols correspond to the four favored (the two disfavored) orientations in the ordered zigzag structure.

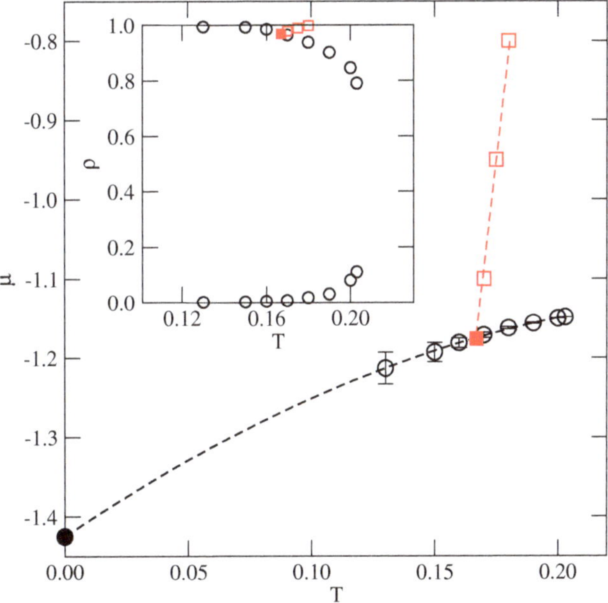

Figure 5. The estimated phase diagram for the system with $u^{|ast} = -0.1$. The main part shows the $T - \mu$ projection, and the inset gives the $T - \rho$ projection. Circles and squares represent the coexistence points of the first−order and continuous transitions, respectively. The filled square marks the location of the critical end point, while the filled circle (in the main part) shows the location of the first-order transition in the ground state.

The situation changes when u^* becomes higher than about -0.084, since the order–disorder transition temperatures, $T_o(u^*)$, exceed the expected critical temperatures (cf. Figure 3). This does not exclude the possibility that the phase diagrams may still look like those shown in Figure 5, however. The calculations carried out for the systems with $u^* = -0.05$ and -0.03 demonstrated a different behavior. The isotherms (see the main part of Figure 6) and the isothermal changes of the ratio $\langle\rho_k\rangle/\langle\rho\rangle$ (see the inset to Figure 6) obtained for the system with $u^* = -0.05$, demonstrated that, at the temperature $T = 0.14$, the first-order transition leads from the gas phase directly to the well developed Z structure.

At higher temperatures of $T = 0.15$ and 0.17, which are still lower than $T_o(-0.05) \approx 0.19$, the isotherms do not show the first-order transition between the gas and liquid phases, and the density smoothly increases with μ. However, at sufficiently high densities, a continuous transition, associated with the development of orientationally ordered Z phase, takes place at temperatures up to $T_o(-0.05)$. At the temperatures above $T_o(-0.05)$, we did not observe the formation of the Z phase at all.

The orientational order–disorder transition is not accompanied by any visible density anomalies along the isotherms, but it leads to large changes of the ratio $\langle\rho_k\rangle/\langle\rho\rangle$ (cf. the inset to Figure 6). In the orientationally disordered phase, all six orientations are equally probable, and hence $\langle\rho_k\rangle/\langle\rho\rangle \approx 1/6$. In the orientationally ordered phase, four orientations are favored, while the remaining two are disfavored. The transition is also accompanied by the appearance of heat capacity peaks of the height and location of maxima depending on the simulation cell size.

In Figure 7, we present examples of heat capacity curves obtained at $T = 0.15$ and for different sizes of the simulation cell. Here, we should recall that, in the close-packed system, the orientational order–disorder transition belongs to the universality class of the three-state Potts model [12,20]. Therefore, the observed transition also belongs to the same universality class. However, to confirm this prediction, one would need to evaluate the size dependence of the joint distribution of density and energy fluctuations, since the scaling fields comprise mixtures of temperature and chemical potential [25,26].

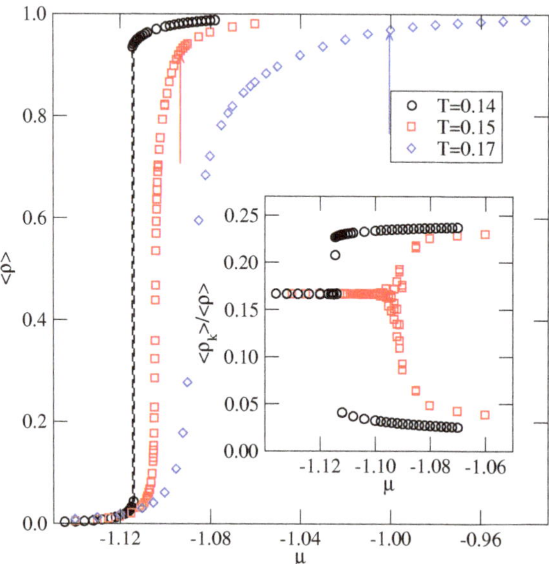

Figure 6. The main part shows the isotherms recorded at different temperatures for the system with $u^* = -0.05$, while the inset presents the changes of the ratio $\langle\rho_k\rangle/\langle\rho\rangle$ along the isotherms at $T = 0.14$ and 0.15. The arrows in the main part mark the locations of the orientational order−disorder transition.

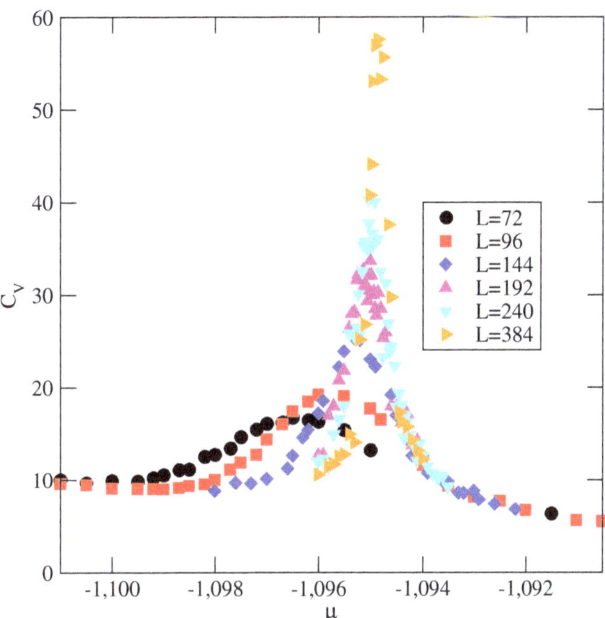

Figure 7. The heat capacity curves for the system with $u^* = -0.05$ at $T = 0.15$, recorded using the simulation cells of different sizes (shown in the figure).

The constructed phase diagrams for the systems with $u^* = -0.05$ and -0.03 are shown in parts a and b of Figure 8, respectively. Taking into account that, in both systems, the coexistence lines of the first-order transition smoothly meet the lines of the continuous transition, we conclude that the first-order transition terminates in the tricritical point, $T_{trc}(u^*)$, which replaces the critical point. The estimated tricritical point temperatures in these two systems are $T_{trc}(-0.05) \approx 0.144$ and $T_{trc}(-0.03) \approx 0.12$.

The calculations carried out for the system with $u^* = -0.01$, i.e., quite close to zero, demonstrated the behavior quite similar to that found in the case of $u^* = 0.0$ [14]. Figure 9 shows a series of isotherms recorded for this system, which exhibit two discontinuous density jumps at temperatures between 0.07 and 0.09, indicating the presence of two first-order transitions. The first transition occurs between the orientationally disordered low density phase and the partially ordered phase, Z_2, in which two orientations are favored. This is illustrated by the changes of $\langle \rho_k \rangle$ along the isotherm at $T = 0.09$ (see Figure 10a).

The phase Z_2 consists of long kinked zigzag clusters, with a large number of 120° kinks, predominantly oriented along with two out of three axes of the triangular lattice (see Figure 11). The second transition, which takes place at higher densities, leads to the development of the ordered Z phase. At temperatures $T > 0.09$, only one continuous transition takes place. Figure 10b shows the changes of $\langle \rho_k \rangle$ along the isotherm at $T = 0.10$ and demonstrates that the only transition occurs between the orientationally disordered (lamellar) fluid and the orientationally ordered Z phase.

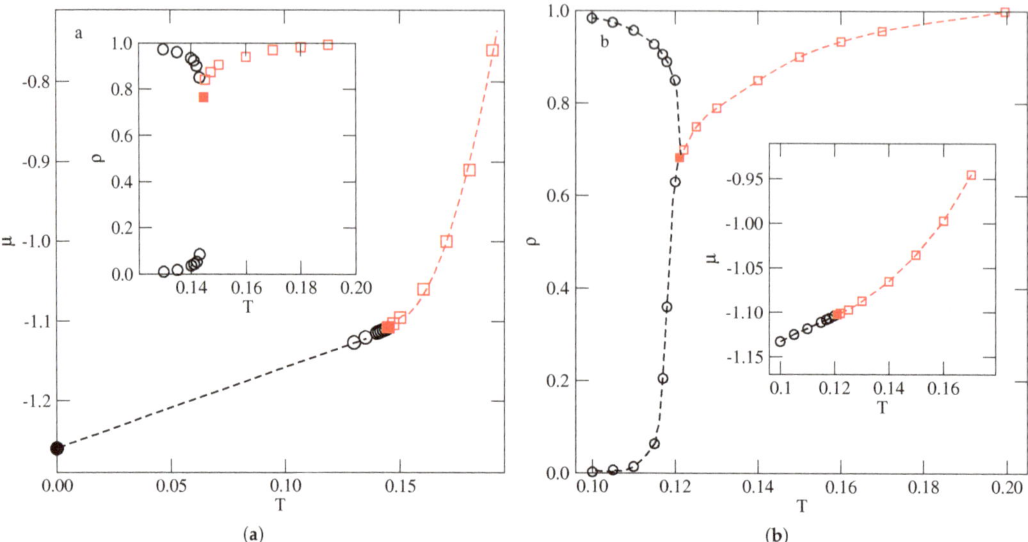

Figure 8. The phase diagrams evaluated for the systems with $u^* = -0.05$ (**a**), and -0.03 (**b**). Open circles and squares show the phase boundaries for the first−order and second−order transitions, respectively. The filled squares mark the locations of the tricritical points. In the main panel, the location of the gas–zigzag transition in the ground state is marked by the filled circle.

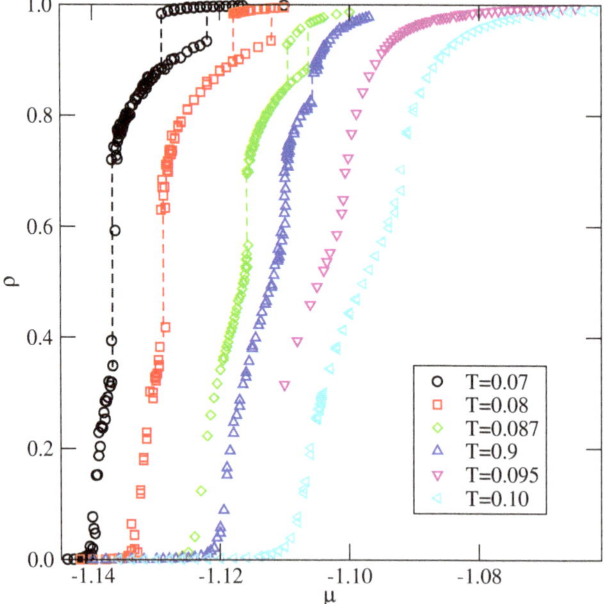

Figure 9. The isotherms obtained for the system with $u^* = -0.01$ at different temperatures.

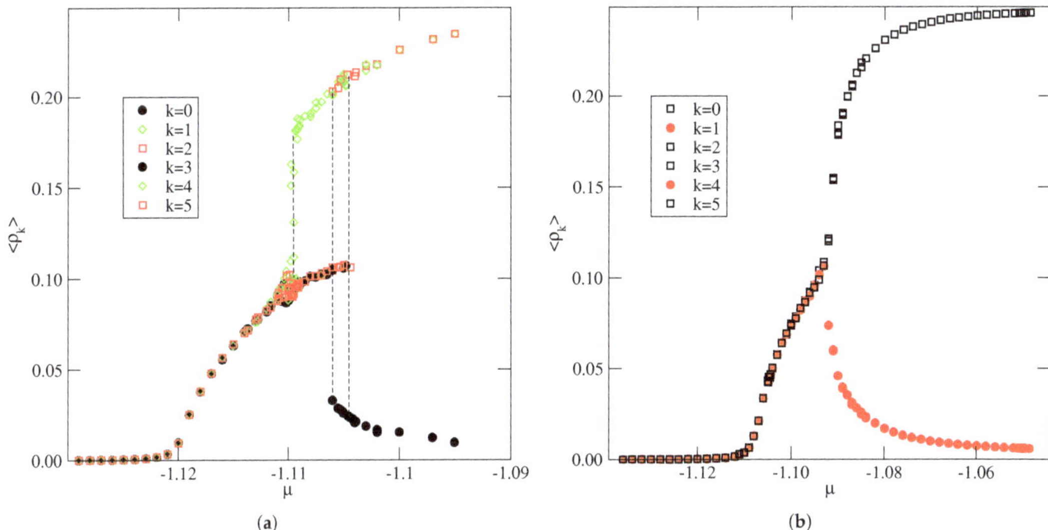

Figure 10. The changes of densities of differently oriented particles along the isotherm at $T = 0.09$ (**a**) and $T = 0.10$ (**b**), obtained for the system with $u^* = -0.01$. The dashed vertical lines in part (**a**) mark the locations of density jumps due to the transitions. One should note the hysteresis loop accompanying the transition between Z_2 and Z phases.

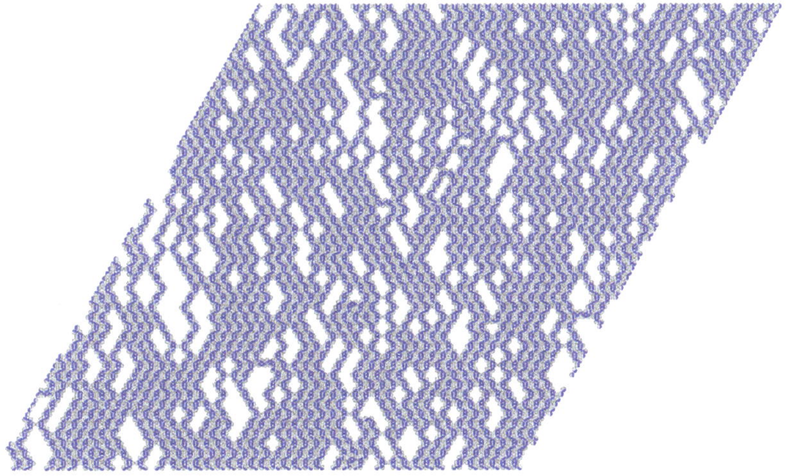

Figure 11. The snapshot recorded for the system with $u^* = -0.01$, at $T = 0.08$ and $\mu = -1.126$, which shows the structure of the phase Z_2.

The recorded isotherms, heat capacities, and density susceptibilities, allowed us to construct $T - \mu$ and $T - \rho$ projections of the phase diagram as shown in Figure 12. Taking into account that, in the ground state, this system exhibits only one transition, between a gas-like and the ordered Z phases, we conclude that the triple point, $T_{tr,1}$, in which a very dilute gas-like phase coexists with the Z_2 and Z phases, must exist at a certain temperature below 0.07.

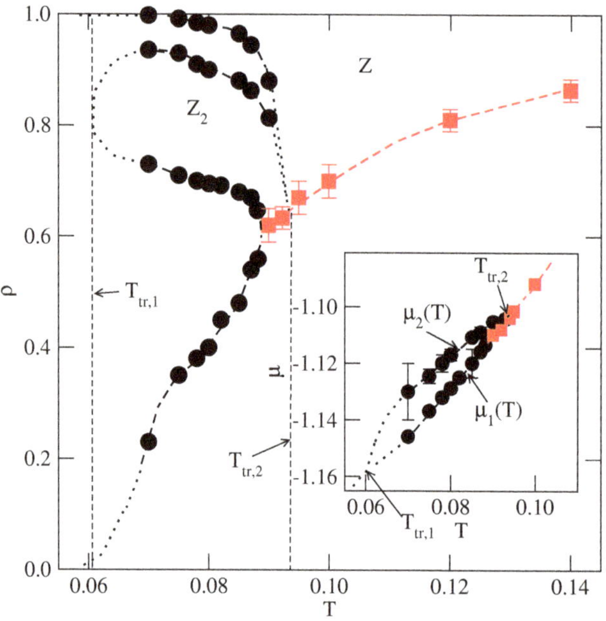

Figure 12. The proposed phase diagram for the system with $u^* = -0.01$. The main part and the inset show the $T - \rho$ and the $T - \mu$ projections, respectively. The circles mark the first−order transitions, while the squares mark the locations of continuous transitions. The dotted vertical lines in the main panel mark the expected locations of the triple points.

On the other hand, the lines of low and high density transitions, $\mu_1(T)$ and $\mu_2(T)$, are expected to meet at a temperature of about 0.093 and a density of about 0.65. At this point, the LF, Z_2, and Z phases coexist.

The change of the phase behavior between the systems with $u^* = -0.03$ and -0.01 results from the weakening of attractive AB and BB interactions. In the Z and Z_2 phases, every particle enjoys attractive interactions with all neighboring particles; however, the AB and BB attraction is weaker when $u^* = -0.01$. In the Z phase, the A half of each particle has contact with the A parts of four neighboring particles, and the formation of straight zigzag stripes is enhanced by the AB and BB attractions.

In the Z_2 phase, the A half of each particle interacts with either three or four A halves of neighboring particles, while the AB and BB attractive interactions are of lesser importance. On the other hand, the Z phase has a negligible residual entropy per particle in the thermodynamic limit [3], and this is stabilized by strong attraction. In the Z_2 phase, with large regions of empty sites, the entropy is higher than in the well-ordered Z phase. Thus, the phase Z_2 is stabilized by entropic effects.

However, the contribution of entropy to the free energy decreases when the temperature is lowered, and hence the structure of the condensed phase becomes dominated by the potential energy. This explains the appearance of the triple point $T_{tr,1}$. As the temperature increases, the ordering in both the Z and Z_2 phases is gradually destroyed by thermal fluctuations; however, the effect of these fluctuations is considerably stronger in the case of the already not well-ordered Z_2 phase. Therefore, the disordering of the Z_2 phase takes place at lower temperatures than the disordering of the Z structure, thus, leading to the presence of another triple point at the temperature $T_{tr,2} \approx 0.93$.

Now, we turn to the results obtained for a series of systems with $u_{AB} = 0$, which are the most closely related to the usual models of Janus particles [15]. The close-packed systems with $u_{AB} = 0$ and $u_{BB} < 0$ undergo a continuous orientational transition, however, at temperatures increasing linearly from about 0.204, when $u_{BB} = 0$, up to about 0.408, when $u_{BB} = -1.0$. Similarly, to the series with $u_{AB} = u_{BB}$, the phase behavior is expected to depend on the stability of the ordered Z phase.

The systems with u_{BB} very close to zero were found to exhibit qualitatively the same behavior as the already discussed system with $u^* - 0.01$ and the system with $u^* = 0.0$. For lower u_{BB}, equal to -0.05 and -0.1, the phase diagrams were qualitatively the same as found for the systems with u^* equal to -0.03 and -0.05. Thus, the dilute gas-like phase condenses directly into the zigzag ordered structure at the temperatures up to the tricritical point temperature, $T_{trc}(-0.05) \approx 0.135$ and $T_{trc}(-0.1) \approx 0.175$.

At higher temperatures, the disordered fluid undergoes a continuous order–disorder transition, up to the temperatures corresponding to the order–disorder transition in close-packed systems. Upon a further lowering of u_{BB} below about -0.116, the phase behavior changes, and the first order transition between the gas-like phase, and the condensed phase was found to lead directly to the ordered zigzag structure only at temperatures below the critical end point, which was the onset of the continuous order–disorder transition.

Above the critical end point temperature, the gas-like phase condenses into the disordered lamellar liquid, and the transition terminates in the critical point. The critical end point temperature and the critical temperature were found to increase when u_{BB} decreased. Figure 13 presents the examples of phase diagrams obtained for $u_{BB} = -0.1$ (part a) and -0.2 (part b), which demonstrate the changes of the topology with u_{BB}, while Figure 14 shows the estimated changes of T_c, T_{cep}, T_{trc} and T_o with u_{BB}.

It is well seen that T_c and T_{cep} meet T_{trc} at $u_{BB} \approx -0.116$. Thus, for any u_{BB} lower than about -0.116, the phase diagram topology is expected to remain unchanged. When u_{BB} decreases, the critical end point temperature gradually approaches the temperature at which the close-packed systems undergo the order–disorder transition. We limited the calculations to u_{BB} down to -0.3, since the estimation of the critical end point temperatures for lower values of u_{BB} was quite difficult. Already for $u_{BB} = -0.3$, $T_{cep}(-0.3)$ is quite close to $T_o(-0.3)$, and the density at the critical end point is quite high and equal to about 0.98.

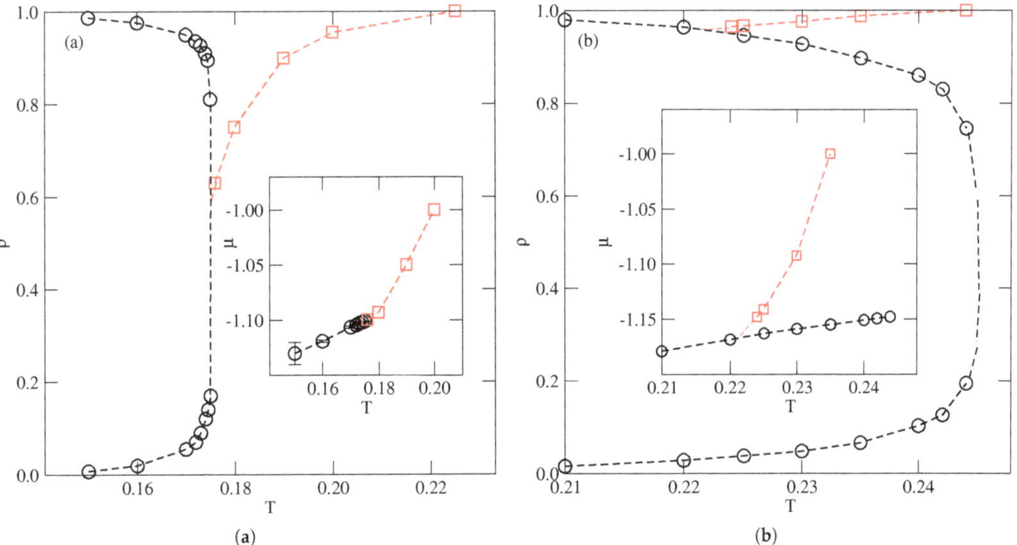

Figure 13. The estimated phase diagrams for the systems with $u_{AB} = 0$ and u_{bB} equal to -0.10 (**a**) and -0.2 (**b**). The main panels and insets show the $T - \rho$ and the $T - \mu$ projections, respectively. The circles and squares mark coexistence points of first−order and continuous transitions, respectively

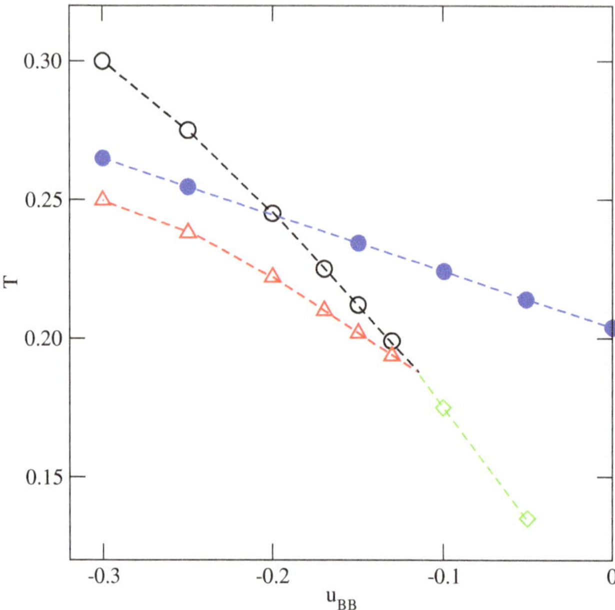

Figure 14. The changes of the critical temperature, $T_c(u_{BB})$, (circles), the tricritical point temperatures, $T_{trc}(u_{BB})$ (diamonds), the critical end point temperatures, $T_{cep}(u_{BB})$ (squares), and the orientational order−disorder transition temperature in close-packed systems, $T_o(u_{BB})$ (filled circles), for the systems with $u_{AB} = 0$.

In the series with $u_{BB} = 0$, the nature of the order–disorder transition in dense systems is different for u_{AB} lower and higher than -0.5. Namely, when $u_{AB} > -0.5$, this transition belongs to the universality class of the three-states Potts model, while, for $u_{AB} < -0.5$, it belongs to the universality class of the four-state Potts model [20]. In the particular case of $u_{AB} = -0.5$, the orientational order–disorder transition does not occur at all. In the close-packed systems, the temperature of the order–disorder transition decreases from 0.204, when $u_{AB} = 0$ to zero, when $u_{AB} = -0.5$. Then, for $u_{AB} < -0.5$, the transition temperature increases from zero up to about 0.14, when u_{AB} decreases from -0.5 to -1.0.

Here, we studied only the systems with u_{AB} between -0.05 and -0.25, and Figure 15 presents three phase diagrams obtained for $u_{AB} = -0.05$, -0.1, and -0.15. In the case of $u_{AB} = -0.05$, the phase behavior is qualitatively the same as in the already discussed systems with $u_{AB} = u_{BB} = -0.03$ and -0.05, as well as in the systems with $u_{AB} = 0$ and $u_{BB} = -0.05$ and -0.1. Thus, the fluid phase condenses directly into the ordered zigzag structure at any temperature, between zero and the temperature at which the close-packed system undergoes the order–disorder transition.

In the case of $u_{AB} = -0.1$, the dilute gas-like phase condenses into the ordered zigzag phase, only at temperatures up to the triple point temperature, $T_{tr}(-0.1) \approx 0.1$, At slightly higher temperatures, up to about 0.103, the gas condenses into the disordered liquid phase. The transition terminates at the usual critical point. However, at temperatures between about 0.1 and 0.102, the disordered liquid undergoes the first-order transition to the ordered zigzag phase. At $\approx$0.102, the tricritical point appears, and, at still higher temperatures, the disordered fluid undergoes a continuous transition to the zigzag phase.

The system with $u_{AB} = -0.15$ shows qualitatively different phase behavior, and the onset of a continuous order–disorder transition is located in the critical end point. Here, again, at the temperatures above the critical end point and up to the critical point, the gas-like phase condenses into the disordered liquid. Thus, the behavior is the same as in the already discussed systems with $u^* < -0.076$ and with $u_{AB} = 0$ and $u_{BB} < -0.116$.

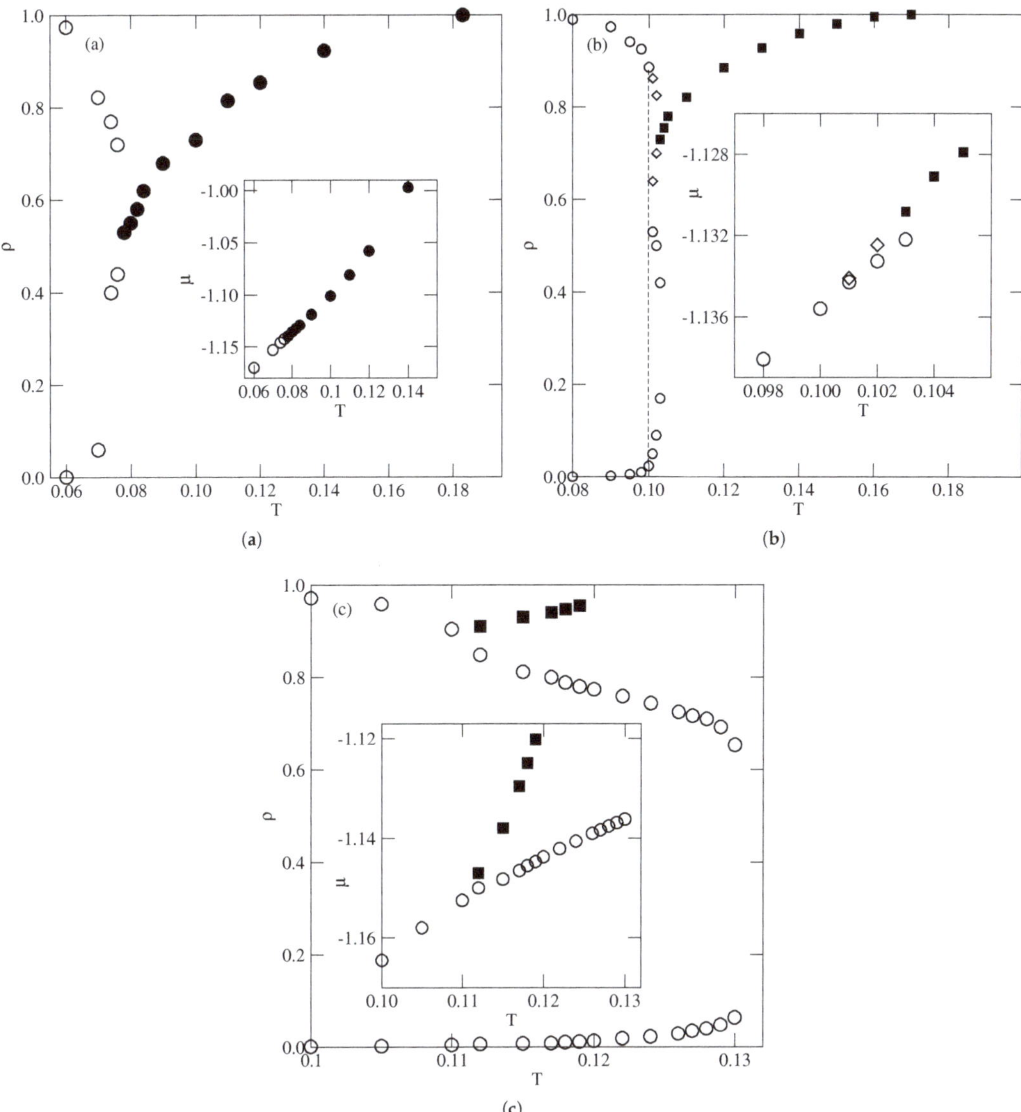

Figure 15. The estimated phase diagrams for the systems with $u_{BB} = 0$ and different values of u_{AB}, equal to -0.05 (**a**), -0.1 (**b**) and -0.15 (**c**). The main figures and insets show the $T-\rho$ and the $T-\mu$ projections, respectively. The vertical line in part b marks the location of the triple point. Open and filled symbols correspond to the first−order and continuous transitions, respectively. In part b, the low and high density transitions are marked by circles and diamonds, respectively

From the calculations carried out for several systems, we can estimate the changes of T_c, T_{cep}, T_{trc} and T_{tr} with u_{AB}, shown in Figure 16. Similarly to previously discussed systems, the tricritical point and the critical point temperatures increase when the AB attraction becomes stronger, and these two regimes meet when $u_{AB} \approx -0.116$. However, the critical end point temperature exhibits non-monotonous changes with u_{AB}. The temperature of the order–disorder transition in a close-packed system. $T_o(u_{AB})$, decreases when the AB attraction becomes stronger, and $T_{cep}(u_{AB})$ is bound to be lower than $T_o(u_{AB})$.

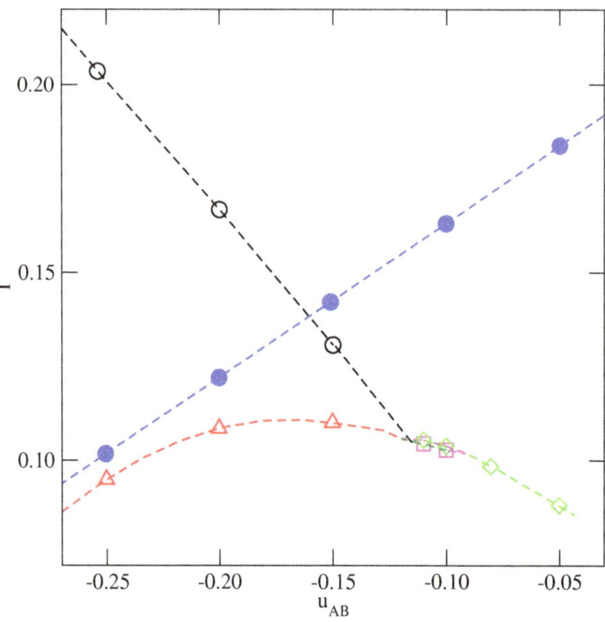

Figure 16. The changes of the $T_c(u_{AB})$, (circles), $T_{trc}(u_{AB})$ (triangles), $T_{cep}(u_{AB})$ (diamonds), $T_{tr}(u_{AB})$ (squares), and $T_o(u_{AB})$ (filled circles) estimated for the series with $u_{BB} = 0$.

4. Summary

We studied the phase behavior of two-dimensional systems of Janus-like particles on a triangular lattice. Here, we assumed that all, AA, AB, and BB, interactions are attractive. The AA interaction energy was fixed, while the AB and/or BB interaction energies were varied, and assumed to be less attractive than the AA interaction. We assumed that the particles can take on only six different orientations and that the interaction energy between a pair of nearest neighbors depends on their mutual orientations. Using the grand canonical Monte Carlo simulation method, we considered three series of systems with $u_{AB} = u_{BB}$, $u_{AB} = 0$ and with $u_{BB} = 0$.

We demonstrated that the phase behavior of all systems strongly depends on the stability of the high density zigzag (Z) phase. The stability of the Z phase is determined by the anisotropy of interactions and increases when the AB and/or BB attractions become weaker. As a consequence, in the systems with sufficiently strong anisotropy of interactions, the liquid phase does not appear, and the dilute fluid condenses directly into the zigzag ordered phase. The transition terminates in the tricritical point. At temperatures above the tricritical point, the disordered fluid undergoes a continuous transition into the zigzag phase.

When the AB and/or BB attraction increases, the stability of the zigzag phase becomes weaker, and its formation is possible only at sufficiently high densities and at sufficiently low temperatures. This means that the dilute phase condenses into the zigzag phase only at the temperatures lower than the critical end point temperature. At temperatures above the critical end point, the dilute phase condenses into the disordered liquid-like phase, and the transition terminates in the usual critical point.

The above scenario was found in all three considered series of systems. Whenever the critical point appears, the critical temperature increases when the attraction between AB and/or BB halves becomes stronger. In the particular series with $u_{AB} = u_{BB} = u^*$, the critical temperature went up to the value corresponding to the critical point of the uniform system when $u^* = -1.0$. In the series with $u_{AB} = 0$ and with $u_{BB} = 0$, the critical temperatures reached lower values when u_{BB} or u_{AB} went to -1.0.

In the case of the series with $u_{AB} = 0$, the phase diagram topology remained the same for any u_{BB} lower than about -0.116. Thus, the onset of the continuous order–disorder transition in the dense fluid meeting the bulk coexistence in the critical end point, $T_{cep}(u_{BB})$, and $T_{cep}(u_{BB})$ gradually increased when u_{BB} is lowered. On the other hand, the series with $u_{BB} = 0$ is expected to show different behavior, when u_{AB} decreases. In this paper, we discussed only the systems with $u_{AB} \geq -0.25$.

In this series, the critical end point temperature, $T_{cep}(u_{AB})$, is bound to go to zero for $u_{AB} = -0.5$, since this particular system does not undergo any orientational order–disorder transition [20]. However, a further decrease of u_{AB} below -0.5 means that the order–disorder transition reappears; however, now, this transition belongs to the universality class of the four-state Potts model. Therefore, it is expected that the continuous order–disorder transition should occur at sufficiently high densities and at sufficiently low temperatures. The onset of this transition is also expected to be located at the critical end point, $T_{cep}(u_{AB})$.

Here, we recall the results obtained for symmetric mixtures [27,28], which show qualitatively the same changes in the phase diagram topology when the tendency toward demixing becomes weaker. In that case, the demixed fluid is an ordered state, and by lowering its stability, the same sequence of phase diagram topologies appears.

Funding: This research received no external funding.

Institutional Review Board Statement: Not applicable.

Informed Consent Statement: Not applicable.

Data Availability Statement: Not applicable.

Conflicts of Interest: The authors declare no conflict of interest.

References and note

1. de Gennes, P. G. Soft matter (Nobel Lecture). *Angew. Chem. Int. Ed. Engl.* **1992**, *31*, 842. [CrossRef]
2. Jiang, S.; Granick, S. (Eds.) *Janus Patricles Synthesis, Self-Assembly and Applications*; RCS Publishing: Cambridge, UK, 2012.
3. Shin, H.; Schweizer, K.S. Theory of two-dimensional self-assembly of Janus colloids: crystalization and orientational ordering. *Soft Matter* **2014**, *10*, 229. [CrossRef] [PubMed]
4. Iwashita, Y.; Kimura, Y. orientational order of one-patch colloidal particles in two dimensions. *Soft Matter* **2014** *10*, 7135. [CrossRef]
5. Jiang, S.; Yan, Y.; Whitmer, J.K.; Anthony, S.M.; Luijten, E.; Granick, S. Orientationally glassy crystals of Janus spheres. *Phys. Rev. Lett.* **2014**, *112*, 218301. [CrossRef]
6. Vanakaras, A.G. Self-organization and pattern formation of janus particles in two dimesions by computer simulations. *Langmuir* **2006**, *22*, 88. [CrossRef] [PubMed]
7. Chen, Q.; Bae, S.C.; Granick, S. Directed self-assembly of a colloidal kagome lattice. *Nature* **2011**, *469*, 181. [CrossRef] [PubMed]
8. Borówko, M.; Rżysko, W. Phase transitions and self-organization of Janus disks in two dimensions studied by Monte Carlo simulations. *Phys. Rev. E.* **2014**, *90*, 062308. [CrossRef] [PubMed]
9. Sciortino, F.; Giacometti, A.; Pastore, G. Phase diagram of janus particles. *Phys. Rev. Lett.* **2009**, *103*, 237801. [CrossRef] [PubMed]
10. Preisler, Z.; Vissers, T.; Smallenburg, F.; Munaó, G.; Sciortino, F. Phase diagra of one-patch colloids forming tubes and lamellae. *J. Phys. Chem. B* **2013**, *117*, 32. [CrossRef] [PubMed]
11. Vissers, T.; Preisler, Z.; Smallenburg, F.; Dijkstra, M.; Sciortino, F. Predicting crystals of Janus colloids. *J. Chem. Phys.* **2013**, *138*, 164505. [CrossRef] [PubMed]
12. Mitsumoto, K.; Yoshino, H. Orientational ordering of closely packed Janus particles. *Soft Matter* **2018**, *14*, 3919. [CrossRef] [PubMed]
13. Patrykiejew, A.; Rżysko, W. Order–disorder transitions in systems of Janus particles on a triangular lattice. *Phys. A* **2020**, *548*, 123883. [CrossRef]
14. Patrykiejew, A.; Rżysko, W. Two-dimensional Janus particles on a triangular lattice. *Soft Matter* **2020**, *16* 6633. [CrossRef] [PubMed]
15. Kern, N.; Frenkel, D. Fluid-fluid coexistence in colloidal systems with short-ranged strongly directional interaction. *J. Chem Phys.* **2003**, *118*, 9882. [CrossRef]
16. Jiang, S.; Granick, S. Controlling the geometry (Janus balance) of amphiphilic colloidal particles. *Langmuir* **2008**, *24*, 2438. [CrossRef] [PubMed]
17. Wu, F.Y. The Potts model. *Rev. Mod. Phys.* **1982**, *54*, 235. [CrossRef]
18. Vanderzande, C. *Lattice Models of Polymers*; Cambidge University Press: Cambridge, UK, 1998.

19. Patrykiejew, A.; Sokołowski, S.; Pizio, O. Statistical Surface Thermodynamics. In *Surface and Interface Science: Solid-Gas Interfaces II*; Vandelt, K., Ed.; Wiley-VCH Verlag: Weinheim, Germany, 2015; Volume 6.
20. Patrykiejew, A.; Rżysko, W. Order–disorder transitions in systems of Janus particles on a triangular lattice:revisited. *Phys. A* **2021**, *570*, 125819. [CrossRef]
21. Landau, D.P.; Binder, K. *A Guide to Monte Carlo Simulations in Statistical Physics*; Cambridge University Press: Cambridge, UK, 2000.
22. Privman, V. (Ed.) *Finite Size Scaling and Numerical Simulation of Statistical Mechanics*; World Scientific: Singapore, 1990.
23. Baxter, R.J. *Exactly Solved Models in Statistical Mechanics*; Academic Press: London, UK, 1989.
24. Note that in the case of a lattice gas model, $T_c(1)$ is equal to $1/4$ of the critical temperature of the corresponding Ising model.
25. Wilding, N.B.; Bruce, A.D. Density fluctuations anf field mixing in the critical fluid. *J. Phys. Condens. Matter* **1992**, *4*, 3087. [CrossRef]
26. Wilding, N.B. *Computer Simulation Strudies in Condensed Matter Physics VIII*; Landau, D.P.K., Mon, K., Schüttler, H.-B., Eds.; Springer: Heidelberg, Germany, 1995.
27. Wilding, N.B.; Schmid, F.; Nielaba, P. Liquid-vapor phase behavior of a symmetrical binary fluid mixture. *Phys. Rev. E* **1998**, *58*, 2201. [CrossRef]
28. Patrykiejew, A. Effects of geometrical and energetic nonadditivity on the phase behavior of two-component symmetric mixtures. *Phys. Rev. E* **2017**, *95*, 012145. [CrossRef] [PubMed]

International Journal of
Molecular Sciences

MDPI

Article

Can We Predict the Isosymmetric Phase Transition? Application of DFT Calculations to Study the Pressure Induced Transformation of Chlorothiazide

Łukasz Szeleszczuk [1,*], Anna Helena Mazurek [2], Katarzyna Milcarz [1], Ewa Napiórkowska [1] and Dariusz Maciej Pisklak [1]

1 Department of Physical Chemistry, Chair and Department of Physical Pharmacy and Bioanalysis, Faculty of Pharmacy, Medical University of Warsaw, Banacha 1 Street, 02-093 Warsaw, Poland; kmilcarz@wum.edu.pl (K.M.); enapiorkowska@wum.edu.pl (E.N.); dpisklak@wum.edu.pl (D.M.P.)

2 Department of Physical Chemistry, Chair and Department of Physical Pharmacy and Bioanalysis, Faculty of Pharmacy, Doctoral School, Medical University of Warsaw, Banacha 1 Street, 02-093 Warsaw, Poland; anna.mazurek@wum.edu.pl

* Correspondence: lszeleszczuk@wum.edu.pl; Tel.: +48-501-255-121

Abstract: Isosymmetric structural phase transition (IPT, type 0), in which there are no changes in the occupation of Wyckoff positions, the number of atoms in the unit cell, and the space group symmetry, is relatively uncommon. Chlorothiazide, a diuretic agent with a secondary function as an antihypertensive, has been proven to undergo pressure-induced IPT of Form I to Form II at 4.2 GPa. For that reason, it has been chosen as a model compound in this study to determine if IPT can be predicted in silico using periodic DFT calculations. The transformation of Form II into Form I, occurring under decompression, was observed in geometry optimization calculations. However, the reverse transition was not detected, although the calculated differences in the DFT energies and thermodynamic parameters indicated that Form II should be more stable at increased pressure. Finally, the IPT was successfully simulated using ab initio molecular dynamics calculations.

Keywords: DFT; CASTEP; aiMD; ab initio molecular dynamics; phase transition; polymorphism

check for updates

Citation: Szeleszczuk, Ł.; Mazurek, A.H.; Milcarz, K.; Napiórkowska, E.; Pisklak, D.M. Can We Predict the Isosymmetric Phase Transition? Application of DFT Calculations to Study the Pressure Induced Transformation of Chlorothiazide. *Int. J. Mol. Sci.* **2021**, *22*, 10100. https://doi.org/10.3390/ijms221810100

Academic Editor: Małgorzata Borówko

Received: 27 July 2021
Accepted: 15 September 2021
Published: 18 September 2021

1. Introduction

Polymorphism, commonly defined as the ability of a substance to exist as two or more crystalline phases that have different arrangements or conformations of the molecules in the crystal lattice [1] is a phenomenon with particular importance in the pharmaceutical sciences and industry. The differences between polymorphs at the molecular level can manifest themselves in different properties, important in this field, such as hygroscopicity, solubility, thermal stability, rate of dissolution, hardness, chemical reactivity, and many others [2,3]. For the correct design of a pharmaceutical compound, it is, therefore, crucial to control its solid-state form to guarantee its properties.

The application of high pressure has been shown as a route to access new phases of solid-state materials—possibly the most famous example being the transformation of graphite into diamond [4]. Under increased pressure, the geometry of both inter and intramolecular bonds can often be altered, new hydrogen bonds can be formed, and existing ones broken or symmetrized. In most cases, pressure-induced phase transition can occur in a single step between higher- and lower-symmetry space groups (type I), through a low-symmetry transition state between relatively higher- symmetry initial and final structures (type II), or via the transformation in which the mechanism is more complex (type III). Isosymmetric structural phase transition (IPT, type 0), in which there are no changes in the occupation of Wyckoff positions, the number of atoms in the unit cell, and the space group symmetry are relatively uncommon [5]. However, there are some well-known and recently discovered examples of such transitions: sodium oxalate [6], 1,3-cyclohexandione [7] L-serine [8], sulfamic acid [9], biurea [10], and α-glycylglycine [11].

The density functional theory (DFT) methods are commonly used to model the structure and properties of organic molecules. However, the uniqueness of each polymorphic form property arises mostly from short- and long-distance intermolecular interactions. Therefore, DFT-based methods in which a single molecule in vacuum or in solution is being modeled were found to be inappropriate and inaccurate to study the polymorphism-related phenomena. While those "single molecules" methods are generally successfully applied in other aspects of pharmaceutical sciences, i.e., to study drug–biomolecule interactions or to predict the formation of complexes, in order to study the solid-state pharmaceutics, other types of calculations, sometimes called "periodic DFT calculations", should be used [12–14]. In this case, the adjective "periodic" is an abbreviation of "performed under periodic boundary conditions" which is a crucial requirement for accurate modeling of crystals. Further, in such calculations, the pseudopotentials are frequently used to represent an effective interaction that approximates the potential experienced by the valence electrons. Additionally, the plane-wave basis sets are employed instead of the localized ones [15,16].

Noncovalent forces, such as hydrogen bonding and van der Waals interactions, are crucial for the formation, stability, and function of molecules and materials [17,18]. There exists a variety of hybrid semiempirical solutions that introduce dispersion corrections in the DFT formalism [19]. These semiempirical approaches provide the best compromise between the cost of the first principals evaluation of the dispersion terms and the need to improve non-bonding interactions in the standard DFT description. The Tkatchenko–Scheffler (TS) correction [20], used in this study, exploits the relationship between polarizability and volume, and thus accounts to some degree for the relative variation in dispersion coefficients of differently bonded atoms. This is achieved by weighting values taken from the high-quality first-principals database with atomic volumes derived from Hirshfeld partitioning of the self-consistent electronic density. It should be noted, however, that the TS scheme is an atom-pairwise dispersion model. As dispersion interactions are not strictly pairwise additive, many-body dispersion interactions can become important for some systems, especially when large and flexible molecules are involved [21]. Such interactions can be captured by the many-body dispersion (MBD) model [22]. However, the object of this study, chlorothiazide, is a rather small and rigid molecule, therefore, only small differences between the TS and MBD dispersion models would have been expected. The increasing number of studies presenting results of periodic DFT calculations on pharmaceutical solids confirms that those kinds of computations can be successfully used to answer the fundamental questions as well as to provide specific solutions for experimental challenges. Many successful applications of such calculations have been recently reviewed by us [23]. However, among those works, there was no application of periodic DFT calculations to study the phenomenon of isosymmetric phase transition in the manner presented in this study.

To be precise, it must be stated that some of the crystals that have been proven experimentally to undergo IPT were also modeled using the DFT methods. However, in those cases, the computational part was limited to the geometry optimization at the pressure at which the particular structure was obtained, followed sometimes by the thermodynamic parameters calculations [10,11].

However, presently, we have decided to expand the spectrum of the applied computational methods by using the various DFT functionals, empirical dispersion corrections, phonon density of states calculations, and computationally demanding ab initio molecular dynamics calculations (aiMD). This effort has been made to answer the question if such IPT can be predicted and, if yes, how it should be carried out. To the best of our knowledge, there are no published works presenting an application of such combined calculations to study the phenomenon of IPT.

To achieve this aim, chlorothiazide (6-chloro-4H-1,2,4-benzothiadiazine-7-sulfonamide 1,1-dioxide, CT, Figure 1) has been chosen as a model compound. This active pharmaceutical ingredient (API) is used as a diuretic agent with a secondary function as an antihypertensive [24]. It exists in various complex solid-state forms, including solvates and co-crystals. However, at ambient conditions, only one polymorphic form of CT has so far been reported (Form I). More importantly, this API has been chosen as its crystal structure and pressure-induced IPT have been studied experimentally by Oswald et al. [25]. In that work, the authors obtained a series of crystal structures of CT at various pressure conditions and confirmed that the IPT occurs at 4.2 GPa resulting in Form II, which was found to be more stable at pressures higher than 4.2 GPa. For CT, the most noticeable differences between Forms I and II studied at the same conditions can be observed for "a" length of a unit cell, which is clearly visible in Figure 2. Unit cell dimensions of all studied structures can be found in Table 1.

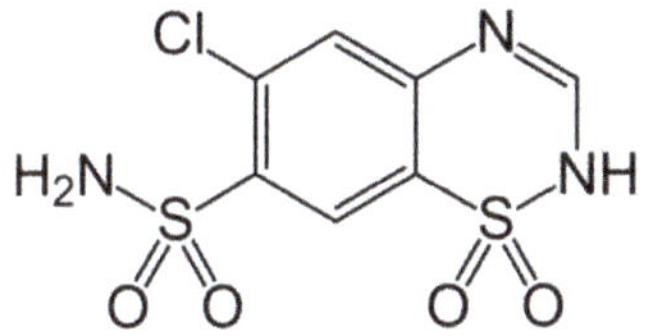

Figure 1. Chemical structure of chlorothiazide (CT).

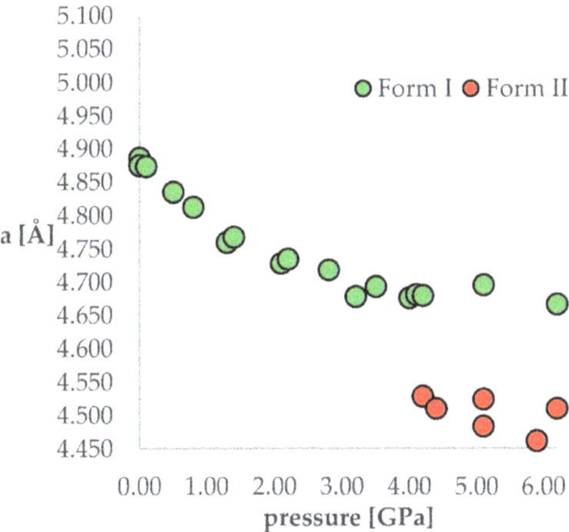

Figure 2. The change of unit cell edge "a" length with respect to pressure. Green circles—Form I, Red circles—Form II.

As the aim of the study was to also check the accuracy of the applied computational methods, it was important that this coherent group of crystal structures was obtained in one study as it allowed the elimination of the possible differences in the unit cell dimensions caused by the application of different diffractometers or diamond anvil cells used for high-pressure experiments. This would be an issue if the structures originated from different works. From Figure 2 and Table 1, it can be observed that some changes of the unit cell dimensions are non-monotonic upon compression. This was probably caused by inaccuracies in their determinations as, at some of the studied pressures, the authors performed solely powder X-ray diffraction (PXRD) measurements followed by pattern refinement.

Table 1. Structural parameters of crystal forms of chlorothiazide. Data for Form II were bolded to increase the clarity.

Form	p [GPa]	a [Å]	b [Å]	c [Å]	α [°]	β [°]	γ [°]	V [Å ^{3}]	XRD
I	0.0	4.886	6.405	8.985	74.05	83.56	80.54	265.01	PXRD
I	0.0	4.875	6.401	8.980	74.05	83.54	80.47	264.01	SCXRD
I	0.1	4.873	6.413	8.941	74.07	83.71	80.59	263.50	PXRD
I	0.5	4.835	6.310	8.895	74.62	84.12	80.67	256.84	SCXRD
I	0.8	4.812	6.278	8.829	75.09	84.07	80.94	253.18	PXRD
I	1.3	4.760	6.146	8.737	76.06	84.56	81.25	244.06	SCXRD
I	1.4	4.768	6.185	8.767	75.66	84.53	81.24	246.43	PXRD
I	2.1	4.728	6.050	8.684	76.54	84.73	81.67	238.02	SCXRD
I	2.2	4.734	6.074	8.691	76.30	84.83	81.70	239.29	PXRD
I	2.8	4.718	5.998	8.656	76.64	84.89	82.05	235.11	PXRD
I	3.2	4.678	5.910	8.593	77.22	85.05	82.54	228.90	PXRD
I	3.5	4.693	5.901	8.599	77.48	84.97	82.35	229.53	SCXRD
I	4.0	4.676	5.812	8.543	77.95	85.11	82.77	224.41	SCXRD
I	4.1	4.681	5.895	8.587	77.34	85.12	82.36	228.31	PXRD
I	4.2	4.680	5.848	8.592	77.63	84.70	81.77	226.34	PXRD
II	**4.2**	**4.529**	**5.957**	**8.540**	**76.52**	**76.52**	**83.27**	**216.35**	**PXRD**
II	**4.4**	**4.510**	**5.929**	**8.503**	**76.53**	**85.62**	**83.20**	**218.91**	**SCXRD**
II	**5.1**	**4.483**	**5.893**	**8.465**	**76.23**	**85.84**	**83.28**	**215.16**	**SCXRD**
I	5.1	4.696	5.804	8.538	77.98	84.85	81.80	224.37	PXRD
II	**5.1**	**4.524**	**5.933**	**8.524**	**76.64**	**85.54**	**83.48**	**220.50**	**PXRD**
II	**5.9**	**4.461**	**5.859**	**8.427**	**75.99**	**86.06**	**83.35**	**211.77**	**SCXRD**
I	6.2	4.666	5.793	8.618	77.62	85.25	82.03	224.58	PXRD
II	**6.2**	**4.510**	**5.897**	**8.472**	**76.31**	**85.65**	**83.30**	**216.80**	**PXRD**

Form—either I or II polymorph, according to [21]; p—pressure at which the structure was studied; a, b, c, α, β, γ, V—unit cell dimensions; XRD-type of X-ray diffraction experiment applied to obtain the structure information (PXRD—powder X-ray diffraction; SCXRD—single-crystal X-ray diffraction).

Having the appropriate amount of structural data, clearly showing the IPT of CT, there was nothing left but to check how the periodic DFT calculations performed in such cases. Our motivation was that if we succeed with this model compound, this method can be further validated on other solid organics undergoing IPT and could finally be used as a screening method in order to predict if the pressure-driven IPT would occur for a particular compound, assuming that only the low-pressure structure is known. Such a method would surely help to design the demanding high-pressure experiments.

2. Results and Discussion

2.1. Choice of the DFT Functionals

The first set of calculations (Table S1) was performed to find out how the choice of the DFT functional would affect the accuracy of geometry optimization and to choose the most accurate one for the subsequent calculations. This was carried out by the optimization of the experimental crystal structure obtained at normal conditions (refcode QQQAUG09). This step is usually omitted in the studies presenting the results of DFT calculations on crystals. This is because, similarly to B3LYP for in vacuo calculations, the GGA PBE functional with TS dispersion correction has not been proved many times to be the most accurate in the case of solid-phase modeling. Unsurprisingly, also in this study, the most accurate results have been obtained by the GGA PBE TS approach. Using this functional, the differences between the experimental and calculated unit cell dimensions were lower than 0.05 Å for lengths and 0.5 ° for angles, while for the other functionals those differences were in some cases found to be larger than 1 Å and 10.

However, in addition to the GGA PBE TS, we have also decided to choose the PBESOL. Although, without any dispersion correction, this functional was specially designed and validated for the densely packed solids, and was in some cases shown to be more accurate than GGA PBE TS, especially for the calculations performed under increased pressure [26].

2.2. Geometry Optimization of Forms I and II under External Pressure—Unit Cell Dimensions Analysis

The next stage of this work was the optimization of both polymorphic forms (I and II) under external pressure to find out if the IPT can be observed. More specifically, we have chosen two crystal structures of CT—the structure of Form I obtained at normal conditions (QQQAUG09) and Form II obtained at 5.9 GPa (QQQAUG17). These structures were obtained at the most distinct pressure conditions. Both structures were optimized at 19 different values of external pressure, exactly those that had been applied experimentally, listed in Table 1. Due to the large number of calculations (two crystal structures, two DFT functionals, 18 values of pressure = 72 optimizations), the obtained unit cell dimensions are listed in Tables S2 and S3 for clarity reasons. Additionally, the visual representation of some of the results are presented in Figure 3. Below, firstly, the changes in the unit cell dimensions upon compression (when geometry optimization starting from Form I) or decompression (when geometry optimization starting from Form II) will be discussed. Then, the RMSD of the calculated structures will be presented and analyzed. Finally, the differences in the energies calculated at the same pressure and using the same DFT functionals but different structures will be discussed.

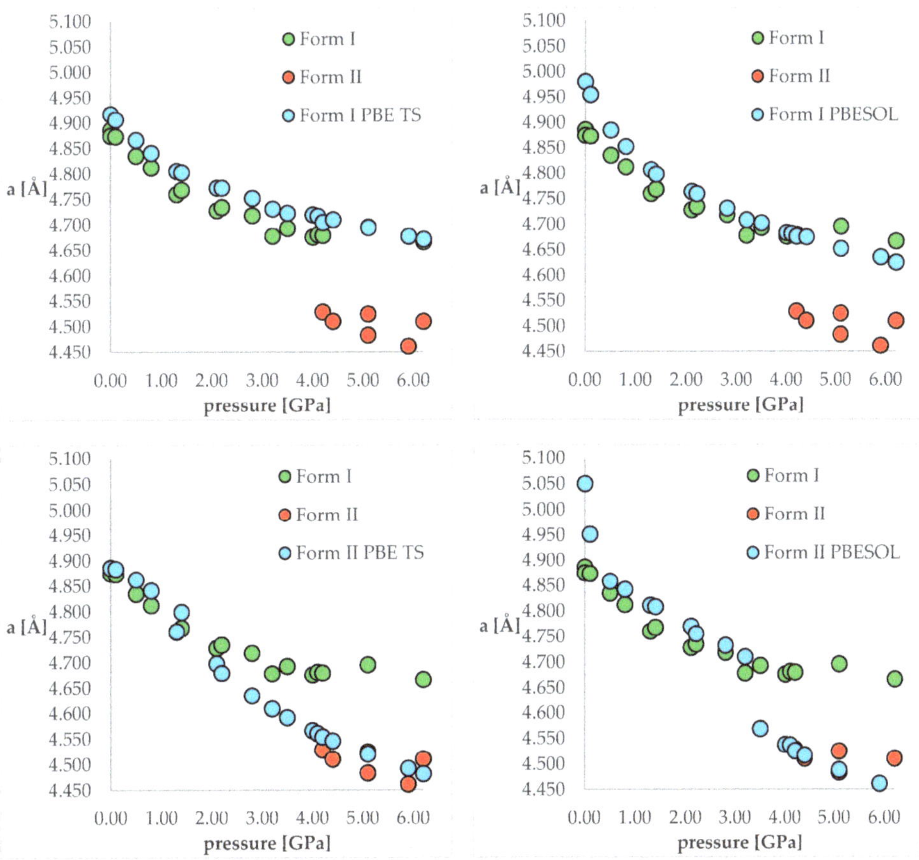

Figure 3. The change of unit cell edge "a" length with respect to pressure. Green circles—experimental Form I; red circles—experimental Form II; blue circles—calculated ones. **Top left**—using PBE TS and starting from Form I; **top right**—using PBESOL and starting from Form I; **bottom left**—using PBE TS and starting from Form II; **top right**—using PBESOL and starting from Form II.

Looking at Figure 3, it is clearly visible that both choice of the DFT functional and of the initial structure (polymorphic form) subjected to geometry optimization had an influence on the obtained results. For the calculations of Form I (top left and top right), the results obtained using PBE TS were more accurate than those obtained using PBESOL, especially for lower pressure values (0 and 0.1 GPa). Application of neither of those functionals enables prediction of the IPT during the compression process because even at higher pressures Form I was still preserved after optimization.

More interesting were the results obtained through modeling of the decompression, which is when Form II had been chosen as the starting one (bottom left and bottom right). For the PBE TS calculations, Form II was preserved, though it is impossible to assess the accuracy of those calculations as Form II was not obtained at pressures lower than 4.2 GPa. However, the changes in the unit cell dimensions were continuous, which suggested that there should be no IPT. However, when the calculations were performed using PBESOL, Form II was preserved only for pressures higher than 3.5 GPa. At this pressure, a jump discontinuity can be observed, which would suggest that at this point one can expect IPT. As an experimentally determined pressure at which IPT occurs was found to be 4.2 GPa, there was a 0.7 GPa difference between the experimental and theoretical values. However, it should be noticed that the experiments were carried out at 293 K while the geometry optimization calculations were performed at 0 K. Therefore, neglecting the thermal motions could be the main reason for this inaccuracy. Nevertheless, DFT calculations of the decompression of Form II using PBESOL enable the prediction that IPT would occur. Unfortunately, in most cases, the low-pressure crystal structure of a compound is known while the question remains if compression would result in the IPT. In the case of CT, this question could not have been answered using solely energy minimization. However, this did not discourage us to look for another solution which will be described in one of the next paragraphs (2.5).

2.3. Geometry Optimization of Forms I and II under External Pressure—RMSD Analysis

So far, only the differences between the experimental and calculated unit cell dimensions were discussed. However, to confirm the results discussed above, the analysis of the conformational changes should also be performed. Though CT is an API with rather limited conformational space, there are of course some conformational differences between the molecules found in Form I and Form II. Instead of presenting the comparison of the lengths, angles, and dihedrals, we have decided to calculate the root mean square deviation (RMSD) values between the calculated and experimental crystal structures. This analysis was, however, hampered by the lack of some experimental data. As is presented in Table 1, only part of the structural parameters originated from the SCXRD measurements while the rest from PXRD. The crystal structures have been fully solved, including the determination of the positions of the atoms, only for structures studied by SCXRD. Besides, there was no pressure at which the SCXRD has been done for both Form I and Form II. Therefore, the RMSD analysis was limited to nine examples (Table 2 A–I).

The coloring scheme for the values in Table 2A–I was applied to facilitate their categorization into five groups (A and B; C and D; E and F; G, H, and I) in order of increasing pressure. In the first group (A and B), the most accurate results were obtained when PBE TS functional was used for optimization, which is also consistent with results presented in Table S1. Besides, the choice of the initial structure had no influence on the obtained results. In other words, no matter whether Form I or Form II was used as the initial structure, after geometry optimization, an accurate structure of Form I was obtained. However, in the case of PBESOL calculations, the choice of the initial structure had an influence on the obtained results. Though PBESOL, as described previously, was found to be less accurate than PBE TS for the calculations at 0 GPa, more accurate results were obtained for Form I than II (RMSD 0.1249 vs. 0.1724).

Table 2. A. RMSD of the structures at 0 GPa. **B.** RMSD of the structures at 0.5 GPa. **C.** RMSD of the structures at 1.3 GPa. **D.** RMSD of the structures at 2.1 GPa. **E.** RMSD of the structures at 3.5 GPa. **F.** RMSD of the structures at 4.0 GPa. **G.** RMSD of the structures at 4.4 GPa. **H.** RMSD of the structures at 5.1 GPa. **I.** RMSD of the structures at 5.9 GPa. To facilitate the analysis of the data in Table 2 a three-color scale was applied. In this scale, the 50th percentile (midpoint) was calculated, and the cell that holds this value was colored yellow. The cell that holds the minimum value was colored green, and the cell that holds the maximum value was colored red.

A **p = 0 GPa, Form I**	**Exp**	**PBESOLII**	**PBESOL I**	**PBE TS II**	**PBE TS I**
Exp	0	0.1724	0.1249	0.1159	0.1010
PBESOL II	0.1724	0	0.0854	0.1138	0.1467
PBESOL I	0.1249	0.0854	0	0.0836	0.0862
PBE TS II	0.1159	0.1138	0.0836	0	0.0572
PBE TS I	0.1010	0.1467	0.0862	0.0572	0
B **p = 0.5 GPa, Form I**	**Exp**	**PBESOLII**	**PBESOL I**	**PBE TS II**	**PBE TS I**
Exp	0	0.1398	0.1079	0.1028	0.1039
PBESOL II	0.1398	0	0.0663	0.1153	0.1257
PBESOL I	0.1079	0.0663	0	0.0585	0.0653
PBE TS II	0.1028	0.1153	0.0585	0	0.0209
PBE TS I	0.1039	0.1257	0.0653	0.0209	0
C **p = 1.3 GPa, Form I**	**Exp**	**PBESOL II**	**PBESOL I**	**PBE TS II**	**PBE TS I**
Exp	0	0.1115	0.1093	0.1447	0.1070
PBESOL II	0.1115	0	0.0241	0.1255	0.0478
PBESOL I	0.1093	0.0241	0	0.1024	0.0367
PBE TS II	0.1447	0.1255	0.1024	0	0.0921
PBE TS I	0.1070	0.0478	0.0367	0.0921	0
D **p = 2.1 GPa, Form I**	**Exp**	**PBESOL II**	**PBESOL I**	**PBE TS II**	**PBE TS I**
exp	0	0.1192	0.1166	0.1703	0.1169
PBESOL II	0.1192	0	0.0169	0.1635	0.0325
PBESOL I	0.1166	0.0169	0	0.1478	0.0230
PBE TS II	0.1703	0.1635	0.1478	0	0.1428
PBE TS I	0.1169	0.0325	0.0230	0.1428	0
E **p = 3.5 GPa, Form I**	**Exp**	**PBESOL II**	**PBESOL I**	**PBE TS II**	**PBE TS I**
exp	0	0.2278	0.1224	0.2566	0.1212
PBESOL II	0.2278	0	0.2165	0.0407	0.2120
PBESOL I	0.1224	0.2165	0	0.2507	0.0149
PBE TS II	0.2566	0.0407	0.2507	0	0.2456
PBE TS I	0.1212	0.2120	0.0149	0.2456	0
F **p = 4.0 GPa, Form I**	**Exp**	**PBESOL II**	**PBESOL I**	**PBE TS II**	**PBE TS I**
exp	0	0.2646	0.1231	0.2767	0.1254
PBESOL II	0.2646	0	0.2577	0.0262	0.2738
PBESOL I	0.1231	0.2577	0	0.2738	0.0215
PBE TS II	0.2767	0.0262	0.2738	0	0.2898
PBE TS I	0.1254	0.2738	0.0215	0.2898	0

Table 2. *Cont.*

G p = 4.4 GPa, Form II	Exp	PBESOL II	PBESOL I	PBE TS II	PBE TS I
exp	0	0.1408	0.2952	0.1422	0.3058
PBESOL II	0.1408	0	0.2846	0.0226	0.2973
PBESOL I	0.2952	0.2846	0	0.3002	0.0194
PBE TS II	0.1422	0.0226	0.3002	0	0.3127
PBE TS I	0.3058	0.2973	0.0194	0.3127	0
H p = 5.1 GPa, Form II	**Exp**	**PBESOL II**	**PBESOL I**	**PBE TS II**	**PBE TS I**
exp	0	0.1411	0.3171	0.1402	0.3243
PBESOL II	0.1411	0	0.3125	0.0216	0.3216
PBESOL I	0.3171	0.3125	0	0.3245	0.0179
PBE TS II	0.1402	0.0216	0.3245	0	0.3334
PBE TS I	0.3243	0.3216	0.0179	0.3334	0
I p = 5.9 GPa, Form II	**Exp**	**PBESOL II**	**PBESOL I**	**PBE TS II**	**PBE TS I**
exp	0	0.1453	0.3372	0.1425	0.3419
PBESOL II	0.1453	0	0.3361	0.0194	0.3421
PBESOL I	0.3372	0.3361	0	0.3462	0.0182
PBE TS II	0.1425	0.0194	0.3462	0	0.3520
PBE TS I	0.3419	0.3421	0.0182	0.3520	0

Quite opposite results could be observed in the second group (C and D). This time, the accuracy of the results obtained using PBESOL and PBE TS was almost the same, with one important observation. For the PBESOL, the choice of the initial form had no influence on the results, while for the PBE TS the accurate results were obtained only if the proper form (I) had been chosen as the initial structure.

For the third group (E and F), the accuracy of PBE TS and PBESOL was almost the same. However, this time, the choice of the correct initial form (I) was crucial to obtain accurate results for both functionals. This is also elegantly reflected in Figure 3, which shows that the length of "a" edge obtained after optimization at 3.5 and 4.0 GPa depends significantly on the initial structure but not on the DFT functional used. Those results correspond nicely with experimental ones, as for those pressure values Forms I and II were found to coexist. This has been proven by the PXRD analysis, as in the diffractograms recorded at those conditions peaks from both Form I and Form II could have been observed. This may suggest that the Gibbs free energies of those two polymorphs are very similar, and thus if the phase transition had occurred during optimization, it would have been associated with a negligible change of the free energy.

The last analyzed group (G, H, and I) is also very consistent in terms of the received RMSD values. For this group, unlike for the others, Form II was the experimentally obtained one, which resulted in small RMSD values for comparisons between the experimental and modeled structures when Form II was used as the initial structure for calculations. As previously stated, based on the results presented in Tables S2 and S3 and Figure 3, the application of neither PBE TS nor PBESOL enabled the prediction of the ISP to obtain Form II while optimizing Form I at the increased pressure. This is also reflected in the large RMSD values between the experimental structure (Form II) and calculated ones, using Form I as the initial for both PBESOL and PBE TS. For this group, the differences between the results obtained using PBE TS and PBESOL were found to be negligible when either Form I or Form II was used as the initial.

2.4. Geometry Optimization of Forms I and II under External Pressure—Energy and Thermodynamic Parameters Differences Analysis

The next step of the analysis was the comparison of the energies obtained while applying the same pressure values and DFT functionals but using different initial structures

(Forms I and II). Results are presented in Tables S4 and S5, Figures 4 and 5. In those figures, the calculated unit cell lengths "a", obtained starting either from Form I or II, have also been shown as they are relevant for the discussion below. The positive values of differences indicate that form II is energetically preferred.

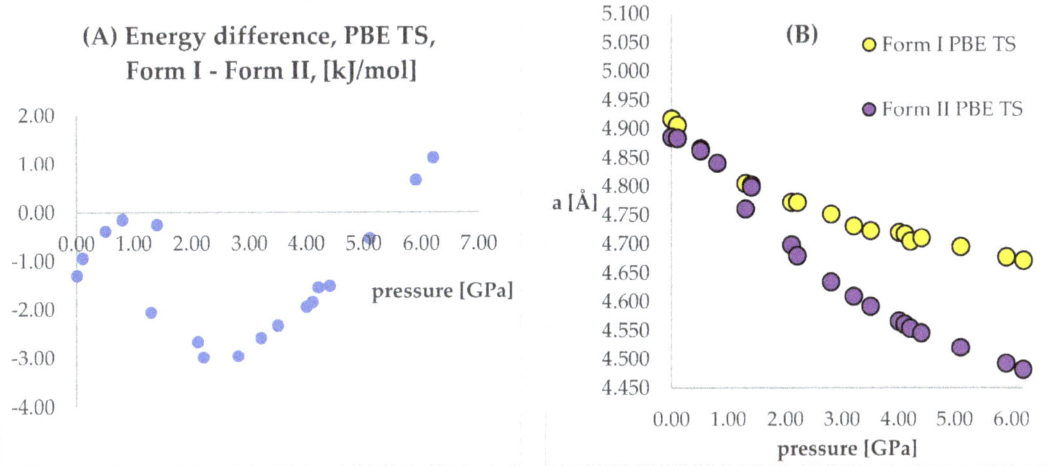

Figure 4. (**A**): Differences between the energies (Form I—Form II) of the structures modeled using PBE TS functional, with respect to pressure. (**B**): The change of unit cell edge "a" length obtained from calculations using PBE TS functional, with respect to pressure. Yellow circles—using Form I as initial; violet circles—using Form II as initial.

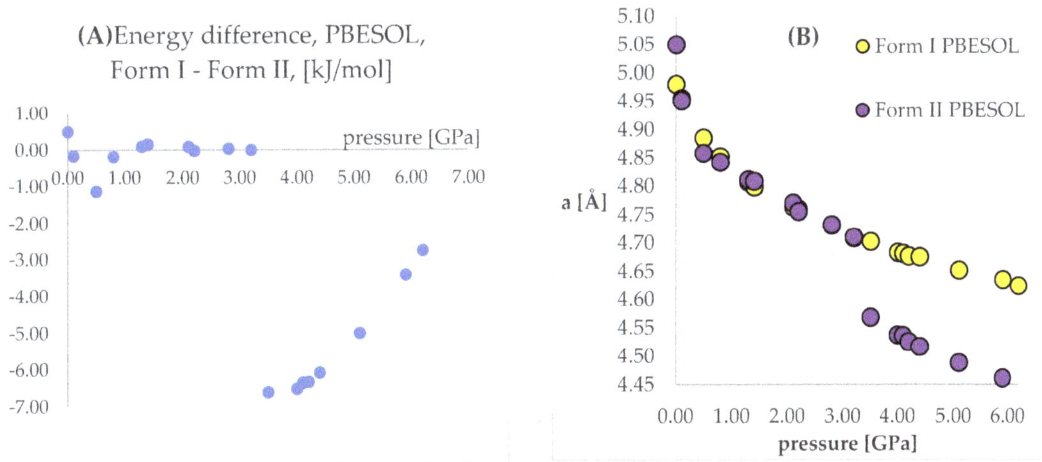

Figure 5. (**A**): Differences between the energies (Form I—Form II) of the structures modeled using PBESOL functional, with respect to pressure. (**B**): The change of unit cell edge "a" length obtained from calculations using PBE TS functional, with respect to pressure. Yellow circles—using Form I as initial; violet circles—using Form II as initial.

Looking at Figures 4 and 5, some similarities and differences between those results can be observed. For the lower pressure values (lower than 1.40 GPa and 3.20 GPa for PBE TS and PBESOL, respectively), the calculated differences between the energies were found to be very small. It is not surprising that for those calculations the transition of Form II into Form I has been observed. This is clearly visible, by looking at the right parts of those figures presenting the values of "a" lengths. The absolute value of the energy difference corresponds with the difference between the "a" values calculated using different Forms as the initial. Then, above a certain value of pressure (2.20 GPa and 3.50 GPa for PBE TS and PBESOL, respectively), the values of energy differences begin to increase monotonically, meaning, that according to both PBE TS and PBESOL calculations results, Form II is becoming more stable with the increase in pressure, which is in agreement with experimental observations. However, in the whole studied pressure range, the differences obtained using PBESOL were found to be negative, meaning that according to the PBESOL calculations Form I should be more energetically favorable than Form II under those conditions. Those results were not in agreement with experimental observations. However, for the results obtained using PBE TS, the change of the sign of the calculated differences was observed at a pressure higher than 5.50 GPa. Therefore, according to the calculations obtained using PBE TS functional, Form II should be more stable than Form I at higher pressure, which was confirmed experimentally. For the PBE TS calculations, if the difference between the forms was larger than 3 kJ/mol the transition of Form II into Form I was observed (at c.a. 2.00 GPa). In the case of PBESOL, the difference had to be larger than 6.60 kJ/mol to force the transition which occurred at 3.20 GPa.

The most important conclusion from the results observed and discussed above was that the results obtained using the PBE TS functional suggested that Form II should be more stable under increased pressure; however, the IPT was not observed during geometry optimization. This conclusion encouraged us to try to overcome the energy barrier between Form I and Form II using ab initio molecular dynamics (aiMD) which will be described in detail in Section 2.5.

More accurately, it is not the electronic energy difference discussed above, but the difference between the Gibbs free energy (ΔG) that decides which polymorphic form is more stable at certain conditions. Therefore, to obtain the thermodynamic parameters of the studied forms, the calculations of phonon density of states were performed. The results of those calculations are presented and discussed below (Figure 6, Table S6). The entropy (ΔS) values were multiplied by the temperature (T = 293 K) to facilitate the analysis, as ΔG = ΔH–TΔS.

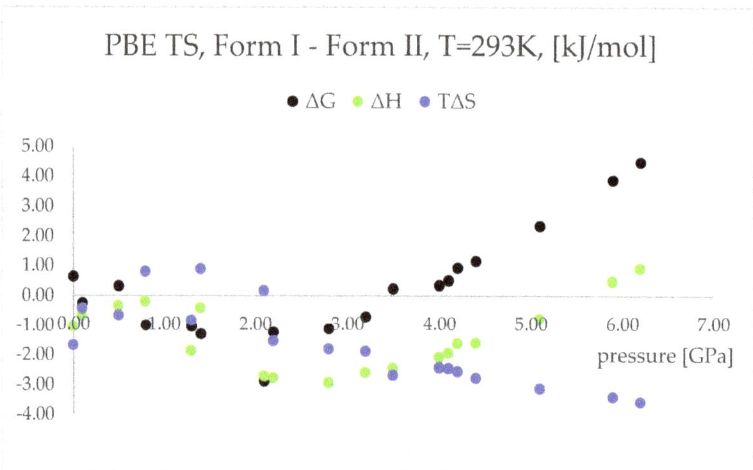

Figure 6. Differences between the thermodynamic parameters: free energy (ΔG, black dots) enthalpy (ΔH, green dots), temperature times entropy (TΔS, blue dots); Form I—Form II; of the structures modelled using PBE TS functional at 293 K, with respect to pressure.

The differences between the ΔH values were found to change similarly to the changes in energy values (Figure 4), suggesting that the IPT of Form I to Form II should be exothermic at pressures higher than 5.5 GPa. However, the differences between the TΔS in the pressure range 2.20–6.20 GPa were found to be negative and monotonically decreasing. This indicated that the studied pressure-induced IPT should be entropy-driven transformation. The calculated changes in the free energy (ΔG) suggested that the studied Forms should coexist at pressure range 3.5–4.1 GPa and above this pressure Form II should be more stable and dominant, which is in agreement with the experimental data.

Concluding this section, the geometry optimization and thermodynamic properties calculations enabled the prediction of Form II to Form I IPT that occurs upon decompression. Further, the free energy calculations results agreed with experimental observations, indicating that Form II is more stable at higher pressure. However, probably due to the large energy barrier between Form I and Form II at higher pressure, the geometry optimization of Form I did not result in obtaining Form II at any of the studied pressure values. In order to achieve this aim, the aiMD dynamics calculations were performed.

2.5. *Ab Initio Molecular Dynamics Simulations*

As stated above, the geometry optimization of Form I at increased pressure did not result in Form II, which was the major aim of this study. However, encouraged by the results of energy (Figure 4) and thermodynamic parameters (Figure 6) calculations, showing that Form II is indeed more stable at increased pressure, we decided to perform the computationally demanding ab initio molecular dynamics (aiMD) simulations. Due to the small differences between the energies of the studied Forms, even at 6.2 GPa, being in the order of 1 kJ/mol, we have not performed the "classical" molecular dynamics simulations based on the molecular mechanics' framework as their accuracy was expected to be insufficient.

For the aiMD simulations, GGA PBE TS functional was chosen, opposite to PBESOL, and Form II was found to be energetically favorable (Figure 4) at increased pressure. The simulations were performed at T = 293 K and p = 6.20 GPa, as this was the largest value of pressure at which the structural information for CT was obtained and, at the same time, the difference between the energies of the studied Forms at this pressure was the largest, in favor of Form II (Figure 4).

The geometry optimized at 6.20 GPa Forms I (QQQAUG09) and II (QQQAUG17) structures were used as starting for aiMD simulations. Form I was the obvious choice as the aim was to observe the IPT and obtain Form II. Additionally, simulations with the same parameters were performed also using Form II as initial for several reasons. First, we wanted to confirm that under those conditions no other phase transition would occur as well as to confirm the stability of this form under this pressure condition. Though experimentally Form II was found to be stable at 6.20 GPa, we wanted to ensure that the results of the calculations would be in agreement with this experimental observation. Secondly, since the introduction of kinetic energy associated with temperature always results in structural parameter fluctuations, it was necessary to determine the magnitude of such fluctuations. The results of aiMD are presented in Figures 7 and 8 and Figures S1–S4.

The results of aiMD showed that it was a proper method to simulate the pressure-induced IPT of CT Form I into Form II. All the structural unit cell parameters of Form I changed into those of Form II during the simulation. Using previously discussed "a" edge as an example, the value of Form II exhibits only thermal fluctuations while the value of Form I decreases monotonically for the first 15 ps, reaching the experimental and computational values obtained for Form II. The simulation time needed for the structural parameters of Form I to convert into those of Form II was not common, i.e., in the case of angle α (Figure 8) it was shorter than for the edge "a". Nevertheless, if only Form I is known, performing aiMD simulations at 6.20 GPa, preceded by the geometry optimization, would suggest that IPT may occur and would allow the estimation of the unit cell parameters of the new form. The reason why the aiMD simulations were required to observe the IPT that was not achieved in the geometry optimization is surely connected with the energy barrier between those two forms. The results of the thermodynamic calculations, presented in Figure 6, suggested that this transition is entropy-driven. Therefore, by adding kinetic energy in the aiMD simulations, it was possible for the studied system (Form I) to overcome this energy barrier and reach the deeper minimum (Form II). MD simulations are complementary to lattice-dynamical calculations in the sense that the latter are better suited to low temperatures, whereas the former are subject to ergodicity problems. Lattice dynamics are by definition limited to the (quasi)harmonic regime, while molecular dynamics naturally account for all the anharmonic effects occurring at high temperatures.

Figure 7. Running average of the unit cell edge length "a" obtained from aiMD simulation at T = 293 K and p = 6.2 GPa using PBE TS functional. Horizontal lines represent the experimental values.

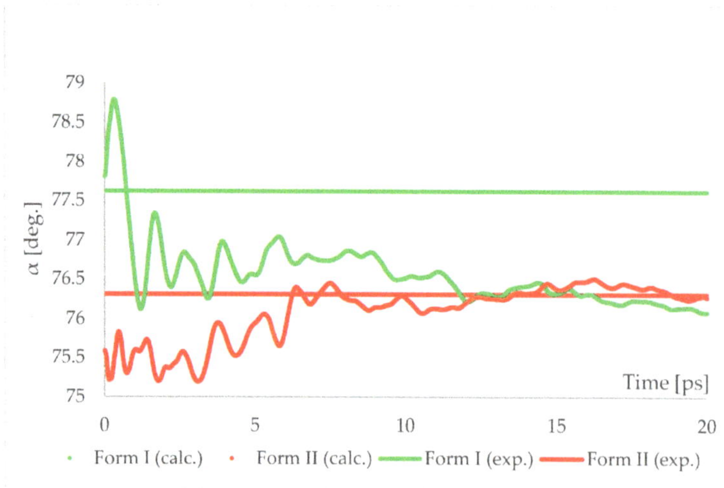

Figure 8. Running average of the unit cell angle "α" obtained from aiMD simulation at T = 293 K and p = 6.2 GPa using PBE TS functional. Horizontal lines represent the experimental values.

Therefore, answering the question stated in the title of this work, the DFT-based calculations can predict the pressure-induced IPT of CT, though only through the aiMD simulations.

3. Computational Methods

The density functional theory (DFT) calculations of geometry optimization, ab initio molecular dynamics (aiMD), phonon dispersion, and density of states were carried out with the CASTEP program [27] implemented in the Materials Studio 2017 software [28] using the plane wave pseudopotential formalism. On the fly generated (OTFG) norm-conserving pseudopotentials (NCP) were generated using the Koelling–Harmon (KH) scalar relativistic approach [29].

For comparison of the conformations of structures under investigation, the direct root mean square deviation (RMSD) of atomic positions for single molecules was calculated (Table 2) according to the Equation:

$$RMSD = \sqrt{\frac{\Sigma_i d_i^2}{n}}$$

where d is the distance between each of the n pairs of equivalent atoms in two optimally superposed structures.

To facilitate the analysis of the data in Table 2 a three-color scale was applied. In this scale, the 50th percentile (midpoint) was calculated, and the cell that holds this value was colored yellow. The cell that holds the minimum value was colored green, and the cell that holds the maximum value was colored red.

3.1. DFT Functionals and Dispersion Correction Methods

The Perdew–Burke–Ernzerhof (PBE) [30] pure or with Tkatchenko–Scheffler (TS) [20] or Gimme [31] dispersion correction, Perdew–Wang (PW91) [32] pure or with Ortmann–Bechstedt–Schmidt (OBS) [33] dispersion correction, revised Perdew–Burke–Ernzerhof (RPBE) [34], Wu–Cohen (WC) [35], solid-design version of the PBE (PBESOL) [36] exchange-correlation functionals, defined within the generalized gradient approximation (GGA) as well as the local exchange-correlation functional of Perdew and Zunger [37] with the parameterization of the numerical results of Ceperley and Alder [38] (LDA CA-PZ), with or without the OBS method of dispersion correction were used in the calculations.

3.2. Geometry Optimization

Geometry optimization was carried out using the Broyden−Fletcher−Goldfarb−Shanno (BFGS) [39] optimization scheme and smart method for finite basis set correction. The electronic parameters—kinetic energy cutoff for the plane waves (E_{cut}) and number of Monkhorst–Pack k-points during sampling for a primitive cell Brillouin zone integration [40] were optimized and set to 990 eV and $3 \times 3 \times 2$, respectively.

The experimental X-ray structure of chlorothiazide Form I (refcode QQQAUG09) and Form II (QQQAUG17) from the Cambridge Structure Database (CSD) were used as initial for calculations. During geometry optimization, all atom positions and the cell parameters were optimized, with no constraints. The convergence criteria were set at 5×10^{-6} eV/atom for the energy, 1×10^{-2} eV/Å for the interatomic forces, 2×10^{-2} GPa for the stresses, and 5×10^{-4} Å for the displacements. The fixed basis set quality method for the cell optimization calculations and the 5×10^{-7} eV/atom tolerance for SCF were used.

3.3. Thermodynamic Parameters Calculations

Phonon frequencies were obtained by diagonalization of dynamical matrices computed using linear response methodology (also known as density functional perturbation theory, DFPT) [41]. DFPT is the most commonly used ab initio calculation method of phonons. This method is different from a direct method since DFPT calculates the change in the Hamiltonian under a given perturbation of charge density or wavefunction, rather than directly displacing atoms in a direct method. The q-point separation parameter, which represents the average distance between Monkhorst–Pack mesh q-points used in the real space dynamical matrix calculations, was set to 0.05 $Å^{-1}$. The convergence criterion for the force constants during a phonon properties run was set to 1×10^{-5} eV/$Å^2$. This model was employed to calculate band structure, DOS, phonon spectrum, and phonon DOS properties. For the dispersion calculations, the separation between consecutive q-vectors on the reciprocal space path was set to 0.015 $Å^{-1}$. For the DOS calculations, the $3 \times 3 \times 2$ Monkhorst–Pack k-points grid has been chosen, resulting in the q-vector separation of 0.04 $Å^{-1}$.

The results of a calculation of phonon spectra have been used to compute, in the quasi-harmonic approximation, zero-point vibrational energy (E_{zp}), entropy (S), Gibbs free energy (G), and enthalpy (H) as functions of temperature, using the Formulas (1)–(4) below that are based on the work by Baroni et al. [41]. In those formula, ω represents phonon frequency, $F(\omega)$ represents the vibrational density of states for a phonon spectrum, E_{tot} is the total electronic energy at 0 K, k is Boltzmann's constant, and $\hbar$ is the Dirac's constant.

$$E_{zp} = \frac{1}{2}\int F(\omega)\hbar\omega d\omega \quad (1)$$

$$S(T) = k\left\{\int \frac{\frac{\hbar\omega}{kT}}{\exp\left(\frac{\hbar\omega}{kT}\right)-1}F(\omega)d\omega - \int F(\omega)\ln\left[1-\exp\left(-\frac{\hbar\omega}{kT}\right)\right]d\omega\right\} \quad (2)$$

$$G(T) = E_{tot} + E_{zp} + kT\int F(\omega)\ln\left[1-\exp\left(-\frac{\hbar\omega}{kT}\right)\right]d\omega + pV \quad (3)$$

$$H = G + TS \quad (4)$$

3.4. Ab Initio Molecular Dynamic Simulations (aiMD)

Born–Oppenheimer ab initio molecular dynamics (aiMD) [42] simulations were run in CASTEP using an NPT ensemble maintained at a constant temperature of 293 K and pressure of 6.20 GPa, using Nosé thermostat, Parinello barostat, and PBE TS functional. The kinetic energy cutoff for the plane waves (E_{cut}) was set to 990 eV and the integration time step was set to 0.5 fs. No symmetry constraints were applied during the simulations. The total time of the simulation was set to 20 ps.

4. Conclusions

In this work, the pressure-induced IPT of chlorothiazide was studied using DFT methods. First, the accuracy of the calculations using different DFT functionals was evaluated, resulting in the choice of the PBE TS and PBESOL for future calculations. Then, the geometry optimization calculations of Form I and Form II at all experimentally studied pressure conditions were performed. It was observed that the choice of DFT functional had a significant influence on the received results. In particular, the dispersion correction (TS) was found to be crucial for achieving accurate results, and thus for the thermodynamic and aiMD calculations only the PBE TS functional has been chosen. Through the geometry optimization, the IPT of Form II into Form I upon decompression was achieved, however, the opposite pressure-induced transition was not observed, regardless of the chosen functional. However, both the comparison of the electronic energies and chosen thermodynamic parameters (ΔG, ΔH, ΔS) indicated that Form II is more stable at the increased pressure. Finally, in order to observe the pressure-induced IPT of Form I into Form II, ab initio molecular dynamics simulations were successfully applied.

Since the DFT calculations enabled to predict the IPT of CT, we will continue to study the other IPT using the methodology described in this work. We hope that such an approach, when successfully validated on the reasonable number of examples, will be used in the future as a screening method to predict the isosymmetric phase transition, lowering the costs and increasing the efficiency of the studies designed for searching of new polymorphic forms.

Supplementary Materials: The following are available online at https://www.mdpi.com/article/10.3390/ijms221810100/s1.

Author Contributions: Conceptualization, Ł.S., A.H.M. and D.M.P.; methodology, Ł.S., A.H.M. and D.M.P.; software, Ł.S., K.M. and E.N.; validation, Ł.S., A.H.M., K.M. and E.N.; formal analysis, Ł.S., A.H.M. and D.M.P.; investigation, Ł.S., A.H.M., K.M. and E.N.; resources, Ł.S.; data curation, Ł.S.; writing—original draft preparation, Ł.S. and A.H.M.; writing—review and editing, Ł.S., A.H.M. and D.M.P.; visualization, Ł.S., A.H.M. and D.M.P.; supervision, Ł.S.; project administration, Ł.S.; funding acquisition, Ł.S. All authors have read and agreed to the published version of the manuscript.

Funding: This work was supported by a research grant (Project GW/F/16) from the Medical University of Warsaw.

Institutional Review Board Statement: Not applicable.

Informed Consent Statement: Not applicable.

Data Availability Statement: The data presented in this study are available on request from the corresponding author.

Conflicts of Interest: The authors declare no conflict of interest. The funders had no role in the design of the study; in the collection, analyses, or interpretation of data; in the writing of the manuscript, or in the decision to publish the results.

References

1. Purohit, R.; Venugopalan, P. Polymorphism: An overview. *Resonance* **2009**, *14*, 882–893. [CrossRef]
2. Censi, R.; Di Martino, P. Polymorph Impact on the Bioavailability and Stability of Poorly Soluble Drugs. *Molecules* **2015**, *20*, 18759–18776. [CrossRef]
3. Chistyakov, D.; Sergeev, G. The Polymorphism of Drugs: New Approaches to the Synthesis of Nanostructured Polymorphs. *Pharmaceutics* **2020**, *12*, 34. [CrossRef]
4. Guerain, M. A Review on High Pressure Experiments for Study of Crystallographic Behavior and Polymorphism of Pharmaceutical Materials. *J. Pharm. Sci.* **2020**, *109*, 2640–2653. [CrossRef] [PubMed]
5. Christy, A.G. Isosymmetric structural phase transitions: Phenomenology and examples. *Acta Crystallogr. Sect. B Struct. Sci.* **1995**, *51*, 753–757. [CrossRef]
6. Goryainov, S.V.; Boldyreva, E.V.; Smirnov, M.B.; Ahsbahs, H.; Chernyshev, V.V.; Weber, H.-P. Isosymmetric Reversible Pressure-Induced Phase Transition in Sodium Oxalate at 3.8 GPa. In *Doklady Physical Chemistry*; Kluwer Academic Publishers-Plenum Publishers: Dordrecht, The Netherlands, 2003; Volume 390, pp. 154–157. [CrossRef]
7. Kartusiak, A. Structure and phase transition of 1,3-cyclohexanedione crystals as a function of temperature. *Acta Crystallogr. Sect. B Struct. Sci.* **1991**, *47*, 398–404. [CrossRef]
8. Fisch, M.; Lanza, A.; Boldyreva, E.; Macchi, P.; Casati, N. Kinetic Control of High-Pressure Solid-State Phase Transitions: A Case Study on L-Serine. *J. Phys. Chem. C* **2015**, *119*, 18611–18617. [CrossRef]
9. Li, Q.; Li, S.; Wang, K.; Li, X.; Liu, J.; Liu, B.; Zou, G.; Zou, B. Pressure-induced isosymmetric phase transition in sulfamic acid: A combined Raman and X-ray diffraction study. *J. Chem. Phys.* **2013**, *138*, 214505. [CrossRef]
10. Bull, C.L.; Funnell, N.P.; Ridley, C.J.; Pulham, C.R.; Coster, P.; Tellam, J.P.; Marshall, W.G. Pressure-Induced Isosymmetric Phase Transition in Biurea. *CrystEngComm* **2019**, *21*, 5872–5881. [CrossRef]
11. Clarke, S.M.; Steele, B.A.; Kroonblawd, M.P.; Zhang, D.; Kuo, I.-F.W.; Stavrou, E. An Isosymmetric High-Pressure Phase Transition in α-Glycylglycine: A Combined Experimental and Theoretical Study. *J. Phys. Chem. B* **2019**, *124*, 1–10. [CrossRef]
12. Roques, J.; Veilly, E.; Simoni, E. Periodic Density Functional Theory Investigation of the Uranyl Ion Sorption on Three Mineral Surfaces: A Comparative Study. *Int. J. Mol. Sci.* **2009**, *10*, 2633–2661. [CrossRef]
13. Chilukuri, B.; Mazur, U.; Hipps, K.W. Structure, Properties, and Reactivity of Porphyrins on Surfaces and Nanostructures with Periodic DFT Calculations. *Appl. Sci.* **2020**, *10*, 740. [CrossRef]
14. Medvedev, A.G.; Churakov, A.V.; Prikhodchenko, P.V.; Lev, O.; Vener, M.V. Crystalline Peroxosolvates: Nature of the Coformer, Hydrogen-Bonded Networks and Clusters, Intermolecular Interactions. *Molecules* **2021**, *26*, 26. [CrossRef]
15. Szeleszczuk, Ł.; Pisklak, D.M.; Zielińska-Pisklak, M. Does the choice of the crystal structure influence the results of the periodic DFT calculations? A case of glycine alpha polymorph GIPAW NMR parameters computations. *J. Comput. Chem.* **2018**, *39*, 853–861. [CrossRef]
16. Szeleszczuk, Ł.; Pisklak, D.M.; Gubica, T.; Matjakowska, K.; Kaźmierski, S.; Zielińska-Pisklak, M. Application of combined solid state NMR and DFT calculations for the study of piracetam polymorphism. *Solid State Nucl. Magn. Reson.* **2019**, *97*, 17–24. [CrossRef] [PubMed]
17. Dolgonos, G.A.; Hoja, J.; Boese, A.D. Revised values for the X23 benchmark set of molecular crystals. *Phys. Chem. Chem. Phys.* **2019**, *21*, 24333–24344. [CrossRef]

18. Buchholz, H.K.; Stein, M. Accurate lattice energies of organic molecular crystals from periodic turbomole calculations. *J. Comput. Chem.* **2018**, *39*, 1335–1343. [CrossRef]
19. Grimme, S. Density functional theory with London dispersion corrections. *Wiley Interdiscip. Rev. Comput. Mol. Sci.* **2011**, *1*, 211–228. [CrossRef]
20. Tkatchenko, A.; Scheffler, M. Accurate Molecular Van Der Waals Interactions from Ground-State Electron Density and Free-Atom Reference Data. *Phys. Rev. Lett.* **2009**, *102*, 073005. [CrossRef] [PubMed]
21. Reilly, A.M.; Tkatchenko, A. van der Waals dispersion interactions in molecular materials: Beyond pairwise additivity. *Chem. Sci.* **2015**, *6*, 3289–3301. [CrossRef]
22. Xu, P.; Alkan, M.; Gordon, M.S. Many-Body Dispersion. *Chem. Rev.* **2020**, *120*, 12343–12356. [CrossRef]
23. Mazurek, A.H.; Szeleszczuk, Ł.; Pisklak, D.M. Periodic DFT Calculations—Review of Applications in the Pharmaceutical Sciences. *Pharmaceutics* **2020**, *12*, 415. [CrossRef] [PubMed]
24. Carpenter, R.J.; Kouyoumjian, S.; Moromisato, D.Y.; Lieu, L.; Amirnovin, R. Lower-Dose, Intravenous Chlorothiazide Is an Effective Adjunct Diuretic to Furosemide Following Pediatric Cardiac Surgery. *J. Pediatr. Pharmacol. Ther.* **2020**, *25*, 31–38. [CrossRef]
25. Oswald, I.D.H.; Lennie, A.R.; Pulham, C.R.; Shankland, K. High-pressure structural studies of the pharmaceutical, chlorothiazide. *CrystEngComm* **2010**, *12*, 2533. [CrossRef]
26. Szeleszczuk, Ł.; Pisklak, D.M.; Zielińska-Pisklak, M. Can we predict the structure and stability of molecular crystals under increased pressure? First-principles study of glycine phase transitions. *J. Comput. Chem.* **2018**, *39*, 1300–1306. [CrossRef] [PubMed]
27. Clark, S.J.; Segall, M.D.; Pickard, C.J.; Hasnip, P.J.; Probert, M.J.; Refson, K.; Payne, M.C. First principles methods using CASTEP. *Z. Krist. Cryst. Mater.* **2005**, *220*, 567–570. [CrossRef]
28. BIOVIA Materials Studio. Available online: http://accelrys.com/products/collaborative-science/biovia-materials-studio/ (accessed on 15 July 2021).
29. Koelling, D.D.; Harmon, B.N. Technique for relativistic spin-polarized calculations. *J. Phys. C Solid State Phys.* **1977**, *10*, 3107–3114. [CrossRef]
30. Perdew, J.P.; Burke, K.; Ernzerhof, M. Generalized Gradient Approximation Made Simple. *Phys. Rev. Lett.* **1996**, *77*, 3865–3868. [CrossRef]
31. Grimme, S. Semiempirical GGA-type density functional constructed with a long-range dispersion correction. *J. Comput. Chem.* **2006**, *27*, 1787–1799. [CrossRef]
32. Perdew, J.P.; Chevary, J.A.; Vosko, S.H.; Jackson, K.A.; Pederson, M.R.; Singh, D.J.; Fiolhais, C. Atoms, molecules, solids, and surfaces: Applications of the generalized gradient approximation for exchange and correlation. *Phys. Rev. B* **1992**, *6*, 6671–6687. [CrossRef]
33. Ortmann, F.; Bechstedt, F.; Schmidt, W.G. Semiempirical van der Waals correction to the density functional description of solids and molecular structures. *Phys. Rev. B* **2006**, *73*, 205101. [CrossRef]
34. Hammer, B.; Hansen, L.B.; Norskov, J.K. Improved adsorption energetics within density-functional theory using revised Perdew-Burke-Ernzerhof functionals. *Phys. Rev. B* **1999**, *59*, 7413–7421. [CrossRef]
35. Wu, Z.; Cohen, R.E. More accurate generalized gradient approximation for solids. *Phys. Rev. B* **2006**, *73*, 235116. [CrossRef]
36. Perdew, J.P.; Ruzsinszky, A.; Csonka, G.I.; Vydrov, O.A.; Scuseria, G.E.; Constantin, L.A.; Zhou, X.; Burke, K. Restoring the Density-Gradient Expansion for Exchange in Solids and Surfaces. *Phys. Rev. Lett.* **2008**, *100*, 136406. [CrossRef] [PubMed]
37. Perdew, J.P.; Zunger, A. Self-interaction correction to density-functional approximations for many-electron systems. *Phys. Rev. B* **1981**, *23*, 5048–5079. [CrossRef]
38. Ceperley, D.M.; Alder, B.J. Ground State of the Electron Gas by a Stochastic Method. *Phys. Rev. Lett.* **1980**, *45*, 566–569. [CrossRef]
39. Pfrommer, B.G.; Cote, M.; Louie, S.G.; Cohen, M.L. Relaxation of Crystals with the Quasi-Newton Method. *J. Comput. Phys.* **1997**, *131*, 233–240. [CrossRef]
40. Monkhorst, H.J.; Pack, J.D. "Special points for Brillouin-zone integrations"—A reply. *Phys. Rev. B* **1977**, *16*, 1748–1749. [CrossRef]
41. Baroni, S.; de Gironcoli, S.; dal Corso, A.; Giannozzi, P. Phonons and related crystal properties from density-functional perturbation theory. *Rev. Mod. Phys.* **2001**, *73*, 515–562. [CrossRef]
42. Arias, T.A.; Payne, M.C.; Joannopoulos, J.D. Ab initiomolecular dynamics: Analytically continued energy functionals and insights into iterative solutions. *Phys. Rev. Lett.* **1992**, *69*, 1077–1080. [CrossRef] [PubMed]

International Journal of
Molecular Sciences

MDPI

Article

Structural and Dynamical Behaviour of Colloids with Competing Interactions Confined in Slit Pores

Horacio Serna [1], Wojciech T. Góźdź [1] and Eva G. Noya [2,*]

1 Institute of Physical Chemistry, Polish Academy of Sciences, Kasprzaka 44/52, 01-224 Warsaw, Poland; hserna@ichf.edu.pl (H.S.); wtg@ichf.edu.pl (W.T.G.)
2 Instituto de Química Física Rocasolano, CSIC, C/ Serrano 119, 28006 Madrid, Spain
* Correspondence: eva.noya@iqfr.csic.es

Abstract: Systems with short-range attractive and long-range repulsive interactions can form periodic modulated phases at low temperatures, such as cluster-crystal, hexagonal, lamellar and bicontinuous gyroid phases. These periodic microphases should be stable regardless of the physical origin of the interactions. However, they have not yet been experimentally observed in colloidal systems, where, in principle, the interactions can be tuned by modifying the colloidal solution. Our goal is to investigate whether the formation of some of these periodic microphases can be promoted by confinement in narrow slit pores. By performing simulations of a simple model with competing interactions, we find that both the cluster-crystal and lamellar phases can be stable up to higher temperatures than in the bulk system, whereas the hexagonal phase is destabilised at temperatures somewhat lower than in bulk. Besides, we observed that the internal ordering of the lamellar phase can be modified by changing the pore width. Interestingly, for sufficiently wide pores to host three lamellae, there is a range of temperatures for which the two lamellae close to the walls are internally ordered, whereas the one at the centre of the pore remains internally disordered. We also find that particle diffusion under confinement exhibits a complex dependence with the pore width and with the density, obtaining larger and smaller values of the diffusion coefficient than in the corresponding bulk system.

Keywords: colloids with competing interactions; periodic microphases; confinement

Citation: Serna, H.; Góźdź, W.T.; Noya, E.G. Structural and Dynamical Behaviour of Colloids with Competing Interactions Confined in Slit Pores. *Int. J. Mol. Sci.* **2021**, *22*, 11050. https://doi.org/10.3390/ijms222011050

Academic Editor: Małgorzata Borówko

Received: 16 September 2021
Accepted: 10 October 2021
Published: 13 October 2021

1. Introduction

Competing attractive and repulsive interactions can be found in a wide variety of systems, ranging from block copolymers, proteins, or colloids, just to mention a few examples [1]. Even though the physical origin of the interactions are different in these systems, theory predicts that they all exhibit similar phase diagrams in which periodic microphases (cluster-crystal, hexagonal, bicontinuous gyroid and lamellar phases) are stable at low temperatures [2–6]. One might think that colloidal systems, in which the attractive and repulsive interactions can be tuned by modifying the colloidal solution, could be a good playground to experimentally study the formation of periodic microphases. Still, these periodic microphases have not yet been experimentally observed in colloidal systems [7], which has been attributed to the particle size polidispersity [8], to the slow kinetics of the fluid [9–11], or to the inability of a simple effective potential to capture the behaviour of the colloidal solution [7]. Additionally, recent studies have suggested that apart from the strength of the interactions, the attractive and repulsive ranges play an important role and their variation can induce different phase behaviours [1,12]. In this regard, the ranges of interaction can be easily tuned in block copolymers by varying the length of the chains composed of one or another monomer [13]. Block copolymers self-assembly has many potential applications in nanotechnology and industry, such as separation and ion conduction in batteries, templating for nanomaterial synthesis and sensing [14,15].

Obtaining the proper ranges in experimental colloidal systems has been challenging, but some recent approaches in which the colloidal particles are functionalised with hydrophobic molecules have shown promising results [16]. In a previous work [17], we showed how the Lennard-Jones plus Yukawa (LJY) potential, with the proper ranges and strengths of the interactions, can form ordered microphases in bulk. Here we want to stress the importance of simulations in predicting new physical phenomena. In particular, simulation is useful to guide the design of experiments that can lead finally to new discoveries. Regarding the applications, confined colloidal particles in channels of different geometries have been used to build wave-guide devices useful in sensing [18].

There are several ways in which the ordering of the periodic microphases can be induced, for example, by applying shear [19] or by confining the fluid in pores with the appropriate geometry [20]. In this work, we will explore this second route. It is known that confinement of simple and complex fluids can change the phase behaviour by shifting coexistence lines to lower or to higher temperatures than in bulk, depending on the shape and size of the pores and on the nature of the interactions of the fluid with the pore walls [21]. It can also induce significant changes on the dynamic behaviour, in some cases finding a nonmonotonous variation of the diffusion coefficient with the pore size [22–24]. In the particular case of systems with competing interactions, previous theory and simulation studies showed that confinement can promote or inhibit the formation of periodic microphases depending on whether the pore size is commensurate or not with the periodicity of the bulk microphase. For example, we showed in a previous work that confinement in channels with triangular and hexagonal cross-sections favour the formation of the hexagonal phase, as well as by introducing wedges in pores with cylindrical cross-sections (which otherwise promote the formation of helical structures) [25,26]. We also found that new phases that are not stable in bulk can be stabilised when confined by the appropriate pore geometry. In this way, cluster-crystals with different symmetries were obtained by confinement in bicontinuous porous materials [27]. Surprisingly, the study of confinement in simple geometries, such as a slit pore, has not been sufficiently explored. Indeed we are only aware of a few studies in which fluids with competing interactions confined between parallel plates were studied, but those were restricted to two- and one-dimensional cases [20,28,29]. The literature on the dynamic behaviour of fluids with competing interactions in confinement is also scarce [30].

In this work we undertake a simulation study to investigate the effects of confinement on the structural and dynamic behaviour of fluids with competing interactions in narrow slit pores as a function of the pore width. The study is performed under conditions at which the cluster-crystal, the hexagonal and the lamellar phases are stable in bulk.

Our goal is to determine whether the formation of periodic microphases can be thermodynamically and/or kinetically favoured by confinement.

2. The Model and the Simulation Method

The colloidal particles interact with each other via an effective short-range attraction long-range repulsion (SALR) model potential resulting from the addition of a Lennard-Jones potential plus a Yukawa repulsive term:

$$u_{SALR}(r_{ij}) = 4\epsilon\left[\left(\frac{\sigma}{r_{ij}}\right)^{2\alpha} - \left(\frac{\sigma}{r_{ij}}\right)^{\alpha}\right] + \frac{A}{(r_{ij}/\zeta)}\exp\left(-r_{ij}/\zeta\right) \tag{1}$$

The parameters of the model were assigned the same values as in our previous work in which the bulk phase diagram was investigated [17]. In particular, we chose $\epsilon = 1.6$, $\sigma = 1.0$, $\alpha = 6$, $A = 0.65$, and $\zeta = 2.0$. For computational efficiency, the potential is truncated and shifted at $r_c = 4.0\sigma$. In what follows, all the magnitudes are reduced taking σ and ϵ as units of length and energy, respectively.

The confinement is implemented along the z direction by placing two parallel walls at $z_w = \pm W/2$, so that the separation between them is W. Periodic boundary conditions are imposed along the x and y directions. The walls are structureless and repulsive. Particles interact with the walls via a Lennard-Jones model truncated and shifted at the energy minimum:

$$\mathcal{V}_{z_w}(z_{iw}) = \begin{cases} 4\epsilon_w \left[\left(\frac{\sigma_w}{z_{iw}} \right)^{12} - \left(\frac{\sigma_w}{z_{iw}} \right)^{6} \right] + \epsilon_w & : z_{iw} < 2^{1/6}\sigma_w \\ 0 & : z_{iw} \geq 2^{1/6}\sigma_w \end{cases}, \tag{2}$$

where $\epsilon_w = 1.0$, $\sigma_w = 1.0$, and z_{iw} is the distance from particle i to the pore wall z_w. The Lennard-Jones plus Yukawa interaction potential used to model the interactions between the particles and the truncated Lennard-Jones potential that accounts for the interactions between the particles and the walls are plotted in Figure 1.

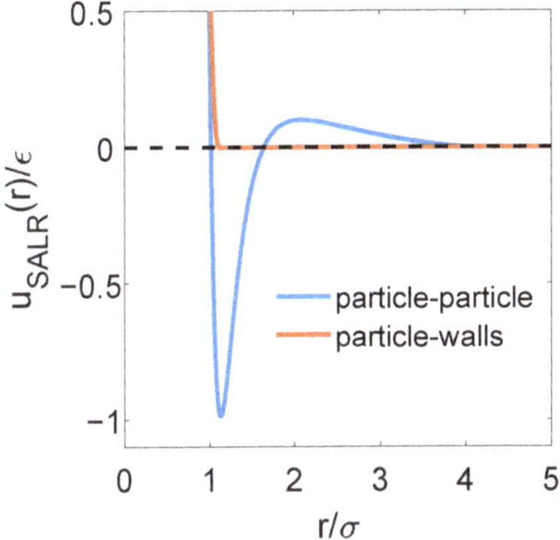

Figure 1. The Lennard-Jones plus Yukawa potential used to model the interactions between colloidal particles and the Lennard-Jones potential truncated and shifted at the energy minimum used to model the interactions between the slit walls and the particles

Thus, the total energy of the system is given by:

$$U_{tot} = \sum_{i=1}^{N-1} \sum_{j>i}^{N} u_{SALR}(r_{ij}) + \sum_{i=1}^{N} (\mathcal{V}_{W/2}(z_{iw}) + \mathcal{V}_{-W/2}(z_{iw})). \tag{3}$$

The slit width is given by the centre–centre separation between the confining walls, W. Taking into account that the energy of a particle becomes very repulsive for distances to the wall shorter than σ_w, the available width for the particle volume is actually $2\sigma_w$ smaller than the pore width. The edges of the simulation box were set to $L_x = L_y = 40\sigma$, and $L_z = W$, with $W^* = W/\sigma = 5.0, 7.0, 9.0$ and 11.0. Thus, the number density is calculated as $\rho^* = (N\sigma^3)/(L_x L_y W)$. Given the large dimensions of the simulation box along the x and y directions, we expect that finite-size effects will be small.

The phase behaviour of the confined SALR fluid was explored by performing a series of Monte Carlo (MC) simulations in the grand canonical ensemble at $T^* = k_B T/\epsilon = 0.30$ for several values of the chemical potential within $-1.2 \leq \mu^* = \mu/\epsilon \leq 0.5$. For each wall separation, we chose three states at which the cluster-crystal, the hexagonal and the lamellar phases exhibit the most ordered structure (as compared to those obtained at other chemical potentials). The numbers of particles confined in the slit pore at each considered state are given in Table 1.

Table 1. Average number of particles confined in the slit pores at which the fluid organises into ordered structures at $T^* = 0.3$ at densities at which the bulk fluid assembles into a cluster-crystal, a cylindrical and a lamellar phase. Note that the chemical potential for the more dense lamellar phase might not be reliable due to the low acceptance probability of the insertion/deletion MC moves. In any case, we only used these simulations to generate the initial configurations for the NVT MD runs.

Phase	$W^* = 5.0$	$W^* = 7.0$	$W^* = 9.0$	$W^* = 11.0$
Cluster-Crystal	$N = 1030$ $\mu^* = -1.20$	$N = 1748$ $\mu^* = -1.00$	$N = 2403$ $\mu^* = -1.00$	$N = 2807$ $\mu^* = -1.00$
Cylindrical	$N = 1868$ $\mu^* = -0.60$	$N = 2923$ $\mu^* = -0.40$	$N = 3845$ $\mu^* = -0.40$	$N = 4065$ $\mu^* = -0.60$
Lamellar	$N = 3432$ $\mu^* = 0.50$	$N = 5399$ $\mu^* = 0.20$	$N = 6124$ $\mu^* = 0.50$	$N = 8058$ $\mu^* = 0.40$

* denotes reduced variables.

Starting from these configurations, the confined fluid was then heated and cooled using Molecular Dynamics (MD) simulations in the canonical ensemble (NVT). The MD simulations were performed with the LAMMPS simulation package [31], in which the truncated and shifted SALR model described above was implemented in an external subroutine coded by us. The time step was set to $dt = 0.005\sqrt{m\sigma^2/\epsilon}$. Temperature was controlled with the Nose-Hoover thermostat with a relaxation time of $100dt$. Simulations were evolved for 10^6 MD steps for equilibration, followed by another 10^6 MD steps for taking averages.

The structure of the fluid was identified mainly by visual inspection of local density plots. These plots were built by dividing the simulation box in small cubic cells of approximate edge length σ, measuring the particle density in each of these cells and averaging over 10,000 independent configurations, so that we can evaluate the local density function:

$$\rho_{xyz}(x,y,z) = \frac{\langle N(x,y,z)\rangle}{\Delta V}, \tag{4}$$

where $\langle N(x,y,z)\rangle$ is the average number of particles in a cubic cell of edge σ and centred at the point (x,y,z), and ΔV is the volume of each small cubic cell, in our case $\Delta V = \sigma^3$. Isosurfaces of these density maps were visualised using OpenDX software. Clusters were also identified by performing a cluster size analysis [32], adopting the convention that two particles are nearest neighbours if the distance between them is lower than $r_{cut} = 1.6\sigma$ for the cluster-crystal and hexagonal phases and lower than $r_{cut} = 1.4\sigma$ for the lamellar phase, i.e., roughly the distance to the first minimum in the radial distribution function of each periodic microphase [17]. This information was used to calculate the cluster size distribution (CSD).

The spatial distribution of the particles along the direction perpendicular to the pore walls was investigated by measuring the density profiles, calculated by dividing the pore volume in small slabs of width $\Delta z = 0.1\sigma$ and averaging the number density in each of these slabs:

$$\rho_z(z) = \frac{\langle N(z+\Delta z)\rangle}{L_x L_y \Delta z}. \tag{5}$$

Here $\langle N(z+\Delta z)\rangle$ is the ensemble average of the number of particles in the slab between $z - \Delta z/2$ and $z + \Delta z/2$, and L_x and L_y are the two periodic edges of the simulation box.

Following our preliminary study of the bulk system [17], we also investigated the internal ordering of the clusters at the particle scale as a function of temperature. For that purpose, for each periodic microphase, we chose an order parameter that is able to discriminate particles within local ordered environments from those within local disordered environments. As in the bulk system, the spherical and cylindrical clusters that form the cluster-crystal and hexagonal phases at low temperatures have local icosahedral symmetries. In this case, a common neighbour analysis (CNA) [33] allows us to distinguish particles with local icosahedral environments from liquid environments. The CNA analysis was made with the OVITO visualisation tool [34], using a fixed cutoff radius of 1.6σ, that corresponds to the distance to the first minimum in the pair distribution function in the bulk cluster-crystal and hexagonal phases [17]. On the contrary, in the lamellar phase, particles are arranged in stacks of hexagonally-packed layers. Therefore, it is more convenient to use the Lechner and Dellago order parameter, that is able to effectively distinguish particles in ordered local environments (i.e., solid-like, including particles in the frozen lamellae [17]) from those in disordered local environments. For the lamellar phase, first neighbours were defined using a slightly shorter cutoff distance than for the cluster-crystal and hexagonal phases of 1.4σ, corresponding to the first minimum in the pair distribution function of the bulk lamellar phase [17].

Finally, we also measured the mean squared displacement (MSD) that provides information on the single particle dynamics:

$$\left\langle \Delta r(t)^2 \right\rangle = \left\langle \frac{1}{N} \sum_{i=1}^{N} (\mathbf{r}_i(t) - \mathbf{r}_i(0))^2 \right\rangle, \tag{6}$$

where $\mathbf{r}_i(t)$ and $\mathbf{r}_i(0)$ are the positions of particle i at times t and zero, respectively. The diffusion coefficient, D, is estimated from Einstein's relation:

$$D = \frac{1}{2d} \lim_{t\to\infty} \frac{\partial \langle \Delta r(t)^2 \rangle}{\partial t} \tag{7}$$

where d is the dimensionality of the system. For the confined systems, we calculated the diffusion coefficient in the direction parallel to the walls ($D_{\|}$), because the particle displacement in the perpendicular direction is limited by the narrow width of the pores. In this case, the MSD is calculated using only the x and y coordinates, and d is set to 2. For the bulk system, movements in the three directions of space are considered and $d = 3$. To calculate the diffusion coefficient, we divide the MSD data into 10 independent blocks. Then, we fit the MSD to a straight line and calculate the diffusion coefficient in each block following Equation (7). We discard the first steps in which the systems usually exhibit ballistic behaviour. We only calculate the diffusion coefficient when the MSD scales linearly with t, i.e., in the diffusive regime. The diffusion coefficients are averaged over the independent blocks and the errors are estimated as the standard deviation of the sample of blocks.

3. Results

3.1. Equilibrium Properties

The qualitative phase diagram of the bulk system was published in previous work [17] and is sketched in Figure 2. The three isochores studied for each pore width W are marked with symbols in this diagram. These isochores correspond to values at which the cluster-crystal, the hexagonal and the lamellar phases are stable at low temperatures in bulk. The structures of the confined fluid in the slit pores at $T^* = 0.3$ obtained from the MD simulations are shown in Figure 3. The stability of these structures with temperature was studied by performing simulations in the NVT ensemble at T^* = 0.20–0.50, and the results are summarised in Figure 2, where the colors of the symbols represent the various structures formed. The results obtained in each density region are described in detail in what follows.

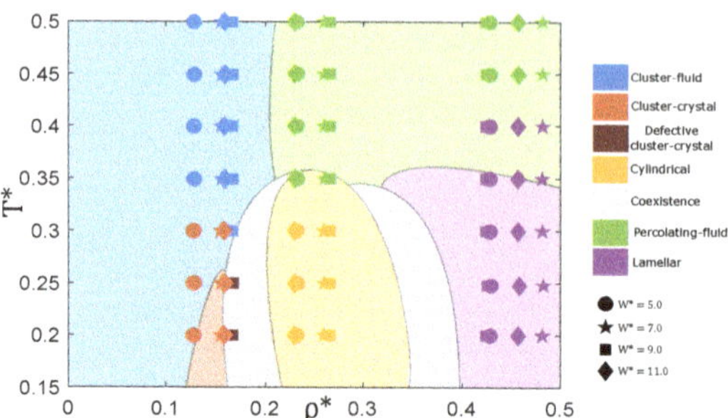

Figure 2. Sketch of the bulk phase diagram, using data from Ref. [17]. The state points studied for each pore size $W^* = 5, 7, 9$ and 11 are marked with different symbols, and their colours indicate the structure adopted by the confined fluid in each thermodynamic state, as provided in the legend.

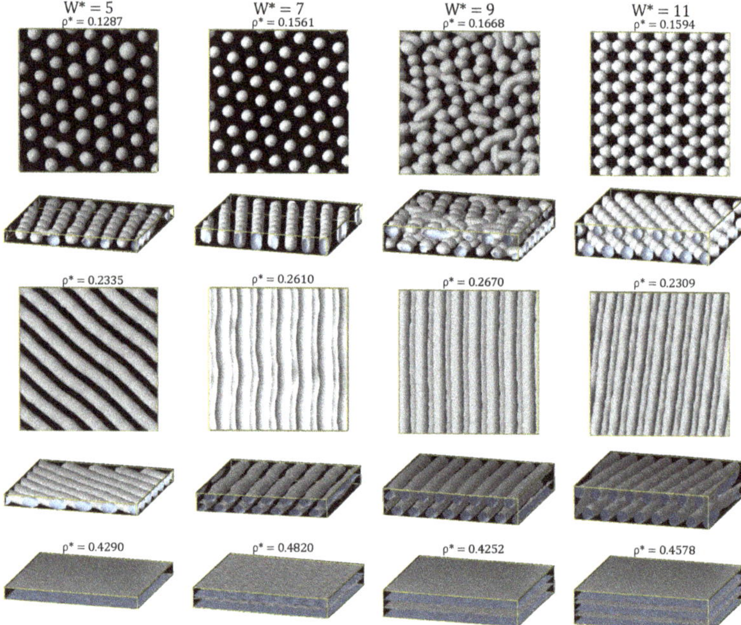

Figure 3. Local density isosurfaces $\rho^*_{iso} = 0.30$ for all the ordered microphases obtained at different slit widths, W^*. Note that the density chosen for the isosurfaces is somewhat lower than that in our previous work on SALR systems modelled with the square-well linear model (in which we chose $\rho^*_{iso} = 0.40$) [25–27]. The reason for this new choice is that the clusters obtained with the Lennard-Jones plus Yukawa model used in this work are appreciably smaller [17]. Two views are presented for cluster-crystal and hexagonal phases and one for the lamellar phase. The number densities in reduced units, ρ^*, are specified and the temperature is $T^* = 0.30$.

3.1.1. Low Density: The Cluster-Crystal

Let us start discussing the structures obtained by the MD simulations at $T^* = 0.3$ and low densities, $\rho^* \approx 0.12 - 0.16$ (Figure 3). Under these conditions the fluid is still able to organise into an ordered cluster-crystal under confinement, except for the pore size $W^* = 9$. In particular, the fluid forms one layer of hexagonally-packed clusters for $W^* = 5$ and $W^* = 7$, the difference being that clusters are roughly spherical in the narrowest pore and adopt a spherocylindrical shape in the $W^* = 7$ pore. At $W^* = 11$, the pore is wide enough to host a stack of two hexagonally-packed layers of nearly spherical clusters. Finally, at

$W^* = 9$, the system assembles into a structure composed of two layers formed by a mixture of spherical and spherocylindrical clusters, in which some local hexagonal ordering can be observed, but that is globally disordered.

The diameter of the clusters ($d_0^* = d_0/\sigma$) measured from the local density isosurfaces projected on the plane perpendicular to the pore walls, as well as the distance ($l_0^* = l_0/\sigma$) between nearest neighbour clusters are given in Table 2. For the three pore widths for which ordered cluster structures are observed, l_0^* and d_0^* adopt values relatively close to those of the bulk system, with a maximum deviation of about 5%–6%. Larger differences can be seen in the CSD (Figure 4, left panel). At $T^* = 0.30$, the bulk CSD is bimodal, exhibiting two peaks of similar probability at $n = 19$ and $n = 23$ [17]. For $W^* = 5$ the CSD is slightly narrower than in the bulk system and peaks at slightly smaller sizes, whereas for the remaining pore widths, the CSD is shifted to larger sizes. This effect is especially pronounced for $W^* = 7$, in which the maximum is located at $n = 31$, consistently with the formation of elongated clusters (Figure 3).

Table 2. Estimation of the distance between clusters ($l_0^* = l_0/\sigma$) and the average cluster size ($d_0^* = d_0/\sigma$) in bulk and in the confined systems. The cluster size d_0^* corresponds to the average cluster diameter in spherical and cylindrical clusters, and to the width of the lamellae in the lamellar phase.

W^*	5.0	7.0	9.0	11.0	Bulk
Cluster-crystal	$l_0^* = 5.9$ $d_0^* = 3.3$ $\rho^* = 0.1287$	$l_0^* = 5.8$ $d_0^* = 3.1$ $\rho^* = 0.1561$	‒ ‒ ‒ ‒ $\rho^* = 0.1668$	$l_0^* = 6.0$ $d_0^* = 3.3$ $\rho^* = 0.1594$	$l_0^* = 5.6$ $d_0^* = 3.3$ $\rho^* = 0.155$
Hexagonal	$l_0^* = 5.3$ $d_0^* = 3.0$ $\rho^* = 0.2335$	$l_0^* = 5.5$ $d_0^* = 3.0$ $\rho^* = 0.2610$	$l_0^* = 5.8$ $d_0^* = 3.2$ $\rho^* = 0.2670$	$l_0^* = 6.2$ $d_0^* = 3.0$ $\rho^* = 0.2309$	$l_0^* = 5.6$ $d_0^* = 2.8$ $\rho^* = 0.252$
Lamellar	‒ ‒ $d_0^* = 2.9$ $\rho^* = 0.4290$	$l_0^* = 3.7$ $d_0^* = 1.5$ $\rho^* = 0.4820$	$l_0^* = 4.7$ $d_0^* = 2.2$ $\rho^* = 0.4252$	$l_0^* = 3.9$ $d_0^* = 1.7$ $\rho^* = 0.4578$	$l_0^* = 4.6$ $d_0^* = 2.2$ $\rho^* = 0.407$

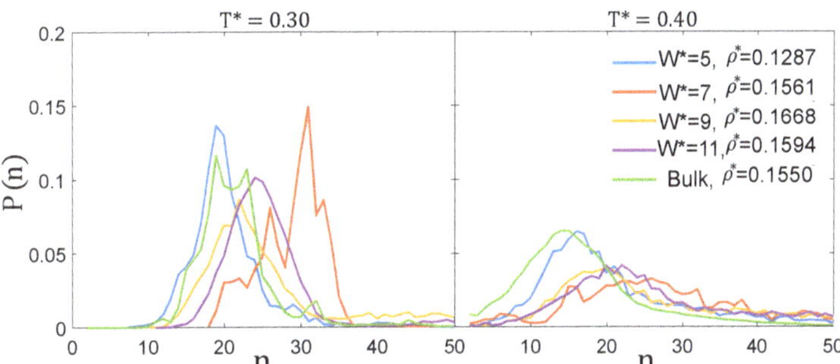

Figure 4. Cluster-size distributions of the cluster-crystal in bulk and in the slit pores of width W^* at $T^* = 0.3$ and $T^* = 0.4$.

The reason why the cluster-crystal becomes incommensurate for the $W^* = 9$ slit pore can be rationalised from the values of d_0^* and l_0^*. This pore is too wide for the fluid to organize into a single layer of spherocylinders with their axial directions aligned perpendicularly to the pore walls. In spherocylindrical clusters that are 7σ high, i.e., of height comparable to the accessible pore width in which the particles can move, many particles experience repulsive interactions, making this configuration energetically unfavourable. A possible alternative would be to form two layers of hexagonally-packed spherical clusters. Taking the values of d_0^* and l_0^* from the bulk system, these two layers can be nicely accommodated in a slit pore of width $W^* \approx 2 + d_0^* + \sqrt{2/3}l_0^* = 9.9$, where d_0^* is the average cluster diameter, $\sqrt{2/3}l_0^*$ is the z-distance between two hexagonally-packed layers in which the distance between the nearest clusters is l_0^*, and the factor 2 takes into account that the centre of the particles can not get closer than σ to the pore walls. Thus, in order to accommodate these two layers in the $W^* = 9$ pore, either the distance between the two layers and/or the shape of the clusters would have to be modified from those in the bulk phase. Our simulations indicate that in these conditions the system is not able to find an ordered cluster phase, forming instead a structure in which spherical and elongated clusters coexist and exhibiting only local order in some regions. Finally, in the pore size $W^* = 11$, two layers of the bulk cluster-crystal can be fitted leaving some extra room in the pore. In this case, the cluster-crystal is somewhat expanded in order to occupy the whole accessible volume within the pore, as evidenced by the larger values of d_0^* (as well as in the shift of the CSD to larger sizes, Figure 4) and of l_0^* as compared to in bulk.

Focusing now on the stability of the assembled structures with temperature, our simulations indicate that the confined cluster-crystal phase remains stable up to higher temperatures than in bulk, except for the pore width $W^* = 9$, which, as we have just seen, is incommensurate with the bulk cluster-crystal (Figure 2). In particular, the cluster-crystal is able to survive up to $T^* = 0.30$ when confined in pores of sizes $W^* = 5, 7$ and 11, whereas in bulk it melts at $T^* = 0.20 - 0.25$ depending on the density. Shifts in the coexistence lines between two phases are common under confinement and have been observed either in complex [26,27,35] and simple [36] fluids. On the contrary, for the pore size $W^* = 9$, the structure remains only partially ordered down to $T^* = 0.20$.

As can be seen in Figure 5 (first row), the particle density profiles measured along the direction perpendicular to the pore walls, exhibit pronounced maxima and minima, indicating the ordering of the particles in layers for all the investigated pore widths and at all temperatures, becoming particularly sharp at low temperatures. At $T^* = 0.20$, the density profiles in the $W^* = 5$ and $W^* = 7$ pores are larger than zero anywhere within the pore, except for distances shorter than the particle repulsive core radius, which reflects that a single layer of clusters has been formed. However, the number of maxima in the density profiles differs in the two pores: four for $W^* = 5$ and six for $W^* = 7$. In both cases, two of these maxima form at the walls and are less pronounced than those in the pore central region. The distance between two adjacent maxima is of the order of the particle diameter, being somewhat shorter at the pore central region (about 0.7σ for $W^* = 5$ and 0.9σ for $W^* = 7$) than at the pore walls (about 1.0σ in both cases). For the two widest pores, $W^* = 9$ and $W^* = 11$, the formation of two clearly different layers of clusters is reflected in the pronounced decrease of the local density at the pore central region. Curiously, this decrease is more pronounced for $W^* = 9$ than for $W^* = 11$ at all temperatures, despite the fact that the confined fluid is more ordered at $W^* = 11$. Especially remarkable, is that at $T^* = 0.5$, the two layers are significantly smoothed for $W^* = 11$ but are still visible for $W^* = 9$. As temperature is lowered, smaller peaks develop within each of these two layers, and their interdistance is again of the order of the particle diameter (about 0.95σ for $W^* = 9$ and 0.90σ for $W^* = 11$).

The presence of very sharp maxima and minima in the density profiles at low temperatures indicates internal ordering of the clusters, a phenomenon already observed in the bulk system [17]. This motivated us to monitor the internal ordering of the clusters as a function of temperature by measuring the fraction of particles within local icosahedral environments as identified with the CNA analysis. As can be seen in Figure 6 (top panel), at these low densities the internal ordering exhibits a similar temperature behaviour as in the bulk system. Clusters are internally ordered at low temperatures (as only the inner particles have icosahedral local environments, a fraction of particles in icosahedral environments of around 0.1 indicates that almost nearly all the clusters are internally ordered [17]), and gradually become disordered as the temperature increases. Visual inspection of the configurations (see Figure 6) reveals that, at the lowest temperature, the clusters exhibit well defined geometries, often consisting in interpenetrated icosahedra sharing a five-fold axis (two, three or even four icosahedra can be merged to form clusters with $n = 19$, $n = 25$ and $n = 31$ particles, which appear with relatively large probabilities, as shown in Figure 4 (left)) or formed by adding particles at the surface of an icosahedral cluster (e.g., the cluster with $n = 24$ shown in Figure 6).

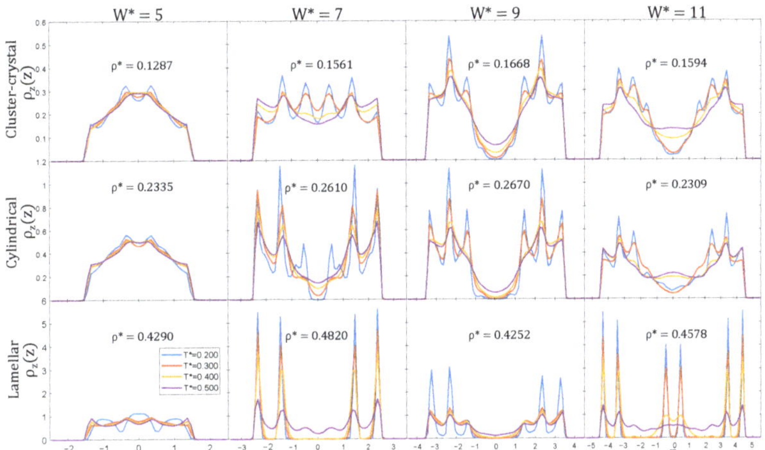

Figure 5. Density profiles along the z-direction at different temperatures and slit widths.

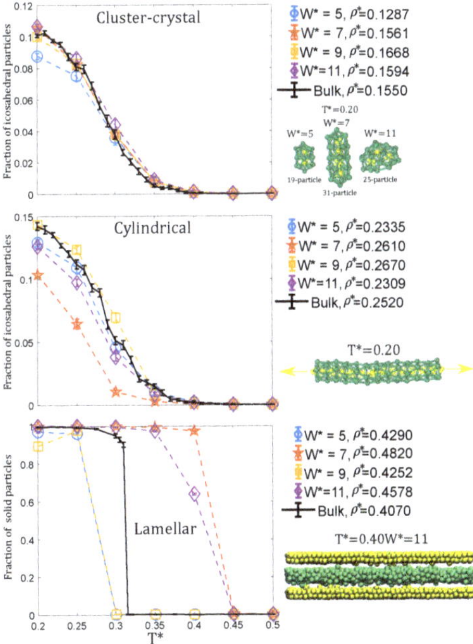

Figure 6. Top panel:the fraction of particles with local icosahedral environments as a function of temperature for the bulk and the confined systems at a low density at which the cluster-crystal phase is stable. Central panel: the fraction of particles with local icosahedral environments as a function of temperature for the bulk and the confined systems at an intermediate density at which the cylindrical phase is stable. Bottom panel: the fraction of particles within a hexagonal local environment as a function of temperature for the bulk and the confined systems at a high density at which the lamellar phase is stable. Note that the classification of particles in these plots is based solely on analysis of the local structure around each particle; the dynamics of the particles was not taken into account in this analysis.

3.1.2. Intermediate Density: The Hexagonal Phase

At intermediate densities ($\rho^* \approx 0.23 - 0.27$) and $T^* = 0.3$, the confined fluid organises into cylindrical clusters for the four pore sizes, as in the bulk system (two different views for each pore are shown in Figure 3), but there are clear structural differences depending on the pore width. For $W^* = 5$, one layer of cylindrical aggregates with roughly circular cross-sections is formed. This is hardly surprising because the bulk cylinder diameter ($d_0 = 2.75\sigma$) is comparable to the accessible pore width (which is approximately equal to $W^* - 2\sigma$). The diameter of the confined cylinders becomes somewhat larger than in the bulk system, probably to occupy as much as possible of the available pore space (see Table 2). For a wider pore ($W^* = 7$), the fluid organises into two layers of cylindrical clusters. The cylinders in this structure are deformed with respect to those in the bulk phase, adopting ellipsoidal (instead of circular) cross-sections. This suggests certain incommensurability of the pore size with the bulk cylindrical phase that can be overcome with a small deformation of the cylinders. As a rough estimation, two layers of the bulk cylindrical phase can fit in a pore of width $W^* = 2 + d_0^* + \sqrt{3}/2l_0^* \approx 9.5$, where the term $\sqrt{3}/2l_0^*$ accounts for the distance in the z direction between the centres of two layers of cylinders in which the distance between the nearest cylinders is l_0^*. Surprisingly, this pore size estimate is significantly larger than the actual pore width ($W^* = 7$), but the cylindrical phase is still able to survive by deformations of the cross-section of the cylinders and probably also by adjusting the distance between them. As a consequence of this, the average energy of the confined fluid in the $W^* = 7$ pore ($\langle u^* \rangle = -1.904$) is higher than in the bulk system ($\langle u^* \rangle = -2.065$).

Following the same reasoning, the pore size $W^* = 9$ has almost the appropriate size to fit two layers of the bulk hexagonal phase and, thus, one would expect that the fluid would be less compressed in this case. This is exactly what we observe in the simulations. The fluid still organises into two layers of cylindrical clusters, which now adopt nearly circular cross-sections as in bulk. On the contrary, for $W^* = 11$, the pore size is somewhat wider than needed for hosting two perfect layers of the bulk cylindrical phase. This is partially offset by forming slightly thicker cylinders and increasing the distance between them as compared to the bulk phase (see Table 2). Note that the orientation of the cylinders with the simulation box is different depending on the pore size to adjust the separation between the nearest cylinders to a value similar to that of the bulk system.

The cylindrical phase is destabilised at temperatures slightly lower than in the bulk system, in particular it remains stable up to $T^* \approx 0.30$, whereas in bulk it survives up to $T^* \approx 0.35$ (Figure 2). The number of layers of cylinders can again be easily inferred from the local density profiles $\rho(z)$ and, except for the narrower pore, the layering of particles at low temperatures is enhanced as compared to the cluster-crystal (see Figure 5). For the pore size $W^* = 5$, one single layer of cylinders is formed, and the density profiles exhibit two rounded peaks of enhanced density, indicating a mild tendency of the particles in the cylinders to sit preferentially in these two planes. For $W^* = 7$ and $W^* = 9$, two layers of cylinders are formed, and this is reflected in the density profiles by a region of low density between the two layers. At low temperatures, each of these two layers of cylindrical clusters exhibits three peaks, the intensities of which vary with the pore width. For the two pore sizes, the central peak in each of the two layers is the highest. The two edge peaks away from the centre of the cylindrical clusters are equal in the $W^* = 9$ pore, but in the $W^* = 7$ pore the peak closer to the pore centre is smaller.

These profiles are consistent with the previous observation from the 3D local density plots shown in Figure 3, in that the cross-sections of the cylinders is circular in the $W^* = 9$ pore, but are significantly deformed in the $W^* = 7$ pore. The two peaks closer to the pore centre become shoulders of the central peak of each layer of cylindrical clusters at $T^* = 0.3$ in the $W^* = 7$ pore, but they survive up to $T^* = 0.4$ in the $W^* = 9$ pore. The same occurs for the region of very low density between the two layers of cylinders that persists up to $T^* = 0.3$ in the $W^* = 9$ pore, but that becomes a region of small density in the $W^* = 7$ pore at this same temperature. For $W^* = 11$, the two layers of cylinders are not separated by a region of low density even at the lowest considered temperature. The reason is that the cylinders adopt a sinusoidal shape in the direction perpendicular to the walls to use as much as possible of the free pore volume that remains by fitting two layers of cylinders in this wide pore.

Visual inspection of the configurations reveals that, as in the bulk system [17], at low temperatures the cylindrical clusters adopt ordered configurations consisting in decagonal tubes made by interpenetration of isosahedra sharing a five-fold axis. As can be seen in Figure 6, central panel, the evolution of the fraction of particles with icosahedral symmetry with temperature exhibits a similar behaviour to the bulk system. Unsurprisingly, the results are almost exactly the same as those of the bulk system for the most commensurate pore $W^* = 9$, and the larger reduction of order is observed for the most incommensurate pore $W^* = 7$.

3.1.3. High Density: The Lamellar Phase

At high densities ($\rho^* \approx 0.42$–0.49) and $T^* = 0.3$, the fluid organises into lamellar structures for the three larger considered pore sizes, as in bulk. This is the expected behaviour, as the geometry of the pores is fully compatible with the lamellar phase. At $W^* = 5$, the fluid occupies the whole pore volume, that is wider than the size of bulk lamellae ($d_0^* = 2.9$ to be compared to $d_0^* = 2.2$ in bulk, see Table 2). For $W^* = 7$, two lamellae are formed, this time narrower than in the bulk system ($d_0^* = 1.5$ to be compared to $d_0^* = 2.2$ in bulk), and with a slightly shorter separation between them ($l_0^* = 3.7$ versus $l_0^* = 4.6$ in bulk). For $W^* = 9$, the system assembles into two lamellae, the thickness and interdistance of which are comparable to those in bulk, indicating that this pore size matches very well the periodicity of the bulk lamellar phase. For $W^* = 11$, a third lamella is formed, although at this temperature, $T^* = 0.3$, both the width of the lamellae and especially the distance between them are reduced with respect to the bulk phase to adjust to the available pore volume.

The lamellar phase is able to survive up to $T^* = 0.4$ for all pore widths, i.e., at temperatures higher than the bulk system in which the transition occurs at $T^* \approx 0.35$ [17] (see Figure 2). Again this is not entirely surprising, as the geometry of the pores is compatible with the lamellar phase. The density profiles $\rho(z)$ further reveal that the lamellar phase is particularly stable with temperature, especially for the pore sizes $W^* = 7$ and $W^* = 11$, in which the fluid organises into two and three lamellae, respectively. Each lamella is made of two molecular layers, as evidenced by the two sharp peaks observed in each layer up to relatively high temperatures (up to $T^* = 0.4$ for $W^* = 7$ and up to $T^* = 0.3$ for $W^* = 11$). The sharpness of these peaks indicates that the internal structure of the lamellae remains ordered at these temperatures. Curiously, for $W^* = 11$ and $T^* = 0.4$ the two lamellae close to the pore walls are still very structured (they are internally ordered), but the middle one exhibits two much more rounded peaks (it is internally disordered). This indicates that the proximity to the walls induces the internal ordering of the lamellae, and such internally ordered lamellae can coexist with disordered lamellae away from the walls. For the pore width $W^* = 9$, only two lamellae are formed, but each one is now composed of two equally populated molecular layers (next to the pore walls) and by an incomplete third hexagonally-packed layer facing the centre of the pore, signalled by a less sharp peak. In this case, the peaks in each lamella are already rounded at $T^* = 0.3$, indicating that in this case the internal ordering of the lamellae is less significant at a high temperature. As found for the cluster-crystal and the cylindrical phases, the tendency of the particles to arrange in layers parallel to the walls is significantly reduced in the narrowest pore ($W^* = 5$), which only exhibits fairly rounded peaks even at low temperatures.

The internal ordering of the lamellae was investigated by measuring the fraction of particles with solid-like environments using the Lechner and Dellago local order parameter [37]. As can be seen in Figure 6 (bottom panel), the fraction of solid-like particles as a function of temperature is strongly dependent on the pore size, differing also from the bulk phase for all pore sizes. At $T^* = 0.2$ the vast majority of the particles have local solid environments (close to 100%) for all pores, except for $W^* = 9$, in which case it drops to 90%. This is due to the growth of a third incomplete molecular layer facing the centre of the pore in each lamella. For the $W^* = 7$ pore, the transition from ordered to disordered lamellae is discontinuous as in bulk, but occurs at a higher temperature than in the homogeneous system. For $W^* = 11$, the decay of the number of particles in ordered environments is more gradual, e.g., at $T^* = 0.4$ about 60%–65% of the particles have solid-like environments. The reason is that, as mentioned before, the two lamellae at the pore walls become disordered at a higher temperature ($T^* = 0.4$) than the one at the centre of the pore ($T^* = 0.3$) (see Figure 6). These results indicate that the walls promote the internal ordering of the lamellae (i.e., they remain ordered up to a higher temperature than in bulk), but those lamellae that are not next to the walls get ordered at similar temperatures as in bulk. Finally, for $W^* = 5$ and $W^* = 9$, the lamellae become disordered at lower temperatures than in bulk, which is attributed to some incommensurability of the bulk lamellar phase with these pore sizes.

3.2. Dynamic Properties

Once we had characterised the equilibrium phase behaviour, we also analysed the dynamics at temperatures around which the periodic microphases start to form. The MSD measured in the four pore sizes at densities at which the cluster-crystal, the cylindrical and lamellar phases are formed are collected in Figure 7, and the diffusion coefficients obtained from these data are plotted in Figure 8. The comparison of the diffusion constant is not made either at constant density nor at constant chemical potential for all the pores, as it is often done in the literature. Instead, in this work we choose to make the comparison under those conditions at which the most ordered structure was obtained in each pore size (which are those shown in Figure 3), as our aim is to investigate if the diffusion of the particles is altered by confinement under the temperature and density conditions at which the periodic microphases start to form from the fluid phase. Note that it is not always possible to obtain different ordered structures at the same density or the same chemical potential for different sizes of the pores, because, due to incommensurability, these ordered structures may be destabilised and become disordered.

3.2.1. Low Density: Cluster-Crystal

The behaviour of the MSD at low densities and temperatures just above those at which the cluster-crystal starts to form, $T^* = 0.4$, is qualitatively similar in the four considered pore sizes and also in the bulk system. The MSD at long times is diffusive for all the pore sizes and in the bulk system. The diffusion coefficient, calculated using Einstein's relation (Equation (7)), as a function of pore size is shown in Figure 8. As can be seen, the diffusion coefficient has a nonmonotonic behaviour with the pore size. The maximum diffusion is achieved for the narrowest pore, in which case the diffusion coefficient is even higher than for the bulk system. The minimum diffusion corresponds to the $W^* = 7$ pore, and then diffusion increases with pore size until it reaches the bulk behaviour. Note that the density in the narrowest pore is somewhat lower than in the remaining pores and than in bulk, and this might partly explain why diffusion is faster in this system. However, it is also worth mentioning that the narrowest pore is the only one in which the fluid is still organised in intermediate size clusters at $T^* = 0.4$, as in the bulk system (see Figure 4, right panel). On the contrary, in the remaining pore sizes, the CSD distributions are shifted to larger sizes. It is important to highlight that the nonmonotonic variation of the diffusion constant with the pore size cannot be explained solely based on the density of the fluid. For example, the diffusion constant for $W^* = 7$ ($\rho^* = 0.1561$) is lower than that for $W^* = 9$ ($\rho^* = 0.1668$), in spite of the density of the fluid being higher in the latter case.

On the contrary, at a temperature just below that at which the periodic microphases start to be seen, $T^* = 0.3$, the behaviour of the MSD at long times depends on the pore size. For the narrowest pore, the particle movement is diffusive as in the bulk system, but, for the remaining pores, it becomes subdiffusive, this effect being more pronounced for the pore size $W^* = 7$. Our hypothesis is that the lower diffusion in the pore $W^* = 7$ is a consequence of the higher ordering of the clusters as compared to the defective structures found for $W^* = 5$ and especially for $W^* = 9$, as can be seen in the density isosurface plots shown in Figure 3.

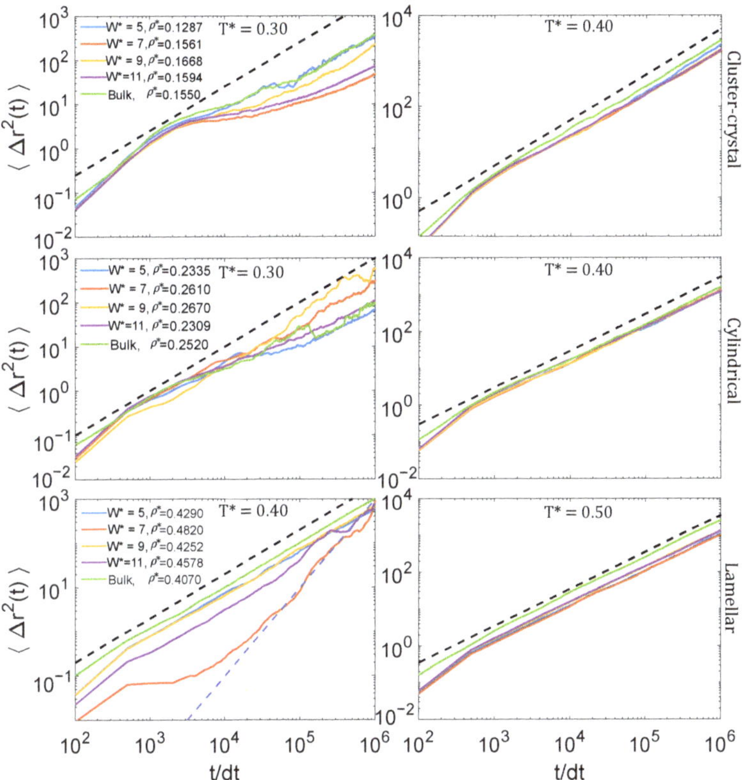

Figure 7. Particle mean squared displacement (MSD) of the fluid confined in slit pores of different width W^* at different temperatures at densities at which the cluster-crystal (**top** row), the cylindrical (**middle** row) and the lamellar (**bottom** row) phases are stable. For comparison, the MSD for the bulk system under similar thermodynamic conditions are also included. The dashed black and blue lines show the expected behaviour for diffusive (MSD$\propto t^{\beta}$, $\beta = 1$) and ballistic ($\beta = 2$) behaviour.

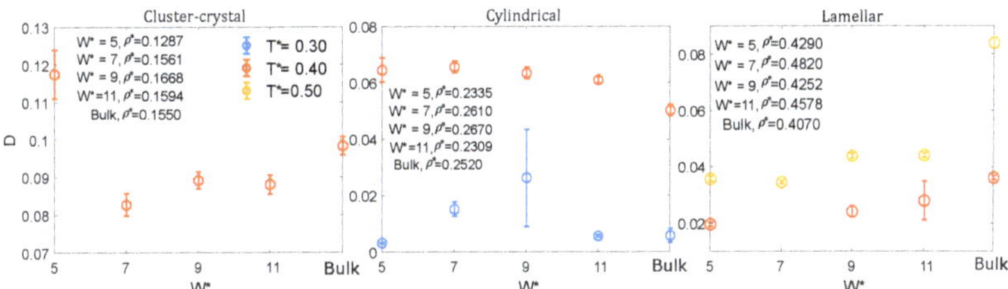

Figure 8. Diffusion coefficient as a function of pore width, for low (**left** panel), medium (**middle** panel) and high (**right** panel) densities, at temperatures around those at which the periodic microphases start to spontaneously form.

3.2.2. Intermediate Density: Cylindrical Phase

In the cylindrical phase the particle movement is diffusive at long times either at temperatures just above ($T^* = 0.4$) and below ($T^* = 0.3$) those at which the cylindrical clusters start to form. At $T^* = 0.4$, the diffusion coefficient adopts similar values for the four pore sizes, being higher in confinement than in the bulk system (Figure 8, central panel). Note that this cannot be explained on the basis of the densities, because, depending on the pore size, the chosen states exhibit both higher and lower densities than the bulk system. Curiously, both in confinement and in bulk, the system is organised in a percolating fluid at this temperature. Our hypothesis is that, consistently with the shift to lower temperatures of the stability region of the cylindrical clusters with respect to the bulk system (Figure 2), confinement partly destroys the clustering at these intermediate densities, thus facilitating the particle diffusion.

At $T^* = 0.3$, the diffusion coefficient is similar in the bulk system and in the $W^* = 5$ and $W^* = 11$ pores, adopting somewhat larger values for $W^* = 7$ and $W^* = 9$. It is important to note, however, that cylinders are parallel to one of the edges of the square section of the simulation box in the $W^* = 7$ and $W^* = 9$ pores, whereas for $W^* = 5$ and $W^* = 11$, they are tilted with respect to one of the edges. Thus, in the former case each cylinder is an independent cluster, but in the latter case all the cylinders are connected to each other due to the periodicity of the system. As a consequence, the cylindrical clusters can move with respect to their neighbours in the narrowest and widest considered pores, but not in the $W^* = 7$ and $W^* = 9$ pores. Thus, it is not possible to make a fair comparison between the diffusion coefficients as a function of the pore sizes in this case.

3.2.3. High Density: Lamellar Phase

In the lamellar case the MSD exhibits diffusive behaviour at long times at temperatures somewhat above ($T^* = 0.5$) those at which the lamellar phase starts to form. At this temperature, the diffusion coefficient under confinement is significantly lower than that of the bulk system for the four considered pore sizes. Our hypothesis is that this could be related to the more efficient packing obtained in the four confined systems as compared to that in the bulk system. Note however, that there is some ambiguity on how to define the density in the confined pores, as one can take into account or not the particle excluded volume close to the pore walls.

At $T^* = 0.4$, the diffusion coefficient of the lamellar phase increases with the pore size, except for $W^* = 7$, the size at which the fluid exhibits the higher diffusion under confinement. As can be seen in Figure 7, the scaling of the MSD with time reveals a superdiffusive behaviour in this case (i.e., the MSD is proportional to t^β, β being an exponent higher than 1). Our hypothesis is that the origin of this enhanced diffusion is related to the smooth surfaces of the formed lamellae in this pore size. For $W^* = 7$, the fluid organises into two lamellae, each one made of two hexagonally-packed layers of particles. As a consequence of the smooth lamellar surfaces and pore walls, the two lamellae can slide with respect to each other due to the thermal movement, leading to a high diffusion coefficient in this case. This effect is not observed for $W^* = 9$ and $W^* = 11$, because the lamellae surfaces are no longer smooth, due to the formation of an incomplete third layer at $W^* = 9$ and to the disordered local structure of the central lamella at $W^* = 11$. At $T^* = 0.4$, the diffusion coefficient is only slightly higher in bulk than in any of the considered confined systems. It is interesting to note that the fluid is organised into a lamellar phase at this temperature in the three largest considered pores, but it forms a percolating fluid in bulk.

4. Discussion and Conclusions

In this work we have studied the assembly of colloids with competing interactions when confined in narrow slit pores at densities at which the cluster-crystal, the hexagonal and the lamellar phases are stable in bulk. We have found that those periodic modulated phases also form under confinement. In particular, our simulations predict that the cluster-crystal and the lamellar phases are often stable up to higher temperatures when confined in slit pores than in the bulk system, but that of the cylindrical phase is lower than in bulk. In the cases in which the pore size is not perfectly commensurate with the pore width, the fluid is often able to adjust the cluster shape, size and interdistance between clusters to fit in the available volume in the pore. One exception to this general behaviour is that the cluster-crystal was not formed even at very low temperatures in the $W^* = 9$ wide slit pore, which can be easily rationalised because this pore size is incommensurate with the corresponding bulk cluster-crystal. Thus, we can conclude that, in general, the presence of walls promotes the formation of the lamellar phase, as expected, but also of the cluster-crystal. The results obtained in this work are similar to those observed in block copolymers confined in slit pores [38]. This suggests that the universality of the phase behaviour in systems with competing interactions observed in bulk can be extended to confined systems.

We have also observed that the presence of walls induces the ordering of the particles within the clusters (spherical, cylindrical or lamellar) in layers of particles parallel to the pore walls. This has already been observed in simple and complex fluids (see, e.g., Refs. [24,39]), and also recently in the adsorption of colloids with competing interactions at an attractive surface [40]. In the case of the lamellar phase, this leads to the internal ordering of the lamellae at higher temperatures than in the bulk system, and to an interesting behaviour in which the lamellae adjacent to the pore walls remain ordered while the lamellae further away from the walls become disordered, a phenomenon observed in the widest considered slit pore. The local ordering in the vicinity of a flat surface has also been observed in experiments of block copolymers under confinement [41]. Both our simulations and the experiments already performed in different systems with competing interactions under confinement might help to design new experiments under the proper conditions to finally obtain ordered microphases in colloidal systems.

Finally, we found that at temperatures just above those at which periodic microphases start to be seen, the diffusion coefficient of the confined fluid can adopt values higher or lower than in bulk depending on the density and on the pore width. In particular, we observed that the diffusion coefficient of the colloidal particles depends on the pore width. This dependence is complex, because the walls induce the internal ordering of the clusters. For a given phase (cluster-crystal, cylindrical or lamellar), the observed nonmonotonic behaviour with the pore size cannot be explained based solely on the density of the confined fluid, as it is often found that the diffusion coefficient does not correlate inversely with the density of the confined fluid, as one would expect for normal fluids. Depending on the shape of the clusters and the separation between the pore walls, it is possible to obtain either larger and smaller values of the diffusion coefficient than in the corresponding bulk systems.

The main conclusion of our work is that the formation of the lamellar and cluster-crystal phases appears to be favoured by confinement in simple slit pores. Although in some cases, the diffusion might be lower in confinement than in bulk, the formation of the periodic microphases was observed in relatively short times in our simulations, indicating that there should not be important kinetic bottlenecks that hinder their formation. We suggest that confining colloids with competing interactions in simple slit pores might be a promising route for the experimental observation of those phases.

Author Contributions: Conceptualisation, H.S., W.T.G. and E.G.N.; methodology, H.S., W.T.G. and E.G.N.; software, H.S.; validation, H.S.; formal analysis, H.S.; investigation, H.S.; resources, H.S.; data curation, H.S.; writing—original draft preparation, E.G.N.; writing—review and editing, H.S., W.T.G. and E.G.N.; visualisation, H.S.; supervision, W.T.G. and E.G.N.; project administration, W.T.G. and E.G.N.; funding acquisition, W.T.G. and E.G.N. All authors have read and agreed to the published version of the manuscript.

Funding: This publication is part of a project that has received funding from the European Union's Horizon 2020 research and innovation programme under the Marie Skłodowska-Curie grant agreement No. 711859. Scientific work was funded from the financial resources for science in the years 2017–2021 awarded by the Polish Ministry of Science and Higher Education for the implementation of an international co-financed project. We would like to acknowledge the support from NCN grant

No 2018/30/Q/ST3/00434 and from the Agencia Estatal de Investigación and the Fondo Europeo de Desarrollo Regional (FEDER), Grant No FIS2017-89361-C3-2-P.

Data Availability Statement: The data that support the findings of this study are available from the corresponding author upon reasonable request.

Acknowledgments: In this section you can acknowledge any support given which is not covered by the author contribution or funding sections. This may include administrative and technical support, or donations in kind (e.g., materials used for experiments).

Conflicts of Interest: The authors declare no conflict of interest. The funders had no role in the design of the study; in the collection, analyses, or interpretation of data; in the writing of the manuscript, or in the decision to publish the results.

References

1. Ruiz-Franco, J.; Zaccarelli, E. On the role of competing interactions in charged colloids with short-range attraction. *Annu. Rev. Condens. Matter Phys.* **2021**, *12*, 51–70. [CrossRef]
2. Ciach, A.; Pękalski, J.; Góźdź, W. Origin of similarity of phase diagrams in amphiphilic and colloidal systems with competing interactions. *Soft Matter* **2013**, *9*, 6301–6308. [CrossRef]
3. Ciach, A. Universal sequence of ordered structures obtained from mesoscopic description of self-assembly. *Phys. Rev. E* **2008**, *78*, 061505. [CrossRef] [PubMed]
4. Pini, D.; Parola, A. Pattern formation and self-assembly driven by competing interactions. *Soft Matter* **2017**, *13*, 9259–9272. [CrossRef] [PubMed]
5. Zhuang, Y.; Zhang, K.; Charbonneau, P. Equilibrium phase behavior of a continuous-space microphase former. *Phys. Rev. Lett.* **2016**, *116*, 098301. [CrossRef] [PubMed]
6. Zhuang, Y.; Charbonneau, P. Equilibrium phase behavior of the square-well linear microphase-forming model. *J. Phys. Chem. B* **2016**, *120*, 6178–6188. [CrossRef]
7. Royall, C.P. Hunting mermaids in real space: Known knowns, known unknowns and unknown unknowns. *Soft Matter* **2018**, *14*, 4020–4028. [CrossRef]
8. Zhang, T.H.; Kuipers, B.W.M.; de Tian, W.; Groenewold, J.; Kegel, W.K. Polydispersity and Gelation in Concentrated Colloids with Competing Interactions. *Soft Matter* **2015**, *11*, 297–302. [CrossRef]
9. Toledano, J.C.F.; Sciortino, F.; Zaccarelli, E. Colloidal systems with competing interactions: From an arrested repulsive cluster phase to a gel. *Soft Matter* **2009**, *5*, 2390–2398. [CrossRef]
10. Klix, C.L.; Royall, C.P.; Tanaka, H. Structural and dynamical features of multiple metastable glassy states in a colloidal system with competing interactions. *Phys. Rev. Lett.* **2010**, *104*, 165702. [CrossRef]
11. Campbell, A.I.; Anderson, V.J.; van Duijneveldt, J.S.; Bartlett, P. Dynamical arrest in attractive colloids: The effect of long-range repulsion. *Phys. Rev. Lett.* **2005**, *94*, 208301. [CrossRef]
12. Zhuang, Y.; Charbonneau, P. Recent advances in the theory and simulation of model colloidal microphase formers. *J. Phys. Chem. B* **2016**, *120*, 7775–7782. [CrossRef]
13. Khandpur, A.K.; Foerster, S.; Bates, F.S.; Hamley, I.W.; Ryan, A.J.; Bras, W.; Almdal, K.; Mortensen, K. Polyisoprene-polystyrene diblock copolymer phase diagram near the order-disorder transition. *Macromolecules* **1995**, *28*, 8796–8806. [CrossRef]
14. Hu, H.; Gopinadhan, M.; Osuji, C.O. Directed self-assembly of block copolymers: A tutorial review of strategies for enabling nanotechnology with soft matter. *Soft Matter* **2014**, *10*, 3867–3889. [CrossRef]
15. Doerk, G.S.; Yager, K.G. Beyond native block copolymer morphologies. *Mol. Syst. Des. Eng.* **2017**, *2*, 518–538. [CrossRef]
16. Guo, Y.; van Ravensteijn, B.G.P.; Kegel, W.K. Self-assembly of isotropic colloids into colloidal strings, Bernal spiral-like, and tubular clusters. *Chem. Commun.* **2020**, *56*, 6309–6312. [CrossRef] [PubMed]
17. Serna, H.; Díaz Pozuelo, A.; Noya, E.G.; Góźdź, W.T. Formation and internal ordering of periodic microphases in colloidal models with competing interactions. *Soft Matter* **2021**, *17*, 4957. [CrossRef]
18. Míguez, H.; Yang, S.M.; Ozin, G.A. Optical properties of colloidal photonic crystals confined in rectangular microchannels. *Langmuir* **2003**, *19*, 3479–3485. [CrossRef]
19. Pękalski, J.; Rzadkowski, W.; Panagiotopoulos, A.Z. Shear-induced ordering in systems with competing interactions: A machine learning study. *J. Chem. Phys.* **2020**, *152*, 204905. [CrossRef] [PubMed]
20. Imperio, A.; Reatto, L. Microphase morphology in two-dimensional fluids under lateral confinement. *Phys. Rev. E* **2007**, *76*, 040402(R). [CrossRef]
21. Alba-Simionesco, A.; Coasne, B.; Dosseh, G.; Dudziak, G.; Gubbins, K.; Radhakrishnan, R.; Sliwinska-Bartkowiak, M. Effects of confinement on freezing and melting. *J. Phys. Condens. Matter* **2006**, *18*, R15. [CrossRef]
22. Nygård, K. Colloidal diffusion in confined geometries. *Phys. Chem. Chem. Phys.* **2017**, *19*, 23632. [CrossRef]
23. Zangi, R. Water confined to a slab geometry: A review of recent computer simulation studies. *J. Phys. Condens. Matter* **2004**, *16*, S5371. [CrossRef]
24. Martí, J.; Calero, C.; Franzese, G. Structure and dynamics of water at carbon-based interfaces. *Entropy* **2017**, *19*, 135. [CrossRef]

25. Serna, H.; Noya, E.G.; Góźdź, W.T. Assembly of Helical Structures in Systems with Competing Interactions under Cylindrical Confinement. *Langmuir* **2018**, *35*, 702–708. [CrossRef]
26. Serna, H.; Noya, E.G.; Góźdź, W.T. The influence of confinement on the structure of colloidal systems with competing interactions. *Soft Matter* **2020**, *16*, 718–727. [CrossRef] [PubMed]
27. Serna, H.; Noya, E.G.; Góźdź, W.T. Confinement of Colloids with Competing Interactions in Ordered Porous Materials. *J. Phys. Chem. B* **2020**, *124*, 10567–10577. [CrossRef] [PubMed]
28. Pękalski, J.; Almarza, N.G.; Ciach, A. Effects of rigid or adaptive confinement on colloidal self-assembly. Fixed vs. fluctuating number of confined particles. *J. Chem. Phys.* **2015**, *142*, 204904. [CrossRef] [PubMed]
29. Almarza, N.G.; Pękalski, J.; Ciach, A. Effects of confinement on pattern formation in two dimensional systems with competing interactions. *Soft Matter* **2016**, *12*, 7551. [CrossRef] [PubMed]
30. Schwanzer, D.F.; Coslovich, D.; Kahl, G. Two-dimensional systems with competing interactions: Dynamic properties of single particles and of clusters. *J. Phys. Condens. Matter* **2016**, *28*, 414015. [CrossRef] [PubMed]
31. Plimpton, S. Fast parallel algorithms for short-range molecular dynamics. *J. Comput. Phys.* **1995**, *117*, 1–19. [CrossRef]
32. Allen, M.P.; Tildesley, D.J. *Computer Simulation of Liquids*; Oxford University Press: Oxford, UK, 2017.
33. Faken, D.; Jónsson, H. Systematic analysis of local atomic structure combined with 3D computer graphics. *Comput. Mater. Sci.* **1994**, *2*, 279–286. [CrossRef]
34. Stukowski, A. Visualization and analysis of atomistic simulation data with OVITO-the Open Visualization Tool. *Model. Simul. Mater. Sci. Eng.* **2010**, *18*. [CrossRef]
35. Bores, C.; Almarza, N.G.; Lomba, E.; Kahl, G. Inclusions of a two dimensional fluid with competing interactions in a disordered, porous matrix. *J. Ournal. Phys. Condens. Matter* **2015**, *27*, 194127. [CrossRef]
36. Qiao, C.; Zhao, S.; Liu, H.; Dong, W. Connect the Thermodynamics of Bulk and Confined Fluids: Confinement-Adsorption Scaling. *Langmuir* **2019**, *35*, 3840–3847. [CrossRef] [PubMed]
37. Lechner, W.; Dellago, C. Accurate determination of crystal structures based on averaged local bond order parameters. *J. Chem. Phys.* **2008**, *129*, 114707. [CrossRef] [PubMed]
38. Yu, B.; Li, B.; Jin, Q.; Ding, D.; Shi, A.C. Confined self-assembly of cylinder-forming diblock copolymers: Effects of confining geometries. *Soft Matter* **2011**, *7*, 10227–10240. [CrossRef]
39. Mittal, J.; Truskett, T.M.; Errington, J.R.; Hummer, G. Layering and position-dependent diffusive dynamics of confined fluids. *Phys. Rev. Lett.* **2008**, *100*, 145901. [CrossRef]
40. Litniewski, L.; Ciach, A. Effect of aggregation on adsorption phenomena. *J. Chem. Phys.* **2019**, *150*, 234702. [CrossRef]
41. Liu, Y.; Zhao, W.; Zheng, X.; King, A.; Singh, A.; Rafailovich, M.; Sokolov, J.; Dai, K.; Kramer, E. Surface-induced ordering in asymmetric block copolymers. *Macromolecules* **1994**, *27*, 4000–4010. [CrossRef]

MDPI
St. Alban-Anlage 66
4052 Basel
Switzerland
Tel. +41 61 683 77 34
Fax +41 61 302 89 18
www.mdpi.com

International Journal of Molecular Sciences Editorial Office
E-mail: ijms@mdpi.com
www.mdpi.com/journal/ijms

www.ingramcontent.com/pod-product-compliance
Lightning Source LLC
LaVergne TN
LVHW070200170726
843515LV00007B/1963

* 9 7 8 3 0 3 6 5 2 7 1 0 9 *